5 STEPS TO A 5™

AP Physics 2: Algebra-Based

2020

AP Physics 2: Algebra-Based

2020

Chris Bruhn

New York Chicago San Francisco Athens London Madrid
Mexico City Milan New Delhi Singapore Sydney Toronto

1 2 3 4 5 6 7 8 9 LHS 24 23 22 21 20 19

ISBN 978-1-260-45478-9
MHID 1-260-45478-9

e-ISBN 978-1-260-45479-6
e-MHID 1-260-45479-7

The series editor was Grace Freedson, and the project editor was Del Franz.

Series design by Jane Tenenbaum.

ABOUT THE AUTHOR

Chris Bruhn began his career as an aerospace engineer with Bell Helicopter before choosing teaching as his calling. Since becoming an educator, he has taught all varieties of AP Physics in schools, including inner-city, suburban, and charter schools, and he has shown a particular talent at constructing successful educational programs. Chris was honored as L.D. Bell High School Teacher of the Year and earned the O'Donnell Texas AP Teacher Award for "remarkable contributions to his students and school, as well as to the teaching profession." The *Dallas Morning News* ran a front-page article featuring Chris and his students on January 26, 2010. (*Dallas Morning News* article: "AP teacher sparks students' love of physics and wins $30,000.") Chris is an educational trainer. He continues to create and share curriculum and educational resources as well as lead institutes and study sessions for teachers and students around the country. Outside of teaching, Chris likes building things, sports, painting, travel, watching superhero movies with his kids, and generally having fun. And now he is writing this book!

CONTENTS

INTRODUCTION: THE FIVE-STEP PROGRAM

Welcome!

You are in AP Physics 2, which means you probably just completed—or survived—AP Physics 1 last year. You already have a good idea of what is involved in getting ready for an AP Physics exam. AP Physics 2 builds upon the material you've already learned in AP Physics 1. Be sure to keep your *5 Steps to a 5 AP Physics 1* book handy for reviewing material you learned last year. I'm Chris, and I am going to be your friendly guide throughout the process of getting prepared for your AP exam.

Why This Book?

To have a good understanding of what makes this book special, it would probably be helpful if you knew a little bit about me. I've been an aerospace engineer, so I have experience in how physics is used in the real world. I have run workshops for teachers, helping them become successful with their own students. I've been teaching all varieties of AP Physics 1, 2, B, and C for more than 20 years, to students just like you in all types of schools—inner-city schools, suburban schools, magnet schools, charter schools—schools big and small. In every one of these schools, there are students, just like you, who learn to love and excel in AP Physics because it makes such beautiful sense.

That's where I come in—helping you make sense of it all. I've got a pretty good idea what is needed for you to do well on the AP exam. Throughout this book you will find:

- Clear, simple-to-understand text
- A thorough explanation of all the topics you will need to know
- Problem-solving tips and solutions
- An introduction to the student-tested Five-Step Program to mastering the AP Physics 2 exam

Organization of the Book: The Five-Step Program

You will be taking a comprehensive AP exam this May. Since this is not your first AP Physics exam, you already understand the structure of the exam and how thoroughly you need to understand the material. When you walk into that exam, you want to feel excited but calm, as if you are looking forward to showing off what you can do, not apprehensive or uncertain. Following the Five-Step Program is the best way to prepare and give yourself the best chance to earn that 5.

Step 1: Set Up Your Study Program

As you probably already know from having taken AP Physics 1, you cannot memorize your way through physics, and you can't cram overnight. Success on the AP exam is the result of diligent practice over the course of months, not the result of an all-nighter on the evening before the exam. Step 1 gives you the background structure you need before you even start exam preparation.

Step 2: Determine Your Test Readiness

A problem on the AP exam usually requires conceptual development, considerable problem solving, or critical thinking skills. AP questions do not ask for straightforward facts that you can memorize. The AP Physics exam is designed to test the depth of your understanding of physics concepts and how well you can apply them. But you're going to have to know those basic physics concepts in order to solve more difficult problems, including the ones you learned in AP Physics 1, so keep last year's *5 Steps to a 5* book handy!

You can't learn physics passively; you have to be actively engaged to truly understand the concepts. A good place to start preparing for the test is by quizzing yourself. This way you'll know your areas of strength and weakness. The 5 Steps fundamentals quiz in Chapter 4 will diagnose your areas of strength and weakness. Once you can answer every question on the fundamentals quiz quickly and accurately, you're ready for deeper questions that will challenge you on the AP exam.

Step 3: Develop Strategies for Success

Since this isn't your first AP Physics exam, you already know that an AP Physics test requires a much different approach than just about any other exam you've taken. You will be getting a reference table with equations on it for the entire exam, but it is still helpful to know your equations and understand what they mean. Memorizing equations is not as important as it was for the old AP Physics B exam, but it is very helpful. Having the equations stored in your "mental toolbox" means you never have to take the time to search around the reference table looking for an equation or wonder what relationship is needed to solve a problem. You will always have the reference table as a backup, but you will be surprised how many equations you know without even trying.

An important tool in your arsenal for succeeding on the AP exam is the ability to predict the behavior of a system based upon a physics equation, such as ranking things from largest to smallest. I'll discuss some methods to help you become successful with those problems.

Step 4: Review the Knowledge You Need to Score High

This is a comprehensive review of all the topics on the AP exam. Now, you've probably been in an AP Physics class all year; you've likely read[1] your textbook. Our review is meant to be just that—*review*, in a readable format, and focused exclusively on the AP exam.

These review chapters are appropriate both for quick skimming, to remind yourself of key points, and for in-depth study, with plenty of practice problems for you to work through. I do not go into nearly as much detail as a standard textbook, but the advantage of the lack of detail is that you can focus on only those issues specific to the AP Physics 2 exam.

Step 5: Build Your Test-Taking Confidence

Here is your full-length practice test. Unlike other practice tests you may take, this one comes with thorough explanations. One of the most important elements in learning physics is making, and then learning from, mistakes. I don't just tell you what you got wrong; I explain why your answer is wrong, and how to do the problem correctly. It's okay to make a mistake here, because if you do, you won't make that same mistake again on the "big day" in May.

[1]Or at least tried to read.

The Graphics Used in This Book

To emphasize particular skills and strategies, icons are used throughout this book. An icon in the margin will alert you that you should pay particular attention to the accompanying text. These three icons are used:

1. This icon points out a very important concept or fact that you should not pass over.

2. This icon calls your attention to a problem-solving strategy that you may want to try.

3. This icon indicates a tip that you might find useful.

ACKNOWLEDGMENTS

Well, I just have to say this wouldn't have been possible without my family who has always believed in me. You are the best family I've ever had! Thank you from the bottom of my heart. You are the best!

Oh, yeah! Thank you, Todd, for starting me down the path to this gig. I still owe you a steak . . . or is it two by now?

Special Acknowledgment and Thank You to Greg Jacobs

Everyone who reads this book will quickly notice the similarities to Greg Jacobs' outstanding book *5 Steps to a 5—AP Physics 1*. *5 Steps to a 5—AP Physics 2* is shamelessly modeled after Greg's work for two reasons:

1. It doesn't take a rocket scientist to know that AP Physics 2 follows AP Physics 1. I felt it was important to follow Greg's lead and offer students a smooth transition in both content, style, and format from one book to the next.
2. Greg's previous publication, *5 Steps to a 5—AP Physics B* has helped thousands of students successfully prepare for the AP exam. When AP Physics B was split into AP Physics 1 and 2, Greg took some of the original "B" material and utilized it in his new AP Physics 1 book. I followed this pattern. Material from Greg's AP Physics B book can easily be seen throughout the study guide.

In short, this book stands on the foundation that Greg Jacobs built. I cannot thank him enough for providing a blueprint for success. (You can find out more about Greg Jacobs by visiting his blog: http://jacobsphysics.blogspot.com/.)

5 STEPS TO A 5™

AP Physics 2: Algebra-Based

2020

Set Up Your Study Program

How to Approach Your AP Physics Course

IN THIS CHAPTER

Summary: Recognize the difference between truly understanding physics and just doing well in your physics class.

Key Ideas

- Focus on increasing your knowledge of physics, not on pleasing your teacher.
- Don't spend more than 10 minutes at one time on a problem without getting anywhere—come back to it later if you don't get it.
- Form a study group; your classmates can help you learn physics better.
- If you don't understand something, ask your teacher for help.
- Don't cram; although you can memorize equations, the skills you need to solve physics problems can't be learned overnight.

Before I get started, keep in mind that with physics it is impossible to "snow the exam grader." There are no shortcuts to doing well on a physics exam; you simply have to know your physics![1] Your grade on the exam is going to be based on the *quality of your work.* You will need to know physics so well that you not only are able to solve the problems given to

[1]Which you already know from taking Physics 1!

you but can also explain what you did to solve them. So this book is designed to help you in two ways:

1. to teach you the ways in which the AP exam tests your physics knowledge, and
2. to give you a review of the physics topics that will be tested—and to give you some hints on how to approach these topics.

Everyone who takes the AP Physics 2 exam has just completed an AP Physics course. **Recognize that your physics course is the place to start your exam preparation!** Whether or not you are satisfied with the quality of your course or your teacher, the best way to start preparing for the exam is by doing careful, attentive work in class all year long.

Okay, for many readers, I'm preaching to the choir. You don't want to hear about your physics class; you want the specifics about the AP exam. If that's the case, go ahead and turn to Chapter 2, and get started on your exam-specific preparation. But I think that you can get even more out of your physics class than you think you can. Read these pieces of time-tested advice, follow them, and I promise you'll feel more comfortable about your class and about the AP exam.

Ignore Your Grade

This must be the most ridiculous statement you've ever read. But it may also be the most important of these suggestions. Never ask yourself or your teacher, "Can I have more points on this assignment?" or "Is this going to be on the test?" You'll worry so much about giving the teacher merely what she or he wants that you won't learn physics in the way that's best for you. Whether your score is perfect or near zero, ask, "Did I really understand all aspects of these problems?"

Remember, the AP exam tests your physics knowledge. If you understand physics thoroughly, you will have no trouble at all on the AP test. But, while you may be able to argue yourself a better grade in your physics *class*, even if your comprehension is poor, the AP readers are not so easily moved.

If you take this advice—if you really, truly ignore your grade and focus on physics—your grade will come out in the wash. You'll find that you got a very good grade after all, because you understood the subject so well. But you *won't care*, because you're not worried about your grade!

Don't Bang Your Head Against a Brick Wall

My meaning here is figurative, although there are literal benefits also. Never spend more than 10 minutes or so staring at a problem without getting somewhere. If you honestly have no idea what to do at some stage of a problem, STOP. Put the problem away. Physics has a way of becoming clearer after you take a break.

On the same note, if you're stuck on some algebra, don't spend forever trying to find what you know is a trivial mistake, say a missing negative sign or some such thing. Put the problem away, come back in an hour, and start from scratch. This will save you time in the long run. If you've put forth a real effort, you've come back to the problem many times, and you still can't get it: relax. Ask the teacher for the solution, and allow yourself to be enlightened. You will not get a perfect score on every problem. But you don't care about your grade, remember?

Developing this mentality of "don't bang your head on the wall" is very important for AP exam day. Picture this: You are taking the AP exam. You are under stress. Your brain just went blank, time is short, and you're stuck on a question . . . REMAIN CALM and move on to something you know. Remember, you are not trying to get a perfect score on the AP exam; your goal is to earn as much credit as possible. The student who earns 75% of the points on the AP exam will probably get the same score as a student who earns 95%. You might even pass the exam with a score as low as 40%! Earn half the points and you are sure to earn a qualifying score. It's kind of liberating. Don't beat yourself up.

Work with Other People

When you put a difficult problem aside for a while, it always helps to discuss the problem with others. Form study groups. Have a buddy in class with whom you are consistently comparing solutions.

Although you may be able to do all your work in every other class without help, I have never met a student who is capable of solving every physics problem on his or her own. It is not shameful to ask for help. Nor is it dishonest to seek assistance—as long as you're not copying, or allowing a friend to carry you through the course. Group study is permitted and encouraged in virtually every physics class around the globe.

Just one quick warning about group studying: you still need to spend time on your own solving problems and answering questions. Way too often, students *only* study with a group, and then when it is time to take the exam, they are so used to having somebody bring up hints and facts that nudge them along in problem solving that they have trouble during the test. Make sure you spend some time on your own working on physics. There is no substitution for solo work.

Ask Questions When Appropriate

I know your physics teacher may seem mean or unapproachable, but in reality, physics teachers do want to help you understand their subject. If you don't understand something, don't be afraid to ask. Chances are that the rest of the class has the same question. If your question is too basic or requires too much class time to answer, the teacher will tell you so.

Sometimes the teacher will not answer you directly, but will give you a hint, something to think about so that you might guide yourself to your own answer. Don't interpret this as a refusal to answer your question. You must learn to think for yourself, and your teacher is helping you develop the analytical skills you need for success in physics.

Keep an Even Temper

A football team should not give up because they allow an early field goal. Similarly, you should not get upset at poor performance on a test or problem set. No one expects you to be perfect. Learn from your mistakes, and move on—it's too long a school year to let a single physics assignment affect your emotional state.

On the same note, however, a football team should not celebrate victory because it scores a first-quarter touchdown. You might have done well on this test, but there's the rest of a nine-month course to go. Congratulate yourself, and then concentrate on the next assignment.

Don't Cram

Yes, I know that you got an "A" on your history final because, after you slept through class all semester, you studied for 15 straight hours the day before the test and learned everything. And, yes, I know you are willing to do the same thing this year for physics. I warn you, both from our and from others' experience: *it won't work*. Physics is not about memorization and regurgitation. Sure, there are some equations you need to memorize. But problem-solving skills cannot be learned overnight.

Additionally, physics is cumulative. For AP Physics 2, not only do you need to have a good grasp of the materials you learned in September of this year, but the course builds on the materials you learned in Physics 1. If there was an area in Physics 1 that gave you a problem, it isn't a bad idea to look it over early in the year to get a better grasp of it. Physics 2 is not about mechanics, but there is plenty of material in fluids, thermodynamics, and static electricity that is based upon the concepts you learned last year. It is a good idea to look it over. If you are not sure if something from Physics 1 is important to Physics 2, ask your physics teacher; he or she will be thrilled that you are taking the initiative to get yourself prepared.

So, keep up with the course. Spend some time on physics every night, even if that time is only a couple of minutes, even if you have no assignment due the next day. Spread your "cram time" over the entire semester.

What You Need to Know About the AP Physics 2 Exam

IN THIS CHAPTER

Summary: Learn what topics are tested, how the test is scored, and basic test-taking information.

Key Ideas

- Most colleges will award credit for a score of 4 or 5, some for a 3.
- Multiple-choice questions account for half of your final score.
- There is no penalty for guessing on the multiple-choice questions. You should answer every question.
- Free-response questions account for half of your final score.
- Your composite score on the two test sections is converted to a score on the 1-to-5 scale.
- You have to know your physics to do well on the exam. You can't "sweet talk" your way to a good score.
- The test focuses on deep conceptual understanding and the ability of students to explain things using the language of physics.

Some Frequently Asked Questions About the AP Physics 2 Exam

Why Should I Take the AP Physics Exam?

Many of you take the AP Physics exam because you are seeking college credit. The majority of colleges and universities will award you some sort of credit for scoring a 4 or a 5. A smaller number of schools will even accept a 3 on the exam. This means you are one or two courses closer to graduation before you even start college!

Therefore, one compelling reason to take the AP exam is economics. How much does a college course cost, even at a relatively inexpensive school? You're talking several thousand dollars. If you can save those thousands of dollars by paying less than a hundred dollars now, why not do so?

Even if you do not score high enough to earn college credit, the fact that you elected to enroll in AP courses tells admission committees that you are a high achiever and serious about your education.

You'll hear a whole lot of misinformation about AP credit policies. Don't believe anything a friend (or even an adult) tells you; instead, find out for yourself. A good way to learn about the AP credit policy of the school you're interested in is to look it up on the College Board's official Web site, at http://collegesearch.collegeboard.com/apcreditpolicy/index.jsp. Even better, contact the registrar's office or the physics department chairman at the college directly.

How Do the New AP Physics 1 and 2 Courses Compare to the Old AP Physics B Exam?

Unless you are taking both Physics 1 and 2 in a single year, you have already taken the AP Physics 1 exam, so you have an idea what the exam is like. The new physics exams are going to be relying less on test calculations and more on conceptual understanding of the material and verbal explanations of physics principles. You will still be required to perform calculations, but the exam will have less of them. So, you have to do more than just work lots of mathematical physics problems to be ready for the exam.

What's the Deal with All the Different AP Physics Courses (Physics 1, Physics 2, and Physics C)?

There are four AP Physics courses you can take, Physics 1, Physics 2, Physics C—Mechanics, and Physics C—Electricity and Magnetism, and they are different in the depth of the material and mathematical modeling. Here's the rundown.

Physics 1 and 2

As survey courses, Physics 1 and 2 cover a broad range of topics. The tables of contents of this book and of *5 Steps to a 5: Physics 1* list them all. These courses are algebra-based—no calculus is necessary. In fact, the most difficult math you will encounter will be algebraic equations and basic trigonometry, which is probably something you did by the tenth grade.

The Physics 1 and 2 courses are ideal for *all* college-bound high school students. For those who intend to major in math or the heavy-duty sciences, Physics 1 and 2 serve as perfect introductions to college-level work. For those who want nothing to do with physics after high school, Physics 1 and 2 are terrific terminal courses—you get exposure to many facets of physics at a rigorous yet understandable level.

Most important, for those who aren't sure in which direction their college careers may head, the Physics 1 and 2 courses can help you decide: "Do I like this stuff enough to keep studying it or not?"

Physics C

These courses are ONLY for those who have already taken a solid introductory physics course and are considering a career in math or science. Some schools teach Physics C as a follow-up to Physics 1 and 2, but as long as you've had a rigorous introduction to the subject, that introduction does not have to be at the AP level.

Physics C is two separate courses: (1) Newtonian Mechanics, and (2) Electricity and Magnetism. Of course, the Physics 1 and 2 courses cover these topics as well. However, the C courses go into greater depth and detail. The problems are more involved, and they demand a higher level of conceptual understanding. You can take either or both 90-minute Physics C exams.

The C courses require some calculus. Although much of the material can be handled without it, you should be taking a good calculus course concurrently.

There Are So Many Different AP Physics Exams. Which One Should I Take?

This is a no-brainer. Take the exam that your high school class is preparing you for! If you are taking AP Physics 2 at your school right now, sign up for the AP Physics 2 exam.

Is One Exam Better Than the Other? Should I Take Them Both?

I strongly recommend taking only one exam—the one your high school AP course prepared you for. Physics C is not considered "better" than Physics 1 and 2 in the eyes of colleges and scholarship committees. They are different courses with different intended audiences. It is far better to do well on the one exam you prepared for than to do poorly on both exams.

What Is the Format of the Exam?

Table 1.1 summarizes the format of the AP Physics 2 exam, which is identical to the AP Physics 1 exam.

Table 1.1 AP Physics 2 Exam

Section	Number of Questions	Time Limit
1. Multiple-Choice Questions	45 with a single correct answer	90 minutes
	5 with two correct answers	
Total	50 Multiple-Choice Questions	
2. Free-Response Questions	1 Experimental Design Question	90 minutes
	1 Qualitative/Quantitative Translation Question	
	2 Short-Answer Questions (including a paragraph-length response)	
Total	4 Free-Response Questions	

You are probably asking yourself, what does that all mean?

- In the multiple-choice section, there will be a total of 50 questions; the first 45 will have only one correct answer. The last 5 will have two correct answers. They will be clearly marked so you won't have to guess which is which. I know you are asking yourself: "Can I get half credit if I get only one right on these last five?" Unfortunately not. You have to get both correct to receive credit.
- There will be one laboratory question where you will have to design a laboratory experiment to solve a problem to determine or discover a pattern of behavior.
- The quantitative/qualitative question is a problem you will have to solve mathematically and be able to explain in writing how you went about solving the problem.
- The short-response questions can require a verbal, algebraic, graphical, or numerical answer.
- Included in one of the questions will be a statement like: "In a clear, coherent paragraph-length response, explain . . ." followed by a large expanse of open paper. When students see this, their brains tend to go into "English class mode" and they begin writing, and writing, and writing. This is a physics exam. If you need a lot of words, it usually means that your understanding of the answer is not as good and your score will be lower. Your paragraph should be concise (short and to the point) and coherent (easy to understand in only one quick reading). What the test graders want here is called technical writing: explaining something in an economy of words. If you can clearly and completely answer the question in three sentences, please do so!

Who Writes the AP Physics Exam?

Development of each AP exam is a multiyear effort that involves many education and testing professionals and students. At the heart of the effort is the AP Physics Development Committee, a group of college and high-school physics teachers who are typically asked to serve for three years. The committee and other physics teachers create a large pool of multiple-choice questions. With the help of the testing experts at Educational Testing Service (ETS), these questions are then pre-tested with college students for accuracy, appropriateness, clarity, and assurance that there is only one possible answer. The results of this pre-testing allow each question to be categorized by degree of difficulty. After several more months of development and refinement, Section I of the exam is ready to be administered.

The free-response questions that make up Section II go through a similar process of creation, modification, pre-testing, and final refinement so that the questions cover the necessary areas of material and are at an appropriate level of difficulty and clarity. The committee also makes a great effort to construct a free-response exam that will allow for clear and equitable grading by the AP readers.

At the conclusion of each AP reading and scoring of exams, the exam itself and the results are thoroughly evaluated by the committee and by ETS. In this way, the College Board can use the results to make suggestions for course development in high schools and to plan future exams.

What Topics Appear on the Exam?

The College Board published the *Curriculum Framework*, which describes both AP Physics 1 and 2 in great detail. Unfortunately, this tome of wisdom is organized around six "Big Ideas," which do not clearly list the content the way you would see it in a textbook. Instead, the content for both courses are mixed together and you have to dig deep into the

Curriculum Framework to get a picture of what you need to know, how thoroughly you need to understand it, and how it will be tested. The bottom line is that there are eight major topic areas for AP Physics 2:

- Fluid Statics and Dynamics
- Ideal Gases
- Thermal Physics and Thermodynamics
- Electric Force, Field, and Potential
- DC Circuits and Resistor/Capacitor Circuits (steady state only)
- Magnetism and Electromagnetic Induction
- Physical and Geometric Optics
- Quantum, Atomic, and Nuclear Physics

Can I Use a Calculator? Will There Be an Equation Sheet?

Yes! You can use a calculator and are given an equation and constant sheet to use on the entire exam. But . . . don't plan on using them very much. There are not that many numerical problem-solving questions on the exam. Roughly two-thirds of the questions on the exam won't have any numbers at all. Many of the problems with numbers will be solved using semi-quantitative reasoning where you need to know what the equation means, but won't be doing any calculator work. For example:

> A scientist has a sample of gas at atmospheric pressure and a temperature of 300 K. The gas is in a sealed container with a movable piston that has an original volume of 30 cm^3. Wanting to study the behavior of the gas, the scientist pushes the piston until the gas has a volume of 15 cm^3 and measures the pressure to be 2 atmospheres. What new temperature will the gas have?
>
> Finding the answer using semi-quantitative reasoning:
>
> - I have a gas that's changing pressure and volume, so let's start with the universal gas law: $PV = nRT$.
> - The gas is in a sealed container so the number of moles will be constant and universal gas constant R is a constant. That leaves us with this relationship: (pressure times volume is proportional to the temperature).
> - Pressure is doubled and volume is cut in half, so the temperature remains unchanged!

You can see that trying to solve this problem with a calculator just makes things harder. The bottom line on calculators: You shouldn't use your calculator much on the exam. Try to use semi-quantitative reasoning first because it's faster and easier. For more information on which calculators are acceptable on the exam, visit https://apstudent.collegeboard.org/apcourse/ap-physics-2/exam-policies.

Concerning the equation sheet: First of all, memorize your equations! I know that is not what you wanted to hear, but if you waste time hunting for an equation during the exam you probably won't do very well. I'll say it again: Memorize your equations! That being said, if your teacher does not give you an official AP Physics 2 equation table to use in class, print one out for yourself from https://secure-media.collegeboard.org/digitalServices/pdf/ap/ap-physics-2-equations-table.pdf. Study it. Make sure you know what is and what is not on it. Not every equation you might need is on the equation table. As you study physics this year, highlight the equations you learn and write notes next to them explaining what they mean and when they are useful. Add any equations that may be missing. I'm telling you to do all these things so that you will memorize the equations and become intimately familiar with the formulas you will be using. That way if your brain goes blank during the exam, you won't waste any time finding what you need.

How Is the Multiple-Choice Section Scored?

The multiple-choice section counts for half of your total score. There is no partial credit on the five multiple-correct questions. There are two correct responses on these and you have to choose them both to get credit.

Should I Guess If I Don't Know the Answer?

Yes. Don't leave anything blank. There is no guessing penalty. Try your best to eliminate at least one response that can't be correct before you guess. This "educated guessing" will improve your score.

How Is the Free-Response Section Graded?

Every June, a group of physics teachers gather for a week to assign grades to your hard work. Each of these "readers" spends a day or so getting trained on one question—and one question only. Because each reader becomes an expert on that question, and because each exam book is anonymous, this process provides a very consistent and unbiased scoring of that question.

During a typical day of grading, a random sample of each reader's scores is selected and cross-checked by other experienced "Table Leaders" to ensure that the consistency is maintained throughout the day and the week. Each reader's scores on a given question are also statistically analyzed, to make sure they are not giving scores that are significantly higher or lower than the mean scores given by other readers of that question. All measures are taken to maintain consistency and fairness for your benefit.

Special note: You want to make the grading of your free-response answers as easy as possible for the reader. Readers do everything in their power to be consistent and fair. But, if your handwriting is poor and you don't write in a coherent, logical, and concise manner, it is hard for you to receive credit. I always tell my students that the readers are very nice people who want you to do well. But they can't grade what they can't read. The readers grade many thousands of papers. They don't have the time to hunt through a maze of your gibberish. Make the reader your friend. Provide a clean, easy-to-read, and orderly response.

Will My Exam Remain Anonymous?

Absolutely. Even if your high-school teacher happens to randomly read your booklet, there is virtually no way he or she will know it is you. To the reader, each student is a number, and to the computer, each student is a bar code.

What About That Permission Box on the Back?

The College Board uses some exams to help train high-school teachers so that they can help the next generation of physics students to avoid common mistakes. If you check this box, you simply give permission to use your exam in this way. Even if you give permission, your anonymity is still maintained.

How Is My Final Grade Determined and What Does It Mean?

As I said, each section counts for 50% of the exam. The total composite score is thus a weighted sum of the multiple-choice and the free-response sections. In the end, when all of the numbers have been crunched, the Chief Faculty Consultant converts the range of composite scores to the 5-point scale of the AP grades. This conversion is not a true curve—it's not that there's some target percentage of 5s to give out. This means you're not competing against other test takers. Rather, the 5-point scale is adjusted each year to reflect the same standards as in previous years. The goal is that students who earn 5s this year are just as strong as those who earned 5s in 2015.

The table below gives a rough idea of the raw percentage scores required for each level of AP score:

AP Score	Percentage of Total Points Earned
5	70%
4	55%
3	40%
2	About 25%

These percentages will be used to score your practice exams in this book. You will receive your actual AP grade in early July.

How Do I Register and How Much Does It Cost?

If you are enrolled in AP Physics in your high school, your teacher will provide all of these details, but a quick summary here can't hurt. After all, you do not have to enroll in the AP course to register for and complete the AP exam. When in doubt, the best source of information is the College Board's Web site: www.collegeboard.com.

Currently, the fee for taking the exam is about $100. Students who demonstrate financial need may receive a refund to offset the cost of testing. The fee and the refund usually change a little from year to year. You can find out more about the exam fee and fee reductions and subsidies from the coordinator of your AP program or by checking information on the official Web site: www.collegeboard.com.

I know that seems like a lot of money just for a test. But, you should think of this $100 as the biggest bargain you'll ever find. Why? Most colleges will give you a few credit hours for a good score. Do you think you can find a college that offers those credit hours for $100? Usually you're talking hundreds of dollars per credit hour! You're probably saving thousands of dollars by earning credits via AP.

To find out if a college gives credit for the exam, visit https://apstudent.collegeboard.org/creditandplacement/search-credit-policies. Or better yet, call the college you want to attend and ask them personally.

There are also several optional fees that must be paid if you want your scores rushed to you or if you wish to receive multiple-grade reports. Don't worry about doing that unless your college demands it. (What, you think your scores are going to change if you don't find them out right away?)

The coordinator of the AP program at your school will inform you where and when you will take the exam. If you live in a small community, your exam may not be administered at your school, so be sure to get this information.

What If My School Doesn't Offer AP Physics 2?

If your school does not offer AP Physics 2, you can study for the exam on your own. This review guide will put you on the right path to success. However, there are two additional things you can do when preparing on your own.

1. Ask the physics teacher (AP Physics 1 teacher, if the course is offered) at your school to supply you with some official College Board AP Physics 2 released questions that you can work through. Don't ask for the key! Just ask for the questions. Looking at the key, instead of actually working the problems, fools you into thinking you know what you are doing when really you're just looking at someone else's work. Once you have worked the problems, have the teacher grade them for you and give you advice. Most teachers are more than happy to help motivated students.
2. If you don't already have one, get a good noncalculus-based, college-level physics textbook. *The College Physics: A Strategic Approach* by Randal Knight, Brian Jones, and Stuart Field, and *College Physics* by Eugenia Etkina, Michael Gentile, and Alan Van Heuvelen are good choices for AP Physics 2. If you can't find one, check one out from the library or ask a physics teacher at your school. Read through all the chapters covering the content you find in this review guide.

What Should I Bring to the Exam?

On exam day, I suggest bringing the following items:

- Several pencils and an eraser that doesn't leave smudges.
- Black- or blue-colored pens for the free-response section.[1]
- A ruler or straightedge.
- A scientific calculator with fresh batteries. (A graphing calculator is not necessary.)
- A watch so that you can monitor your time. You never know if the exam room will have a clock on the wall. Make sure you turn off the beep that goes off on the hour.
- Your photo identification and social security number.
- Tissues.
- Your quiet confidence that you are prepared.

What Should I NOT Bring to the Exam?

Leave the following at home:

- A cell phone, PDA, or walkie-talkie.
- Books, a dictionary, study notes, flash cards, highlighting pens, correction fluid, etc., *including this book*. Study aids won't help you the morning of the exam . . . end your studying in the very early evening the night before.
- Portable music of any kind. No iPods, MP3 players, or CD players.
- Clothing with any physics terminology or equations on it.
- Panic or fear. It's natural to be nervous, but you can comfort yourself that you have used this book well and that there is no room for fear on your exam.

---

[1]You may use a pencil, but you should not erase incorrect work, you should cross it out. Not only does crossing out take less time than erasing, if you erase something important by mistake, you lose all your work. If you happen to change your mind about crossing something out, just circle your work and write the reader a note: "Grade this!"

Building Your Personal Attack Plan

IN THIS CHAPTER

Summary: In order to do well on the exam, you need a plan. This chapter will help you prioritize what to study and how to fit that studying into your schedule.

Key Ideas

- Memorizing your way to a good score on the AP Physics 2 exam is impossible because that is not what the exam tests.
- Cramming your way to a good score is not practical and counter-productive. Unlocking the mysteries of the universe takes time, so make the most of your AP Physics 2 course at school.
- Become familiar with the exam, what types of questions will be asked, and what strategies work well.
- Take a practice exam to help you prioritize your study plan based on your personal needs.

Memorization and Cramming Won't Help

The AP exam is not about facts and equations. Sure you need to memorize important information and you will use equations, but that is only the beginning. The exam concentrates on what you can do with what you know. The exam attempts to see if you can connect the factual pieces together, with a solid conceptual framework to demonstrate a deep understanding of how the world works. This means you can't just read the book and/or cram the week before the exam and expect to do well.

Deep understanding of physics comes only with time. Your AP Physics 2 class is your most useful tool in preparing for the exam. You need to invest time during the school year in your course. Ask questions and study. Don't just look for "the right answer." Search for the reasons why the world behaves the way it does. Get to know your physics over time.

Of course, you may not be satisfied with the quantity or quality of your in-class instruction. And even if your class is the best in the country, you will still need a reminder of what you covered way back at the beginning of the year. That's where this book, and extracurricular AP exam preparation, are useful.

Building a Plan That Is Right for You

Everyone is different and will have different needs when preparing for the AP exam. However, there are six steps that every student needs to take to build a plan that works best for them.

1. Learn how the test is structured (Chapter 2).
2. Make sure you know what content will be tested (Chapter 4).
3. Become familiar with how the questions will be asked (Chapter 5).
4. Learn strategies that work for the different types of questions (Chapters 6–8).
5. Practice the material you are weak in (Chapters 9–15).
6. Take a complete practice exam (Step 5, The AP Physics 2 Practice Exams).

There will be topics in physics that you will understand right away. The topic speaks your language because you already think about the world in that way and it just clicks. And then there are those topics that are—well—let's just say alien. The beauty of AP Physics 2 is that there are many different types of topics. So if you get stuck on one, there will be others that will be simple for you. You may love fluids but hate magnetism. Circuits is a snap but thermodynamics is a struggle. Optics is fun but the nano world is weird. The point is that you will be great at some things and not so great at others.

Use Step 2 (Chapters 4–5) to find your strengths and weaknesses. You shouldn't spend lots of time studying things you are already good at. Your plan should focus on the areas you are weak in because that is where you will make the greatest improvements to your score. If you find that you are having difficulty with magnetism, dedicate a few days to work on it by using the sample problems in this book, reviewing the material from your class, and asking questions of your teacher. Sometimes it just takes a little more time and effort to see how it all fits together. Keep at it. You WILL get it. The cool thing is that when you work on the areas you are struggling in, it helps you understand how the rest of physics fits together as well.

So, your plan will be unique to fit your needs and the time you have remaining before the exam. Some of you acquired this book in September, while others waited until spring break. No matter when you start, the AP exam will be here before you know it—early in May. No matter how much time you have left before the exam, build a plan and stick with it. Remember, success does not come to those who cram.

Choose a plan below that matches how much time you have left. Use it as a guide to build your own personal attack plan.

Plan A: I'm Starting in September

Good job! You are starting early. Here is what to do:

Fall Semester

- Read Chapter 1 on how to approach the AP Physics 2 course.
- Make sure you know how the test is structured (Chapter 2).
- Learn strategies that work for the different types of questions and practice them on questions you work on class (Chapters 6–8).
- As you cover the AP Physics 2 content in class, review the same material in this book (Chapters 9–15). This will help give you a different viewpoint on the same material. Hearing the material in different ways and spending additional time on the content will give you a more robust understanding.

Second Semester

- Keep up the good work of reviewing chapters as you cover them in class. If you see that your class is running out of time to finish the course on schedule, read ahead in this book so that you are already primed to understand material that your teacher may be hurrying through (Chapters 9–15).
- In February, use the self-tests in Chapters 4 and 5 to review your content knowledge and how well you handle the different question types. Identify where your weaknesses are and concentrate your study time on the material and types of test questions you are weakest in. Review the question strategies from Chapters 6–8.

April

- Take both practice exams (Step 5). This is a dry run for the actual exam, so time yourself and pretend it's the real thing. This will help you gauge what the real exam will be like in pacing and difficulty. Grade your exam and concentrate your final time on the areas and the types of questions that you did poorest on.

May

- Have confidence that you are ready for the exam. Don't cram! Go to bed and get a good night's sleep. Eat protein and whole grains before the exam. Give your brain a head start without any high-sugar foods that will cause an energy crash in the middle of the exam.
- Walk into the exam with hope and purpose.

Plan B: I'm Starting in January

If you are like most students, you got this book in the second semester. That's OK. There is plenty of time for you to work on any skills you may be lacking. Here is a plan for you:

January

- Read Chapter 1 on how to approach the AP Physics 2 course.
- Make sure you know how the test is structured (Chapter 2).
- Learn strategies that work for the different types of questions and practice them on questions you work on class (Chapters 6–8).
- Find the material you have already covered in your AP Physics 2 course and review that material ASAP (Chapters 9–15). As you cover new material in your class, review the same content in this book. This will give you a different viewpoint on the same material.

Hearing the material in different ways and spending additional time on the content will give you a more robust understanding.

February

- Use the self-tests in Chapters 4 and 5 to review your content knowledge and how well you handle the different question types. Identify where your weaknesses are, and concentrate your study time on the material and types of test questions you are weakest in. Review the question strategies from Chapters 6–8.

March

- Continue working on your areas of weakness. If you see that your class is running out of time to finish the course on schedule, read ahead in this book so that you are already primed to understand material that your teacher may be hurrying through.

April

- Take both practice exams (Step 5). This is a dry run for the actual exam, so time yourself and pretend it's the real thing. This will help you gauge what the real exam will be like in pacing and difficulty. Grade your exam and concentrate your final time on the areas and the types of questions that you did most poorly on.

May

- Have confidence that you are ready for the exam. Don't cram! Go to bed and get a good night's sleep. Eat protein and whole grains before the exam. Give your brain a head start without any high-sugar foods that will cause an energy crash in the middle of the exam.
- Walk into the exam with hope and purpose.

Plan C: It's Spring Break and I Just Got Started!

Remember the mantra from *The Hitchhiker's Guide to the Galaxy*: "Don't Panic." There is still time to prepare. You should have already covered most of the material and acquired many of the skills for success in your AP Physics 2 class. Now all you have to do is put the finishing touches on the skills you already have. Don't use your remaining time to cram. Use this book to discover where your gaps are and spend your remaining time filling in those gaps.

March

- Read Chapter 1 on how to approach the AP Physics 2 course.
- Make sure you know how the test is structured (Chapter 2).
- Learn strategies that work for the different types of questions and practice them on questions you work on class (Chapters 6–8).
- Find the material you have already covered in your AP Physics 2 course and review that material ASAP (Chapters 9–15). As you cover any new material in your class, review the same content in this book. This will give you a different viewpoint on the same material. Hearing the material in different ways and spending additional time on the content will give you a more robust understanding.
- Use the self-tests in Chapters 4 and 5 to review your content knowledge and how well you handle the different question types. Identify where your weaknesses are and concentrate your study time on the material and types of test questions you are weakest in.

April

- Take both practice exams (Step 5). This is a dry run for the actual exam, so time yourself and pretend it's the real thing. This will help you gauge what the real exam will be like in pacing and difficulty. Grade your exam and concentrate your final time on the areas and the types of questions that you did most poorly on.

May

- Have confidence that you are ready for the exam. Don't cram! Go to bed and get a good night's sleep. Eat protein and whole grains before the exam. Give your brain a head start without any high-sugar foods that will cause an energy crash in the middle of the exam.
- Walk into the exam with hope and purpose.

Determine Your Test Readiness

Test Yourself: AP Physics 2 Fundamentals

IN THIS CHAPTER

Summary: This chapter contains a basic concepts quiz for AP Physics 2 designed to help you judge your areas of genius and areas of need.

Key Ideas

- Discover which topics you know and don't know from the AP Physics 2 content.
- Use this self-evaluation quiz to help guide you in constructing your study plan from Chapter 3.

AP Physics 2 Fundamentals Self-Assessment

The questions in this chapter are not written in the AP exam style. This is just a quiz. It's really a bunch of rapid-fire questions. They are designed to get a quick idea of what you know cold, what is fuzzy, and what is absent from your brain. So rifle through the questions, answering as many as you can. Check your answers with the key at the end. Based on your results, you should now know your content strengths and weaknesses. Return to Chapter 3 with this knowledge and build your study plan.

Fluids

1. How would you determine the density of an object?
2. What is the physical mechanism in a fluid that causes a force on a surface? We are not looking for an equation here!
3. Mathematically how do you calculate force knowing the pressure? OK, now we are looking for an equation.
4. For the equation $P = P_0 + \rho gh$,
   (a) for what kind of situation is the equation valid?
   (b) what does P_0 stand for (careful!)?
5. Write Bernoulli's equation.
6. State Archimedes' principle in words by finishing the following sentence: "The buoyant force on an object in a fluid is equal to . . ."
7. For a flowing fluid, what quantity does Av represent, and why is this quantity the same everywhere in a flowing fluid?
8. Write the alternate expression for mass that is useful when dealing with fluids of known density.

Thermal Physics, Thermodynamics, and Gases

9. How do you determine the internal energy of a gas given the temperature of the gas? Define all variables in your equation.
10. How do you determine the rms speed of molecules given the temperature of a gas? Define all variables in your equation.
11. State the equation for the first law of thermodynamics. What does each variable stand for? What are the units of each term?
12. Sketch two isotherms on the PV diagram below. Label which isotherm represents the higher temperature.

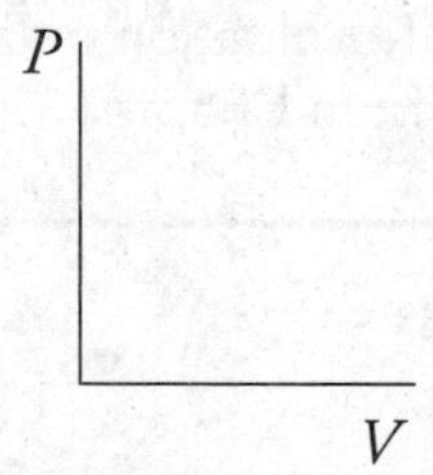

13. Describe a situation in which heat is added to a gas, but the temperature of the gas does *not* increase.
14. Imagine you are given a labeled PV diagram for 1 mole of an ideal gas. *Note that one of the following is a trick question!*
    (a) How do you use the graph to determine how much work is done on or by the gas?
    (b) How do you use the graph to determine the change in the gas's internal energy?
    (c) How do you use the graph to determine how much heat was added to or removed from the gas?

15. How can we estimate absolute zero in a school lab?

16. If the volume of a gas doubles while the temperature is held constant, what happens to the pressure of the gas?

17. Heat is added to a gas in a closed container.
(a) Is there any work done by or on the gas?
(b) What happens to the temperature and pressure of the gas?

18. Why does heat always flow from hot objects to cold objects?

19. What are the three ways heat can be transferred? Explain each process.

Electricity and Magnetism

20. Given the charge of a particle and the electric field experienced by that particle, give the equation to determine the electric force acting on the particle.

21. Given the charge of a particle and the magnetic field experienced by that particle, give the equation to determine the magnetic force acting on the particle.

22. A wire carries a current to the left, as shown below. What is the direction and magnitude of the magnetic field produced by the wire at point P?

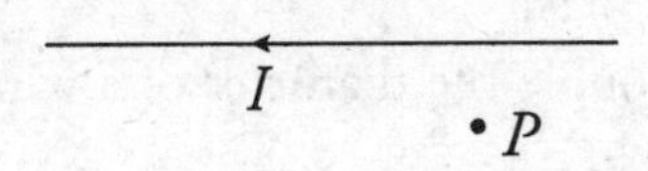

23. When is the equation kQ/r^2 valid? What is this an equation for?

24. The electric field at point P is 100 N/C; the field at point Q, 1 meter away from point P, is 200 N/C. A point charge of +1 C is placed at point P. What is the magnitude of the electric force experienced by this charge?

25. Can a current be induced in a wire if the flux through the wire is zero? Explain.

26. True or false: In a uniform electric field pointing to the right, a negatively charged particle will move to the left. If true, justify with an equation; if false, explain the flaw in reasoning.

27. Which is a vector and which is a scalar: electric field and electric potential?

28. Fill in the blank with either "parallel" or "series":
(a) Voltage across resistors in ______ must be the same for each.
(b) Current through resistors in ______ must be the same for each.
(c) Voltage across capacitors in ______ must be the same for each.
(d) Charge stored on capacitors in ______ must be the same for each.

29. A uniform electric field acts to the right. In which direction will each of these particles accelerate?
(a) proton
(b) positron (same mass as electron, but opposite charge)
(c) neutron
(d) anti-proton (same mass as proton, but opposite charge)
(e) electron

30. A uniform magnetic field acts to the right. In which direction will each of these particles accelerate, assuming they enter the field moving toward the top of the page?
(a) proton
(b) positron (same mass as electron, but opposite charge)
(c) neutron
(d) anti-proton (same mass as proton, but opposite charge)
(e) electron

31. How do you find the potential energy of an electric charge?

32. Describe the processes of conduction, polarization, and induction.

33. What is the smallest unit of charge that you will likely ever find? Why?

34. What do the electric field and electric potential "look like" between two oppositely charged capacitor plates?

35. The force between two charges is *F*. If the size of the charges is doubled and the charges are moved twice as far apart, what will be the new force?

36. What are the similarities and differences between the electric force and gravitational force?

37. Two identical metal spheres, one with a charge of +2 C and the other with a charge of –4 C, touch. What is the new charge of each sphere?

38. If you double the diameter of a wire, what happens to its resistance?

39. How would you determine if a resistor is ohmic?

40. I want to increase the capacitance of a capacitor. How could I change the geometry of the capacitor to accomplish this?

41. An uncharged capacitor is connected to a battery. Compare the current of an uncharged capacitor immediately after it is connected to a battery to its current after it has been connected for a long time. Explain.

42. Describe how ferromagnetic, paramagnetic, and diamagnetic materials behave when placed in an external magnetic field.

Physical and Geometric Optics

43. Describe what each variable means in the equation $d \sin \theta = m\lambda$.

44. Describe how the interference pattern for a single slit is the same and different from an interference pattern for a double slit where the width of the single slit is the same as the distance between the centers of the double slits.

45. Explain how dark and light locations are created when light shines through a double slit.

46. Describe the differences between electromagnetic and mechanical waves.

47. Only 80% of the light that strikes a particular surface reflects from it. What happened to the rest of the light?

48. Why does light refract (bend) when it enters a new medium?

49. (a) When light travels from water ($n = 1.3$) to glass ($n = 1.5$), which way does it bend?
(b) When light travels from glass to water, which way does it bend?
(c) In which of the above cases may total internal reflection occur?
(d) Write (but don't solve) an equation for the critical angle for total internal reflection between water and glass.

50. Describe two principal rays drawn for a convex lens. Be careful to distinguish between the *near* and *far* focal points.

1.

2.

51. Describe two principal rays drawn for a concave lens. Be careful to distinguish between the *near* and *far* focal points.

1.

2.

Quantum, Atomic, and Nuclear

52. After sitting on a shelf for 3 years, a radioactive sample has only one-eighth of its original beta particle emissions. What does this tell you about the sample?

53. Green light shines on a photosensitive material, causing it to eject electrons.
(a) What can you do to cause more electrons to be ejected?
(b) What can you do to cause the ejected electrons to have more energy?

54. What conservation laws are obeyed in the nano-world?

55. A photon collides with a stationary electron giving it a velocity. What has happened to the wavelength and frequency of the photon? Explain.

56. Why does gas in a neon bulb emit only discrete wavelengths of light?

57. What does a wave function tell us?

58. Uranium decays into thorium by ejecting an alpha particle. How does the mass of the uranium compare to the thorium and alpha particle? Explain.

59. We often use two different equations for wavelength:

$$\lambda = \frac{hc}{\Delta E}, \text{ and } \lambda = \frac{h}{mv}$$

When is each used?

60. Name the only decay process that affects neither the atomic number nor the atomic mass of the nucleus.

Answers to AP Physics 2 Fundamentals Self-Assessment

1. $\rho = m/V$. So, all we need to do is find mass and volume. Mass can be found using a balance. If the object has a geometric shape, measure it with a ruler and calculate its volume. If it is irregular, fill a container to the rim with water. Submerge the object, and collect and measure the overflow volume of water.

2. The simple answer is pressure, but what causes the pressure? Fluids are made of a bajillion tiny molecules in constant random motion. These individual molecules are colliding with and recoiling from any surface touching the fluid. This imparts an individual molecular impulse to the surface that is perpendicular to that surface. Add up all these individual impulses and you get pressure.

3. $F = PA$. Remember that the force caused by fluid pressure is always perpendicular to the surface.

4. (a) This is valid for a static (not moving) column of fluid.
   (b) P_0 stands for pressure at the top of the fluid; not necessarily, but sometimes, atmospheric pressure.

5. $P_1 + \rho g y_1 + \frac{1}{2}\rho v_1^2 = P_2 + \rho g y_2 + \frac{1}{2}\rho v_2^2$

6. . . . the weight of the fluid displaced.

7. Av is the volume flow rate. Fluid can't be created or destroyed; so, unless there's a source or a sink of fluid, total volume flowing past one point in a second must push the same amount of total volume past another downstream point in the same time interval.

8. mass = density · volume

9. $U = \frac{3}{2}Nk_BT$. Internal energy is $\frac{3}{2}$ times the number of molecules in the gas times Boltzmann's constant (which is on the constant sheet) times the absolute temperature, in kelvins. Or, $U = \frac{3}{2}n\,RT$ is correct, too, because $Nk_B = nR$. (Capital N represents the number of molecules; small n represents the number of moles.)

10. $v_{rms} = \sqrt{\dfrac{3k_BT}{m}}$

    k_B is Boltzmann's constant, T is absolute temperature in kelvins, and m is the mass of each molecule in kilograms (*NOT* in amu!). This can also be expressed as $\sqrt{\frac{3RT}{\mu}}$ where m is the molar mass of the gas and R is the ideal gas constant.

11. $\Delta U = Q + W$

    Change in internal energy is equal to (say it in rhythm, now) "heat added to, plus work done on" a gas. Each term is a form of energy and so has units of joules.

12. The isotherm labeled as "2" is at the higher temperature because it's farther from the origin.

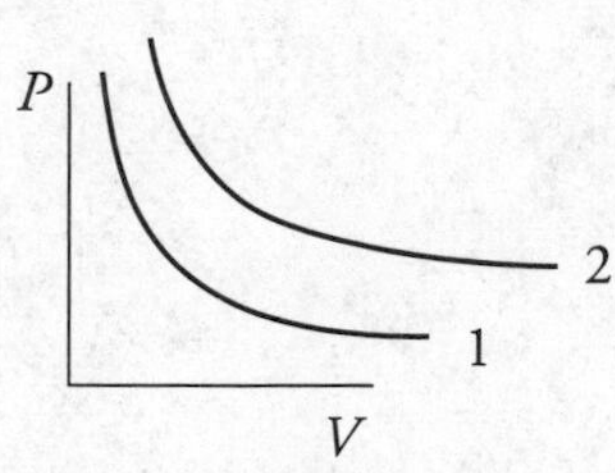

13. Let's put the initially room-temperature gas into a boiling water bath, adding heat. But let's also make the piston on the gas cylinder expand, so that the gas does work. By the first law of thermodynamics, if the gas does as much or more work than the heat added to it, then ΔU will be zero or negative, meaning the gas's temperature stayed the same or went down.

14. (a) Find the area under the graph. (b) Use $PV = nRT$ to find the temperature at each point; then, use $U = \frac{3}{2}nRT$ to find the internal energy at each point; then subtract to find ΔU. (c) You can NOT use the graph to determine heat added or removed. The only way to find Q is to find ΔU and W.

15. An estimate for absolute zero can be found by plotting the relationship between gas pressure and its temperature while keeping the volume constant. For example, confine a gas in a container. While measuring the temperature and pressure, cool the gas. Plot the data and you will see that as temperature goes down, pressure goes down (a direct relationship). Draw a line through your data and extrapolate backward until you get to zero pressure, because you can't go any lower than no pressure. The temperature at zero pressure is absolute zero.

You can also find absolute zero by measuring gas volume as a function of temperature and repeating the extrapolation process to zero volume. The temperature at zero volume is absolute zero.

16. The model for an ideal gas is $PV = nRT$. Temperature is being held constant, R is a constant, and assuming we don't gain or lose any gas in the process, the number of moles is a constant. Therefore: $PV =$ a constant. So, if volume doubles, the pressure will be cut in half.

17. Heat is added ($+Q$). The volume of the container does not appear to be changing ($\Delta V = 0$):

(a) $W_{done by gas} = P\Delta V = 0$.

(b) $\Delta U = Q + W$. Since work is zero and heat is positive, the internal energy of the gas goes up. From the kinetic theory of gases, we know that the internal energy of a gas is directly related to the temperature of the gas. Therefore, as ΔU goes up, so does temperature. Finally, from the ideal gas law $PV = nRT$, we see that volume and the number of molecules aren't changing. Pressure is directly related to temperature. Therefore, the pressure increases as well.

18. Objects are made up of a bajillion atoms moving in a distribution of random speeds. Some fast, some slow. Temperature measures the average motion of these atoms. In hotter objects, on average, the atoms are moving faster. If you put a hot object in contact with a cold object, on average, the faster-moving atoms in the hot object will transfer energy to the slower-moving ones in the cold object by elastic collisions.

19. Conduction is a process by which thermal energy is transferred by physical contact from atom to atom as we just discussed in #18. Convection is a process of thermal energy transfer through the movement of hot fluids as they rise due to changes in density. Radiation is the emission of electromagnetic radiation due to the random vibration of the charged particles. The higher the temperature of the object, the faster the particles vibrate and the more they radiate energy at higher frequencies.

20. $F = qE$

21. $F = qvB \sin \theta$

22. Point your right thumb in the direction of the current (i.e., to the left). Your fingers point in the direction of the magnetic field. This field wraps around the wire, pointing into the page above the wire and out of the page below the wire. Since point P is below the wire, the field points out of the page.

23. This equation is only valid when a point charge produces an electric field. (Careful—if you just said "point charge," you're not entirely correct. If a point charge experiences an electric field produced by something else, this equation is irrelevant.) It is an equation for the electric field produced by the point charge.

24. Do *not* use $E = kQ/r^2$ here, because the electric field is known. So, the source of the electric field is irrelevant—just use $F = qE$ to find that the force on the charge is (1 C)(100 N/C) = 100 N. (The charge is placed at point P, so anything happening at point Q is irrelevant.)

25. Yes! Induced emf depends on the *change* in flux. So, imagine that the flux is changing rapidly from one direction to the other. For a brief moment, flux will be zero; but flux is still changing at that moment. (And, of course, the induced current will be the emf divided by the resistance of the wire.)

26. False. The negative particle will be *forced* to the left. But the particle could have entered the field while moving to the right . . . in that case, the particle would continue moving to the right, but would slow down.

27. Electric field is a vector, so fields produced in different directions can cancel. Electric potential is a scalar, so direction is irrelevant.

28. Voltage across resistors in parallel must be the same for each.
Current through resistors in series must be the same for each.
Voltage across capacitors in parallel must be the same for each.
Charge stored on capacitors in series must be the same for each.

29. The positively charged proton will accelerate with the field, to the right.
The positively charged positron will accelerate with the field, to the right.
The uncharged neutron will not accelerate.
The negatively charged anti-proton will accelerate against the field, to the left.
The negatively charged electron will accelerate against the field, to the left.

30. Use the right-hand rule for each:
The positively charged proton will accelerate into the page.
The positively charged positron will accelerate into the page.
The uncharged neutron will not accelerate.
The negatively charged anti-proton will accelerate out of the page.
The negatively charged electron will accelerate out of the page.

31. If you know the electric potential experienced by the charge, $PE = qV$.

32. Conduction is when a charged object is physically touched to another. The charge migrates between the objects, and the two share the original net charge. Polarization is when you bring a charged object close to, but not touching, something else. The charged object repels like charges and attracts opposite charges in the second object, causing it to have a charge separation, but not a net charge. One side becomes more positive, and the other side becomes more negative. Induction is one step beyond polarization. With induction, the charged object is brought close to, but does not touch, a second object. This creates a charge separation—polarization. Now an escape path, or "ground," is attached to the polarized object, which allows the repelled charge to escape. The ground is removed, and the object is now charged the opposite sign.

33. 1.6×10^{-19} C. This is the charge of an electron/proton. Unless you have a particle accelerator in your classroom, this is the smallest unit of charged matter you will ever encounter. Everything else you will encounter is made up of protons and electrons, so 1.6×10^{-19} C is the smallest you will see.

34. The electric field will be uniform in strength and direction, pointing away from the positive plate and toward the negative one. The electric potential will have isolines that are parallel to the capacitor plates (making them perpendicular to the E-Field lines). Near the edges of the capacitor plates, things begin to get "curvy" as E-Field and isolines of potential spill out.

35. $|\vec{F}_E| = \frac{1}{4\pi\varepsilon_0}\frac{|q_1 q_2|}{r^2}$ Double both charges and double the radius and you get 4 on top and 4 on the bottom, which cancel: $|\vec{F}_E| = \frac{1}{4\pi\varepsilon_0}\frac{|2q_1 2q_1|}{(2r_2)}$. Thus the force stays the same.

36. $|\vec{F}_E| = \frac{1}{4\pi\varepsilon_0}\frac{|q_1 q_2|}{r^2}$ versus $F_G = G\frac{m_1 m_1}{r_2}$

Similarities: both have the same inverse squared relationship with the radius. Both exert a force on a line between the two objects. Both forces extend to infinity. Differences: mass is always positive, while charge can be negative or positive. Gravity only attracts, while electric force attracts and repels. Gravity is very weak in comparison to the electric force. Most objects have a net charge of zero, so the electric forces cancel out in most cases. This is not true with gravity; all objects with mass will have an attraction to all other masses.

37. Since both are conductors, by contact, or conduction, the two spheres will share the net charge of –2 C. Since they are the same size, each sphere will end up with –1 C.

38. $R = \frac{\rho l}{A} = \frac{\rho l}{\pi r^2}$. So if you double the diameter of the wire, the new resistance is one-fourth the original ($R/4$).

39. An ohmic resistor will have the same resistance even though the voltage and current are changing: $R = \frac{\Delta V}{I}$ = constant. So, all you have to do is connect the resistor to different voltages, measure the current in each case, and calculate the resistance in each case to see if it is constant. Better yet, plot voltage versus current on a graph and make sure the graph is linear. If it is curved, the resistor is nonohmic.

40. $C = \kappa\varepsilon_0\frac{A}{d}$. Here are your options for increasing the capacitance: (1) κ—insert an insulator between the plates with a higher dielectric constant; (2) A—increase the area of the plates; and (3) d—decrease the distance between the plates.

41. When the uncharged capacitor is first connected to the battery, it has no potential difference between the plates and offers no resistance to current. Thus, at first, the capacitor behaves like a wire. After a long time, the capacitor fills with opposite charges on the plates and builds a potential difference equal to, but in the opposite direction of, the battery. No more charge can move to the capacitor and it is "full." Thus, in the end, the capacitor acts like a "disconnect" or "open switch" in the circuit and current flow stops.

42. The magnetic domains in a ferromagnetic material line up with the external B-Field amplifying the field. The magnetic field in a paramagnetic material also lines up with the external B-Field, but the effect is weak. Diamagnetic materials generate a magnetic field that is opposite to the external B-Field, partially cancelling the field. Superconductors exhibit diamagnetism and can completely cancel out the magnetic field inside themselves.

43. d is the distance between the slits in the case of a double slit interference pattern, or the width of the slit, in the case of a single slit interference pattern.

m is the "order" of the points of constructive or destructive interference. In the case of a double slit interference pattern, it represents the difference in path length from one source to a point on the screen, and the path length from the other source in terms of wavelengths.

θ is the angle between the central maximum "m" order maxima or minima.

λ is the wavelength of the wave traveling through the slits.

44. A double slit interference pattern has fairly evenly spaced maxima and minima in the interference pattern, while the single slit has a wide central maximum. If the distance between the double slits is the same as the width of the single slit, the maxima on the double slit will be minima for the single slit and vice versa. The wide central maximum of the single slit pattern will be twice as wide as that of the double slit pattern.

45. The interference pattern is created because light behaves as a wave. Light passing through each slit must travel a distance toward the screen. This distance can be measured in wavelengths. When the difference in distance from the slits to the screen is in whole multiples of a wavelength, the waves line up crest to crest to form constructive interference. When the travel distance from the two slits to the screen is off by a ½ wavelength, the waves strike the screen crest to trough for destructive interference.

46. Both exhibit wave behaviors: interference, diffraction, refraction, reflection, Doppler effect, etc. However, electromagnetic waves do not require a medium to travel. They are a self-propagating electric and magnetic wave. The energy of a mechanical wave is dictated by its amplitude. The energy of an EM wave depends on its frequency $\left(E = hf = \frac{hc}{\lambda}\right)$. EM waves also exhibit the particle properties (photons).

47. Well, it depends. For clear substances, the remaining 20% of the light energy could have passed through and refracted. For opaque substances, the remaining 20% of the light energy is mostly absorbed. The key idea here is conservation of energy. Some light energy may reflect, some may transmit, and some may be absorbed, but it all adds up to the original 100% of the light energy.

48. When light enters a new medium, the wave fronts move at a new speed. This new speed causes the light to change direction. The only way light can travel into a new medium and not change direction is if it enters directly into the medium perpendicular to the surface, or if the speed of light in the new medium is the same as in the old medium.

49. (a) Light bends **toward** the normal when going from low to high index of refraction.

(b) Light bends **away from** the normal when going from high to low index of refraction.

(c) Total internal reflection can occur only when light goes from high to low index of refraction.

(d) $\sin \theta_c = 1.3/1.5$

50 For a convex (converging) lens:
- The incident ray parallel to the principal axis refracts through the *far* focal point.
- The incident ray through the *near* focal point refracts parallel to the principal axis.
- The incident ray through the center of the lens is unbent.

(Note that you don't necessarily need to know this third ray for ray diagrams, but it's legitimate.)

51. For a concave (diverging) lens:
- The incident ray parallel to the principal axis refracts as if it came from the *near* focal point.
- The incident ray toward the *far* focal point refracts parallel to the principal axis.
- The incident ray through the center of the lens is unbent.

(Note that you don't necessarily need to know this third ray for ray diagrams, but it's legitimate.)

52. One-eighth the original emission means it has gone through 3 half-lives ($\frac{1}{2} \times \frac{1}{2} \times \frac{1}{2} = \frac{1}{8}$) So the sample is 9 years old.

53. Photoelectric effect and photons

(a) If all you want is more electrons, just shine more green light on the surface.

(b) To get higher-energy electrons, you will need higher-energy light striking the surface. Use blue light, or UV light.

54. Conservation of charge, conservation of momentum, conservation of mass/energy (remember that mass and energy are two aspects of the same thing: $E = mc^2$), conservation of nucleon number.

55. The photon has transferred energy and momentum from itself to the electron. The frequency decreases and the wavelength increases.

$$\left(E = hf = \frac{hc}{\lambda}\right)$$

56. The electrons in the neon atom exhibit the wave properties of constructive and destructive interference. Thus, the elections can only exist in specific energy level locations. To jump up to the next energy level, the electron must absorb the specific amount of energy required to get there. It's kind of like walking up steps. When electrons fall downward from one energy level to another, they must emit a photon of the exact energy difference between the levels: hence the bands of light.

57. The wave function of a particle is a probability distribution that tells us the likelihood of finding the particle at a specific location. The larger the positive or negative amplitude of the wave, at a specific location, the higher probability of finding the particle. If the amplitude is zero at a location, the particle will not be found at that location.

58. The mass of the uranium is larger than the combined mass of the thorium and alpha particle. The loss in mass is due to the kinetic energy gained by the alpha particle in the decay ($E = mc^2$). Spontaneous nuclear reactions generate by-products that have less mass than the original.

59. $\lambda = \dfrac{hc}{\Delta E}$

is used to find the wavelength of a photon only. You can remember this because of the c, meaning the speed of light—only the massless photon can move at the speed of light.

$\lambda = \dfrac{h}{mv}$

is the de Broglie wavelength of a massive particle. You can remember this because of the m—a photon has no mass, so this equation can never be used for a photon.

60. Gamma decay doesn't affect the atomic mass or atomic number. In gamma decay, a photon is emitted from the nucleus, but because the photon carries neither charge nor an atomic mass unit, the number of protons and neutrons remains the same.

What Do I Know, and What Don't I Know?

I'll bet you didn't get every question on all of the fundamentals quizzes correct. That's okay. The whole point of these quizzes is for you to determine where to focus your study.

It's a common mistake to "study" by doing 20 problems on a topic with which you are already comfortable. But that's not studying . . . that's a waste of time. You don't need to drill yourself on topics you already understand! It's also probably a mistake to attack what for you is the toughest concept in physics right before the exam. Virtually every student has that one chapter they just don't get, however hard they try. That's okay.

The fundamentals quizzes that you just took can tell you exactly what you should and should not study. Did you give correct answers with full confidence in the correctness of your response? In that case, you're done with that topic. No more work is necessary. The place to focus your efforts is on the topics where either you gave wrong answers that you thought were right, or right answers that you weren't really sure about.

Now that you have a decent idea of your content knowledge, it's time to take a look at what the AP Physics 2 questions will be like. Turn the page to begin working on your AP question skills.

Test Yourself: AP Physics 2 Question Types

IN THIS CHAPTER

Summary: In the last chapter you analyzed your basic content knowledge. In this chapter you will test yourself with the types of questions you will encounter on the AP Physics 2 exam itself. If you have taken the AP Physics 1 exam, you will already have some experience with these types of questions. Practice makes perfect. A complete key to the practice questions can be found at the end of the chapter.

Key Ideas

- Content knowledge is only the first step. You need to know how questions will be organized and what they expect.
- Some question types will be new to you. Become familiar with their format.
- This self-assessment, and the fundamentals self-assessment from Chapter 4, will help you develop a study plan that matches your unique needs (Chapter 3).

AP Physics 2 Question-Types Self-Assessment

The AP Physics 2 exam has many different question types including:

- Descriptive problems
- Calculation problems
- Ranking tasks
- Experimental description and analysis questions
- Multiple questions referring to the same stem
- Multiple-correct questions
- Experimental design free-response questions
- Qualitative-quantitative free-response questions

Some of these may look familiar, while others will seem new or perhaps even odd. I don't want you to be surprised while taking the AP exam. Practice these problems to find which are easy for you and which ones are a struggle. Adjust your study plan accordingly (Chapter 3). Chapters 6–8 give strategies for handling the question types you will encounter.

Descriptive Problems (No Numbers)

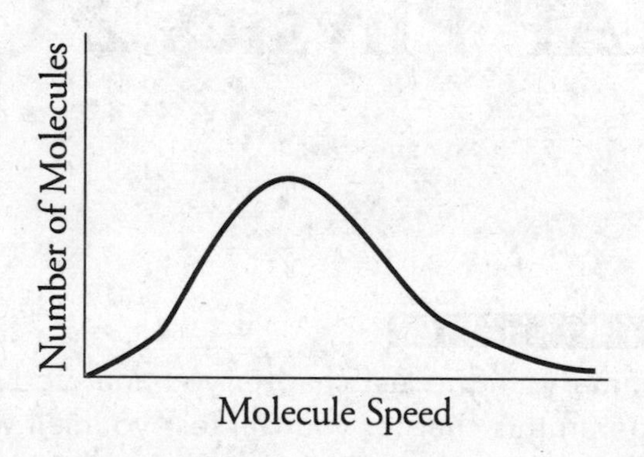

1. The graph shows the distribution of speeds for 1 mole of helium at temperature T, pressure P, and volume V. How would the peak of the graph change if the sample was changed from 1 mole of helium to 1 mole of hydrogen at the same temperature, pressure, and volume?

(A) The peak will shift to the left.
(B) The peak will shift upward and to the left.
(C) The peak will shift to the right.
(D) The peak will shift downward and to the right.

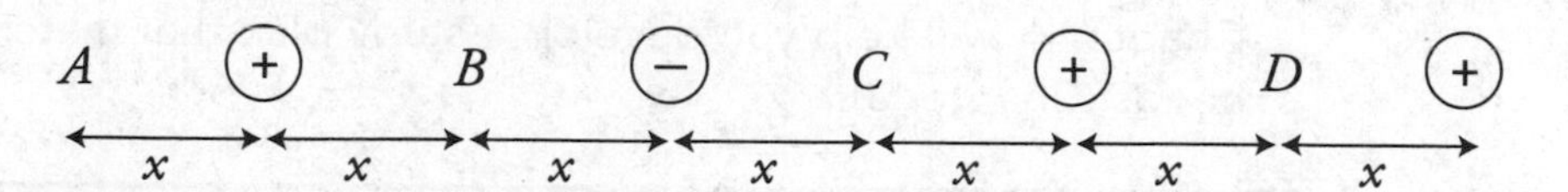

2. Four equally spaced charges of the same magnitude are arranged as shown in the figure. At which location is the electric field directed to the right?

Semiquantitative Reasoning (No Numbers but the Relationships in an Equation Will Show You the Way to the Solution)

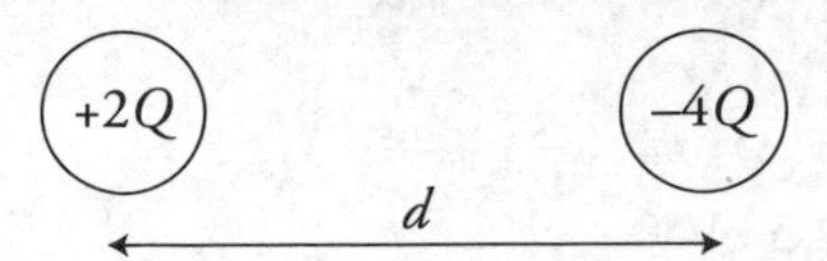

3. Two metal spheres with the same radius have charges of $+2Q$ and $-4Q$. They are separated by a distance d, which is much larger than their radius, as shown in the figure. The two spheres exert a force F on each other. The spheres are brought into contact and again are placed a distance d apart. What is the new force between the spheres?

(A) $\frac{1}{8}F$
(B) $\frac{1}{4}F$
(C) $\frac{1}{3}F$
(D) $\frac{1}{2}F$

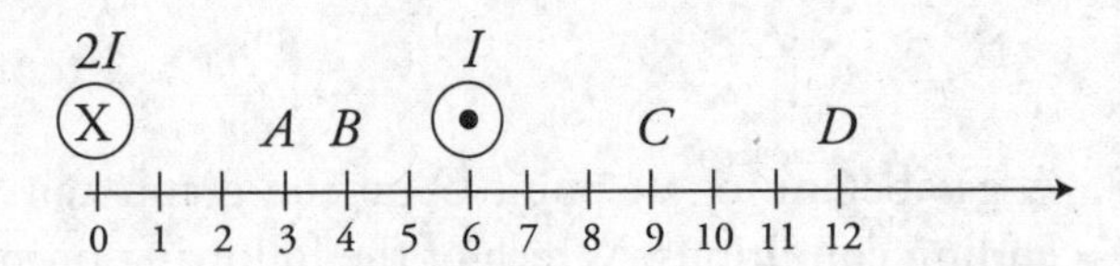

4. Two long wires carry current perpendicular to the page in opposite directions as shown in the figure. The left wire has twice the current of the right wire. At which location will the magnetic field be closest to zero?

(A) A
(B) B
(C) C
(D) D

Calculation Problems

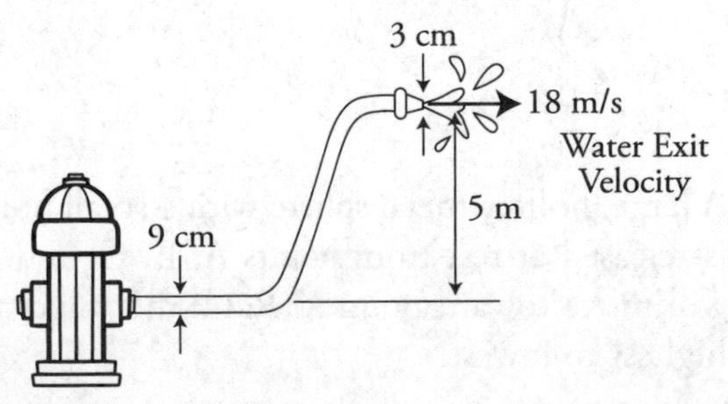

5. A firehose has a 3-cm diameter exit nozzle and a 9-cm diameter supply line from the hydrant that is 5 m below the nozzle as shown in the diagram. What pressure must be supplied at the hydrant to produce an exit velocity of 18 m/s? (Assume the density of water is 1000 kg/m^3 and the exit pressure is 1×10^5 Pa.)

(A) 1.6×10^5 Pa
(B) 2.1×10^5 Pa
(C) 2.9×10^5 Pa
(D) 3.1×10^5 Pa

6. In an LED, electrons are made to drop across an electric potential and emit a photon of light. What must be the voltage drop to produce a blue light of 500 nm?

(A) 1 eV
(B) 1.5 eV
(C) 2 eV
(D) 2.5 eV

Ranking Tasks

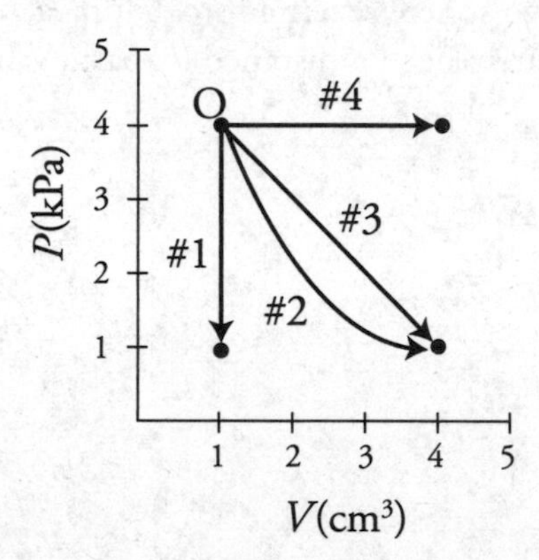

7. A gas beginning at point O on the graph can be taken along four paths to different ending conditions. Which of the following properly ranks the paths for change in temperature of the gas from most positive to most negative?

(A) $\Delta T_4 > \Delta T_3 > \Delta T_2 > \Delta T_1$
(B) $\Delta T_4 > \Delta T_3 = \Delta T_2 > \Delta T_1$
(C) $\Delta T_4 = \Delta T_1 > \Delta T_3 = \Delta T_2$
(D) $\Delta T_3 = \Delta T_2 > \Delta T_4 = \Delta T_1$

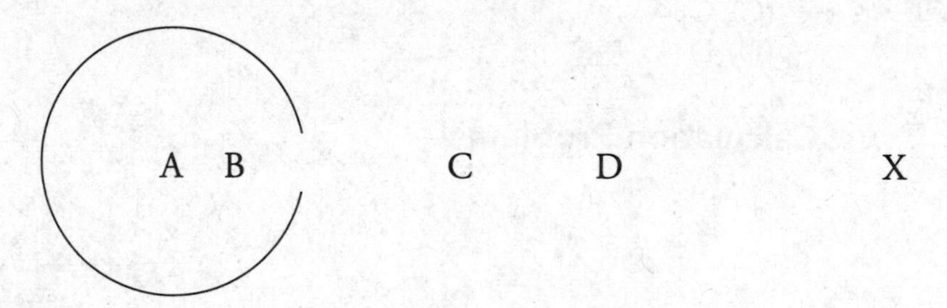

8. A large, hollow metal sphere with a small hole has a positive charge, as shown. A proton is released at rest from points A, B, C, and D. Assume that all protons, if they move, will move toward point X. Rank the velocity of the proton as it reaches point X from highest to lowest.

(A) A > B > C > D
(B) A = B > C > D
(C) D > C > B > A
(D) C > D > A = B

Experimental Description and Analysis

9. Students are asked to determine the number of molecules in a small, sealed pressurized cylinder of some unknown gas. There is a gauge on the cylinder that reads the internal pressure. What additional information and equipment will the students need to accomplish their task?

(A) a ruler and thermometer
(B) a ruler and type of gas inside the cylinder
(C) a thermometer, and type of gas inside the cylinder
(D) the type of gas in the cylinder and the average speed of the molecules

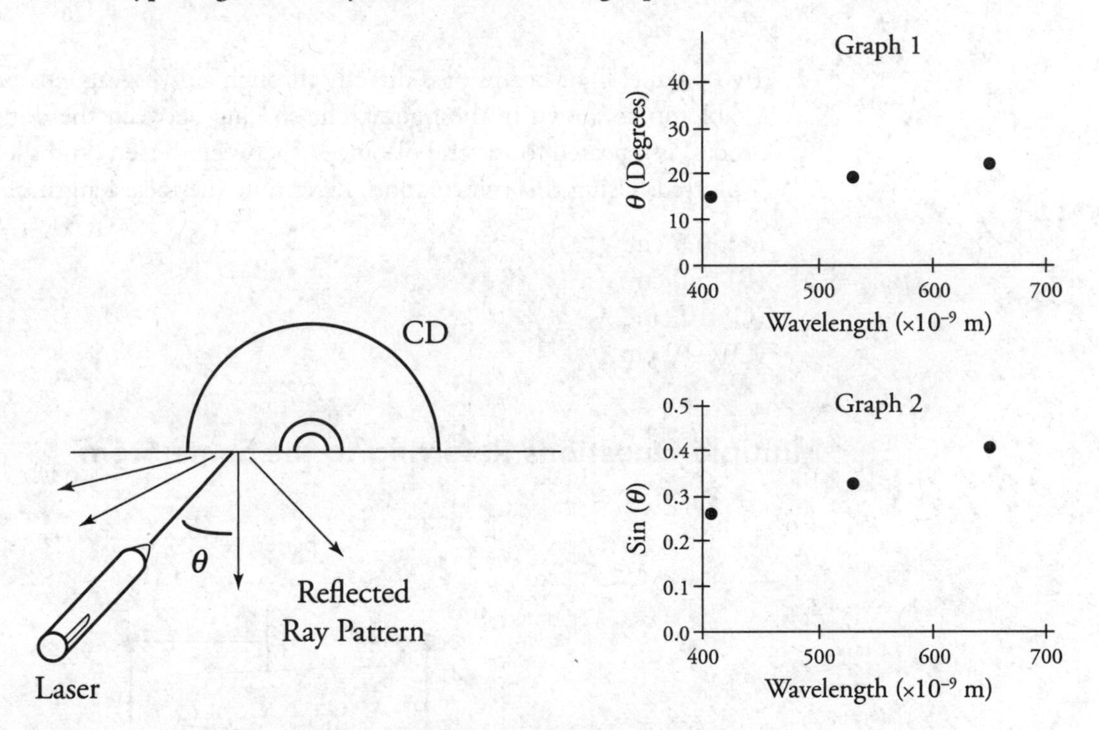

10. Students are attempting to calculate the narrow track spacing etched onto a CD. During the experiment, students shine a laser onto the surface of the CD to produce an interference pattern of reflected rays. The students measure the angle θ between the central ray and the first ray to the side as shown in the figure. This process is repeated for a red, green, and blue laser, and the data is presented in two graphs. Which of the following will be equal to the track spacing of the CD?

(A) the slope of the best-fit line for graph 1
(B) the slope of the best-fit line for graph 2
(C) the inverse of the slope of the best-fit line for graph 1
(D) the inverse of the slope of the best-fit line for graph 2

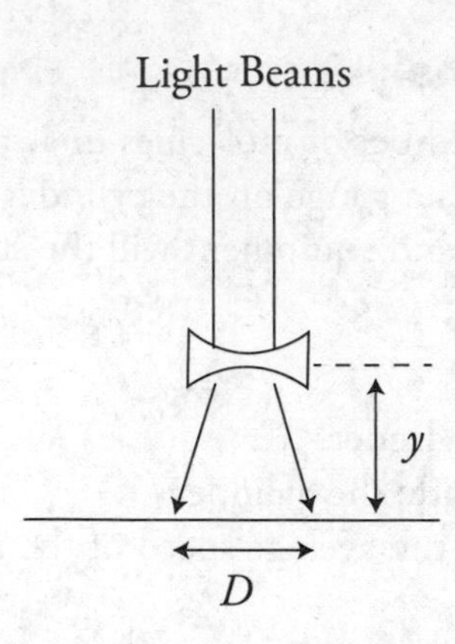

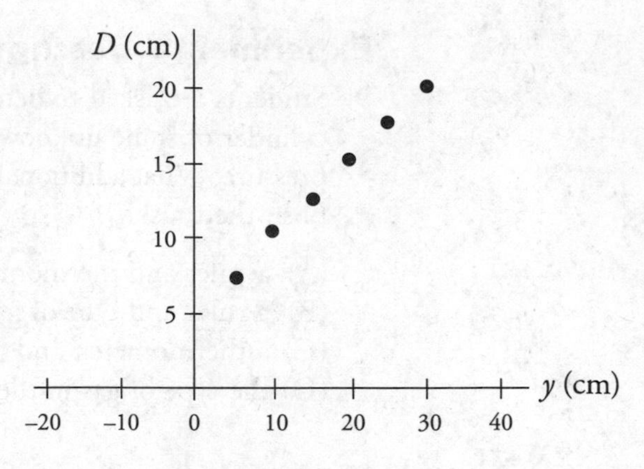

11. Two parallel light beams pass directly through a diverging lens producing two dots on a tabletop as shown in the figure. The spacing between the dots D is measured. This process is repeated for several distances y between the lens and the tabletop and the data is plotted. Using this information, determine the focal length of the lens.

(A) 0.5 cm
(B) 5.0 cm
(C) 10 cm
(D) −10 cm

Multiple Questions Referring to the Same Stem

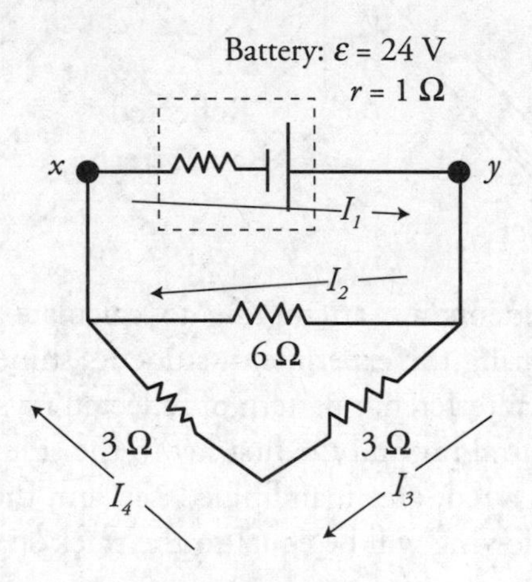

Questions 12–14 refer to the following material.

The circuit shown in the figure has a battery with an emf of 24 V and internal resistance of 1 Ω and three external resistors. The arrows show the currents I_1, I_2, I_3, and I_4 through each resistor.

12. What is the equivalent resistance of the three external resistors?

(A) 12 Ω
(B) 3.0 Ω
(C) 2.3 Ω
(D) 1.2 Ω

13. Which of the following correctly ranks the currents I_1, I_2, I_3, and I_4?

(A) $I_1 = I_2 > I_3 = I_4$
(B) $I_1 > I_2 = I_3 = I_4$
(C) $I_1 > I_2 > I_3 = I_4$
(D) $I_1 > I_2 > I_3 > I_4$

14. When operating as shown, what is the voltage difference V_{XY} of the battery?

(A) ε
(B) $\varepsilon - I_1(1\ \Omega)$
(C) $\varepsilon + I_1(1\ \Omega)$
(D) $I_2(6\ \Omega) + I_3(3\ \Omega) + I_4(3\ \Omega)$

Multiple-Correct Questions

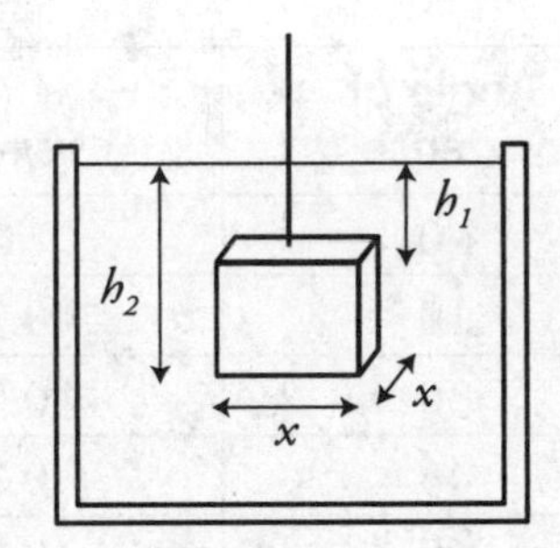

15. A mass m is suspended in a fluid of density ρ by a string as shown in the figure. The tension in the string is F_T. Which of the following are correct mathematical statements? **(Select two answers.)**

(A) $F_b = x^2 \rho g h_2$
(B) $F_b = x^2 \rho g (h_2 - h_1)$
(C) $F_b = mg$
(D) $F_b = mg - F_T$

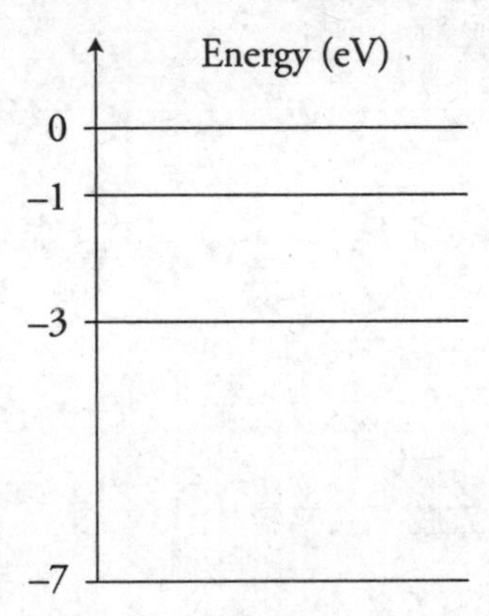

16. An energy-level diagram for an atom is shown in the diagram. Which of the following would be possible for this atom with an electron in the first excited state? **(Select two answers.)**

(A) absorption of an 8 eV photon
(B) absorption of a 1 eV photon
(C) emission of a 2 eV photon
(D) emission of a 4 eV photon

Experimental Design Free Response

17. In a laboratory experiment, your teacher has asked you to experimentally find the index of refraction of a clear rectangular block of glass. On the supply table are laser pointers (red, green, and blue) and other commonly available lab equipment.

(a) List the equipment you would choose to successfully carry out your investigation.
(b) Sketch the setup for your investigation. Indicate all the items you will be using. Clearly label and show the measurements you would take.
(c) Outline the experimental procedure you would use to gather the necessary data. Make sure the outline contains sufficient detail so that another student could follow your procedure.
(d) The table shows data taken by another group for a clear plastic block. Using this data, what quantities would you plot that would produce a straight line that could be used to determine the index of refraction of the plastic block?

Angle of Incidence θ_i	sin θ_i	Angle of Refraction θ_r	sin θ_r
0	0	0	0
10	0.17	8	0.14
30	0.50	20	0.34
45	0.71	31	0.51
60	0.87	35	0.57
80	0.98	46	0.72

(e) Using your answer from (d), plot the data and determine the index of refraction using a best-fit line.

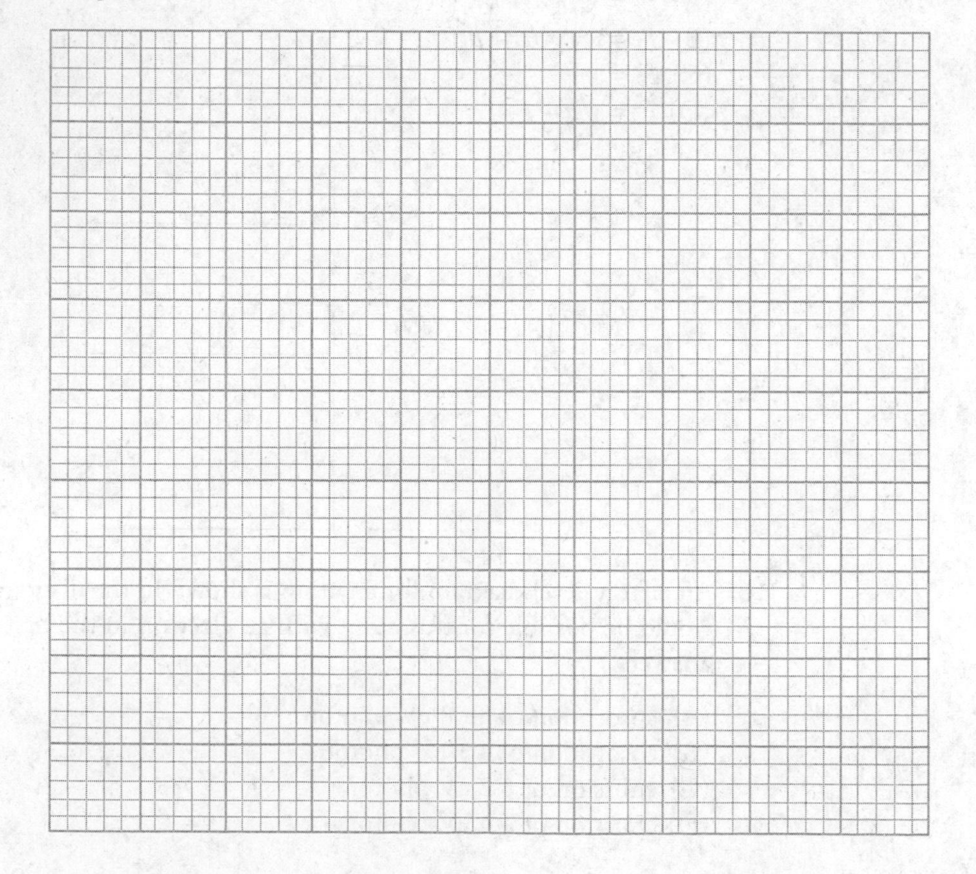

(f) Your teacher tells you that your glass block has a larger index of refraction than the other group's plastic block. Sketch what your data might look like on the graph from part (e).

Qualitative-Quantitative Transition Free Response

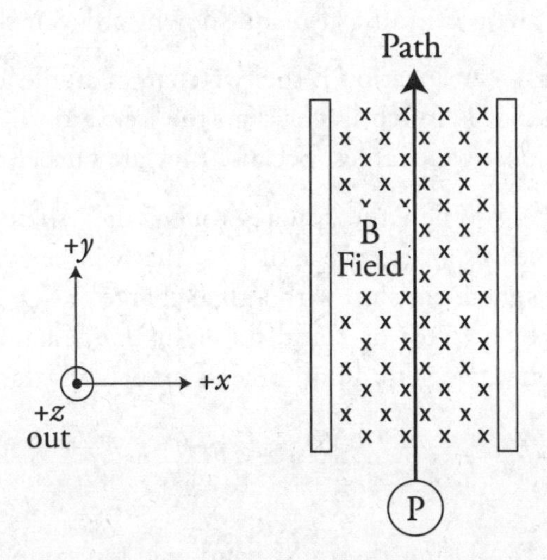

18. A proton traveling at a velocity v in the upward ($+y$) direction passes undeflected through a capacitor as shown. The capacitor plates are separated by a distance $d = 0.05$ m, have an area of $A = 0.004$ m^2, and a constant potential difference of $\Delta V = 20{,}000$ V. A uniform magnetic field $B = 0.40$ T directed into the page ($-z$ direction) completely covers the area between the plates. Assume the effects of gravity are negligible compared to other forces.

(a) Which capacitor plate has the highest electric potential? Explain how you came to this conclusion.

(b) Derive an algebraic expression for the velocity of the proton so that it maintains an undeflected path through the capacitor plates. Use your expression to calculate a numerical value for the velocity.

(c) The proton is replaced with an electron with the same velocity and launched upward through the plates. Compare the motion of the electron with that of the proton. Justify your response using your answer to (b).

(d) The electron is now given twice the velocity of the proton and travels upward through the plates. On the figure, sketch the path of this electron. Justify your answer.

(e) Derive an algebraic expression for the acceleration of the double-speed electron. Use your expression to calculate a numerical value for the acceleration.

Answers to AP Physics 2 Question-Types Assessment

1. **D**—Both samples have the same temperature and average molecular kinetic energy. Hydrogen is less massive and must move faster on average to have the same average kinetic energy. This means the peak will shift to the right. Since both samples have the same number of moles of gas, the area under the graph must be the same. This means the peak must also shift downward as it shifts to the right.

2. **B**—At location B the two charges on the left create an electric field directed to the right that is much larger than the leftward-directed electric fields created by the two other positive charges, because they are farther away.

3. **A**—When the spheres touch, they share the combined charge by conduction. This leaves a net charge of $-2Q$ that will be equally distributed on each sphere. Thus, each sphere ends up with a final charge of $-Q$. This means that the left charge has decreased by a factor of 2 and the right charge has decreased by a factor of 4. Using Coulomb's law, we get a final force of one-eighth the original:

$$F_{New} = k\frac{(\frac{1}{2}q_1)(\frac{1}{4}q_2)}{r^2} = \frac{1}{8}k\frac{q_1 q_2}{r^2} = \frac{1}{8}F$$

4. **D**—Using the right-hand rule for magnetic fields around current carrying wires, we determine that the magnetic field rotates clockwise around the left wire and counterclockwise around the right wire. Thus, choices A and B cannot be correct, as the two B-fields add in the downward direction. Thus, the choice is between C and D. Using the equation: $B = \frac{\mu_0}{2\pi}\frac{I}{r}$ we see that the left wire, with twice the current, must also have twice the radius in order to produce the same size B-field as the right wire: $\frac{\mu_0}{2\pi}\frac{2I}{2r} = \frac{\mu_0}{2\pi}\frac{I}{r}$

   Point D satisfies this condition.

5. **D**—Using the conservation of mass/continuity equation, we see that the water at the hydrant must be going slower than at the nozzle: $A_1 v_1 = A_2 v_2$.

   The area of the hose is proportional to the diameter squared: $\pi r_1^2 v_1 = \pi r_2^2 v_2$. This gives a velocity at the hose of 2 m/s.

   Using conservation of energy/Bernoulli's equation:

$$(p + \rho g y + \frac{1}{2}\rho v^2)_1 = (p + \rho g y + \frac{1}{2}\rho v^2)_2$$

$$p_1 + 0 + \frac{1}{2}(1000 \text{ kg/m}^3)(2 \text{ m/s})^2$$

$$= \left(100{,}000 \text{ Pa} + (1000 \text{ kg/m}^3)(10 \text{ m/s}^2)(5\text{m}) + \frac{1}{2}(1000 \text{ kg/m}^3)(18\text{m/s})^2\right)_2$$

   Plugging in the values, we get a required pressure of 3.1×10^5 Pa at the hydrant.

6. **D**—$E = hf = \frac{hc}{\lambda} = \frac{1240 \text{ eV·nm}}{500 \text{ nm}} = 2.48 \text{ eV}$

7. **B**—Temperature is directly related to the *PV* value. The starting position has a *PV* value of 4*PV*. Path 4 takes the gas to a *PV* value of 16*PV*, so ΔT is positive. Paths 2 and 3 take the gas to a *PV*, value of 4*PV* so ΔT is zero. Path 1 takes the gas to a lower final *PV* value so ΔT is negative.

8. **D**—Inside a charged conductor, the electric field is zero and the electric potential is constant. Thus, the charges inside the sphere do not experience a force and will not move at all. Outside of the sphere, the positive charges are repelled to the right. The electric potential decreases as the charges move away from the sphere. Since charge C will experience the largest change in potential, it will gain more kinetic energy than D.

9. **A**—With a ruler, the volume of the cylinder can be calculated. Knowing the volume, temperature, and pressure, the number of moles and number of molecules of gas can be calculated using the ideal gas law.

10. **D**—The independent variable is wavelength and the dependent variable is angle. Solving the interference pattern equation ($d \sin \theta = m\lambda$) for the independent variable, we get: $\sin \theta = \frac{m}{d}\lambda$, where *m* equals 1 because the students measure the angle to the first order maxima. Thus, to find the track spacing, we need to find the slope of graph 2 and take the inverse of it.

11. **D**—Extrapolating the data to where the parallel light beams would cross, we get a focal length of −10 cm.

12. **B**—The two 3 Ω resistors are in series giving 6 Ω, which adds in parallel with the 6 Ω resistor to give a total resistance of 3 Ω.

13. **B**—I_I is in the main line and must be the largest. I_1 splits into I_2 and I_3, which must be the same because the resistance in each line is 6 Ω. $I_3 = I_4$. They are the same because there are no branching lines.

14. **B**—Between points X and Y the battery contributes $+\varepsilon = +24$ V while the internal resistance consumes $-\Delta V = -Ir = -I_1\ (1\ \Omega)$.

15. **B and D**—$F_b = \rho V g = x^2 \rho g(h_2 - h_1)$ where $x^2(h_2 - h_1)$ is the volume of the mass. Since the mass is stationary, the forces on it must cancel out: $F_b + F_T = mg$.

16. **A and D**—From the first excited state, the electron could absorb the 8 eV photon, which will ionize the atom sending the electron away with 5 eV of kinetic energy. The atom could also emit a 4 eV photon, which would send the electron down to the −7 eV energy level.

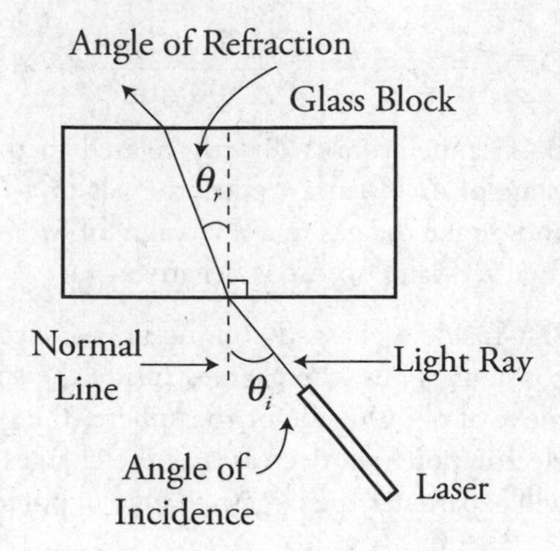

17. (a) Start with the equation that models refractive behavior, Snell's law: $n_i \sin \theta_i = n_r \sin \theta_r$. You will need a way to measure angles. This could be accomplished with a protractor, or just a ruler if you choose to use trig functions to calculate the angle. You will also need a light source (one of the lasers) and a method for tracing the angles, like a piece of paper, pencil, and ruler.

(b) See sketch.

(c) Place the glass block on a piece of paper and trace the outline. Shine a laser into the block and trace its path. Mark the exit location of the beam on the other side of the block. Remove the block from paper and use exit marks to trace the ray's path inside the glass. Draw the normal line at the entry point of the ray into the glass, and then measure the angles of incidence and refraction as shown in the picture. Repeat for multiple angles.

(d) This is called linearizing the graph. Plot $\sin \theta_r$ as a function of $\sin \theta_i$ to create a straight line. The slope will equal $1/n_r$. (You could also swap the axis, but I plotted the independent variable on the *x*-axis and the dependent variable on the *y*-axis. As you should!) We will discuss more about linearization in Chapter 6.

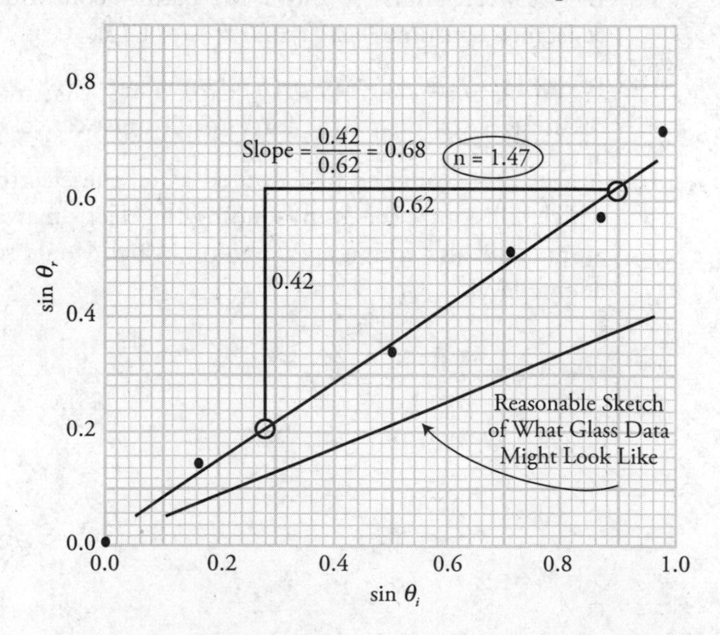

(e) Notice how the best fit line represents the average of the data and does not even go through any of the data points! Do not use the data points to calculate the slope. Pick two convenient points from the best-fit line as shown in the figure. The slope is 0.68 = $1/n_r$. Invert and get $n_r = 1.47$.

(f) Since the slope of the graph is the inverse of the index of refraction, and the glass block has a larger index of refraction, the graph for the glass block should have a shallower slope and still pass through zero. (See the graph.)

18. (a) The left plate. Using the right-hand rule for magnetic forces on moving charges, we see that the magnetic force will be directed to the left. Therefore, we need an electric force to the right. Since the proton is positively charged, we need an electric field to the right to accomplish this, and thus, the left plate must be positive and the right plate negative.

(b) For a charged particle to pass undeflected straight through crossed perpendicular magnetic and electric fields, the electric and magnetic forces on the charge must cancel.

$$F_{Electric} = F_{Magnetic}$$
$$Eq = qvB \sin \theta$$

Note that since the velocity is perpendicular to the magnetic field: $\sin \theta = \sin (90°) = 1$. Therefore, $E = vB$ when the charge travels in a straight line through the fields. $v = \frac{E}{B} = 1.0 \times 10^6$ m/s

(c) The electron also passes through the plates undeflected. The charge canceled out of the equation derived in (b) $v = \frac{E}{B}$.

(d) The electron will curve to the right between the plates. The magnitude of the electric force does not change. However, the increased velocity of the electron causes the magnetic force to become larger. Using the right-hand rule for forces on moving charges, we can see that the increased magnet force causes the electron to accelerate to the right.

(e) $\Sigma F = F_M - F_E = qvB - Eq = qvB - \frac{\Delta V}{d} q = m_e a$

$$a = \frac{qvB - \frac{\Delta V}{d} q}{m_e} = 7.0 \times 10^{16} \text{ m/s}^2$$

Develop Strategies for Success

General Strategies

IN THIS CHAPTER

Summary: This chapter contains general strategies useful for the entire AP Physics 2 exam—multiple-choice and free-response sections. First, let's talk about AP Physics 1 and what you need to remember. Second, I'll discuss the tools you have at your disposal (calculator, a table of information, and equation sheet) and how to use them. Next, I'll investigate what those equations you are given mean, how to relate them to a graph, and how to use a graph to find information. Finally, I'll work on ranking task skills.

Key Ideas

- You should dust off that *5 Steps to a 5 AP Physics 1* book you had last year. The skills you learned in AP Physics 1 are going to be needed.
- Sure you can have a calculator, but it won't help you for most of the exam. Only use it when you actually need it.
- The table of information/equation sheet is good to have in a pinch, but it won't save you if you don't know what it all means.
- Each equation tells a story of a relationship. Graphs are a picture of these relationships. Learn to see the relationships.
- There are three ways to get information from a graph: (1) read it, (2) find the slope, and (3) calculate the area under the graph.

✪ Ranking task questions show up in both multiple-choice and free-response questions. Some require conceptual analysis and others have numbers.

What Do I Need to Remember from AP Physics 1?

The short answer is everything. The prior skills you learned in AP Physics 1 are expected knowledge on the AP Physics 2 exam. Don't panic. You won't be asked any questions about blocks on an incline attached to a pulley. Only the content in AP Physics 2 is tested. However, the information you learned about forces, energy, momentum, motion, graphing, free body diagrams, and all the rest is assumed to be still accessible in your brain. Physics is cumulative. There won't be any roller coasters going around a track, but there will be charged particles that experience forces, accelerate, and convert potential energy into kinetic energy. All the skills you learned last year will help you this year.

So what do you do if all that past information is fuzzy? Ask your teacher to review the concepts and dust off that *5 Steps to a 5 AP Physics 1* book you had last year.

Tools You Can Use

The Calculator

You can use a calculator on both sections of the AP exam. Most calculators are acceptable—scientific calculators, programmable calculators, graphing calculators. However, you cannot use a calculator with a QWERTY keyboard, and you'll be restricted from using any calculators that make noise. You also cannot share a calculator with anyone during the exam.

The real question, though, is whether a calculator will really help you. The short answer is "Yes": You will be asked a few questions on the exam that require you to do messy calculations. The longer answer, though, is "Yes, but it won't help very much."

The majority of the questions on the exam, both multiple choice and free response, don't have any numbers at all.

There are questions that have numbers but don't want a numerical answer. For example:

A convex lens of focal length $f = 0.2$ m is used to examine a small coin lying on a table. During the examination the lens is held a distance of 0.3 m above the coin and is moved slowly to a distance of 0.1 m above the coin. During this process, what happens to the image of the coin?

(A) The image continually increases in size.
(B) The image continually decreases in size.
(C) The image gets smaller at first and then bigger in size.
(D) The image flips over.

The numbers in these problems are only there to set the problem up. (The correct answer is D).

Then there are questions with numerical answers but using a calculator is counterproductive. For example:

A cylinder with a movable piston contains a gas at pressure $P = 1 \times 10^5$ Pa, volume $V = 20$ cm^3, and temperature $T = 273$ K. The piston is moved downward in a slow, steady fashion, allowing heat to escape the gas and the temperature to remain constant. If the final volume of the gas is 5 cm^3, what will be the resulting pressure?

(A) 0.25×10^5 Pa
(B) 2×10^5 Pa
(C) 4×10^5 Pa
(D) 8×10^5 Pa

Using your calculator to solve this one will take too much time. You can do this one in your head: $PV = nRT$, nRT is constant. So, if the volume is four times smaller, the pressure has to be four times greater! Correct answer (C) 4×10^5 Pa. In fact, many times the numerical calculations are simple or involve ratios that don't require a calculator.

Here is the big takeaway—use your calculator only when it is absolutely necessary.

Special Note for Students in AP Physics 2 Classes
Many, if not most, of your assignments in class involve numerical problems. What can you do? Start by trying to solve every problem without a calculator first. Be resourceful. Draw a diagram, sketch a graph, use equations with symbols only, etc. Second, work the conceptual problems from your textbook and ask your teacher for the key. Practice the skills that will make you successful on the AP exam.

The Table of Information and the Equation Sheet

The other tools you can use are the table of information and equation sheet. You will be given a copy of these sheets in your exam booklet. It's a handy reference because it lists all the constants, math formulas, and the equations that you're expected to know for the exam.

However, the equation sheet can also be dangerous. Too often, students interpret the equation sheet as an invitation to stop thinking: "Hey, they tell me everything I need to know, so I can just plug-and-chug through the rest of the exam!" Nothing could be further from the truth.

First of all, you've already *memorized* the equations on the sheet. It might be reassuring to look up an equation during the AP exam, just to make sure that you've remembered it correctly. And maybe you've forgotten a particular equation, but seeing it on the sheet will jog your memory. This is exactly what the equation sheet is for, and in this sense, it's pretty nice to have around. But beware of the following:

- Don't look up an equation unless you know *exactly* what you're looking for. It might sound obvious, but if you don't know what you're looking for, you won't find it.
- Don't go fishing. If part of a free-response question asks you to find an object's velocity, and you're not sure how to do that, don't just rush to the equations sheet and search for every equation with a "V" in it.

If your teacher has not issued you the official AP Physics 2 table of information and equation sheet, download one from the College Board at https://secure-media.collegeboard.org/digitalServices/pdf/ap/ap-physics-2-equations-table.pdf. Exam day shouldn't be the first time you see these tools.

Get to Know the Relationships

Now that you have an official AP Physics 2 equation sheet, let's talk about what the jumble of symbols tell us. Take a look under the "FLUID MECHANICS AND THERMAL PHYSICS" heading. See the equation $PV = nRT$? What does it tell us? It shows us how all these individual quantities are related and what their relationship is. Rearranging the equation for P we get: $P = \frac{nRT}{V}$. T is in the numerator, which means that if T doubles, and all the other variables on the right stay the same, P must also double. Pressure is directly proportional to temperature: $P \propto T$. See graph #2 below. If V doubles, and all the other variables stay the same, P will be cut in half. Pressure is inversely proportional to volume: $P \propto \frac{1}{V}$. On a graph, an inverse relationship looks like #4. What other relationships are we likely to see? Shown below are the six most frequent relationships in AP Physics 2.

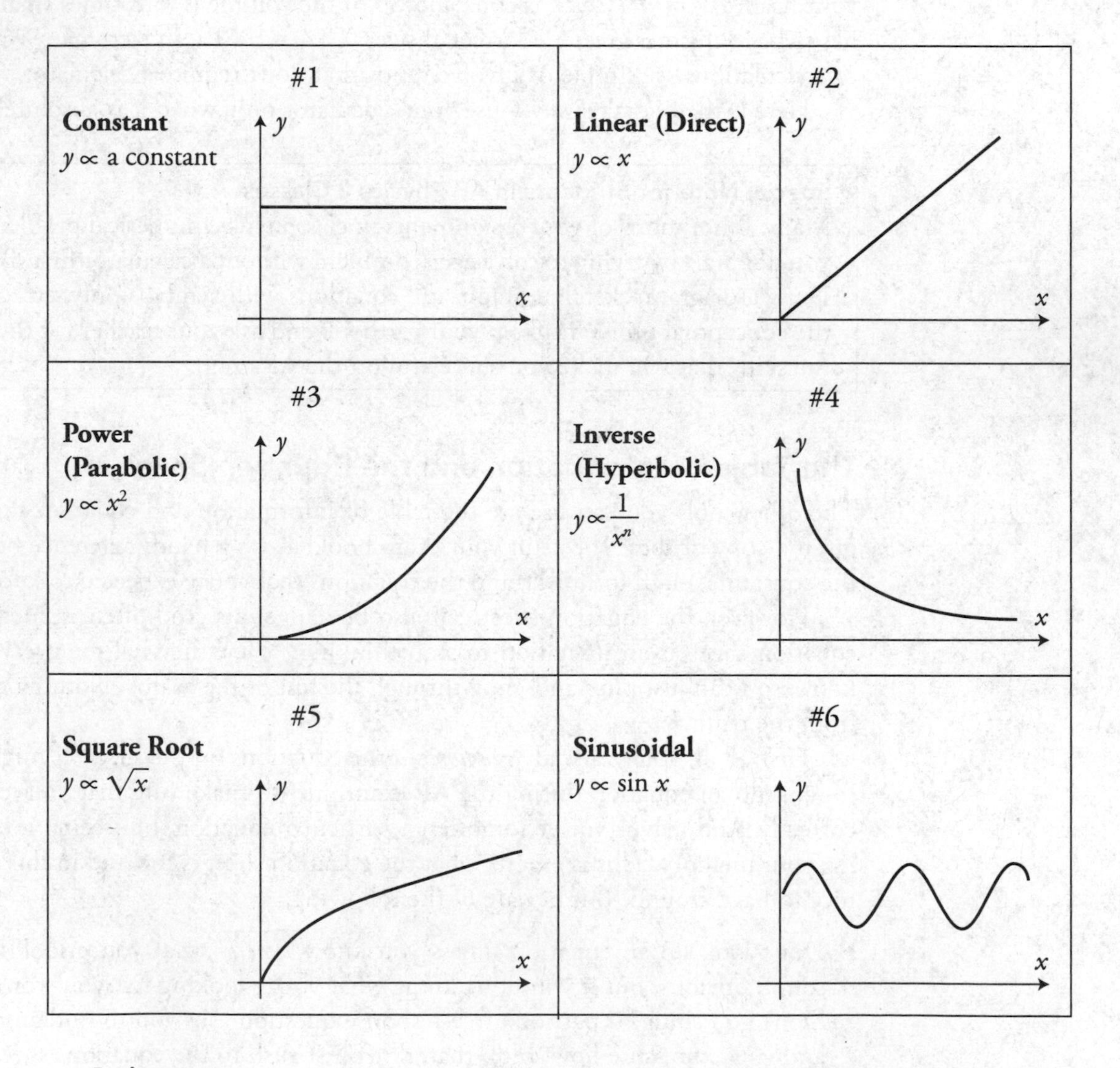

Let's practice.

Kinetic energy $K = \frac{1}{2}mv^2$. Kinetic energy is directly proportional to the velocity squared, $K \propto v^2$. It will have a graph like #3. If you double the velocity, the kinetic energy quadruples: $4K \propto (2v)^2$.

Rearrange the kinetic energy equation to solve for $v = \sqrt{\frac{2K}{m}}$. Velocity is proportional to the square root of the kinetic energy $v \propto \sqrt{K}$; see graph #5. If you double the kinetic energy, the velocity goes up by a factor of $\sqrt{2}$.

Electric force: $|\vec{F}_E| = \frac{1}{4\pi\varepsilon_0}\frac{|qq|}{r^2}$. The electric force is inversely proportional to the radius squared $F_E \propto \frac{1}{r^2}$, but it is directly related to the charge $F_E \propto q$. See graphs #2 and #4. This means that if you double one charge and also double the radius, the force will be cut in half: $\frac{1}{2}F_E \propto \frac{2q}{(2r)^2}$.

There are always questions on the exam that can be solved this way. Learning how to work with these relationships is crucial to doing well on the exam because they don't require a calculator and save you time. Put your calculator away and practice this skill all year long.

What Information Can We Get from a Graph?

Gathering information from a graph is another highly prized skill on the AP exam. Let's spend some time making sure you have it down cold. The good thing is there are only three things you can do with a graph: read it, find the slope, or find the area. But, before you can do that, you need to examine the graph. Look at the *x*-axis and *y*-axis. What do they represent? What are the variables? What are the units? Which physics relationships (equations) relate to this graph?

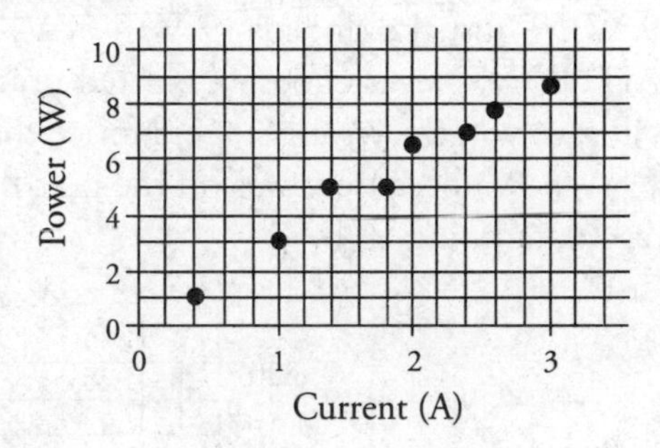

1. Read the Graph

Look at this graph. The data is not in a perfect straight line. This is common on the AP exam, as it includes real data like you would get in an actual lab. If you get data like this, sketch a line or curve that seems to best fit the data.

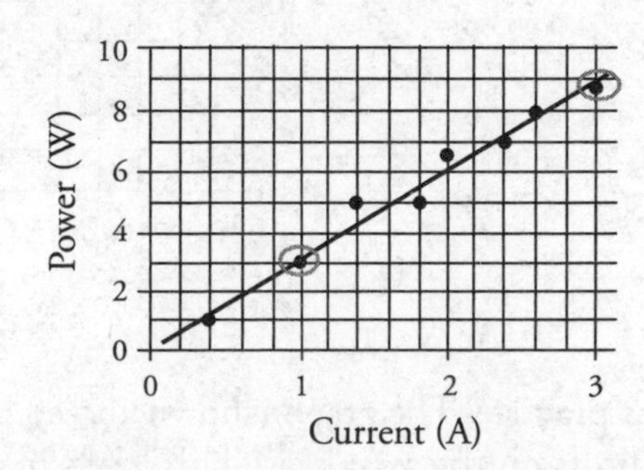

This data seems straight, so draw a "best-fit" line through the data that splits it down the middle. Your line may not touch any of the data points. That's OK. Once you have your best-fit line, forget about the data points and concentrate only on the line you have drawn. The AP exam may ask you to extrapolate beyond the existing data or interpolate between points. For example: at a current of 2 amps the power is approximately 6 watts.

2. Find the Slope

$$P = \Delta V \; I$$
$$y = m \; x + b$$

slope = ΔV

y-intercept = 0

In math class you calculated the slope of lines, but most of the time it didn't have a physical meaning. In physics, the slope can represent something real. Take a look at the axis on the graph. Power is on the y-axis and current on the x-axis. Ask yourself if there is a physics relationship between power and current. $P = I\Delta V$ seems to fit the bill. In math class the equation of a line is $y = mx + b$. Now line up the physics equation with the math equation to find out what the slope's physics meaning is. Turns out that the slope is the potential difference!

This procedure of matching up the physical equation with the math equation of a line will help you find the physics meaning of the slope every time.

Now that we know the slope represents the potential difference, we need to calculate it. Slope is rise over run. Pick two convenient points. I used points (1 A, 3 W) and (3 A, 9 W). Thus, the slope is (9 W – 3 W) / (3 A – 1 A) = 3 W/A = 3 V.

CAUTION! Never choose a plotted point unless it actually falls on your best-fit line. This will give you the wrong slope. Notice that one of the chosen points was an actual data point (1 A, 3 W) because it was on the best-fit line. That's OK. The other point (3 A, 9 W) was not a data point.

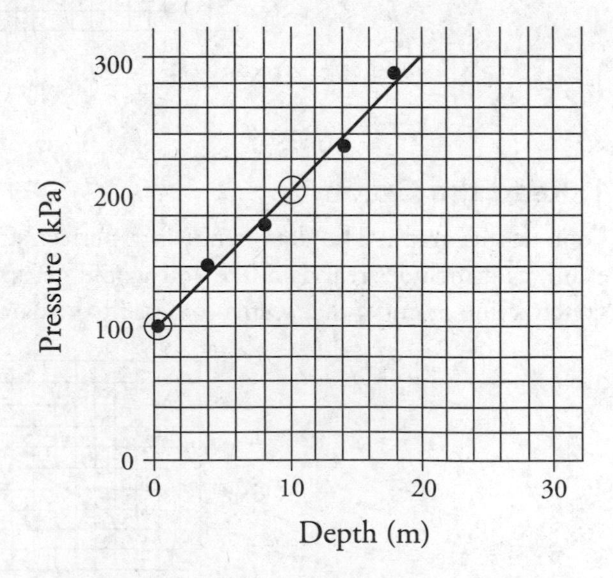

Let's practice. The graph (shown above) has pressure as a function of depth. (Watch out for the units on the axis: kPa.) This looks like a fluids problem. The equation $P = P_0 + \rho gh$ seems to fit. Now let's match the physics equation with the math equation.

$$P = \rho g \; h + P_0$$
$$y = m \; x + b$$

slope = ρg
y-intercept = P_0

So the slope is the density times the acceleration due to gravity. Using the circled points, we get a slope of 10 kPa/m. This time there is a y-intercept, which is the pressure on top of the fluid, $P_0 = 100$ kPa.

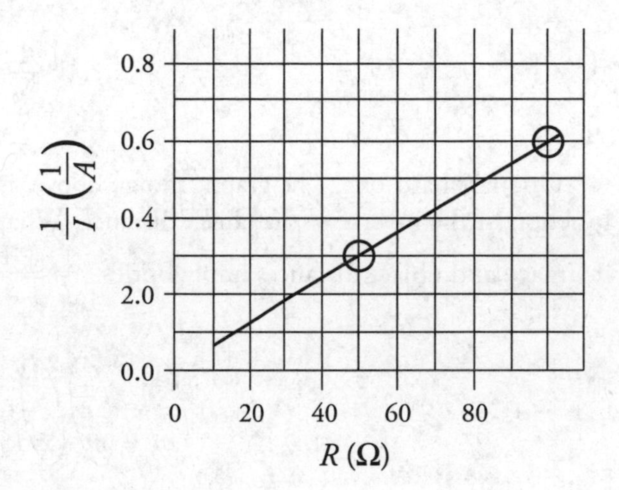

Sometimes the axes are strange. Take a look at the graph, which shows the inverse of current as a function of resistance. Don't fret! What is the physics relationship between current and resistance? $I = \frac{\Delta V}{R}$. Now match up this equation with $y = mx + b$:

$$\frac{1}{I} = \frac{1}{\Delta V} \; R$$
$$y = m \; x + b$$

slope = $\frac{1}{\Delta V}$
y-intercept = 0

The slope is the inverse of the potential difference. That's kind of strange but if that is the graph they give you, just go with it. The slope = 0.006 (1/A)/Ω. Taking the inverse, we get the potential difference of 167 V.

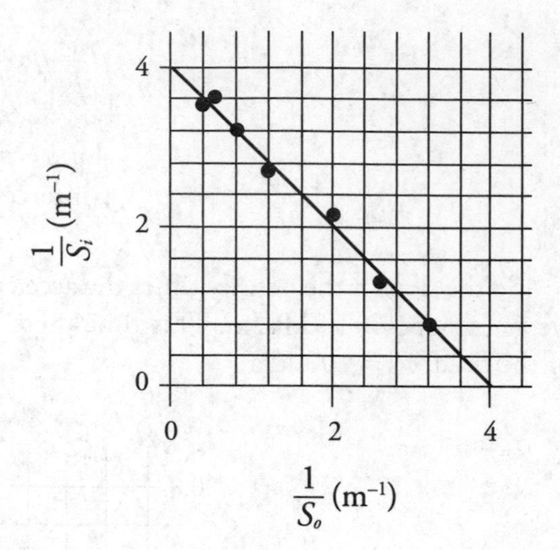

One last hard one! The graph, shown above, is the inverse of the image distance as a function of the inverse of the object distance. What a mess, but who cares? We can handle it. Image and object distances imply optics: $\frac{1}{f} = \frac{1}{S_o} + \frac{1}{S_i}$. Now match the equation up:

$$\frac{1}{S_i} = (-)\frac{1}{S_o} + \frac{1}{f}$$

$$y = m \; x + b$$

slope = −1

y-intercept = $\frac{1}{f}$

The slope turns out to equal −1 and does not have any physical meaning this time. The y-intercept is the inverse of the focal length.

Occasionally the AP exam asks that you to take a graph that is curved and produce a graph that has a straight line. This is called linearization. It is the reverse process of above. Take the equation and match it up with $y = mx + b$ to see what you should graph. Take this equation: $n_1 \sin \theta_1 = n_2 \sin \theta_2$. Let's put the θ_1 on the x-axis and θ_2 on the y-axis. Match the equation. Plot what it tells you to plot, and you get a straight line from your data. Piece of cake.

$$\sin \theta_2 = \frac{n_1}{n_2} \sin \theta_1$$

$$y = m \; x + b$$

plot $\sin \theta_2$ on y-axis

plot $\sin \theta_2$ on x-axis

slope = $\frac{n_1}{n_2}$

y-intercept = 0

3. Find the Area

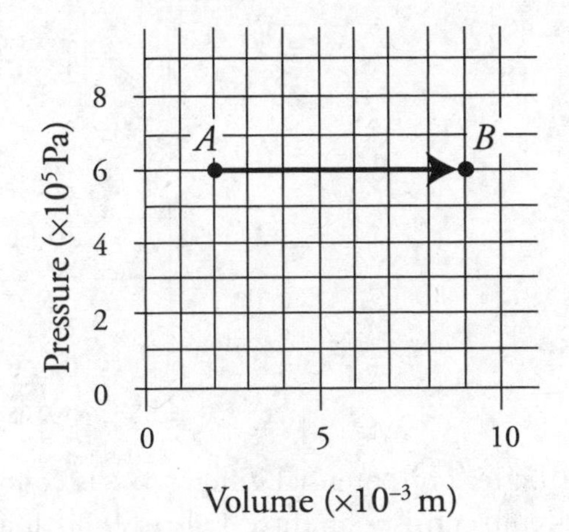

This time we look to see if multiplying the *x*- and *y*-axis variables will produce anything meaningful. If so, the space under the graph has a physical meaning. Take a look at the graph above—pressure times a changing volume. That looks like something to do with gases: $W = -P\Delta V$. The area under the graph is the work. Calculate the area of a graph just like the area of a geometric shape. Keep in mind that the "units" of our graph area will be Pa · m not a geometric unit like meters squared.

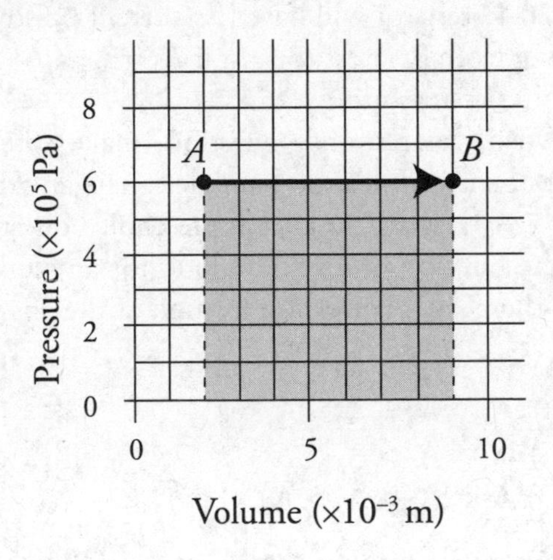

The area of this graph is $(6 \times 10^5\ \text{Pa})(9 \times 10^{-3}\ \text{m} - 2 \times 10^{-3}\ \text{m}) = 4200\ \text{Pa} \cdot \text{m} = 4200\ \text{J}$. Since our equation was $W = -P\Delta V$, our final answer is negative: $W = -4200$ J of work.

Ranking Task Skills

Ranking tasks are an interesting type of question that can show up in both the multiple-choice and free-response portions of the exam. Here is an example:

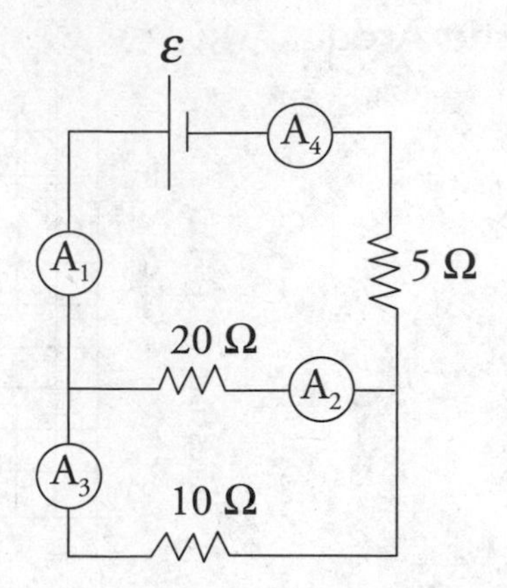

A battery of potential difference ε is connected to the circuit pictured above. The circuit consists of three resistors and four ammeters. Rank the readings on the ammeters from greatest to least.

They are not asking for any numbers, and in most cases, trying to use numbers to solve the problem is much more time-consuming than using conceptual reasoning and semi-quantitative reasoning. In this example, we see that ammeters A_1 and A_4 have the same current because they are in a same single pathway. This is the main pathway that feeds the rest of the circuit. This main current splits before passing through the lower two resistors. The 20-Ω resistor will have less current passing through it than the 10-Ω resistor. Thus, the ranking from greatest to least is $A_1 = A_4 > A_3 > A_2$. No numerical calculations were needed, just physics reasoning.

On a free-response question, make sure you write your answer in a clear way that cannot be misunderstood and designate any that are equal. For example: **Greatest** $(A_1 = A_4) > A_3 > A_2$ **Least**. On a multiple-choice question, look to save time. As soon as you figure out the ranking of any pair, look for any answer choices that don't have that pairing and cross them out. For example, look at the answer choices below:

(A) $A_1 > A_3 > A_2 > A_4$
(B) $A_1 > A_3 = A_2 > A_4$
(C) $A_1 > A_4 > A_3 > A_2$
(D) $A_1 = A_4 > A_3 > A_2$

When you determine that $A_1 = A_4$, a quick look at the answer choices shows that only answer choice (D) will work.

There are students who are more comfortable with numerical thinking. If that is the case with you, you can choose a number for ε and work out the currents. But this will almost always take longer.

Strategies for the Multiple-Choice Questions

IN THIS CHAPTER

Summary: The multiple-choice section of the AP Physics 2 exam contains a wide variety of question types. Some you may be unfamiliar with. This chapter contains strategies to attack this rogues' gallery of questions.

Key Ideas

- There are 50 multiple-choice questions. All of them require deep understanding of physics.
- You should practice a system when working multiple-choice questions. Make sure you know why the correct answer is correct and why the wrong answer is incorrect.
- The last five questions will be "multiple-correct" questions. You will need to choose the two correct responses. There is no partial credit for getting only one correct.
- Time . . . time . . . there is not enough time. Pace yourself so that you have an opportunity to work on every multiple-choice question. Answer every question because there is no guessing penalty.

Multiple-Choice Basics

The multiple-choice section is the first half of the exam. You have 90 minutes to answer 50 questions. Each question has four answer choices. The first 45 questions are the normal "one-correct" response that you are used to. The last five are "multiple-correct" questions where you need to choose both correct responses. There is no penalty for guessing, so be sure to answer every question. I'm going to repeat that so it sinks in—ANSWER EVERY QUESTION because it can only help your score. Don't leave points on the table. Questions not answered = points not earned.

The multiple-choice questions do not involve simple recall of information, facts, or data. You are going to have to put your physics reasoning to work answering these questions. The questions will test all your skills. You will encounter graphs, lab-based questions, numerical problems, proportional and conceptual reasoning, problems with only symbols, ranking tasks, and questions with nothing but words. Some of the questions will seem easy to you and others will be gut-crunching hard.

How to Get Better at Multiple-Choice Problems

Most students read the stem of the question and begin to search for the correct answer. If what they are looking for is not given as a possible response, they guess or give up and move on. I'd like you to practice a different approach that will lead you to better results.

1. Read the stem of the question, paying close attention to any diagrams or graphs that can give you crucial information. (Remember that graphs give us three types of information. See Chapter 6.)
2. Ask yourself what kind of physics is involved. AP Physics 2 has eight major content areas:
   - Fluids
   - Ideal Gases
   - Thermodynamics
   - Electric Force, Field, and Potential
   - Circuits
   - Magnetism and Electromagnetic Induction
   - Waves and Optics
   - Quantum, Atomic, and Nuclear (also called Modern Physics)

   Which content area does this problem fall into? Sometimes more than one content area will apply. This should take only a second to figure out.
3. Think about the tools from physics that would be useful in this situation. I always tell my students to put on their "physics glasses" and tell me what they see. You have six pairs of glasses to view the world:
   - Energy glasses (What energy transformations are occurring? Is there work being done? Is energy conserved?)
   - Force glasses (What are the forces and how are they affecting the system?)
   - Momentum glasses (Are there any interactions of particles?)
   - Kinematics glasses (What kind of motion is occurring?)
   - Wave glasses (Is there any wave nature involved?)
   - Nano glasses (Are the strange and wonderful behaviors of atoms and particles coming into play?)

Viewing the problem through these six "glasses" will help you see how to approach the problem. With practice, you will perform this step naturally.

4. Look at the answer choices. What are they like? Are they numerical, symbolic, graphical, or just words? Looking at your options gives you a clue on how to approach the problem. For instance, there may be numbers in the stem of the question and you start doing calculations only to find the answers are all symbols. A quick look at the answer choices will keep you from launching off in the wrong direction.
5. Eliminate any answer choice that can't be correct by crossing it out. This will keep you from considering it again and wasting time. This will also improve your statistical chances if you have to take an educated guess.

These five steps will help you on exam day. As you prepare for the exam, here is a sixth step that will really amp up your physics skills and improve your exam score.

6. On every multiple-choice question, figure out and explain why the correct answer is correct. Just as important, figure out why each of the incorrect responses is wrong and explain why it is inconsistent with physics. This has two benefits. First, it deepens your physics knowledge to the point that will be expected on the exam. Second, the free-response section of the exam will ask you to justify and explain your reasoning. Practice this skill by writing down your justifications for each answer choice using "physics language." Ask your teacher to review your reasoning if you need help on how to word your justification. This single practice step will improve your skills and score.

Multiple-Correct Questions

The last five multiple-choice questions (46–50) will be of this type of question. To remind you, there will be a short instruction paragraph separating these five multiple-correct from the previous "normal" single-correct questions. In addition, at the end of the stem of each multiple-correct you will see "Select two answers." So, unless you are asleep, you can't miss them.

You have to select both correct answers and none of the incorrect answers to get credit. This should not scare you. These questions are no harder than other multiple-choice questions. They just require a bit more thinking. In many cases you can determine one correct answer quickly. Then if you can eliminate a response that can't be correct, you have a 50-50 shot on the last one. Use the steps outlined previously and you will do great.

Time: Pace Yourself!

Fifty questions in 90 minutes. That gives you 1.8 minutes per question or 9 minutes for every five questions. You don't have time to fool around. Here are some strategies to help you with time management:

- This is not like the SAT where the questions start easy and get harder near the end. The AP exam is generally broken into topic chunks. There will be a few electricity problems, a couple of fluids problems, followed by one or two atomic/nuclear questions, etc. They are mixed up in bite-size pieces.
- You may be a superhero with circuits but a mortal in thermodynamics. The point is that the level of difficulty and variety of topics are sprinkled throughout the exam. Don't get

stuck on a problem you don't know how to do and then run out of time without getting to those easy problems you know at the end of the exam.

- Skip any problems you don't know how to do. Answer every problem you know first. You have secured the easy points. Now go back through the exam again and answer the ones you might know how to do now that your brain is warmed up. Cycle through the multiple-choice questions a third time if you have time.
- When time is running short, be absolutely sure to ANSWER EVERY QUESTION because it can only help your score.
- Take a watch with you to the exam to keep track of time remaining. (Some of you get psyched out by a ticking watch staring at you when you are under stress. If that is you, then skip the watch.)
- Practice makes perfect. You will have access to multiple-choice questions in your textbook or in class. Practice these questions under real testing conditions so that you can find your pace. (Remember the pace of 9 minutes for every five questions.) Ask your teacher for extra AP-style questions if the ones in your book are mostly numerical or not like what you see in this book.

Additional Strategies and Words of Wisdom

- Do not look for patterns in your answers. Every year I'll have a student who says, "I had already picked answer choice (B) four times in a row so I decided to guess (A) since I hadn't used it as much." I like to remind students that they probably missed some of those four (B)s in a row! There is no pattern to the exam. There could be eight (D)s in a row, and 30 of the answers overall could be (A). Don't bother looking for a pattern. Answer each question individually and move on.
- Remember the ranking tasks from Chapter 6. In these problems, you will be given a situation in which you're going to have to rank answers from smallest to largest or largest to smallest. An example might be a circuit in which you're asked to rank the power of the resistors in the circuit from largest to smallest.
- You are able to use a calculator for the entire AP Physics test, but it is not going to be all that important for you. The questions on the test are designed to test your knowledge of physics conceptually. Sure, there will be calculations on the test, but the numbers are not going to be the point of the test. In fact, the majority of the problems won't have numbers.
- You will not find a simple "find the matching equation, plug in the numbers, and solve" problem on the test. Most of your calculation problems will involve multiple steps and will require more physics knowledge than simply plugging numbers into equations.

Strategies for the Free-Response Questions

IN THIS CHAPTER

Summary: The AP Physics 2 exam contains a variety of problem types unique to the free-response section of the exam: laboratory design, qualitative-quantitative transition (QQT), paragraph-length response, and student-contention questions. (If you took the AP Physics 1 exam last year, you have seen these already.) This chapter gives strategies and advice on how to approach these unique questions.

Key Ideas

- The 90-minute free-response section contains four questions: a 12-point lab question, a 12-point qualitative-quantitative transition (QQT) question, and two 10-point short-answer questions. One of the 10-point questions will require a paragraph-length response.
- There are simple tips and a strategy you can use to maximize your score on the lab question.
- This is not an English exam! Keep your paragraph-length response short and sweet. Get to the point and then stop writing.
- Student-contention questions are a unique type of question that can pop up on the exam. They are not that hard once you see how they work.
- The QQT question asks you to use qualitative reasoning and then asks you to use equations and numbers. If you are stronger with numbers, work the problem in reverse.

- You can get partial credit on the free-response portion of the exam. Remember that you need roughly only 70%–75% or so to earn a top score and only about 40%–45% or so to earn a qualifying score. Make sure to attempt every part of each problem to get the maximum partial credit possible.
- The free-response section is read and graded by real people. There are tips at the end of the chapter to help you best communicate your physics understanding to these readers.

The Structure of the Free-Response Section

The free-response section contains only four questions to be completed in 90 minutes. Piece of cake!

- One 12-point laboratory-based question
- One 12-point qualitative-quantitative transition (QQT) question
- Two 10-point short-answer questions of which one will contain within it a requirement for a paragraph-length response

This gives you about two minutes per point or about 25 minutes to answer the longer questions and 20 minutes for the shorter ones.

The best thing about the free-response section of the AP exam is this: you've been preparing for it *all year long*! "Really?" you ask, ". . . I don't remember spending much time preparing for it."

But think about the homework problems you've been doing throughout the year. Every week, you probably answer a set of questions, each of which might take a few steps to solve, and I'll bet that your teacher always reminds you to show your work. This sounds like the AP free-response section to me!

The key to doing well on the free-response section is to realize that, first and foremost, these problems test your *understanding* of physics. The purpose is not to see how good your algebra skills are, how many fancy-sounding technical terms you know, or how many obscure theories you can regurgitate. So all I'm going to do in this chapter is give you a few suggestions about how, when you work through a free-response question, you can communicate to the AP graders that you understand the concepts being tested. If you can effectively communicate your understanding of physics, you will get a good score.

Getting Off to a Good Start

Every year there are horror stories of students getting stuck on a problem and not finishing the free-response questions. "I got stuck on the first one and then I had to rush. But I still ran out of time!" Rule No. 1: Get off to a good start! When you begin the free-response section, remain calm and keep your writing utensil on your desk. Take a quick inventory of four questions and ask yourself: "Which of these looks the easiest?" Start there. This will ensure that you start on a positive note and have as much time as possible to finish the exam. Practice this strategy in class during your exams. Always work your way from what seems easiest to what seems hardest.

Lab Questions

It is all well and good to be able to solve problems and calculate quantities using the principles and equations you've learned. However, the true test of any physics theory is whether or not it WORKS.

The AP development committee is sending a message to students that laboratory work is an important aspect of physics. To truly understand physics, you must be able to design and analyze experiments. Thus, *each free-response section will contain one question that involves experiment design and analysis.*

Here is an example:

In the laboratory, you are asked to investigate the relationship between the temperature and volume of a fixed amount of gas. The following equipment is available for use:

___ a cylinder with a movable piston
___ a cylinder with a fixed volume
___ a pressure sensor
___ a container large enough to hold a cylinder with extra room
___ a source of hot water
___ a source of ice water
___ a ruler
___ a thermometer
___ a stopwatch
___ a set of masses that will fit on the movable piston

(a) Put a check in the blank for each of the items above that you would need for your investigation.
(b) Outline an experimental procedure you would use to gather the necessary data to answer this question. Make sure your procedure contains sufficient detail and information so that another student could follow your directions.

To answer a lab question, just follow these steps:

1. **Follow the directions.** Sounds simple, doesn't it? When the test says, "Draw a diagram," it means they want you to draw a diagram. And when it says, "Label your diagram," it means they want you to label your diagram. You will likely earn points just for these simple steps.

Exam Tip from an AP Physics Veteran
On the AP test, I forgot to label point *B* on a diagram, even though I obviously knew where point *B* was. This little mistake cost me several points!
—*Zack, college senior and engineer*

2. **Use as few words as possible.** Answer the question, then stop. You can lose credit for an incorrect statement, even if the other 15 statements in your answer are correct. The best idea is to keep it simple.
3. **There is no single correct answer.** Most of the lab questions are open-ended. There might be four or more different correct approaches. So don't try to "give them the answer they're looking for." Just do something that seems to make sense—you might well be right!

4. **Don't assume you have to use all the stuff they give you.** It might sound fun to use a force probe while determining the index of refraction of a glass block, but really? A force probe?
5. **Don't over-think the question.** They're normally not too complicated. Remember, you're supposed to take only about 25 minutes to write your answer. You're not exactly designing a subatomic particle accelerator.
6. **Don't state the obvious.** You may assume that basic lab protocols will be followed. So there's no need to tell the reader that you recorded your data carefully, nor do you need to remind the reader to wear safety goggles.
7. **Find a physics relationship/equation to help you.** Determine what physics relationship is being investigated and find an equation that models that behavior. The example above is obviously concerning gases and gas laws. That leads us to the ideal gas law: $PV = nRT$.
8. **Let your physics relationship guide you through the lab.** Now that we have $PV = nRT$, where does it lead us? Well, we are looking for a relationship between temperature and volume. So we will need to change one of these variables and measure the change in the other. The equation also tells us that pressure and number of molecules are involved in the relationship. But, we are not investigating those quantities, so we must keep them constant.
9. **Put your plan together.** Here is one possible plan:
   - Use the movable piston and put a mass on top of the piston to maintain constant pressure. The amount of gas in the cylinder should remain constant.
   - Measure the temperature with the thermometer. Use a ruler to measure the height of the gas in the cylinder and multiply by the area of the piston to find the volume.
   - Starting at room temperature, place the cylinder in the large container with ice water.
   - Record the temperature as it decreases and measure the volume of the gas at every 10°. Collect the data in a table and plot the relationship.

Note that choosing the equipment came after finding a physics relationship and developing a plan. To make sure your explanation is clear, it is helpful to draw a picture even if the question doesn't ask for one. Make sure you describe what measurements you will take. Explain how you will be finding any calculated values, like we just did for volume.

Practice Designing a Lab

In the following table is a list of common lab-based questions that ask you to design a lab. Next to each lab is a list of useful equations that you might use to design your lab around. For example: if you are tasked with finding the density of a fluid, you might use the basic definition of *density* by measuring the mass and volume of the fluid: $\rho = \frac{m}{V}$. Or, you might want to use the buoyancy force by floating an object of known mass on the fluid and measuring the volume displaced: $F_b = \rho V g$. Take some time to review the following list of labs and associated equations. There will be a lab question on the free-response section. Practice designing labs. Ask your teacher to help you perform some of them. Remember, practice makes perfect!

Design a Lab to . . .	**Useful Equations**
Find or investigate the density of an object.	$\rho = \frac{m}{V}$, $P = P_0 + \rho gh$, $F_b = \rho Vg$
Determine if a gas displays ideal gas properties.	$PV = nRT$
Analyze data about thermal conductivity.	$\frac{Q}{\Delta t} = \frac{kA\Delta T}{L}$
Analyze the work done by a gas.	$W = -P\Delta W$
Qualitatively investigate the charge of an object and the induced charge on objects.	$\lvert\vec{F}_E\rvert = \frac{1}{4\pi\varepsilon_0}\frac{\lvert qq\rvert}{r^2}$
The effect of the geometry of an object on its resistance.	$R = \frac{\rho l}{A}$
Analyze the properties of circuits including those with capacitors.	$P = I\Delta V$, $I = \frac{\Delta V}{R}$, $R_S = \Sigma_i R_i$, $\frac{1}{R_p} = \Sigma_i \frac{1}{R_i}$, $C_P = \Sigma_i C_i$, $\frac{1}{C_S} = \Sigma_i \frac{1}{C_i}$
Determine the effect of geometry changes on capacitance.	$C = \kappa\varepsilon_0\frac{A}{d}$, $\Delta V = \frac{Q}{C}$
Investigate the force on moving charges caused by a current carrying wire.	$B = \frac{\mu_0 I}{2\pi r}$, $F_M = qvB\sin\theta$
Investigate reflection of waves and the images formed.	$\frac{1}{f} = \frac{1}{s_0} + \frac{1}{s_i}$, $\lvert M\rvert = \left\lvert\frac{h_i}{h_0}\right\rvert = -\left\lvert\frac{s_i}{s_0}\right\rvert$
Determine the angle relationship between incoming and outgoing rays during refraction.	$n_1 \sin\theta_1 = n_2 \sin\theta_2$
Investigate the refraction of light through lenses and the images formed.	$\frac{1}{f} = \frac{1}{s_0} + \frac{1}{s_i}$, $\lvert M\rvert = \left\lvert\frac{h_i}{h_0}\right\rvert = \left\lvert\frac{s_i}{s_0}\right\rvert$
Investigate the particle properties of light.	$K_{max} = hf - \varphi$
Investigate the wave nature of small particles.	$d\sin\theta = m\lambda$

Paragraph-Length Responses

You are in the middle of the free-response section and see these fateful words: "In a coherent paragraph-length answer, describe . . . ." This is followed by a question and a full page of blank white paper just begging you to write all over it. Your brain automatically switches into English-essay mode. STOP! Before you write anything, remember: Less is better than more. This is not an English exam. The humans grading the free response are physics people. They do not want fluff. They are looking for specific information that tells them you understand the physics of the situation.

Here is an acronym that will help: CLEVeR.

1. **CL**—Make your CLAIM.
2. **EV**—Give your EVIDENCE.
3. **R**—Explain your REASONING.

State your claim and support it with evidence and reasoning. Get to the point. Answer the question and stop writing. You usually earn credit for just having a sequential and logical answer that does not have anything contradictory, wrong, or irrelevant! If you keep writing, you are likely to contradict yourself or use incorrect physics and lose points. If you can answer the question in three sentences, do so. Be CLEVeR! Keep it short and sweet.

"How do I maximize my score on paragraph-length response questions?" First off, there is only one per AP exam. Second, they are always worth 5 points, which means they will only be a part of a 10-point or 12-point question. So don't get stressed out about paragraph-length response questions. Botching your answer won't keep you from passing the exam.

Yeah, yeah . . . but "How do I maximize my score on paragraph-length response questions?" Here is what the graders are looking for:

1. Answer **primarily written in prose form**. (That means you must use words as your main method for explaining your ideas.)
2. Handwriting **must be readable**! (Sloppy handwriting that cannot be read by a reasonable person in a reasonable amount of time with reasonable effort will not receive any points. It is your responsibility to put your ideas on paper so they can be understood by others.)
3. The answer follows an **organized and logical thought process**. (One idea leads to another, and it should make sense after being read through the first time.)
4. The answer **sites physics principals and ideas** by name. (For example: "Charge will move *until equilibrium is established*, which is when the *electric potential is the same* for both metal spheres." Here is another example: "Because there is *only one pathway* in the circuit, *Kirchhoff's current rule* tells us that the current is the same for both resistors.)
5. The answer **avoids extraneous and contradictory information**. (Usually, the more you write, the worse your answer gets.)

"Wait a minute. Lots of stuff in physics is hard to explain without an equation, graph, diagram, or sketch! Can't I use them?!?!" YES, you can use equations, graphs, diagrams, and sketches, BUT your answer must be primarily WORDS.

One last piece of advice. The AP Physics readers consistently tell us that students need to be more precise in their explanations. Remember, you are not writing poetry that is up to the interpretation of the reader. You need to use "physics language" and avoid the use of pronouns. Instead of saying "It gets pushed away." Be more precise! Say something more like this: "The electron receives a force to the right due to the electric field." Or, "The electric force pushes the electron to the right." Speak in "Physics" and your grade will improve.

Student-Contention Questions

The student-contention question is an interesting type of question that sometimes shows up on the exam. It is not hard, just different. Here is an example:

A student makes this claim about the electrons making the two transitions shown in the energy-level diagram:

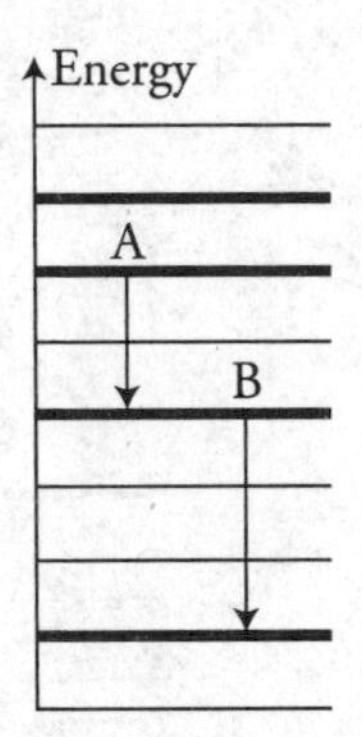

"The electrons will both emit a photon of light during their transition. The photon emitted by transition A will be a shorter wavelength than transition B because the electron begins at a higher-energy level."

(a) What aspects of the student's statement do you agree with and which do you disagree with? Qualitatively explain your reasoning.
(b) Derive an algebraic expression for the emitted photon.
(c) How does your derivation support your argument in part (a)?

In these questions, simply and directly explain your reasoning using physics principles. Here is a sample solution to this problem using the CLEVeR method:

(a) I agree that photons will be emitted in both cases (your claim). The electrons are losing energy as they drop to lower energy levels and will emit a photon (your evidence and reasoning).

I disagree that A will have a shorter wavelength (your claim). B has a larger drop in energy, which will produce a higher-energy photon of shorter wavelength (your evidence and reasoning).

(b) $E = hf = \frac{hc}{\lambda}$, Therefore: $\lambda = \frac{hc}{E}$.

(c) The equation shows the wavelength and photon energy to be inversely related (your claim). Therefore, the larger energy transition in B will produce a smaller wavelength (your evidence and reasoning).

Qualitative-Quantitative Transition (QQT) Questions

Look back at the student-contention question we just completed. Notice how parts (a) and (c) have no numbers. This is a *qualitative* reasoning question asking you to use words to make your argument. Part (b) asks for an algebraic derivation. This is a *quantitative* reasoning question asking you to use math to make your argument. Quantitative reasoning would also include finding a number answer similar to if you had been asked to calculate the wavelength of the emitted photon.

The QQT question will ask you to consider a situation and explain it first qualitatively (with words) and then quantitatively (with equations and numbers). You probably have been solving lots of number problems in your AP Physics 2 class and use your calculator a lot. As a consequence, you may be more comfortable with number problems. If this is the case, play to your strengths and solve the number part of the problem first, and then answer the conceptual part. Here is an example:

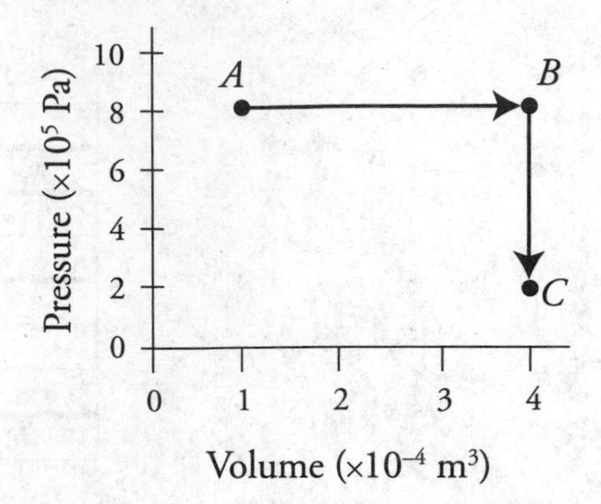

Two moles of gas is taken from its original state at point A through point B to a final state at point C as shown in the graph. A student makes this statement: "The net heat added or subtracted from the gas is zero because the final temperature of the gas is the same temperature as its starting temperature."

(a) Do you agree or disagree with the student's statement? Explain your reasoning with words.
(b) Justify your argument with a calculation.

Which do you feel more comfortable answering (a) or (b)? If you are a numbers person, start by solving the calculation first and then put it into words. Here's how it would work for this question:

Solving part (b) with equations:

$P_A V_A = P_C P_C = nRT$, since the PV value at points A and C are the same, the temperature change in the process is equal to zero.

$\Delta U = \frac{3}{2} nR\Delta T = 0$, since the change in temperature during the process is zero, the change in internal energy is also zero.

$W_{A \to B \to C} = -P\Delta V = -(800{,}000 \text{ Pa})(0.0004 \text{ m}^3 - 0.0001 \text{ m}^3) = -240 \text{ J}$, there is no work done moving from point B to C as there is no change in volume.

$\Delta U = 0 = Q + W$, therefore $Q = -W = -(240 \text{ J}) = 240 \text{ J}$.

Now that you have the math worked out, put it in words to answer (a). Remember to be CLEVeR with your response!

I agree the temperatures at A and C are the same because the pressure times volume is the same at both locations. I disagree that the heat is zero. Since the temperature does not change, the change in the internal energy of the gas is zero. There is negative work done by the gas during its expansion. Therefore, heat must be added to the system.

Choose to approach the QQT question with numbers first and words second, or words first and numbers second. It doesn't matter. The key is to play to your strengths so that you earn as many points as you can. Speaking of earning points, let's talk about how the exam is graded.

What the Exam Reader Looks For

Before grading a single student's exam, the high school and college physics teachers who are responsible for scoring the AP free-response section make a "rubric" for each question. A rubric is a grading guide; it specifies exactly what needs to be included for an answer to receive full credit, and it explains how partial credit should be awarded.

For example, consider part of a free-response question from AP Physics 1:

> A student pulls a 1.0-kg block across a table to the right, applying a force of 8.0 N. The coefficient of kinetic friction between the block and the table is 0.20. Assume the block is at rest when it begins its motion.
>
> (a) Determine the force of friction experienced by the block.
> (b) Calculate the speed of the block after 1.5 s.

Let's look just at part (b). What do you think the AP graders are looking for in a correct answer? Well, we know that the AP free-response section tests your understanding of physics. So the graders probably want to see that you know how to evaluate the forces acting on an object and how to relate those forces to the object's motion.

In fact, if part (b) were worth 4 points, the graders might award 1 point for each of these elements of your answer:

1. Applying Newton's second law, $F_{net} = ma$, to find the block's acceleration.
2. Recognizing that the net force is not 8.0 N, but rather is the force of the student minus the force of friction (which was found in [a]), 8.0 N – 2.0 N = 6.0 N.
3. Using a correct kinematics equation with correct substitutions to find the final velocity of the block; i.e., $v = v_o + at$, where $v_o = 0$ and $a = 6.0\ \text{N}/1.0\ \text{kg} = 6.0\ \text{m/s}^2$.
4. Obtaining a correct answer with correct units, 9.0 m/s.

Now, we're not suggesting that you try to guess how the AP graders will award points for every problem. Rather, we want you to see that the AP graders care much more about your understanding of physics than your ability to punch numbers into your calculator. Therefore, you should care much more about demonstrating your understanding of physics than about getting the right final answer.

Partial Credit

Returning to part (b) from the example problem, it's obvious that you can get lots of partial credit even if you make a mistake or two. For example:

- If you forgot to include friction, and just set the student's force equal to *ma* and solved, you could still get 2 out of 4 points.
- If you solved part (a) wrong but still got a reasonable answer, say 4.5 N for the force of friction, and plugged that in correctly here, you would still get either 3 or 4 points in part (b)! Usually the rubrics are designed not to penalize you twice for a wrong answer. So if you get part of a problem wrong, but your answer is consistent with your previous work, you'll usually get full or close to full credit.
- That said, if you had come up with a 1000 N force of friction, which is clearly unreasonable, you probably will not get credit for a wrong but consistent answer, unless you clearly indicate the ridiculousness of the situation. You'll still get probably 2 points, though, for the correct application of principles!

- If you got the right answer using a shortcut—say, doing the calculation of the net force in your head—you would not earn full credit but you would at least get the correct answer point. However, if you did the calculation *wrong* in your head, then you would *not* get any credit—AP graders can read what's written on the test, but they're not allowed to read your mind. Moral of the story: Communicate with the readers so you are sure to get all the partial credit you deserve.
- Notice how generous the partial credit is. You can easily get 2 or 3 points without getting the right answer and 40%–75% is in the 3–5 range when the AP test is scored!

You should also be aware of some things that will NOT get you partial credit:

- You will not get partial credit if you write multiple answers to a single question. If AP graders see that you've written two answers, they will grade the one that's wrong. In other words, you will lose points if you write more than one answer to a question, even if one of the answers you write is correct.
- You will not get partial credit by including unnecessary information. There's no way to get extra credit on a question, and if you write something that's wrong, you could lose points. Answer the question fully, then stop.

Final Advice About the Free-Response Section

- Always show your work. If you use the correct equation to solve a problem but you plug in the wrong numbers, you will probably get partial credit, but if you just write down an incorrect answer, you will definitely get no partial credit.
- If you don't know precisely how to solve a problem, simply explain your thinking process to the grader. If a problem asks you to find the radius of a charged particle in a magnetic field, for example, and you don't know what equations to use, you might write something like this: "The centripetal force points toward the center of the path, and this force is the magnetic force." This answer might earn you several points, even though you didn't do a single calculation.
- However, don't write a book. Keep your answers succinct.
- If you make a mistake, cross it out. If your work is messy, circle your answer so that it's easy to find. Basically, make sure the AP graders know what you want them to grade and what you want them to ignore.
- If you're stuck on a free-response question, try another one. Question 3 might be easier for you than question 1. Get the easy points first, and then try to get the harder points only if you have time left over.
- Always remember to use units.
- It may be helpful to include a drawing or a graph in your answer to a question, but make sure to label your drawings or graphs so that they're easy to understand.
- No free-response question should take you more than 20–25 minutes to solve. They're not designed to be outrageously difficult, so if your answer to a free-response problem is outrageously complicated, you should look for a new way to solve the problem, or just skip it and move on.
- Use "physics language" and avoid the use of pronouns. Instead of saying "It gets pushed to the right." Say something like "The electron receives a force to the right due to the electric field." Speak in "Physics"!

Review the Knowledge You Need to Score High

Fluid Mechanics

IN THIS CHAPTER

Summary: AP Physics 2 includes the study of both static (stationary) and flowing (dynamic) fluids. Keep in mind that the definition of a fluid is anything that flows. This means both a liquid and a gas have fluid behavior, but for the most part you will be dealing with liquids. Entire courses are based on the study of fluid behavior, but in AP Physics 2 you look only at a few relatively simple concepts associated with an idealized, or frictionless, incompressible fluid. Even though the principles may be divided into static and dynamic, a single problem may have both concepts in it.

Key Ideas

- The behavior of molecules gives rise to pressure and the forces pressure can produce.
- Static fluids—fluids sitting still
  - (a) Pressure at any point in a fluid is caused by the weight of the column of fluid above that point, plus the pressure acting on the surface of that column of fluid. Therefore, the greater the height of that column, the greater the pressure at the bottom of the column.
  - (b) Absolute pressure is the difference between the measured pressure and a vacuum. Gauge pressure is the difference between the measured pressure and atmospheric pressure.
  - (c) "Eureka!" Archimedes' principle states that the buoyant force on a submerged object is based upon the weight of the displaced fluid.

(d) Pascal's principle states that any increase in pressure on the surface of a fluid creates an equal and undiminished increase in pressure in all points throughout the fluid.

- Dynamic fluids—fluids in motion
  (a) Conservation of mass leads us to the continuity equation. Frequently the continuity equation will be needed to determine the speed of a fluid moving through a pipe of changing cross-sectional area. Keep in mind that the continuity equation can be altered to give mass flow rate and volumetric flow rate.
  (b) Conservation of energy leads us to Bernoulli's equation, which is used to relate the velocity, pressure, and height of a flowing fluid from one point in a fluid flow to another.

Relevant Equations

Definition of Pressure

$P = F/A$ where F = force in newtons
A = surface area in m^2

Density

$\rho = m/V$ V = volume in m^3
m = mass in kg

Pressure in a Static Column of Fluid

$P = P_0 + \rho gh$
P = pressure in pascals
P_0 = pressure acting on the surface of the fluid
ρ = density in kg/m^3
g = acceleration due to gravity
h = height of the column of fluid above the point where pressure is being determined

Buoyancy Force

$F_b = \rho Vg$
ρ = density of the displaced fluid in kg/m^3
V = volume of the displaced fluid in m^3

Continuity Equations

$A_1v_1 = A_2v_2$
A = the cross-sectional area of the pipe or flow tube in m^2
v = fluid velocity in m/s

$\Delta V/t = Av$ $\Delta V/t$ = volumetric flow rate in m^3/s
$\Delta m/t = \rho Av$ $\Delta m/t$ = mass flow rate in kg/s
ρ = density of the fluid in kg/m^3

Bernoulli's Equation

$$P_1 + \rho gy_1 + \tfrac{1}{2}\rho v_1^2 = P_2 + \rho gy_2 + \tfrac{1}{2}\rho v_2^2$$

y = the height of a position in a column of fluid in meters

How the Nano-World Influences the Fluid World We Live In

Before we talk about fluids, let's put on our "nano-glasses" and take a look at the world of the very small to see how the behavior of atoms gives us the behavior of fluids that we experience.

So what causes pressure in a fluid? There are two major causes of pressure.

1. Pressure Due to the Thermal Motion of the Molecules

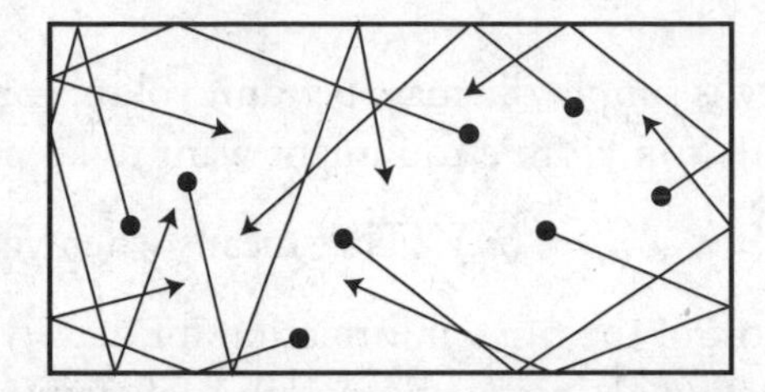

Remember that fluids are made up of lots and lots and lots of molecules. Each of these molecules is vibrating around in a random fashion due to the fluid's thermal energy. The hotter the fluid, the faster the vibrations of the molecules will be. (More about this in Chapter 10.) These vibrating molecules collide with anything the fluid comes in contact with. The above diagram shows the molecules colliding with the walls of a container. Each collision imparts a small impulse on the wall. Because of the large number of random collisions that occur in every direction, any parallel forces to the wall exerted by the collisions will cancel out! However, the perpendicular component of the collision forces will not cancel out. This means that the forces caused by fluid pressure will always be perpendicular to the surface the fluid is in contact with.

2. Pressure Due to Gravity

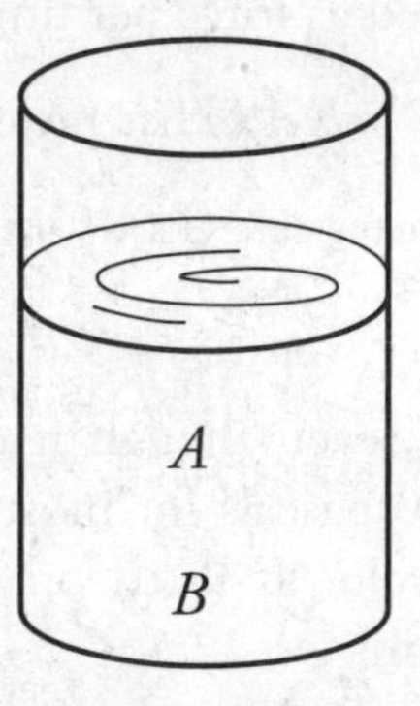

The above diagram shows a stationary liquid in a glass. Since the liquid is stationary, it must be in equilibrium, which means the forces all must be canceling out. Now consider points A and B in the liquid. At each point the forces must cancel out so that they can remain stationary. Each point must support the weight of all the fluid above itself with a counteracting force upward in order to maintain equilibrium. Otherwise the liquid above would accelerate downward due to gravity. Point B has more atoms stacked on top of it than point A does. This means that the pressure from all those molecular collisions at point B must be greater than at point A simply because it has to support more fluid above itself in a gravitational field.

The thermal effect of atomic collisions and the gravitational effect on the atoms combine to cause the overall pressure in a fluid. In gases, the thermal effect causes most of the pressure in a gas. Gases have a very small density, thus the gravitational effect on a gas is very small. The general rule of thumb is that air pressure is assumed to be constant unless there is a lot of vertical height to worry about. Liquids have a much larger density than gases. Thus, the gravitational effect can cause a lot of the pressure in a liquid. Swim just 10 feet to the bottom of a pool and you will notice this pressure increase.

Density

Density is simply the mass per unit volume of an object. In AP Physics 2 you will primarily measure it in kg/m^3. You might want to keep in mind that 1 g/cm^3 = 1000 kg/m^3.

$\rho = m/V$ (That means when you want to find mass: $m = \rho V$.)

Most of the time, information for density will be given to you in the problem, or there will be enough information given to determine the density of the material. It is a good idea to know the density of water, which is 1000 kg/m^3.

Density and Mass

When dealing with fluids, we don't generally talk about mass—we prefer to talk about density and volume. But, of course, mass is related to these two measurements: *mass* = *density* · *volume*.

So the mass of any fluid equals ρV. And because weight equals mg, the weight of a fluid can be written as ρVg.

Pressure

Pressure is defined as force per unit area:

$P = F/A$ (That means when you want to find force: $F = PA$.)

It is measured in pascals (Pa). One pascal equals one newton per meter squared:

$$1 \text{ Pa} = 1 \text{ N/m}^2$$

Air pressure, even though it does vary with weather conditions, is given an average value of 100,000 pascals for most AP Physics problems unless otherwise indicated in the problem. This value is listed on the reference table you will be given when taking the AP Physics 2 exam.

With fluids, we talk a lot about pressure, but remember that forces cause things to accelerate. When you need a force, just multiply the pressure times the areas of contact: $F = PA$.

You need to be aware pressure may be indicated in one of two different ways, absolute pressure and gauge pressure. Absolute pressure is the pressure measured relative to a vacuum, and gauge pressure is the pressure measured relative to atmospheric pressure. So when you measure the gauge pressure in a tire, you are measuring the difference between the pressure in the tire and the atmosphere, or what pressure exists in the tire above that of atmospheric pressure. Absolute pressure means exactly what it sounds like. An absolute pressure of zero describes a complete vacuum. Therefore, absolute pressure equals gauge pressure plus one atmosphere:

$$P_{Absolute} = P_{Gauge} + 100000 \text{ Pa}$$

Static Fluids

Anybody who has gone swimming at the bottom of a deep swimming pool experiences an increase in pressure as they travel deeper and deeper into the pool. A good way to think about it is as you submerge further into the pool, the column of water above you becomes taller and taller, making the weight of that column of water greater and greater. This increased weight of the column of water corresponds to an increase in pressure at the bottom of that column. Any pressure acting on the surface of that column of water causes an equal, undiminished increase in pressure at all points throughout the fluid; this is known as Pascal's law. To find the pressure at any point in the fluid:

$P = P_0 + \rho g h$ where P = pressure
P_0 = pressure on the surface of the fluid
ρ = density in kg/m^3
g = acceleration due to gravity
h = height of the column of fluid above the point where pressure is being determined

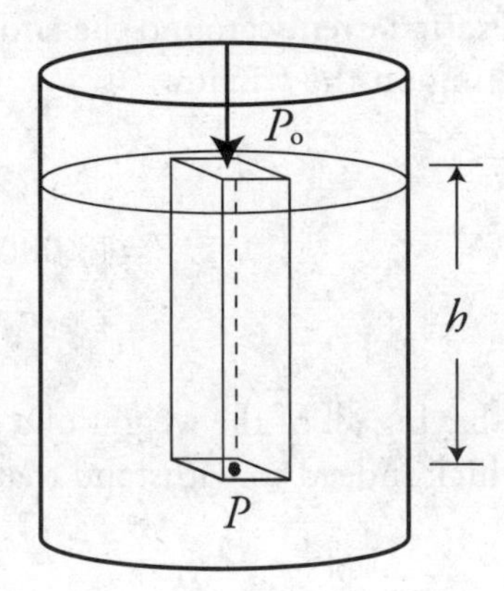

Notice that for a static fluid, the pressure is solely based on the height of the fluid column, the fluid's density, and the pressure acting on the surface of the column. The shape of the container has no effect. In the diagram below, the pressure is the same in the tube along the bottom, so it supports the same height column of fluid in each container, regardless of the shape of the container.

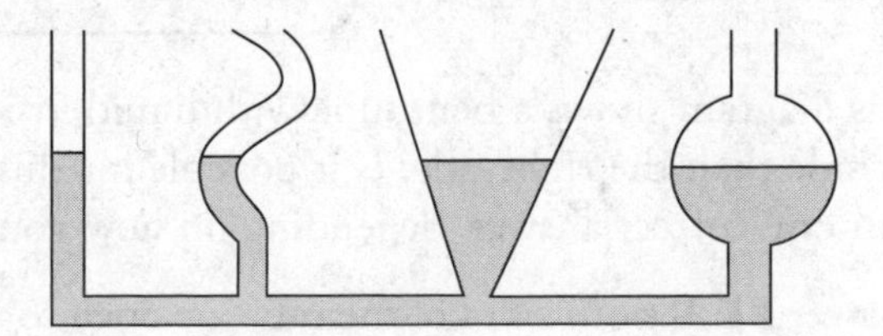

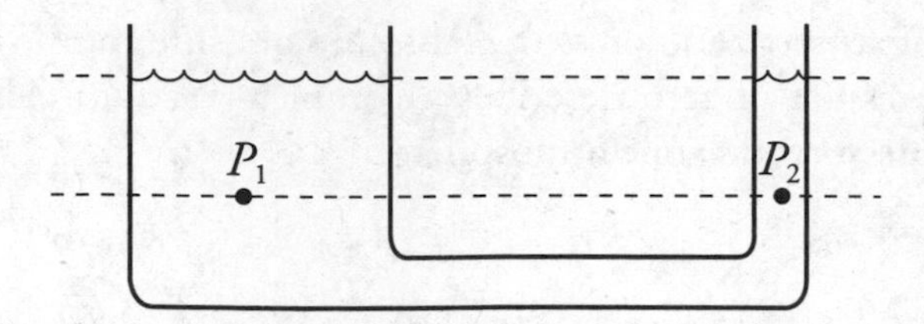

The pressure is the same along any horizontal line drawn through in a stationary connected fluid. In the diagram above, pressure P_1 and P_2 are the same. Horizontal lines higher in the fluid represent lower pressure. Horizontal lines deeper in the fluid represent higher pressure. The width of the fluid does not matter.

Example

A submarine is at a depth of 300 m in fresh water. If it has a window of 0.10 m radius, what is the force of the water on the window?

First you must keep in mind that you are asked for the actual force of the water on the window, so you need to use absolute pressure when determining the pressure:

$$P = P_0 + \rho gh$$

$$P = 100{,}000 \text{ Pa} + (1000 \text{ kg/m}^3)(9.8 \text{ m/s}^2)(300 \text{ m})$$

$$P = 3{,}000{,}000 \text{ Pa}$$

Ignoring the slight height difference from the top and bottom of the vessel's hull, this pressure is exerted evenly around the entire hull. Since you know the pressure, you can find the force acting on the window.

$$F = PA = P\pi r^2$$

$$F = (3{,}000{,}000 \text{ Pa})\pi(0.10 \text{ m}^2)$$

$$F = 94{,}000 \text{ N}$$

That is like having all of the weight of a good-sized truck pressing on the window so it has to be very thick indeed to withstand that force!

Applications of Static Fluids

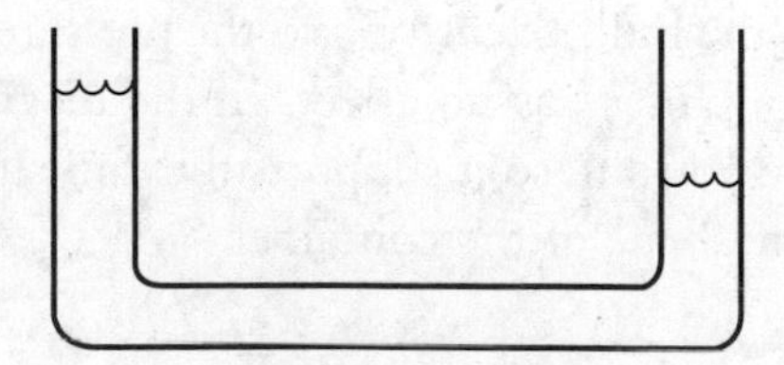

This diagram shows a bent tube with liquid inside. The level of the liquid is higher on the left side than the right side. Is it possible for this to happen in a static fluid? There is more than one correct answer, depending on how you explain yourself.

Answer #1: If both sides of the tube are open to the atmosphere, then the answer is no. The left tube has a higher fluid column, which will produce more pressure than at the bottom than the right tube. Therefore, the fluid will flow to the right until the level of fluid is the same in both vertical tubes.

Answer #2: It could be possible to produce different column heights of fluids if the pressures on top of the fluids in the left and right tubes are not the same. The pressure P_0 in the right tube would need to be higher.

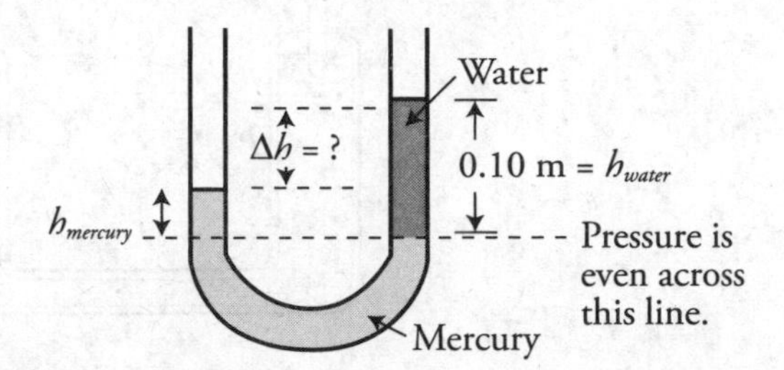

Take a look at this U-tube with two different fluids inside of it and open to the atmosphere on both sides. The fluid is stationary. Since the fluid does not accelerate in one direction or another, you know the forces exerted on the bottom of the U-tube must be equal from both sides of the tube. If this is true, then the pressures exerted on the bottom must be the same from both sides of the tube. When you look at the mercury in the diagram, the pressure, which exists at the mercury-water interface on the right-hand side of the tube, must equal the pressure in the mercury at the same height in the left side of the tube. So that means that the pressure caused by the column of mercury above that height must equal the pressure caused by the column of water on the other side. This makes it relatively easy to find the height of the column of mercury:

$$P_{mercury} = P_{water}$$

$$P_0 + \rho g h_{mercury} = P_0 + \rho g h_{water}$$

Since P_0 and g are the same, you can cancel them out:

$$\rho h_{mercury} = \rho h_{water}$$

$$(13{,}600\ \text{kg/m}^3) h_{mercury} = 1000\ \text{kg/m}^3\ (0.10\ \text{m})$$

$$h_{mercury} = 0.0074\ \text{m}$$

Now it is easy to find the difference in height:

$$0.1000\ \text{m} - 0.0074\ \text{m} = 0.9926\ \text{m}$$

Another application of static pressure is a hydraulic jack. A hydraulic jack consists of two pistons, one with a small area and the other with a large area. If you consider the height of the two pistons to be the same, then the pressure acting on the faces of the pistons will also be the same. When you apply a force to the smaller piston, it yields a larger force on the larger piston due to its larger surface area.

Example

A hydraulic jack is made of two pistons as shown in the next diagram; the smaller piston has an area of 0.01 m^2 and the larger one has an area of 0.20 m^2. What force would be required to hold up a 1500-N (about 300-lb.) football player?

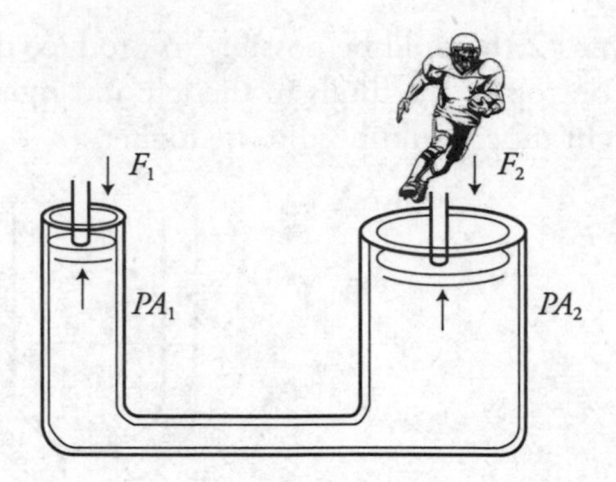

$$P_1 = P_2$$

$$\frac{F_1}{A_1} = \frac{F_2}{A_2}$$

$$\frac{F_1}{0.01\ \text{m}^2} = \frac{1500\ \text{N}}{0.20\ \text{m}^2}$$

$$F_1 = 75\ \text{N}$$

So it takes only 1/20th of the weight of the football player applied to the smaller piston to support the football player.

Barometer

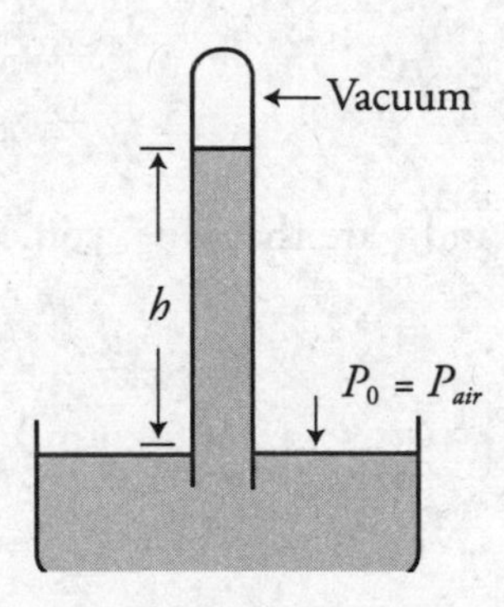

A barometer is a simple device that can measure the pressure acting on the surface of a fluid. A closed-end tube is filled with a fluid (usually water or mercury) and then turned upright inside a container of fluid with the closed end facing upward. The weight of the fluid pulls the fluid down in the tube, creating a vacuum at the top of the tube ($P_{vacuum} = 0$). Now, remember that the pressure at the surface of the fluid must be equal at all points at the same horizontal height; therefore, pressure exerted on the surface of the fluid by atmospheric air pressure equals the pressure inside the column of fluid in the tube at the same horizontal level. This pressure on the surface of the mercury holds up the column of mercury in the tube.

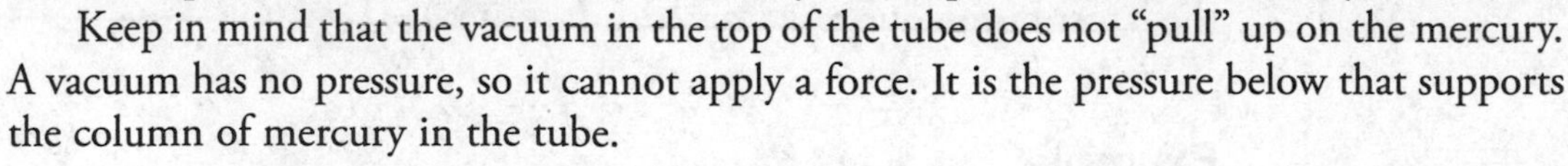

Keep in mind that the vacuum in the top of the tube does not "pull" up on the mercury. A vacuum has no pressure, so it cannot apply a force. It is the pressure below that supports the column of mercury in the tube.

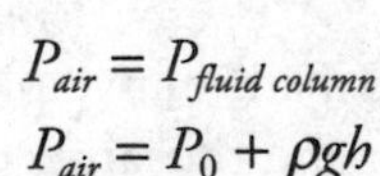

$$P_{air} = P_{fluid\ column}$$

$$P_{air} = P_0 + \rho g h$$

And, since there is a vacuum at the top of the column: $P_0 = 0$.

$$P_{air} = \rho g h$$

Buoyancy and Archimedes' Principle

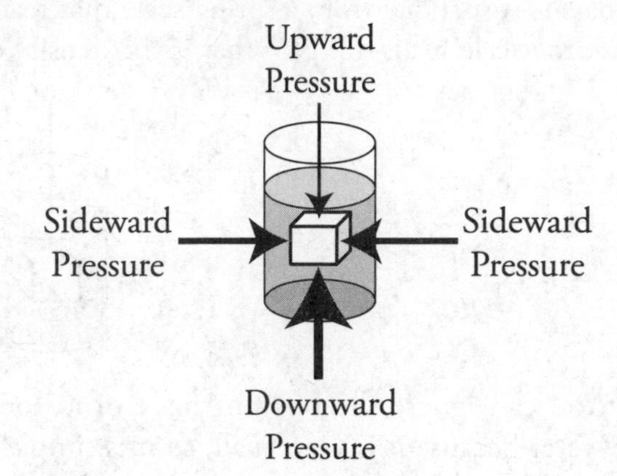

Why do things float? Because there is a gravitational effect on fluids that causes pressure to be greater the deeper you get in a fluid. We know:

1. Pressure in a fluid increases with depth: $P = P_0 + \rho gh$.
2. The bottom of the block is deeper in the fluid than the top of the block.
3. Therefore, the upward pressure on the bottom of the block must be greater than the downward pressure on the top of the block.
4. The pressure on the sides of the block must be identical, thus canceling each other out.
5. This means that every object surrounded by a fluid must receive an upward force due to the difference in pressures on the top and bottom of the object!

Do you realize that a helium balloon will not float in the International Space Station? We don't have the gravity effect pressure in space, only thermal effect pressure, which is the same everywhere in the fluid.

Archimedes discovered the buoyancy principle while taking a bath, so it is named after him—the Archimedes' principle. According to this principle, when a body is immersed in a fluid either fully or partially, the fluid exerts an upward force on the body equal to the weight of the fluid that is displaced by the body:

$$F_b = \textit{Weight of displaced fluid} = mg_{\textit{displaced fluid}} = (\rho_{\textit{fluid}} V_{\textit{fluid}})g$$

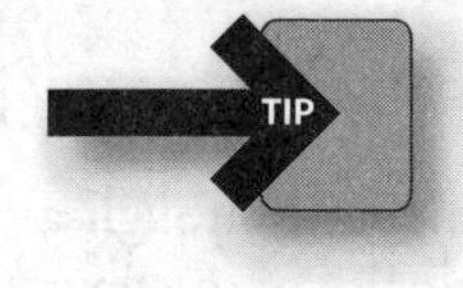

Be careful and remember to use the density and volume of the displaced fluid.

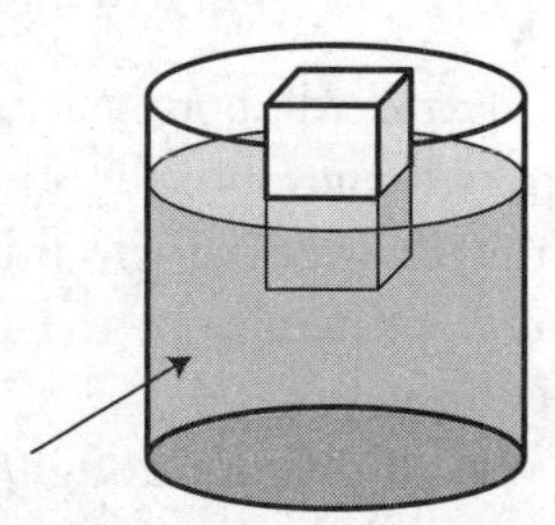

Caution: One common mistake is to assume that the volume of the object is always equal to the volume of the displaced fluid. Objects that float do not necessarily displace the fluid equal to their volume. Only the submerged portion of the object displaces fluid!

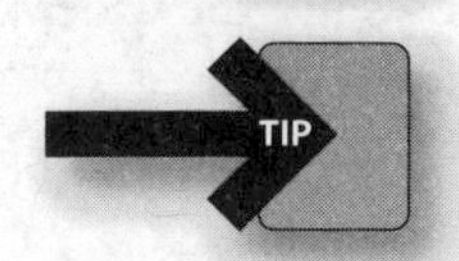

Remember Newton's third law. The fluid is pushing up on the object with a buoyant force. Therefore, the object must be pushing downward on the fluid with an equal but opposite force.

Example

A mass is suspended from a spring scale that reads 5.0 N. When the mass is submerged in water the scale reads 4.2 N. What is the density of the object?

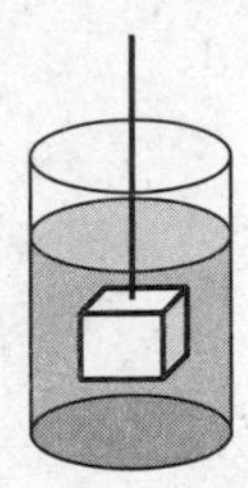

You can ignore the buoyant force of air on the object when it is suspended out of the water because it is very small, so it is simple to find the mass. When out of the water, Newton's second law gives us:

$$F = ma$$
$$T - mg = 0$$
$$5.0\text{ N} - m(9.8\text{ m/s}^2) = 0$$
$$m = 0.51\text{ kg}$$

Now apply Newton's second law when the object is submerged. (Remember to draw your free-body diagram!):

$$F = ma$$
$$T + F_{buoyancy} - mg = 0$$
$$T + \rho V g - mg = 0$$
$$4.2\text{ N} + (1000\text{ kg/m}^3)V(9.8\text{ m/s}^2) - 0.51\text{ kg }(9.8\text{ m/s}^2) = 0$$
$$V = 8.1 \times 10^{-5}\text{ m}^3$$

You are all set; you have the mass and the volume:

$$\rho = m/V$$
$$\rho = 0.51\text{ kg} / 8.1 \times 10^{-5}\text{ m}^3 = 6300\text{ kg/m}^3$$

Dynamic Fluids—Continuity

So far we've only talked about stationary fluids. But what if a fluid is moving through a pipe? There are two questions that then arise: first, how do we find the velocity of the flow, and second, how do we find the fluid's pressure? To answer the first question, we'll turn to the continuity principle. In the next section, we'll answer the second question using Bernoulli's equation.

Consider water flowing through a pipe. Imagine that the water always flows such that the entire pipe is full. Now imagine that you're standing in the pipe[1]—how much water flows past you in a given time?

We call the volume of fluid that flows past a point every second the volume flow rate, measured in units of m^3/s. This rate depends on two characteristics: the velocity of the fluid's flow, v, and the cross-sectional area of the pipe, A. Clearly, the faster the flow, and the wider the pipe, the more fluid flows past a point every second. To be mathematically precise,

$$\text{volume flow rate} = Av.$$

[1]Wearing a scuba mask, of course.

If the pipe is full, then any volume of fluid that enters the pipe must eject an equal volume of fluid from the other end. (Think about it—if this weren't the case, then the fluid would have to compress or burst the pipe. Once either of these things happens, the principle of continuity is void.) So, by definition, *the volume flow rate is equal at all points within an isolated stream of fluid.* This statement is known as the principle of continuity. For positions "1" and "2," for example, we can write $A_1v_1 = A_2v_2$. In reality, continuity is a consequence of conservation of mass. Since no mass is created or destroyed inside the pipe, an equal amount of fluid mass that flows into one end of the pipe must exit out the other end of the pipe.

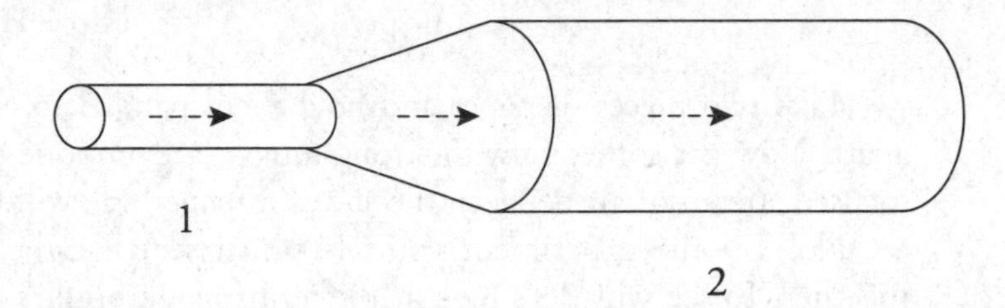

The most obvious physical consequence of continuity is that where a pipe is narrow, the flow is faster. So, the fluid must move faster in the narrower section labeled "1." Most people gain experience with the continuity principle when using a garden hose. What do you do to get the water to stream out faster? You use your thumb to cover part of the hose opening, of course. In other words, you are decreasing the cross-sectional area of the pipe's exit, and since the volume flow rate of the water can't change, the velocity of the flow must increase to balance the decrease in the area.

Dynamic Fluids—Bernoulli's Equation

Bernoulli's equation is probably the longest equation you need to know for the AP exam, but fortunately, you don't need to know how to derive it. What you should know, conceptually, is that it's really just an application of the conservation of energy.

$$P_1 + \rho g y_1 + \tfrac{1}{2}\rho v_1^2 = P_2 + \rho g y_2 + \tfrac{1}{2}\rho v_2^2$$

Bernoulli's equation is useful whenever you have a fluid flowing from point "1" to point "2." The flow can be fast or slow; it can be vertical or horizontal; it can be open to the atmosphere or enclosed in a pipe. (What a versatile equation!) P is the pressure of the fluid at the specified point, y is the vertical height at the specified point, ρ is the fluid's density, and v is the speed of the flow at the specified point.

Too many terms to memorize? Absolutely not! First, we're dealing with pressures, so obviously there should be a pressure term on each side of the equation. Second, we said that Bernoulli's equation is an application of energy conservation, so each side of the equation will contain a term that looks kind of like potential energy and a term that looks kind of like kinetic energy. (Note that the units of every term are N/m^2.)

Principal Consequence of Bernoulli's Equation

The most important result that can be derived from Bernoulli's equation is this: generally, where flow is faster, pressure is lower.

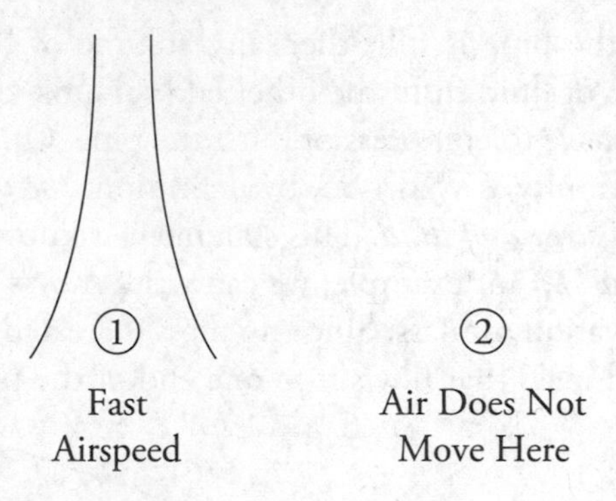

Take two pieces of paper and hold them parallel to each other, a couple centimeters apart. Now get someone with strong lungs—a trombone player, perhaps—to blow a fast, focused stream of air between the sheets of paper. He will blow the sheets apart, right?

No! Try the experiment yourself! You'll see that the pieces of paper actually move together. To see why, let's look at the terms in Bernoulli's equation.[2] Take point "1" to be between the sheets of paper; take point "2" to be well away from the sheets of paper, out where the air is hardly moving, but at the same level as the sheets of paper so that y_1 and y_2 are the same. (See the diagram above.) We can cancel the y terms because they are at the same height and solve for the pressure between the sheets of paper:

$$P_1 = P_2 + \left(\tfrac{1}{2}\rho v_2^2 - \tfrac{1}{2}\rho v_1^2\right)$$

Far away from the pieces of paper, the air is barely moving (so v_2 is just about zero), and the pressure P_2 must be the ambient atmospheric pressure. But between the sheets, the air is moving quickly. So look at the terms in parentheses: v_1 is substantially bigger than v_2, so the part in parentheses gives a negative quantity. This means that to get P_1, we must *subtract* something from atmospheric pressure. The pressure between the pieces of paper is *less than* atmospheric pressure, so the air on the exterior of both pieces of paper push the pieces of paper together.

This effect can be seen in numerous situations. A shower curtain is pushed in toward you when you have the water running past it; an airplane is pushed up when the air rushes faster over top of the wing than below the wing; a curveball curves because the ball's spin causes the air on one side to move faster than the air on the other side.

In general, solving problems with Bernoulli's equation is relatively straightforward. Yes, the equation has six terms: one with pressure, one with height, and one with speed for each of two positions. But in almost all cases you will be able to get rid of several of these terms, either because the term itself is equal to zero, or because it will cancel with an identical term on the other side of the equation. For example:

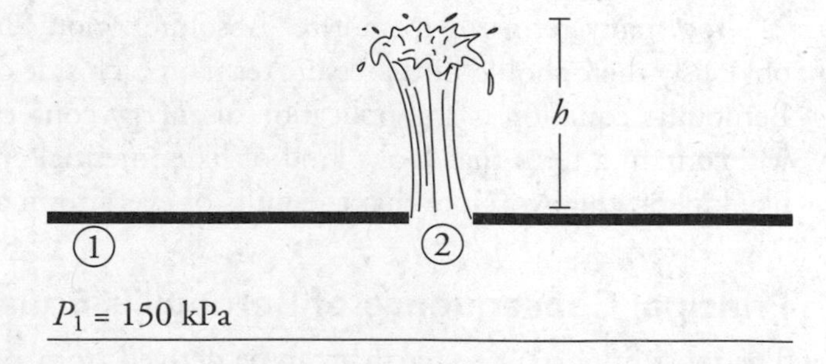

In a town's water system, pressure gauges in still water at street level read 150 kPa. If a pipeline connected to the system breaks and shoots water straight up, how high above the street does the water shoot?

[2]Yes, Bernoulli's equation *can* be applied to gases as well as liquids, as long as the gas is obeying the continuity principle and not compressing. This is a good assumption in most cases you'll deal with on the AP exam.

We know that we should use Bernoulli's equation because we are given the pressure of a flowing fluid, and we want to find the speed of the flow. Then, kinematics can be used to find the height of the fountain.

Let point "1" be in still water where the 150 kPa reading was made, and point "2" will be where the pipe broke. Start by noting which terms in Bernoulli's equation will cancel: both points are at street level, so $y_1 = y_2$ and the height terms go away; the velocity at point "1" is zero because point "1" is in "still" water.

Now the pressure terms: If a *gauge* reads 150 kPa at point "1," this is obviously a reading of *gauge* pressure, which you recall means the pressure over and above atmospheric pressure. The pressure where the pipe broke must be atmospheric.

Of the six terms in Bernoulli's equation, we have gotten rid of three:

$$P_1 + \cancel{\rho g y_1} + \cancel{\tfrac{1}{2}\rho v_1^2} = P_2 + \cancel{\rho g y_2} + \tfrac{1}{2}\rho v_2^2$$

This leaves

$$\tfrac{1}{2}\rho v_2^2 = P_1 - P_2$$

P_1 is equal to P_{atm} + 150 kPa; P_2 is just atmospheric pressure:

$$\tfrac{1}{2}\rho v_2^2 = P_{atm} + 150\,\text{kPa} - P_{atm}$$

The atmospheric pressure terms go away. Plug in the density of water (1000 kg/m^3), change 150 kPa to 150,000 Pa, and solve for velocity:

$$v_2 = 17 \text{ m/s}$$

Finally, we use kinematics to find the height of the water stream. Start by writing out the table of variables; then use the appropriate equation.

v_0	17 m/s
v_f	0 m/s
$x - x_0$	?
a	-10 m/s^2
t	?

$$v_f^2 = v_0^2 + 2a(x - x_0)$$

$x - x_0$ = 15 m high, about the height of a four- or five-story building.

Real-World Examples of Bernoulli

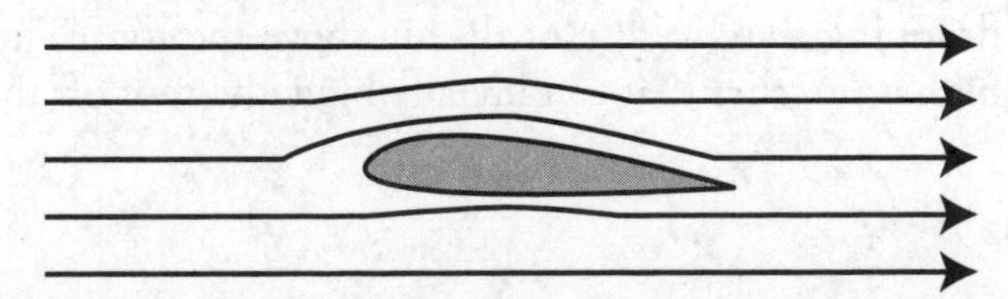

Wings and lift: The shape of the airfoil compresses the flow tube about the wing. This increases the velocity of the air above the wing, creating a pressure difference between the top and bottom of the wing, producing lift.

Spinning ball pulls air past faster on top.

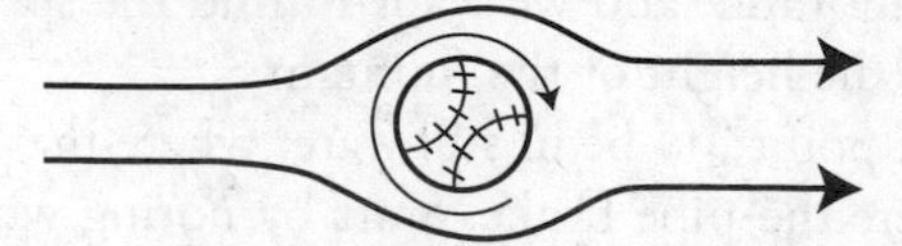

Air moves past slower on the bottom,
causing the ball to curve upward.

Curveball: Friction between the ball and the air causes the ball to drag the air faster past itself on one side while hindering the airflow on the other side. This velocity difference creates a pressure difference and curves the trajectory of the ball. (Try it with a beach ball or Ping-Pong ball!)

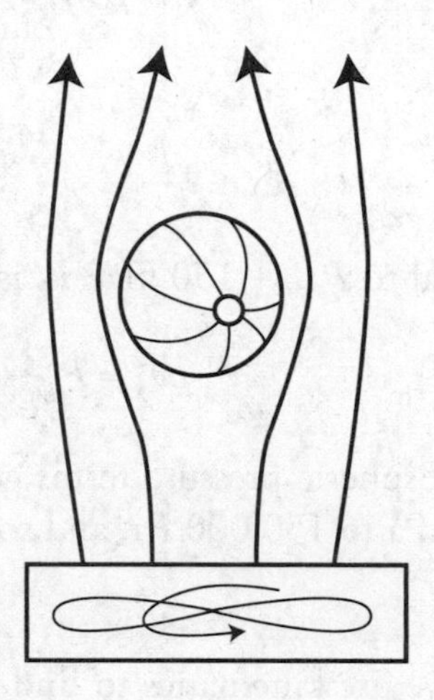

Beach ball trapped above a fan: Position a fan so that it blows air upward. Now release a beach ball in the airstream. When the ball tries to escape from the airstream it will encounter high-pressure stationary air that pushes the ball back into the low-pressure airstream. You can do this with a leaf blower and a tennis ball. Tape a string to a Ping-Pong ball and swing it into a stream of water from a faucet to get the same effect!

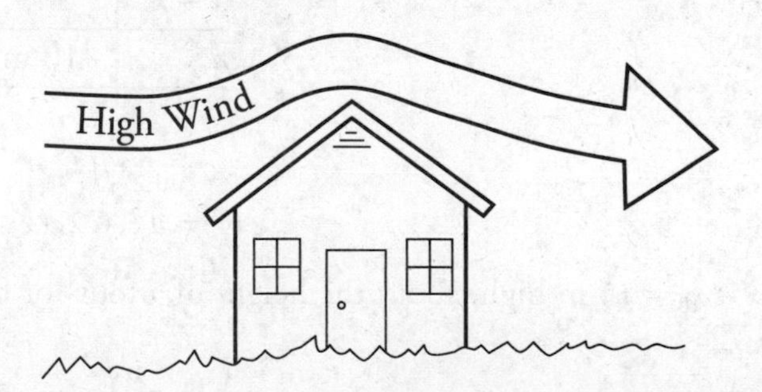

High winds blowing a roof off: During storms, high winds move past the roof of a house dropping the pressure of the air above the roof. The stationary air inside the house has a higher pressure that can literally blow the roof off the house!

❯ Practice Problems

Multiple Choice

1. A dam holds water in a lake that extends 1000 m east of the dam. If it extended 2000 m east instead of 1000 m, the pressure at the base of the dam would:

   (A) increase
   (B) decrease
   (C) remain the same

2. A horizontal 4-cm diameter pipe tapers to a 2-cm nozzle. If the water emerges from the nozzle at 20 m/s, what is the pressure in the 4-cm diameter section of the pipe?

   (A) 107,500 Pa
   (B) 190,000 Pa
   (C) 250,000 Pa
   (D) 290,000 Pa

3. A pirate ship hides out in a small inshore lake. It carries 20 ill-gotten treasure chests in its hold. But lo, on the horizon the lookout spies a gunboat. To get away, the pirate captain orders the heavy treasure chests jettisoned. The chests sink to the lake bottom. What happens to the water level of the lake?

   (A) The water level drops.
   (B) The water level rises.
   (C) The water level does not change.

4. Manny saves 2-liter soda bottles so that he can construct a raft and float out onto the Trinity River. If Manny has a mass of 80 kg, what minimum number of bottles is necessary to support him? The density of water is 1000 kg/m^3, and 1000 L = 1 m^3.

   (A) 1600 bottles
   (B) 800 bottles
   (C) 200 bottles
   (D) 40 bottles
   (E) 4 bottles

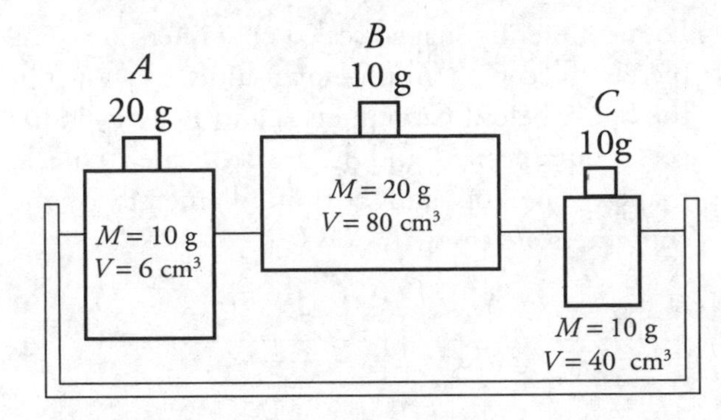

5. Three plastic blocks of different masses and volumes float with weights on top of them. Which of the following correctly ranks the buoyancy force on the plastic blocks?

   (A) $A = B > C$
   (B) $A > B = C$
   (C) $B > A > C$
   (D) $B > A = C$

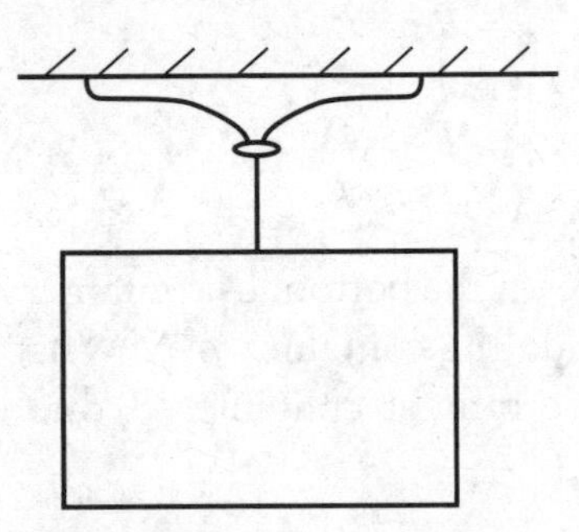

6. A suction cup is used to hang a sign from the ceiling as shown above. Which of the following statements is correct?

   (A) The force from the air upward on the suction cup is greater than the weight of the suction cup and sign.
   (B) The pressure of the air upward on the suction cup is equal to the weight of the suction cup and the sign.
   (C) The difference in pressure between the inside and outside of the suction cup equals the weight of the suction cup and the sign.
   (D) The low pressure inside the suction cup pulls the suction cup upward and holds the sign up, canceling gravity.

Questions 7 and 8

Four differently shaped sealed containers are completely filled with an unknown fluid, as shown in the figure below. Containers A and B are cylindrical. Containers C and D are truncated conical shapes. The top and bottom diameters of the containers are shown below.

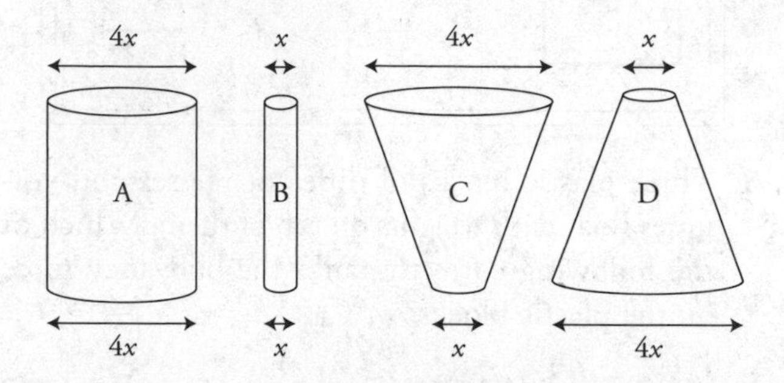

7. Which of the following is the correct ranking of the pressure (P) at the bottom of the containers?

(A) $P_A = P_B = P_C = P_D$
(B) $P_A = P_D > P_C = P_B$
(C) $P_A > P_D > P_C > P_B$
(D) $P_D > P_A > P_C > P_B$

8. The force on the bottom of container A due to the fluid inside the container is F. What is the force on the bottom of container B due to the fluid inside?

(A) F
(B) $F/4$
(C) $F/8$
(D) $F/16$

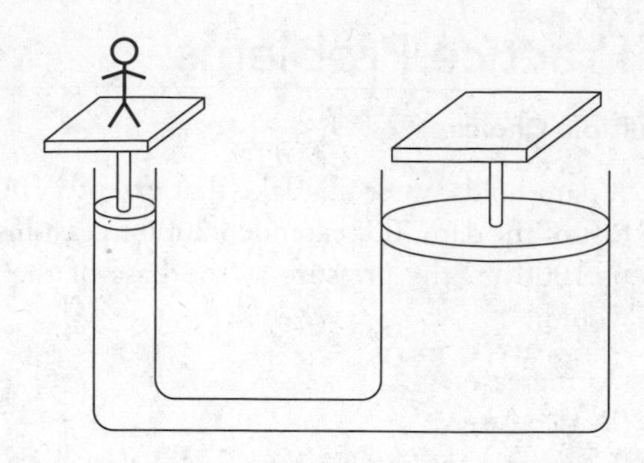

9. Two cylinders filled with a fluid are connected by a pipe so that fluid can pass between the cylinders, as shown in the figure above. The cylinder on the right has 4 times the diameter of the cylinder on the left. Both cylinders are fitted with a movable piston and a platform on top. A person stands on the left platform. Which of the following lists the correct number of people that need to stand on the right platform so neither platform moves? Assume that the platform and piston have negligible mass and that all the people have the same mass.

(A) 16 people
(B) 4 people
(C) 1 person
(D) It is impossible to balance the system because you need 1/16 of a person on the right side

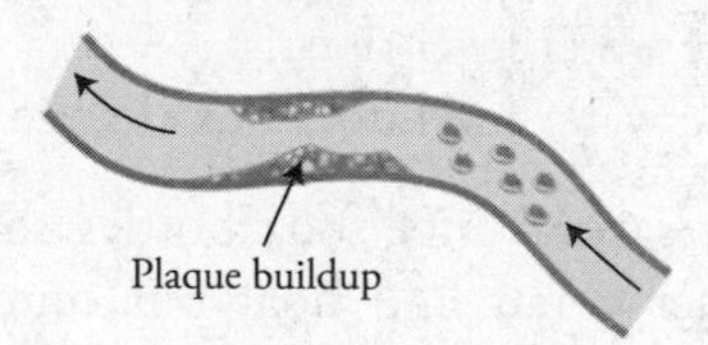

10. Blood cells pass through an artery that has a buildup of plaque along both walls, as shown in the figure. Which of the following correctly describes the behavior of the blood cells as they move from the right side of the figure through the area of plaque? Assume the blood cells can change volume.

(A) The blood cells increase in speed and expand in volume.
(B) The blood cells increase in speed and decrease in volume.
(C) The blood cells decrease in speed and expand in volume.
(D) The blood cells decrease in speed and decrease in volume.

Free Response

11. A person sips a drink from a straw.

(A) Rank the pressure inside the person's mouth A, to the pressure at the surface of the drink B, to the pressure at the bottom of the drink C. Justify your response.
(B) Explain using words how the person is able to drink from a straw.
(C) Justify your explanation in (B) by using a derivation.
(D) The top of the drink is now sealed with an air-tight lid. What affect will this have on the person drinking?

Volume ($\times 10^{-6}$ m^3)	Mass ($\times 10^{-3}$ kg)
50	145
100	161
150	223
200	266
300	334

12. Your teacher has asked you to determine the density of a liquid. You place a graduated cylinder on a balance and measure the mass at different volumes. The data is displayed in the table above.

(A) Graph the data appropriately on the grid provided to the right.

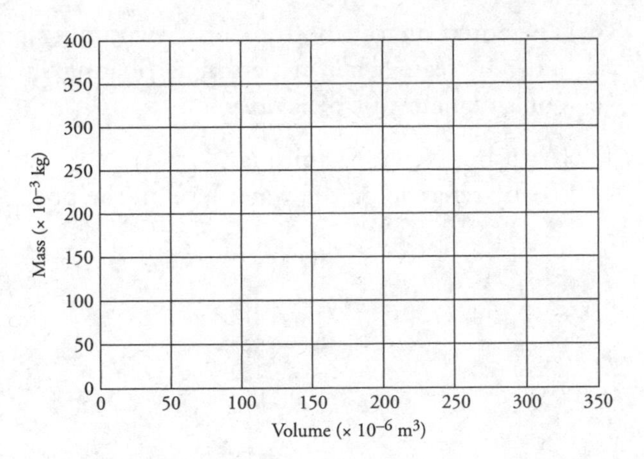

(B) Explain why the best fit line does not go through the origin and explain what the intercept represents.
(C) Use the graph to determine the density of the fluid in kg/m^3. Show your work.
(D) How might you improve the accuracy of the lab?

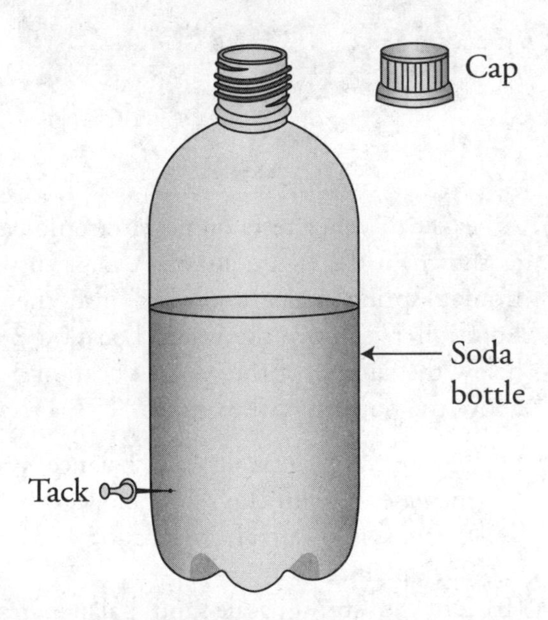

13. A soda bottle is filled partway with water. A tack is poked into the side to create a hole. The cap is removed from the bottle.

(A) When the tack is removed, water flows out. As the water level descends, the velocity of the water leaving the hole decreases. Explain why this happens in terms of pressure and energy.
(B) The bottle is refilled to its original level, and the tack is used to plug up the hole. The cap is tightened, and the tack is again removed. Water comes out of the hole but quickly stops while there is still a large quantity of water above the tack hole. Explain why this occurs.

14. The force on the bottom of a swimming pool must increase when a person is floating in the pool. Explain this behavior

(A) using Newton's third law.
(B) by referencing the water level in the pool.

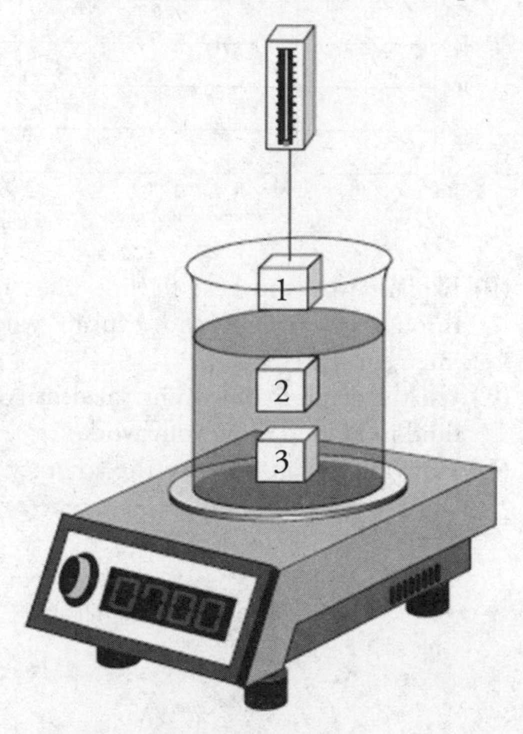

15. A beaker of water rests on an electronic balance, as shown in the figure above. A mass suspended from a spring scale is lowered into the water. Location 1 is above the water. Location 2 is just below the surface of the water. Location 3 is just above the bottom of the beaker.

(A) Do the spring scale and balance readings increase, remain the same, or decrease when the mass is lowered from location 1 to location 2? Explain.
(B) Do the spring scale and balance readings increase, remain the same, or decrease when the mass is lowered from location 2 to location 3? Explain.

16. A hose with a 2-cm radius supplies water to fill a pool. Water flows out of the hose at a rate of 6 m/s. The pool has a length of 12 m, a width of 9 m, and a depth of 2 m.

(A) What is the volume flow rate of the water exiting the hose?
(B) How many hours does it take to completely fill the pool?

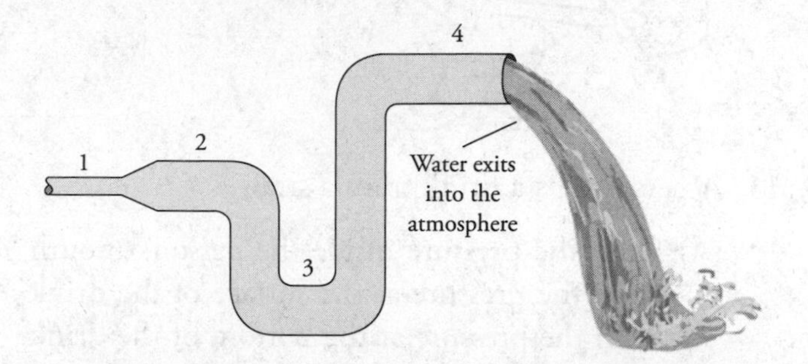

17. Water passes through a closed piping system starting at point 1 and exits to the atmosphere at the highest point near point 4. The pipe has a small diameter at point 1 but then widens out to a larger constant radius for the rest of the pipe.

(A) Rank the velocity of the water in the pipe at the four locations from greatest to least. Justify your answer.
(B) Rank the pressure in the pipe at the four points from greatest to least. Justify your answer.

❯ Solutions to Practice Problems

1. (C) The pressure in a fluid is based upon the height of the column of fluid, the density of the fluid, and pressure exerted on the surface of the fluid. By extending the horizontal dimension of the lake, you are not changing any of these variables, so the pressure remains the same.

2. (D) This is a combination Continuity and Bernoulli's equation problem. The pressure acting on the fluid exiting the pipe is air pressure, so your pressure at this point is atmospheric—100,000 Pa. You will need to find the flow velocity at the wider section of the pipe so you will use the continuity equation:

$$A_1 v_1 = A_2 v_2$$

$$\pi r_1^2 v_1 = \pi r_2^2 v_2$$

You can cancel π out from both sides and have to remember to change your diameters to radii.

$$(0.02\ \text{m}^2)\ v_1 = (0.01\ \text{m}^2)\ (20\ \text{m/s})$$

$$v_1 = 5\ \text{m/s}$$

Now that you have the velocity at the narrow section, you can apply Bernoulli's equation.

$$P_1 + \rho g y_1 + \tfrac{1}{2}\rho v_1^2 = P_2 + \rho g y_2 + \tfrac{1}{2}\rho v_2^2$$

Consider the nozzle point 1 and the wide section point 2. Since the pipe is horizontal, there is no change in height, so the height can be considered zero and those terms can drop out. You are left with:

$$P_1 + \tfrac{1}{2}\rho v_1^2 = P_2 + \tfrac{1}{2}\rho v_2^2$$

$$100{,}000\ \text{Pa} + \tfrac{1}{2}(1000\ \text{kg/m}^3)(20\ \text{m/s})^2 = P_2 + \tfrac{1}{2}(1000\ \text{kg/m}^3)(5\ \text{m/s})^2$$

$$P_2 = 290{,}000\ \text{Pa}$$

3. (A) When the treasure is floating in the boat, it displaces an amount of water equal to its weight. When the treasure is on the lake bottom, it displaces much less water, because the lake bottom supports most of the weight that the buoyant force was previously supporting. Thus, the lake level drops.

4. (D) Since Manny will be floating in equilibrium, his weight must be equal to the buoyant force on him. The buoyant force is $\rho_{\text{water}} V_{\text{submerged}}\, g$, and must equal Brian's weight of 800 N. Solving for $V_{\text{submerged}}$, we find he needs to displace 8/100 of a cubic meter. Converting to liters, he needs to displace 80 L of water, or 40 bottles. (We would suggest that he use, say, twice this many bottles—then the raft would float only half submerged, and he would stay drier.)

5. (A) Since all the objects are floating, the buoyancy force must equal the weight of the combined objects.

6. (A) Pressure is not a force and cannot equal or cancel out the gravity force. Assuming there is still air inside the suction cup, the force from the external air upward would need to be larger than the overall weight to cancel out the extra force from the air still inside the suction cup pushing down from the inside. Even if there was a vacuum inside the suction cup, the force from the air would be greater than the weight of the system as there would be a normal force pushing downward on the suction cup from the ceiling.

7. (A) They are all the same: $P = P_0 + \rho g h$

8. (D) The pressure is the same at the bottom of each container. The area of the bottom surface = $\pi\left(\frac{d}{2}\right)^2$. Thus the force at the bottom on the container is $F = PA = P\pi\left(\frac{d}{2}\right)^2$. Inserting the diameter of each container, we see that the force on the smaller container is 1/16 that of the larger container.

9. (A) The pressure at the top of both pistons must be the same:

$$\left(\frac{F}{A}\right)_{\text{left}} = \left(\frac{F}{A}\right)_{\text{right}}$$

$$\left(\frac{mg}{\pi r^2}\right)_{\text{left}} = \left(\frac{mg}{\pi r^2}\right)_{\text{right}}$$

$$\left(\frac{m}{r^2}\right)_{\text{left}} = \left(\frac{m}{r^2}\right)_{\text{right}} = \frac{m_{\text{right}}}{\left(4r_{\text{left}}\right)^2}$$

$$16m_{\text{left}} = m_{\text{right}}$$

10. (A) The speed increases as it passes through the region of plaque and, as shown by Bernoulli's equation, the pressure in the fluid decreases. Thus, the cell will increase in size.

11. (A) $P_C > P_B > P_A$. The pressure at the bottom of the drink is the highest due to the column of fluid above it and the atmosphere above that. The pressure at the top of the drink is atmospheric. The pressure inside the person's mouth must be less than atmospheric so that fluid rises up the straw.

(B) The person must lower the pressure inside his or her mouth. Then, the higher pressure at the bottom of the straw pushes fluid up toward the person's mouth where the pressure is lower. (Note that the person does not "suck" the drink up to his or her mouth!)

(C) $P_C = P_A + \rho g h_{\text{height of fluid in the straw}}$

$P_A = P_C - \rho g h_{\text{height of fluid in the straw}}$

(D) As the fluid moves up the straw, it reduces the amount of fluid in the cup. When the lid is airtight, this increases the volume of "the air pocket" at the top of the cup, which will reduce the pressure in this air pocket. This will in turn reduce the pressure of the fluid at the bottom of the drink. This makes it harder for the person to produce the pressure differential to push fluid up the straw and drink.

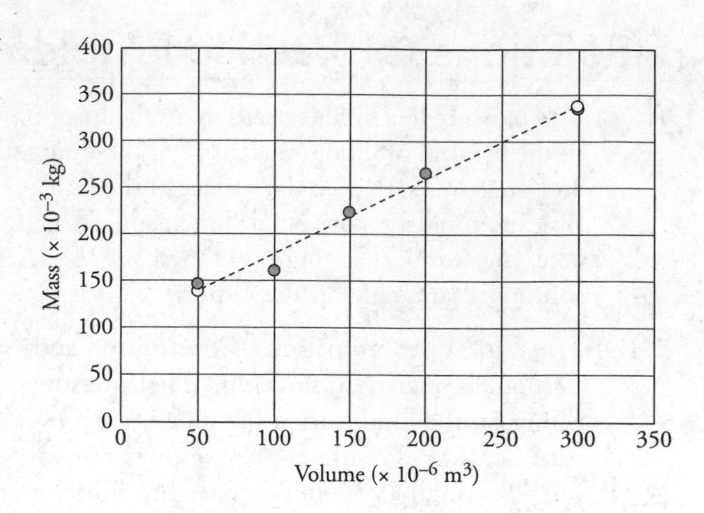

12. (A) See the graph.

(B) You forgot to tare the balance. The intercept is the mass of the graduated cylinder (approximately 100 g).

(C) The density of the fluid equals the slope of the best fit line of the data. The slope is about 800 kg/m^3.

(D) More data points over a wider range almost always helps. Taring the balance won't really help because the slope of the line is independent of the mass of the graduated cylinder.

13. (A) The water pressure at the hole is determined by the height of the water above the hole. As the amount of water decreases, the pressure that pushes the water out of the hole decreases as well. In terms of energy, think of the amount of fluid above the tack as gravitational potential energy that can be converted to kinetic energy in the water exiting the tack hole. As the water level descends, the gravitational potential energy decreases, which decreases the kinetic energy of the exiting water.

(B) When the water level descends, the air pocket at the top of the bottle increases. With the cap on the bottle, air from the outside cannot flow in. The ideal gas law, $PV = nRT$, tells us that as the volume of air above the water increases, the pressure of that air must decrease. The static pressure equation, $P = P_0 + \rho g h$, tells us that as the pressure on top of the water decreases, the water pressure at the tack hole will also decrease. Eventually, the water pressure inside the bottle equals the air pressure outside the bottle, and we have equilibrium. The water stops coming out of the tack hole.

14. (A) When a person floats in a pool, he or she receives a buoyancy force from the water. By Newton's third law, the water must also receive an equal and opposite force downward from the person. This downward force increases the force on the bottom of the pool.

(B) When a person floats in a pool, he or she displaces water, which makes the water level rise. This rise in water height causes an increase in the static fluid pressure on the bottom of the pool, which in turn increases the force on the bottom of the pool.

15. (A) The spring scale reading decreases because the buoyancy force pushes up on the mass. The balance reading increases because the mass pushes down on the water with a Newton's third law force that is equal in magnitude but opposite in direction to the upward buoyancy force on the mass.

(B) Both scales' readings remain the same. The object does not displace any additional water moving from position 2 to position 3. Therefore, the buoyancy force does not change by moving the mass from position 2 to position 3.

16. (A) $\frac{\Delta V}{\Delta t} = Av = \pi r^2 v = 0.0075 \frac{\text{m}^3}{\text{s}}$. Remember to convert your units!

(B) The volume of the pool is 216 m^3. Divide the volume of the pool by the volume flow rate, and you get 28,800 seconds to fill the pool, which is 8 hours.

17. (A) Continuity tells us that the velocity of the water is fastest at location 1 and is uniformly slower at locations 2, 3, and 4.

(B) There is a lot going on here.

- Continuity tells us that the speed is slower at locations 2, 3, and 4.
- Bernoulli tells us that the slower the fluid moves, the higher the pressure. This means the pressure at location 2 is higher than at location 1.
- Static pressure tells us that the lower the pipe is, the higher the pressure. This means that in the large pipe, location 3 has the highest pressure and location 4 has the lowest.
- Combining all these considerations, we can see that the pressure is highest at location 3 followed by location 2. It is impossible to compare locations 1 and 4 accurately without more information about the pipe diameters and the height difference between location 1 and 4. We just know the pressures at locations 1 and 4 are both lower than at location 2.

› Rapid Review

- Pressure is caused by the random thermal motion of the molecules.
- Gravitational force pulling fluids downward increases pressure. The deeper inside a fluid you go, the higher the pressure gets.
- The force caused by pressure is always perpendicular to the surface of contact.
- Pressure equals force divided by the area over which the force is applied.
- Density equals the mass of a substance divided by its volume.
- The absolute pressure exerted by a column of a fluid equals the pressure at the top of the column plus the gauge pressure—which equals ρgh.
- The buoyant force on an object is equal to the weight of the fluid displaced by that object.
- If a force is applied somewhere on a container holding a fluid, the pressure increases everywhere in the fluid, not just where the force is applied.
- The volume flow rate—the volume of fluid flowing past a point per second—is the same everywhere in an isolated fluid stream. This means that when the diameter of a pipe decreases, the velocity of the fluid stream increases, and vice versa.
- Bernoulli's equation is a statement of conservation of energy. Among other things, it tells us that when the velocity of a fluid stream increases, the pressure decreases.

Thermodynamics and Gases

IN THIS CHAPTER

Summary: Energy transfer in gases can do measurable work on large objects. Thermodynamics is the study of heat energy and how it gets transferred. It's different from the other topics in this book because the systems focused on are, for the most part, molecular. Since molecules and gases are at the heart of thermodynamics, this chapter includes a lot of material introducing them.

Key Ideas

- The behavior of molecules gives rise to the thermal behavior of matter.
- Internal energy is the energy possessed by a substance due to the vibration of molecules. A substance's internal energy depends on its temperature.
- Heat is a transfer of thermal energy.
- Heat is transferred by conduction, convection, and radiation.
- When an object's temperature increases, it expands.
- The ideal gas law connects four important thermodynamics variables: the number of gas molecules, gas pressure, temperature, and volume.
- Kinetic theory explains how the random motion of many individual molecules in a gas leads to measurable properties such as temperature.

- The first law of thermodynamics is a statement of conservation of energy. It is usually applied when a gas's state is represented on a pressure vs. volume (*PV*) diagram. From a *PV* diagram:
  (a) Internal energy can be determined using the ideal gas law and *PV* values from the diagram's axes.
  (b) Work done on or by the gas can be determined from the area under the graph.
  (c) Heat added or removed from the gas can NOT be determined directly from the diagram, but rather only from the first law of thermodynamics.
- Absolute zero is where all molecular vibrations stop and a gas would have no pressure or volume.

Relevant Equations

Ideal gas law: $PV = nRT = Nk_BT$

Internal energy of an ideal gas: $U = \frac{3}{2}nRT = \frac{3}{2}Nk_BT$

Average kinetic energy of the molecules in a gas and kinetic energy equation from mechanics: $K_{average} = \frac{3}{2}k_BT$ and $K = \frac{1}{2}mv^2$

RMS speed of a gas molecule: $v_{rms} = \sqrt{\frac{3k_BT}{m}}$

First law of thermodynamics: $\Delta U = Q + W$

Work done by a gas: $W = -P\Delta V$

Rate of heat transfer: $\frac{Q}{\Delta t} = \frac{kA\Delta T}{L}$

Start Small—Atomic Behavior That Produces Thermal Results

Put on your nano-glasses again and peer into the stuff that makes up matter. In Chapter 9, we saw that atoms are in constant motion and their collisions with surfaces give rise to pressure. Let's take a closer look at atomic motion. Atoms are in motion vibrating around nonstop. This motion is completely random. Some atoms are moving very fast and some are moving slowly—some to the right, some to the left. And now that you blinked, they have collided with each other and changed their motion. It's constant motion, constantly changing all the time. What can we gather from all this chaos? Quite a bit actually:

- The temperature of the object tells us the speed of the vibrations. Atoms in hot objects, as a whole, vibrate faster. Atoms in cold objects, on average, vibrate slower. In fact, the temperature of an object is a direct measure of the average kinetic energy of the atoms: $K_{average\ for\ atoms} \propto T$.
- In general, objects get bigger when they are hotter because the atoms are vibrating faster and take up a bit more space. Objects tend to shrink when they become colder.

(There are exceptions, like 0°C water and ice, but we will leave that explanation to your college chemistry teacher.)

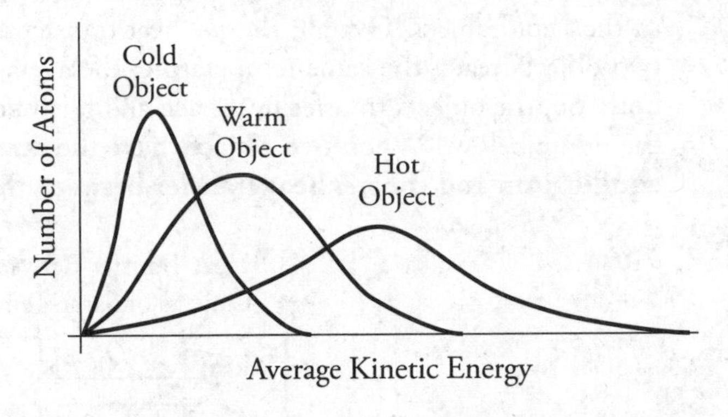

- The motion of atoms is random but has a distinct pattern. The figure above shows the kinetic energy distribution of atoms in an object when it is cold, warm, and hot. (If you are also taking a statistics class right now, you will see that these look similar to "normal distribution curves.") The vertical axis shows the number of atoms while the horizontal axis shows the kinetic energy. Look at the "cold" curve. Most of the atoms have a slow speed and thus a low kinetic energy. A few atoms are moving kind of fast and a few are moving very slowly, but most are in the middle. This is typical of all the curves. If we warm up the object, its curve will shift to the right because the atoms are gaining kinetic energy overall. The peak of the curve will also shift down because the number of atoms in the object is the same. They are just spread over a wider array of kinetic energies now. When we get the object hot, the curve continues to spread out to the right. The area under each graph should be the same because it's the same object with the same number of atoms, just at different temperatures. Notice something interesting: a few of the "cold" atoms have more kinetic energy than some of the atoms in the "hot" object. But, on average, the "hot" atoms have more kinetic energy and are moving faster than the "cold" atoms.
- This random motion distribution of atoms is why thermal energy always moves from hot to cold. Molecules in a hot object tend to collide and transfer more energy to the molecules in a cold object because they are moving faster (conservation of momentum). Is it possible for a "cold" molecule to collide and transfer energy to a "hot" molecule? Sure! But on average it is much more likely for energy to transfer from "hot" to "cold" (see the figure below), just like it is much more likely for a speeding car to transfer energy to a slower moving car in a collision than the other way around.

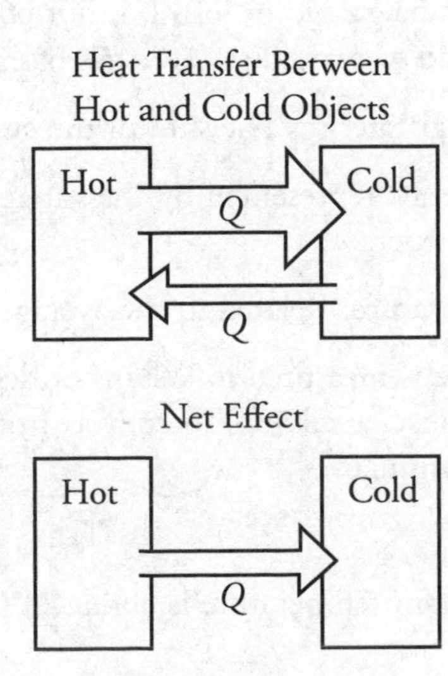

- When does this thermal energy transfer between objects stop? In reality the heat transfer between objects never really stops. Hot objects transfer lots of heat to cold objects. But remember that "cold" objects have a few fast-moving molecules that can transfer heat to the "hot" object. Overall, the *net* heat transfer is always from hot to cold. Once the two objects reach the same temperature, the average molecular motion is the same for both. So, the objects transfer heat back and forth between each other at equal rates. (See the figure below.) **When two objects have the same temperature, they are in thermal equilibrium and the *net* heat transfer between them is zero.**

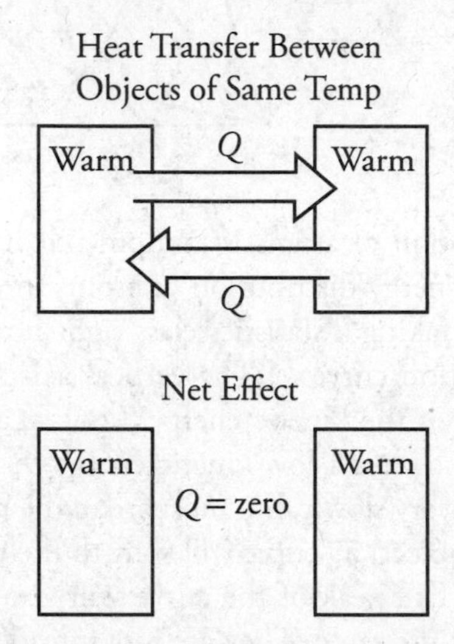

Now that we have a basic understanding of the nanobehavior of matter and how it influences the macrobehavior of what we see around us, let's take a closer look at this behavior on a larger scale.

Heat, Temperature, and Power

Heat, represented by the variable Q, is a type of energy that can be transferred from one body to another. As you might expect, heat is measured in joules, because it is a form of energy. Notice that all energy quantities (KE, PE, work) have units of joules. Heat is no different.

Be careful with phraseology here. Energy must be *transferred* in order to be called heat. So, heat can be gained or lost, but not possessed. It is incorrect to say, "a gas has 3000 J of heat." Instead we would say "the gas has 3000 J of internal energy."

Internal Energy: The sum of the energies of all of the molecules in a substance.

Internal energy, represented by the variable U, is also measured in joules, again because it is a form of energy.

Temperature: Related to the average kinetic energy per molecule of a substance.

Temperature is measured in kelvins or degrees Celsius. There is no such thing as "a degree kelvin"; it's just "a kelvin." To convert from kelvins to degrees Celsius or vice versa, use the following formula:

$$T_{\text{Celsius}} = T_{\text{Kelvin}} - 273$$

So, if room temperature is about 23°C, then room temperature is also about 296 K.

It is important to note that *two bodies that have the same temperature do not necessarily contain the same amount of internal energy*. For example, let's say you have a huge hunk of metal at 20°C and a tiny speck of the same type of metal, also at 20°C. The huge hunk contains a lot more internal energy—it has a lot more molecules moving around—than the tiny speck. By the same logic, you could have two items with very different temperatures that contain exactly the same amount of internal energy.

Power: Work per time

Power is measured in joules/second, or watts:

$$1\text{ W} = 1\text{ J/s}$$

Power is not an idea limited to thermodynamics. In fact, we often talk about power in the study of mechanics.

Heat Transfer

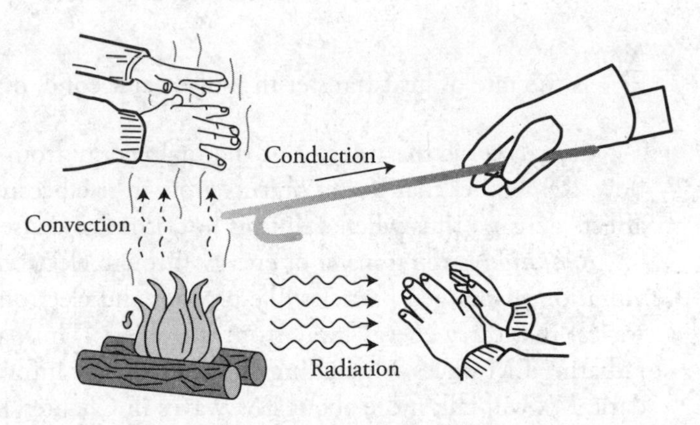

There are three physical methods for heat to transfer: conduction, convection, and radiation.

Conduction: Have you ever noticed that some materials feel colder than others even though they have been sitting in the same room and are at the same temperature? For example, you could have a block of wood and a block of metal sitting on a table in your home. If you picked up the wood, it wouldn't feel particularly warm or cold, but the metal feels cool to the touch. Well, the reason the metal feels colder than the wood is that it is a better conductor of heat than the wood. It draws heat away from your hand at a faster rate than the wood, making it feel colder.

Conduction is the physical transfer of the atomic vibration energy from atom to atom through an object from the hot end to the colder end. Whenever two objects with different temperatures touch each other, heat will transfer from the hotter object to the colder one through atomic collisions until the objects reach the same temperature, or equilibrium. There are a couple of properties that can affect the rate of heat transfer:

- The thermal conductivity, k, of the material. For example, metals tend to conduct heat better and will have a high thermal conductivity compared to nonmetals.
- The difference in temperatures between the cold and hot objects, ΔT. A greater difference will produce a larger rate of heat transfer.

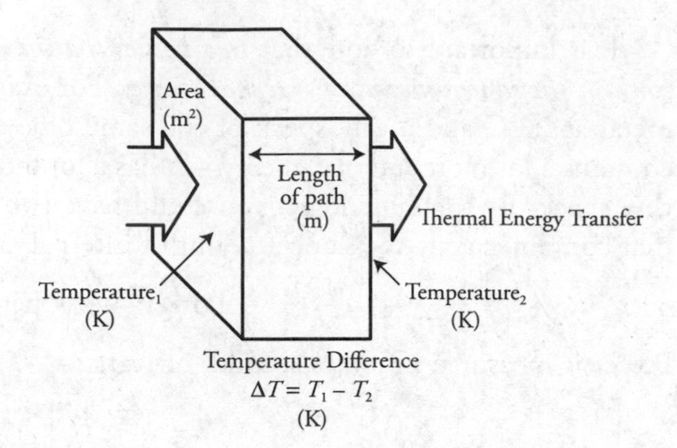

- The cross-sectional area of the material the heat is being transferred along, A—the larger the cross-sectional area of contact, the greater the rate of heat transfer.
- The length, L, of the material the heat is being transferred through—a longer length, the lower the rate of heat transfer.

$$\frac{\Delta Q}{t} = \frac{kA\Delta T}{L}$$

$\frac{\Delta Q}{t}$ is the rate of heat transfer in joules per second, or watts.

Convection is the transfer of thermal energy from one place to another through fluid flow. Remember that hotter objects grow in size because the atoms are moving faster. This affects density; thus when a fluid is hot, it is less dense and will naturally rise.

Radiation is the transfer of energy through electromagnetic waves. It turns out that the vibration of charged particles like protons and electrons create electromagnetic waves (EM waves) that carry energy away from the object. Thus, anything with vibrating molecules is radiating EM waves—including you! This is how infrared vision goggles help you see in the dark. We will talk more about EM waves in Chapter 12.

Kinetic Theory of Gases

We have already looked at how the nano-world affects objects in general. Let's review and see how atoms influence gases in particular.

Kinetic theory makes several sweeping assumptions about the behavior of molecules in a gas. These assumptions are used to make predictions about the macroscopic behavior of the gas. Don't memorize these assumptions. Rather, picture a bunch of randomly bouncing molecules in your mind, and see how these assumptions do actually describe molecular behavior.

- Molecules move in continuous, random motion.
- There are an exceedingly large number of molecules in any container of gas.
- The separation between individual molecules is large.
- Molecules do not act on one another at a distance; that is, they do not exert electrical or gravitational forces on other molecules.
- All collisions between molecules, or between a molecule and the walls of a container, are elastic (i.e., kinetic energy is not lost in these collisions).

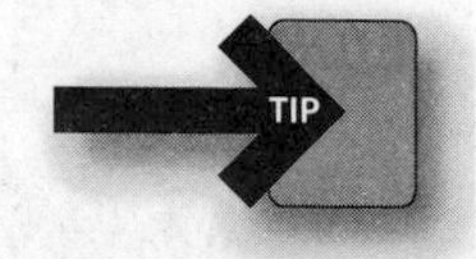

What do you need to know about kinetic theory? You need to understand the above assumptions, and the two important results of kinetic theory. These are derived using

Newtonian mechanics and the assumptions. Don't worry for now about the derivation; just learn the equations and what they mean:

1. The relationship between the internal energy of a gas and its temperature is

$$U = \frac{3}{2}nRT = \frac{3}{2}Nk_BT$$

This equation allows you to find the temperature of a gas given its internal energy, or vice versa.

2. The average kinetic energy of a gas is directly related to the temperature of the gas with the temperature in Kelvin:

$$K_{average} = \frac{3}{2}k_B T$$

Remember that kinetic energy is:

$$K = \frac{1}{2}mv^2$$

Putting this together we get:

$$v_{rms} = \sqrt{\frac{3k_B T}{m_{\text{gas molecule}}}}$$

This means that, for the same temperature, gases of different masses will have the same average kinetic energy but different average velocities. Gases with a smaller molecular mass will travel faster at the same temperature than heavier gas molecules.

The next figure shows the molecule speed distributions of two gases. If the two gases have the same temperature, what can you deduce about the two gases? Since they have the same temperature, they must have different masses. Gas A is a more massive molecule.

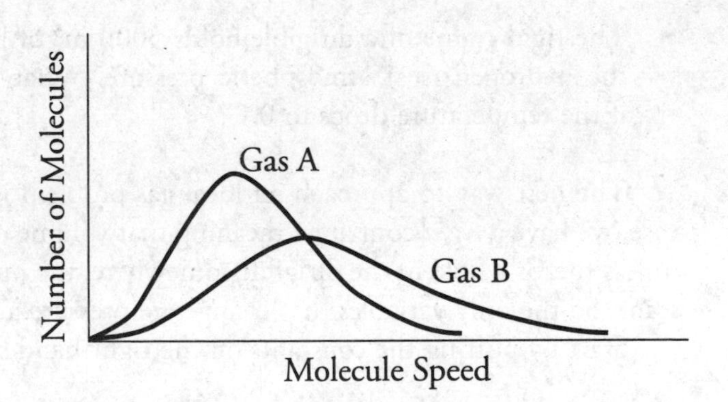

Ideal Gas Law

This is one that you may remember from chemistry class:

$$PV = nRT$$

However, you are probably going to use a different set of units from what you used in chemistry. In physics:

- P is the pressure in pascals. (Remember, a pascal represents a force of 1 newton spread over an area of 1 meter squared.)
- V is the volume of the gas in cubic meters (m^3).

- n is the number of moles of gas.
- R is the ideal gas constant ($R = 8.31$ J/mol·K).
- T is the temperature of the gas in Kelvin (K).

This same relationship can be written as:

$$PV = Nk_BT$$

where:

- N is the number of molecules.
- k_B is Boltzmann's constant. (It is equal to the ideal gas constant, R, divided by Avogadro's number, $k_B = 1.38 \times 10^{-23}$ J/K.)

Be careful about your units. For volume, remember:

- $1\ \text{m}^3 = 1000$ liters $= 1 \times 10^6\ \text{cm}^3$
- the number of moles $= \dfrac{\text{number of molecules}}{\text{Avogadro's number}}$ or $n = \dfrac{N}{N_A}$

  where $N_A = 6.02 \times 10^{23}$ molecules/mole

Keep in mind there is a nice shortcut if you have a constant number of moles of gas and you are changing the volume, temperature, or pressure of that gas:

$$\frac{P_1V_1}{T_1} = \frac{P_2V_2}{T_2}$$

Here's a sample problem.

The rigid frame of a dirigible holds 5000 m³ of hydrogen. On a warm 20°C day, the hydrogen is at atmospheric pressure. What is the pressure of the hydrogen if the temperature drops to 0°C?

The best way to approach an ideal gas problem is to ask: "What is constant?" In this case, we have a *rigid* container, meaning that volume of the hydrogen cannot change. Also, unless there's a hole in the dirigible somewhere, the number of molecules N must stay constant. So the only variables that change are pressure and temperature.

Start by putting the constants on the right-hand side of the ideal gas equation:

$$\frac{P}{T} = \frac{Nk_B}{V}$$

Because the right side doesn't change, we know that the pressure divided by the temperature always stays the same for this dirigible:

$$\frac{10^5\ \text{N/m}^2}{293\ \text{K}} = \frac{P}{273\ \text{K}}$$

Here atmospheric pressure, as given on the constant sheet, is plugged in (note that temperature must always be in units of kelvins in this equation). Now just solve for P. The answer is 93,000 N/m², or just a bit less than normal atmospheric pressure.

Now let's take gases to the extreme. Notice how pressure is directly related to the temperature: $PV = nRT$. If we put a sample of gas in a sealed container so the volume won't change and then cool the gas in the freezer, what happens to the pressure? It goes down. When you plot pressure as a function of temperature, you get the graph shown below. Notice how the pressure keeps going down with temperature, but it can go only so low. The lowest pressure you can get is zero. The temperature where this occurs is absolute zero. This is when the motion of every molecule in the object ceases. No motion. No kinetic energy. Thus the temperature is zero. This same graph can be made by keeping pressure constant and plotting volume as a function of temperature; the result is the same. Zero volume for the gas will occur at absolute zero.

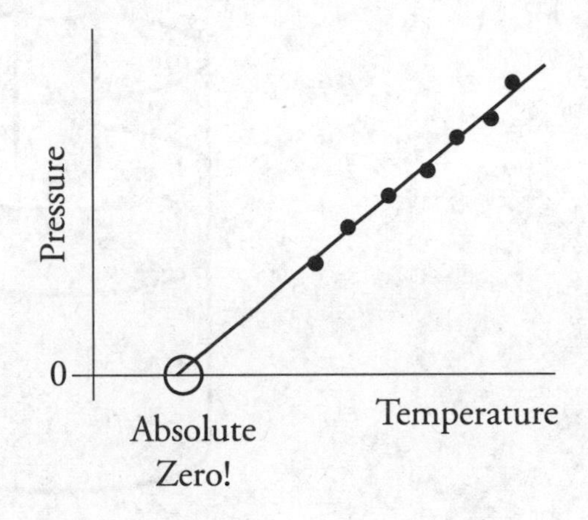

First Law of Thermodynamics

Before you start this section, be sure you understand the difference between heat and internal energy, as described several pages ago. Furthermore, remember that mechanical work is done whenever a force is applied through a distance.

The first law of thermodynamics is simply a statement of conservation of energy.

$$\Delta U = Q + W$$

U is the internal energy of the gas in question. Q is the heat added to the gas, and W is the work done on the gas when the gas is compressed.

The signs of each term in the first law are extremely important. If heat is added to the gas, Q is positive, and if heat is taken from the gas, Q is negative. If the gas's internal energy (and, thus, its temperature) goes up, then ΔU is positive; if the gas's temperature goes down, then ΔU is negative.

The sign of W is more difficult to conceptualize. Positive W represents work done *on* a gas. However, if a gas's volume expands, then work was not done *on* the gas, work was done *by* the gas. In the case of work done *by* a gas, W is given a negative sign.

Here's a sample problem.

A gas is kept in a cylinder that can be compressed by pushing down on a piston. You add 2500 J of heat to the system, and then you push the piston 1.0 m down with a constant force of 1800 N. What is the change in the gas's internal energy?

Define the variables in the first law, being careful of signs: Here $Q = +2500$ J. We're pushing the piston down; therefore, work is done *on* the gas, and W is positive. We remember that work = force × distance; so, the work done on the gas here is 1800 N·1 m, or +1800 J. Now plug this into the first law of thermodynamics: the answer is +4300 J.

PV Diagrams

Imagine a gas in a cylinder with a movable piston on top, as shown below.

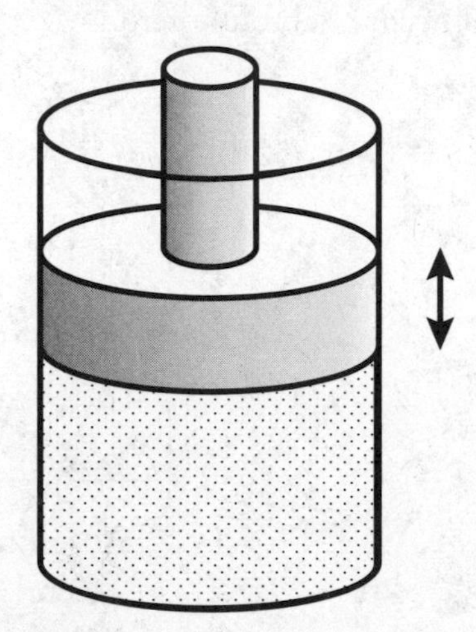

When the gas expands, the piston moves up, and work is done *by* the gas. When the piston compresses, work is done *on* the gas. The point of a *heat engine* is to add heat to a gas, making the gas expand—in other words, using heat to do some work by moving the piston up.

To visualize this process mathematically, we use a graph of pressure vs. volume, called a *PV diagram*. On a *PV* diagram, the *x*-axis shows the volume of the gas, and the *y*-axis shows the pressure of the gas at a particular volume.

Don't be afraid of *PV* diagrams. Once you learn your way around them, they are really quite simple. Let's practice.

Isothermals and How to Find *T* and ΔU on a *PV* Diagram Using: $PV = nRT$ and $\Delta U = \frac{3}{2}nR\Delta T$

How do we find temperature on a *PV* diagram?

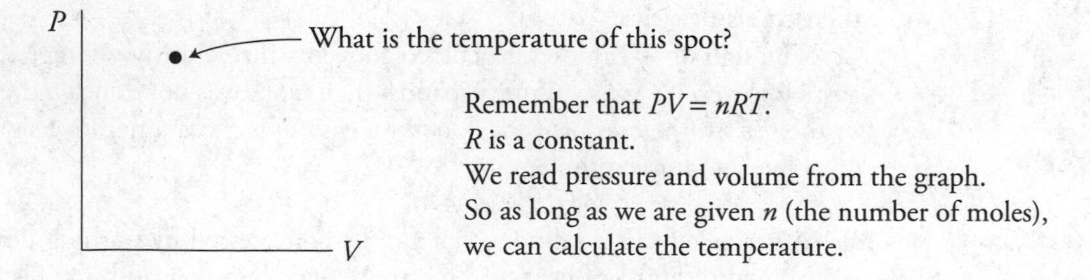

Which of these locations has the highest temperature? At which point is the internal energy highest for the gas?

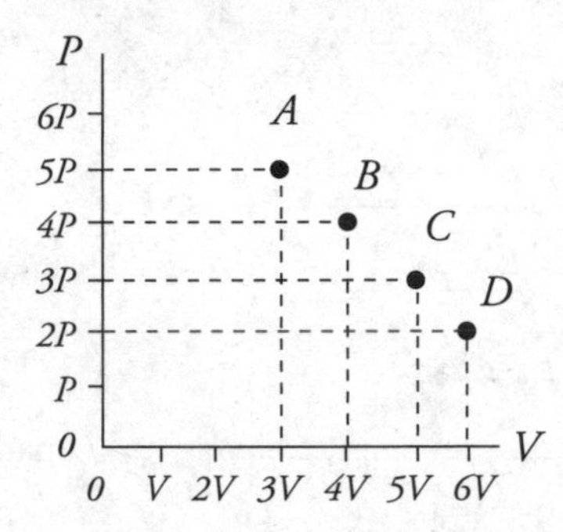

Remember that $PV = nRT$.
So a higher PV value means higher T.

Location	PV Value	
A	15PV	
B	16PV	⟵ Highest T and U (internal energy)
C	15PV	
D	12PV	⟵ Lowest T and U

Which path would you take on this PV diagram to get warmer or colder?

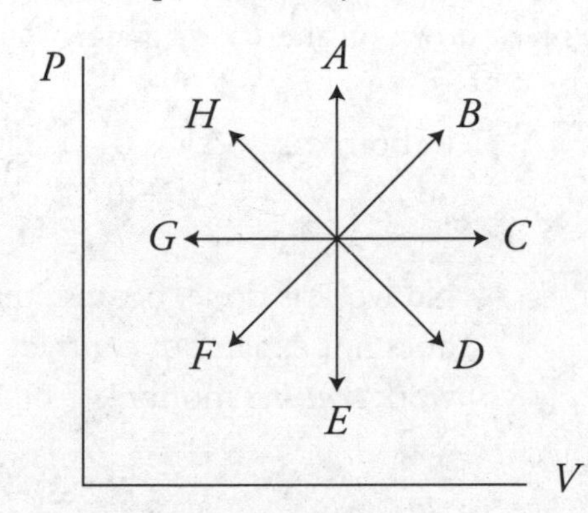

Higher PV = Higher T
Paths A, B, and C take you to higher T.
Lower PV = Lower T
Paths E, F, and G take you to lower T.
It's hard to tell about D and H without more information.
There could be no change in T. It depends on how much the volume and pressure change. So we would need actual numbers to find out.

Every point that has the same PV value has the same temperature. These are called isothermal lines.

If PV is a constant, T is a constant.
Constant PV lines are hyperbolas.
These hyperbolas of constant T are called isothermals.
Lower isothermals are lower temp.
Higher isothermals are higher temp.

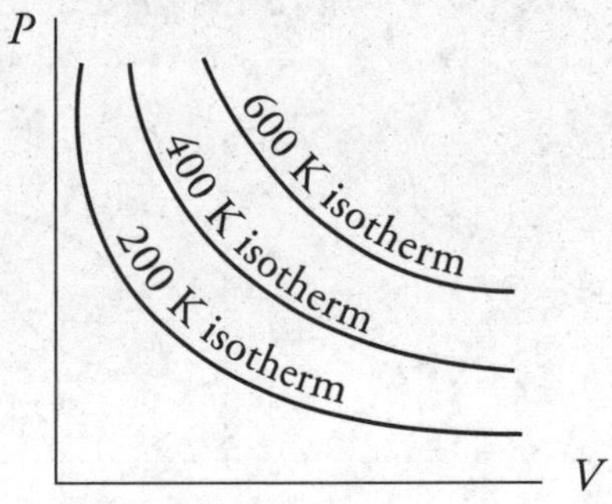

How to Find Work on a *PV* Diagram Using $W_{gas} = -P\Delta V$

Movement to the right is negative work.

An expanding gas moves the environment thus transferring energy from itself to the environment.

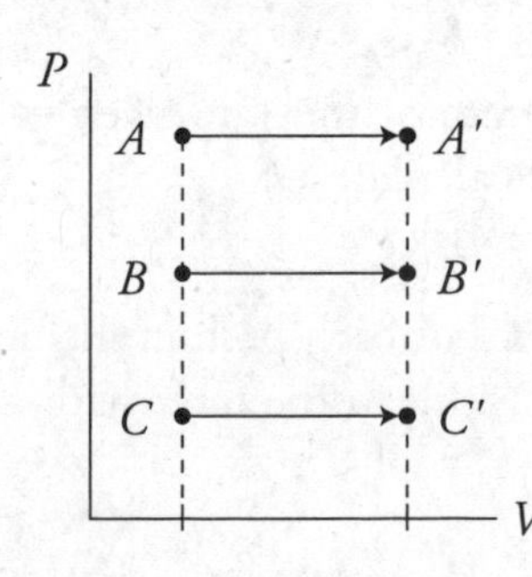

Which path is the largest negative work?

$A \rightarrow A'$ $-W = -P\Delta V$ Largest
$B \rightarrow B'$ $-W = -P\Delta V$ ↓
$C \rightarrow C'$ $-W = -P\Delta V$ Smallest

Even though all three have the same ΔV, the pressure is not the same in each case.

Movement to the left is positive work.

A gas compressed by an outside force receives energy from the outside environment as a consequence.

Which path is the largest positive work?

$D \rightarrow D'$ $W = -3P(-3V) = +9PV$

$E \rightarrow E'$ $W = -2P(-2V) = +4PV$

$F \rightarrow F'$ $W = -P(-V) = +PV$

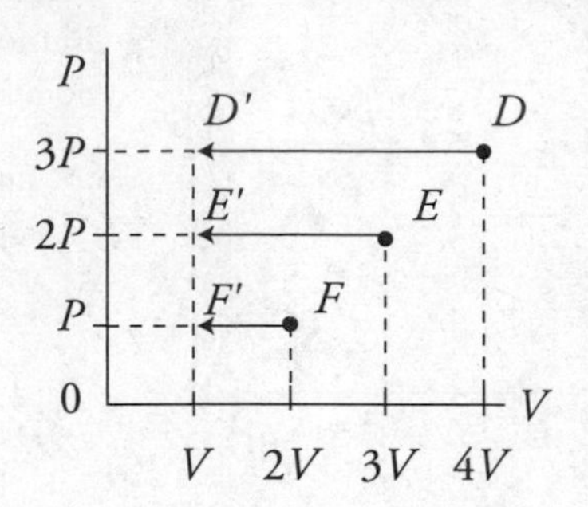

It takes 9 times more work to compress the gas along the D path than the F path.

What happens when you move up or down on the PV diagram?

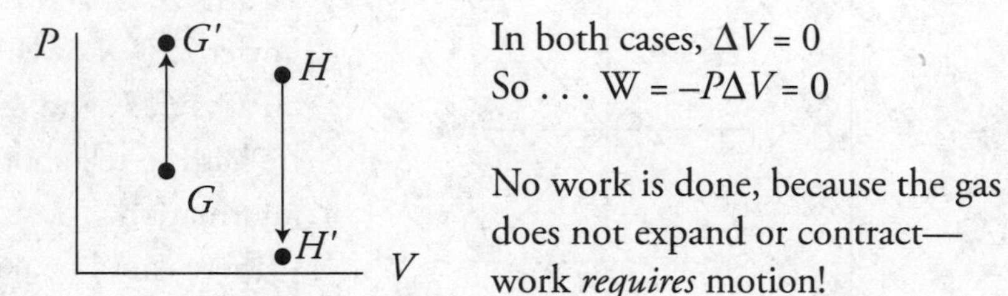

The area under the graph equals the magnitude of the work.

Paths #1 and #2 both have the same starting point and the same ending point, but different amounts of work.

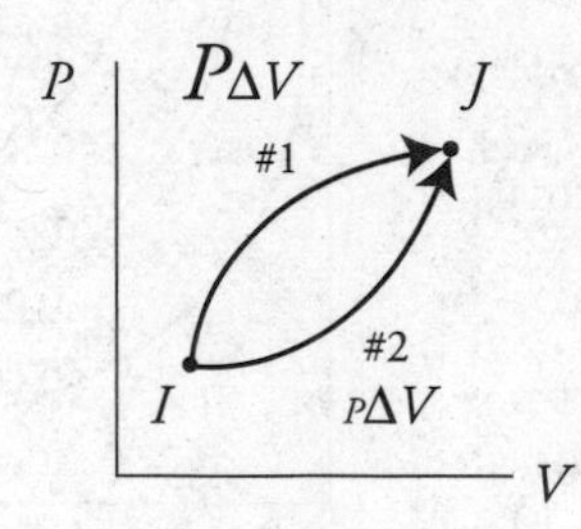

Both paths start and end at the same place. The work is negative in both cases, but the magnitudes are not the same! They don't have the same *average* pressure!

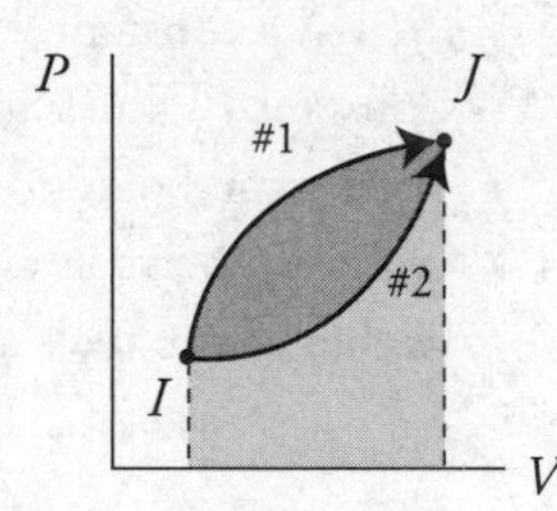

Think area under the curve—Path #1 has more area under it. That means a larger magnitude for work.

How to Find Heat (Q) on a PV Diagram Using $\Delta U = Q + W$

How do we find heat (Q) from a PV diagram? Well . . . we don't. Not directly, anyway. We use the first law of thermodynamics: $\Delta U = Q + W$.

Here is the procedure:

1. We either find W and ΔU from the PV diagram or they are given values in a problem.
2. We plug these values into $\Delta U = Q + W$ and calculate Q.

That's it!

Here is a practice problem with no numbers. For each path in the next diagram, determine if the value is positive, negative, or zero and fill in the table with a +, –, or 0.

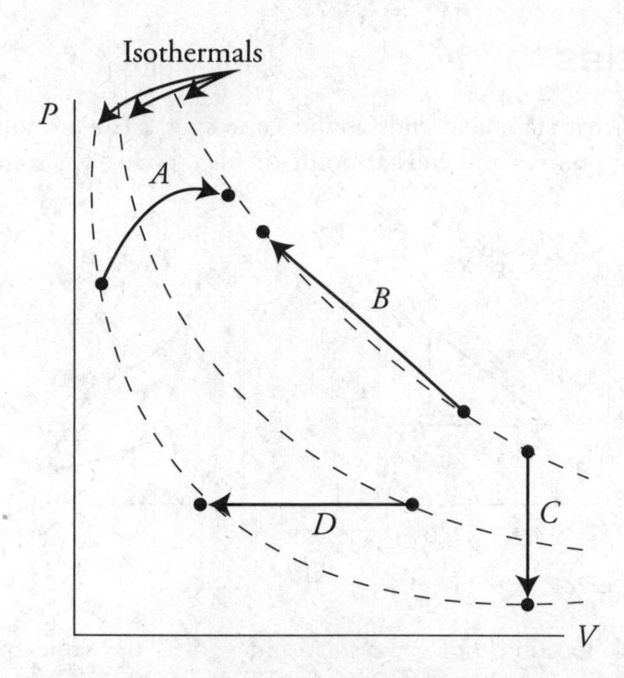

Remember the rules:

- To find ΔT, we see how the path moves through the isothermal lines. Or we see if P times V increases or decreases.
- To find ΔU, we see what ΔT does because ΔU tracks ΔT since $\Delta U = \frac{3}{2}nR\Delta T$.
- To find work, we see if the path moves right ($-W$), left ($+W$), or up and down ($W = 0$). And we use the equation $W = -P\Delta V$.
- To find Q, we use the first law of thermodynamics $\Delta U = Q + W$ and the ΔU and W that we already found.

Path	ΔT	ΔU	W	Q
A				
B				
C				
D				
How to determine the sign	ΔT is found by seeing how the path moves through the isotherms. ΔT and ΔU always have the same sign. $\Delta U = \frac{3}{2}nRT$		Move to the right = – Move to the left = + Up or down = 0 $W_{gas} = -P\Delta V$	Find ΔU and W first. Then use the first law of thermodynamics to calculate Q. $\Delta U = Q + W$

Answer Key

Path	ΔT	ΔU	W	Q
A	+	+	–	+
B	0	0	+	–
C	–	–	0	–
D	–	–	+	–
How to determine the sign	ΔT is found by seeing how the path moves through the isotherms. ΔT and ΔU always have the same sign. $\Delta U = \frac{3}{2}nRT$		Move to the right = – Move to the left = + Up or down = 0 $W_{gas} = -P\Delta V$	Find ΔU and W first. Then use the first law of thermodynamics to calculate Q. $\Delta U = Q + W$

Round and Round We Go: Cycles

A cycle is a path on a *PV* diagram that starts and ends at the same spot. Here are some cycles. Note that each cycle starts at point *A* and ends at point *A*. Also, cycle #1 is sometimes referred to as cycle ABCDA.

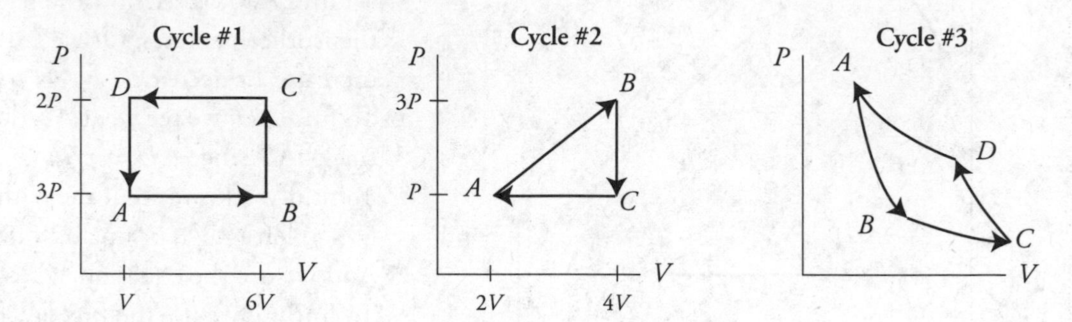

Let's apply what we have learned:

- Temperature depends on the *PV* location. Since we start and end at the same spot $\Delta T_{cycle} = 0$.
- Since ΔU and ΔT are related, when $\Delta T_{cycle} = 0$, $\Delta U_{cycle} = 0$.
- That means ΔU drops out of the first law of thermodynamics and the equation becomes $Q = -W$!

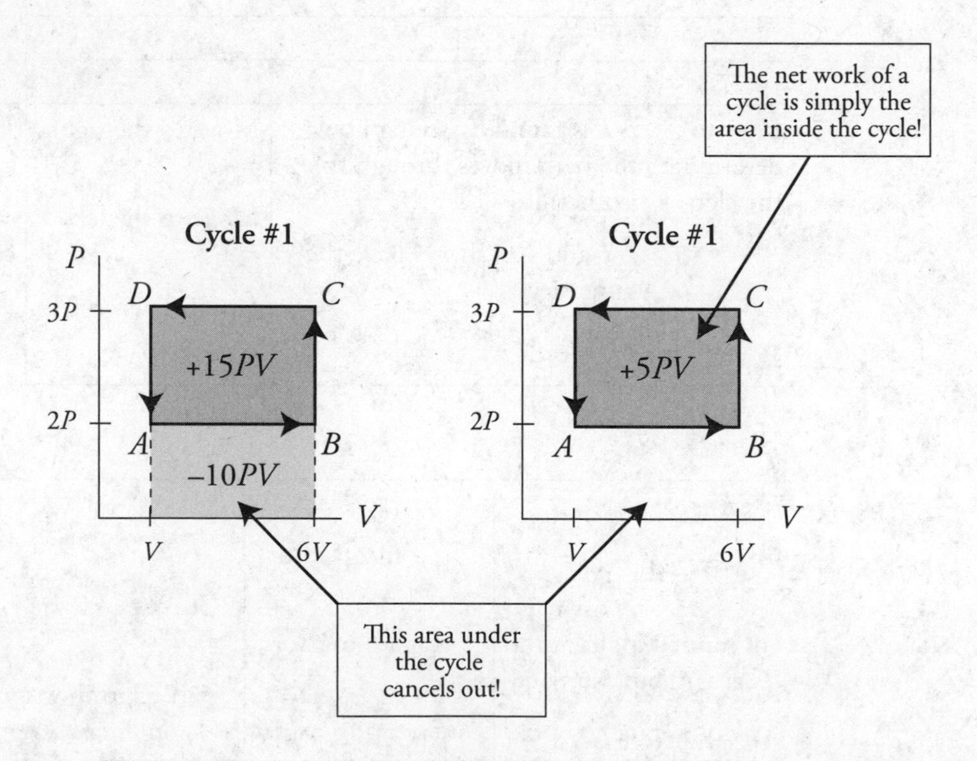

Now look at cycle #1: W_{AB} is negative, W_{BC} and W_{DA} are both zero, and W_{CD} is positive. Note that the positive work of W_{CD} is larger than the negative work of W_{AB}. Thus the net work of cycle #1 is positive! Remember that work = the area under the graph. Look at cycle #1 again. The area under path *CD* is more than the area under the path *AB*. Note that the positive and negative areas under the cycle cancel out, leaving only the area inside the cycle. The net work of a cycle = the area inside the cycle.

Four Special Processes (Paths) on a *PV* Diagram

1. **Constant Pressure—Isobaric**
   $W_{gas} = -P\Delta V$ is easy to calculate since P = constant. These processes move right and left on a *PV* diagram.
2. **Constant Volume—Isochoric/Isovolumetric**
   $W = 0$ because $\Delta V = 0$ thus $\Delta U = Q$. These processes move up and down on a *PV* diagram.
3. **Constant Temperature—Isothermal**
   $\Delta T = 0$; therefore, $\Delta U = 0$ and $Q = -W$. These processes move along the hyperbolic constant *PV* lines.
4. **No Heat Transfer Between the System and the Environment—Adiabatic**
   $Q = 0$ thus $\Delta U = W$. This is a curved path similar to an isothermal but steeper.

What do they look like on the *PV* diagram? *(Note that each of the processes shown next could move in the opposite direction! They just happen to be drawn moving to the right and downward for this example.)*

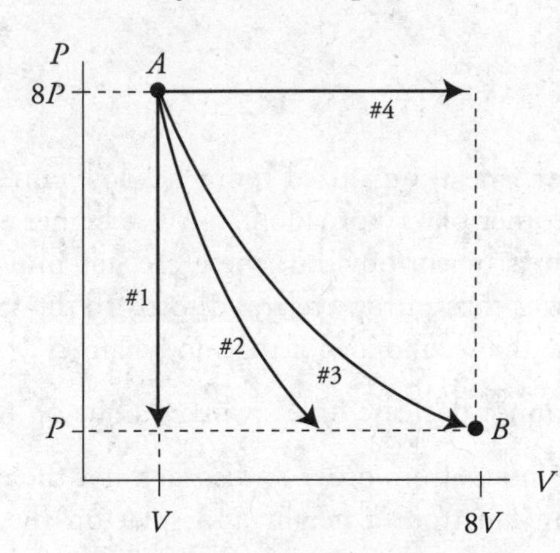

#1 Isochoric/Isovolumetric $W = 0$, $\Delta U = Q$
#2 Adiabatic $Q = 0$, $\Delta U = W$
#3 Isothermal $\Delta T = 0$ and $\Delta U = 0$, $Q = -W$
#4 Isobaric ΔP = Constant, $W = -P\Delta V$
$\Delta U = Q + W$

How do we know #3 is isothermal and that #2 is not? Because *PV* at point $A = 8PV$ and at point $B = 8PV$. Path #3 is a constant *PV* line. Path #2 is not.

Fill in the table below with a +, –, or 0 to show what is happening to each variable for each of the four special processes as shown above.

Path	ΔT	ΔU	W	Q
#1				
#2				
#3				
#4				
How to determine the sign	ΔT is found by seeing how the path moves through the isotherms. ΔT and ΔU always have the same sign. $\Delta U = \frac{3}{2}nRT$		Move to the right = – Move to the left = + Up or down = 0 $W_{gas} = -P\Delta V$	Find ΔU and W first. Then use the first law of thermodynamics to calculate Q. $\Delta U = Q + W$

Answer Key

(If the processes were moving to the left and upward instead of right and downward, all the signs in the table below would be reversed.)

Path	ΔT	ΔU	W	Q
#1	−	−	0	−
#2	−	−	−	0
#3	0	0	−	+
#4	+	+	−	+
How to determine the sign	ΔT is found by seeing how the path moves through the isotherms. ΔT and ΔU always have the same sign. $\Delta U = \frac{3}{2}nRT$		Move to the right = − Move to the left = + Up or down = 0 $W_{gas} = -P\Delta V$	Find ΔU and W first. Then use the first law of thermodynamics to calculate Q. $\Delta U = Q + W$

Entropy

Entropy is a measure of disorder. A neat, organized room has low entropy . . . the same room after your three-year-old brother plays "tornado!" in it has higher entropy.

There are quantitative measures of entropy, but these are not important for the AP exam. What you do need to know is that entropy relates directly to the second law of thermodynamics. In fact, we can state the second law of thermodynamics as follows.

The entropy of a system cannot decrease unless work is done on that system.

This means that the universe moves from order to disorder, not the other way around. For example: A ball of putty can fall from a height and splat on the floor, converting potential energy to kinetic energy and then to heat—the ball warms up a bit. Now, it is not against the laws of conservation of energy for the putty to convert some of that heat back into kinetic energy and fly back up to where it started from. But does this ever happen? Um, no. The molecules in the putty after hitting the ground were put in a more disordered state. The second law of thermodynamics does not allow the putty's molecules to become more ordered without work being done on the system.

Or, a glass can fall off a shelf and break. But in order to put the glass back together—to decrease the glass's entropy—someone has to do work.

If heat flows into a system, the molecules will vibrate around more causing an entropy rise. When heat flows out of a system, the molecules become more ordered and entropy decreases. For example: a refrigerator does work to transfer thermal energy from the food you place inside and deposits this thermal energy into the kitchen. The entropy of the food decreases while the entropy of the kitchen increases.

❯ Practice Problems

Multiple Choice

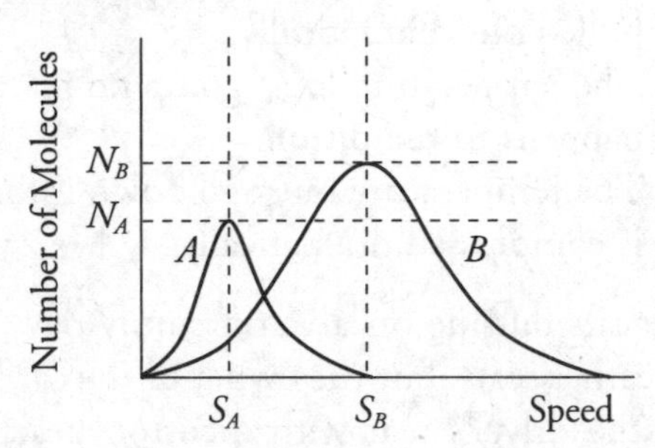

1. The graph above shows the distribution of speeds for two samples of nitrogen gas, each enclosed in separate sealed containers. Sample A has 1 mole of nitrogen molecules and is at a low initial temperature. Sample B has 3 moles of nitrogen molecules and is at a higher initial temperature. (Horizontal and vertical lines N_A, N_B, S_A, and S_B are shown for reference to indicate the initial locations of the peak for each graph.) The two samples are put into thermal contact so that heat can flow between the two containers but they are insulated from the surroundings so that no heat is lost. Which of the following graphs depicts a possible speed distribution of the two gases after they have reached equilibrium?

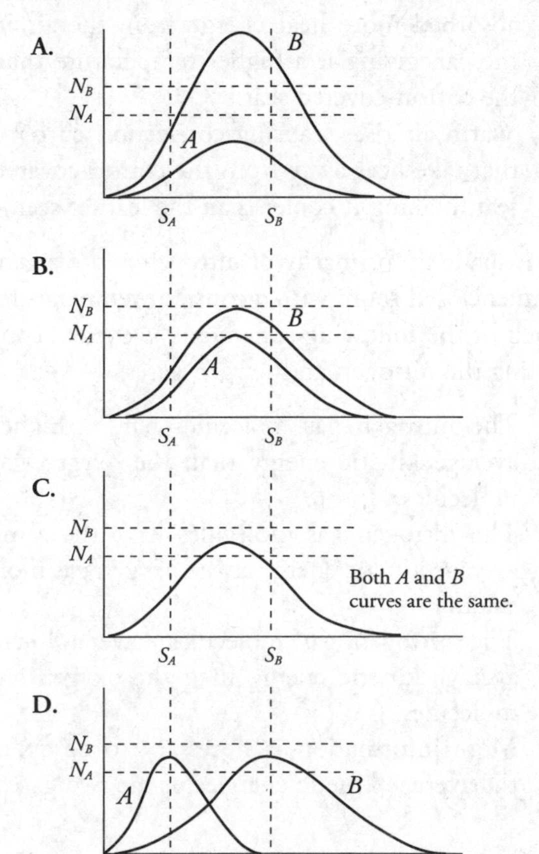

Gas	Temperature (K)	Molar Mass (g/mole)
1	300	2.0
2	400	4.0
3	2400	16

2. Three gas samples with the same number of moles of gas and the same pressure are placed in different temperature environments. Scientists record the molar mass and temperatures of the gases in the table above. What is the correct ranking of the density ρ of the gas samples?

(A) $\rho_1 > \rho_2 > \rho_3$
(B) $\rho_1 = \rho_3 > \rho_2$
(C) $\rho_2 > \rho_1 = \rho_3$
(D) $\rho_3 > \rho_2 > \rho_1$

Questions 3 through 5 refer to the following material.

A gas, beginning at point O on the graph, can be taken along four paths to different ending conditions.

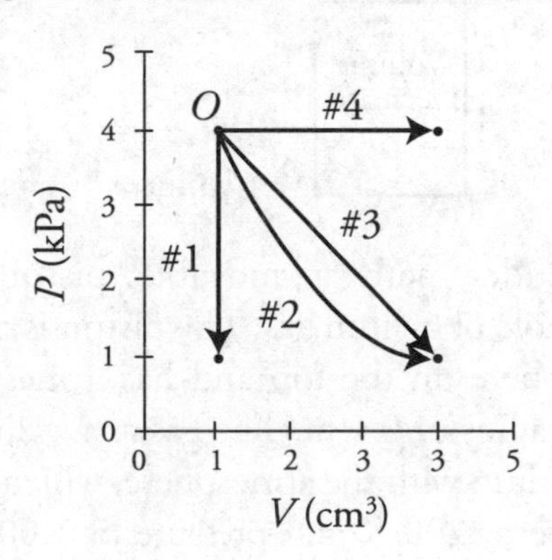

3. Which of the following correctly ranks the paths for work done by the gas?

(A) $W_4 > W_3 > W_2 > W_1$
(B) $W_4 > W_3 = W_2 > W_1$
(C) $W_4 = W_3 = W_2 > W_1$
(D) $W_4 = W_1 > W_3 = W_2$

4. Along which of the paths will the gas absorb the most heat?

(A) 1
(B) 2
(C) 3
(D) 4

5. A gas, beginning at point O on the graph, can be taken along four paths to various ending conditions.

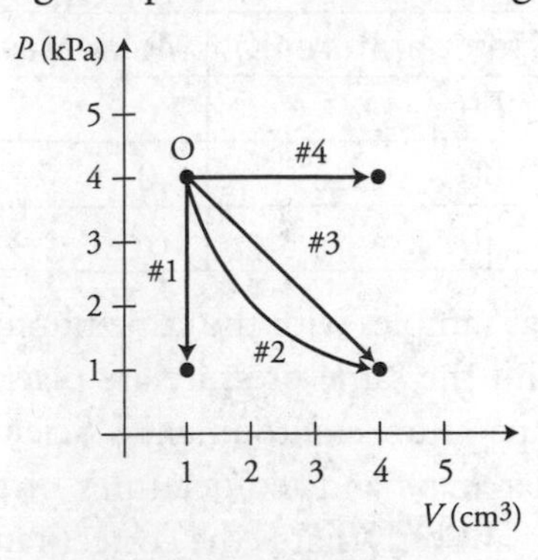

Which of the following quantities are the same for processes 2 and 3? **(Select two answers.)**

(A) Q
(B) ΔT
(C) ΔU
(D) W

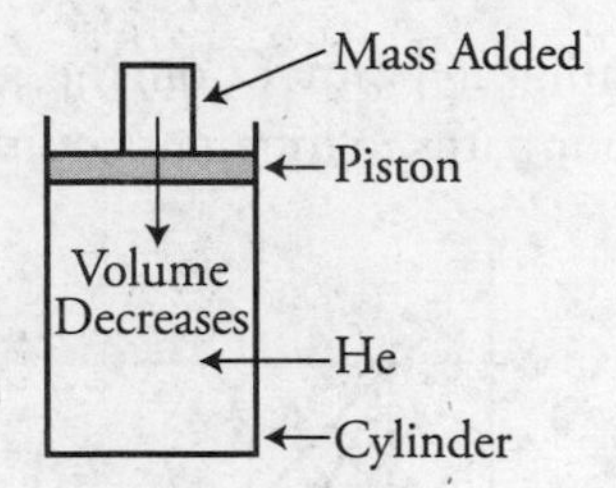

6. A cylinder with a movable piston contains 0.41 mole of helium gas. The piston is open to the atmosphere on the top and has a mass of 30 kg and a radius of 0.1 m. The gas and cylinder are in equilibrium with the atmosphere, which has a temperature of 290 K and pressure of 1.00×10^5 Pa. The initial volume of the helium inside the cylinder is 0.0090 m^3 when a 60-kg mass is placed on top of the piston, which compresses the gas to a smaller volume as seen in the figure above. What is the final pressure of the helium gas?

(A) 2.81×10^4 Pa
(B) 1.28×10^5 Pa
(C) 3.00×10^5 Pa
(D) It is not possible to calculate the new pressure without knowing the final temperature and volume of the gas

7. A gas is enclosed in a metal container with a movable piston on top. Heat is added to the gas by placing a candle flame in contact with the container's bottom. Which of the following is true about the temperature of the gas?

(A) The temperature must go up if the piston remains stationary.
(B) The temperature must go up if the piston is pulled out dramatically.
(C) The temperature must go up no matter what happens to the piston.
(D) The temperature must go down if the piston is compressed dramatically.

8. A car sits outside on a warm sunny day. The car has leather seats, but the owner of the car has covered the driver's seat with a cotton beach towel. The windows of the car are closed. The sun is at an angle so that it shines directly on the driver's seat but not the passenger's seat. After the car has been in the sun for several hours, the owner notices that when she touches the cotton-covered driver's seat it does not feel as hot as the passenger's seat, which was not in direct sunlight. Which of the following are correct statements about the two seats? **(Select two answers.)**

(A) The cotton-covered seat has a higher temperature than the leather seat due to the radiation it has absorbed from the sun.
(B) Cotton conducts heat at a slower rate, making it feel cooler to the owner.
(C) Leather is a better conductor of heat and has absorbed more heat energy from the air in the car, giving it a higher temperature than the cotton-covered seat.
(D) Warm air rises, causing convection currents that take heat away from the cotton-covered seat, making it cooler than the leather seat.

9. Air is made up primarily of nitrogen and oxygen. In an enclosed room with a constant temperature, which of the following statements is correct concerning the nitrogen and oxygen gases?

(A) The nitrogen gas molecules have a higher average kinetic energy than the oxygen gas molecules.
(B) The nitrogen gas molecules have the same average kinetic energy as the oxygen gas molecules.
(C) The nitrogen gas molecules have a lower average kinetic energy than the oxygen gas molecules.
(D) More information is necessary to compare the average kinetic energies of the two gases.

10. Air is made up primarily of nitrogen and oxygen. In an enclosed room with a constant temperature, which of the following statements is correct concerning the nitrogen and oxygen gases?

(A) The nitrogen gas molecules have a higher velocity than the oxygen gas molecules.
(B) The nitrogen gas molecules have the same velocity as the oxygen gas molecules.
(C) The nitrogen gas molecules have a lower velocity than the oxygen gas molecules.
(D) It is impossible to compare the velocity of the two gases without knowing the temperature of the air and the percentage of nitrogen and oxygen in the room.

Free Response

11. A small container of gas undergoes a thermodynamic cycle. The gas begins at room temperature. First, the gas expands isobarically until its volume has doubled. Second, the gas expands adiabatically. Third, the gas is cooled isobarically; finally, the gas is compressed adiabatically until it returns to its original state.

(A) The initial state of the gas is indicated on the *PV* diagram below. Sketch this process on the graph.

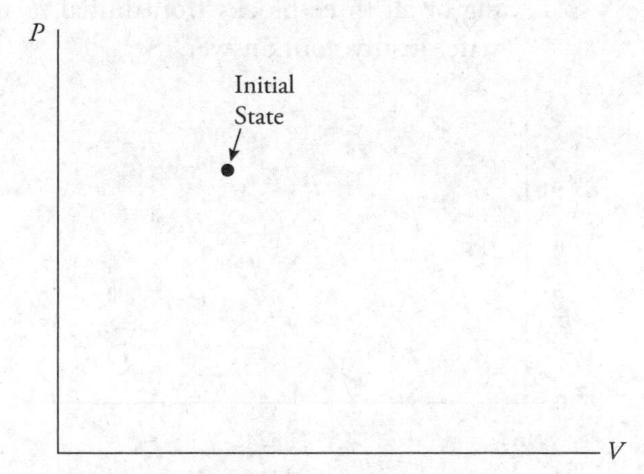

(B) Is the temperature of the gas greater right before or right after the adiabatic expansion? Justify your answer.
(C) Is heat added to or removed from the gas in one cycle?

12. Use the microscopic motion of atoms to explain:

(A) How gases exert pressure on a surface.
(B) Why gas pressure is always perpendicular to the surface with which the gas is in contact.

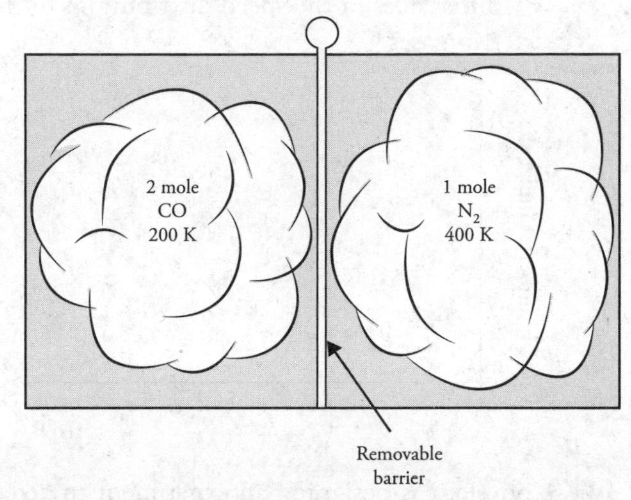

13. Carbon monoxide gas with a molecular mass of 28.0 kg/kmol and an initial temperature of 200 K is confined to the left side of a sealed container. Diatomic nitrogen gas with a molecular mass of 28.0 kg/kmol and an initial temperature of 400 K is confined to the right side of the sealed container, as shown in the figure above. Separating the gases is a removable barrier. When the barrier is removed, the two gases mix.

(A) Describe the temperature changes in the two gases.
(B) Describe any movement of thermal energy over a long period of time. Explain the microscopic process that determines this process.
(C) Sketch the initial molecular speed distributions of the two gases on the axis. Label each gas. Explain any differences in the speed distribution of the two gases.

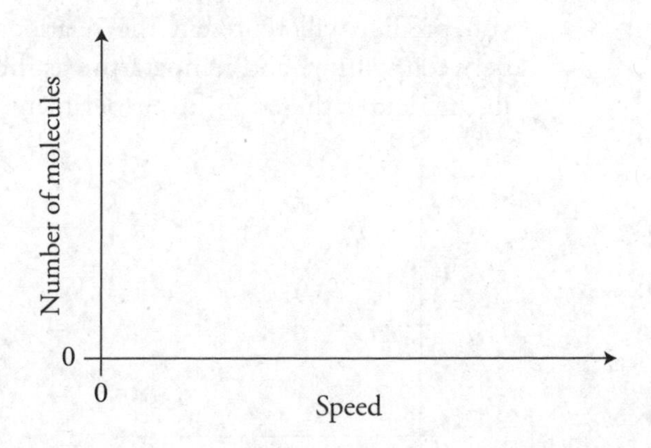

(D) Sketch the final molecular speed distributions of the two gases on the axis. Label each gas. Explain the changes in the graph from the initial condition. Also explain any differences in the speed distributions of the two gases.

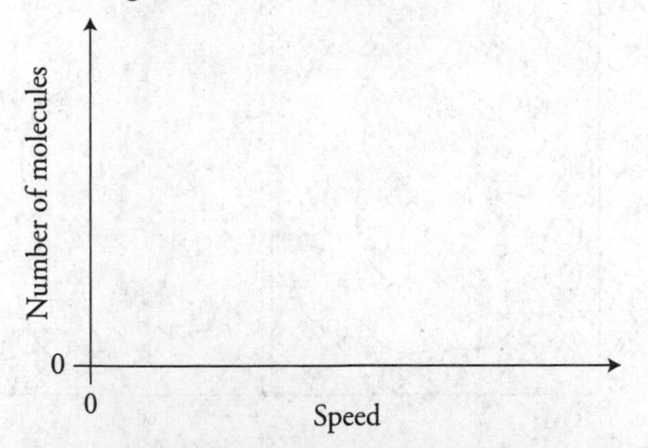

14. A physicist is designing an experiment to determine the relationship between the volume of a gas and the temperature of a gas.

(A) List the items the physicist could use to perform this investigation.
(B) Sketch a simple diagram of the investigation. Make sure to label all items used in the sketch and label any measurements that will be made.
(C) Outline the experimental procedure the physicist could use to gather the necessary data. Indicate the measurements to be taken and how the measurement will be used to obtain the data needed. Make sure your outline contains sufficient detail so that another scientist could follow the procedure and verify the results.
(D) Explain how data from this experiment could be used to determine the value of absolute zero.
(E) On the axis, sketch the line or curve that you predict will represent the relationship between volume and temperature as shown in the data gathered in this experiment.

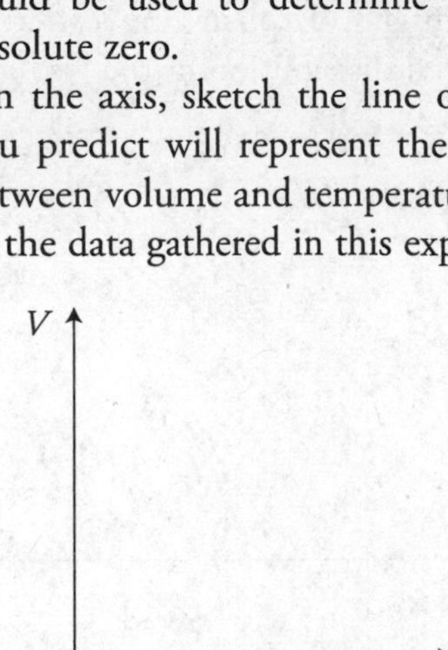

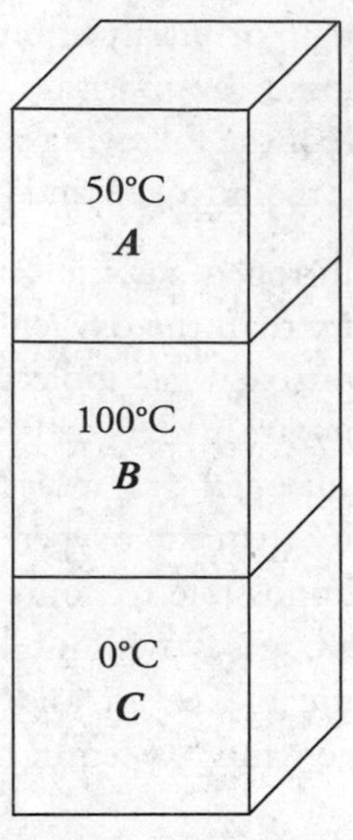

15. Three identical blocks of differing temperatures are stacked on top of one another and insulated from the environment, as shown in the figure above.

(A) Plot the temperature of the blocks versus time. Label each line and indicate all important temperatures on the graph.
(B) Discuss any changes in entropy for block *C*. Explain your answer at the microscopic level.
(C) Discuss the entropy of the system consisting of all three blocks from initial to final state. Justify your answer.

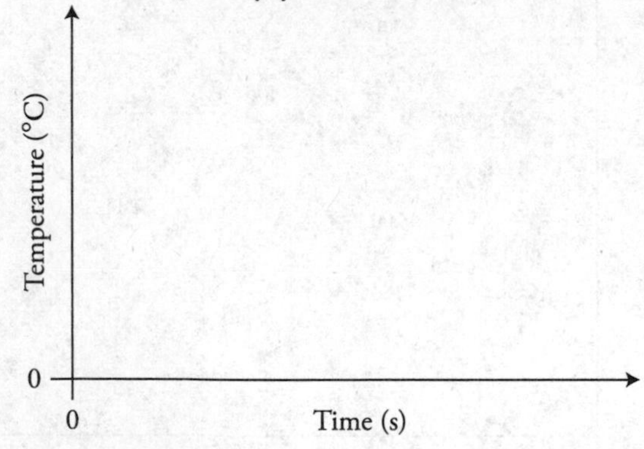

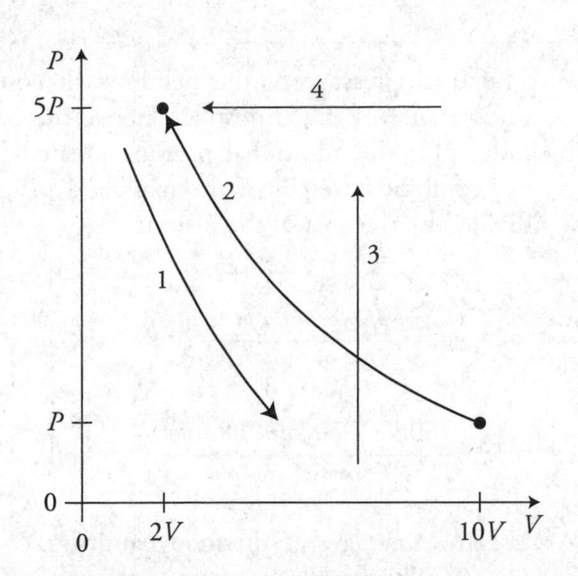

16. A gas is taken through the four processes shown in the figure. For each path, identify the name of the process and determine if the values of ΔT, ΔU, W, and Q are positive, negative, or zero and fill in the table with a +, –, or 0. (The table is begun: Process #1 is adiabatic.) Explain your reasoning.

Name of process	ΔT	ΔU	W	Q
1) Adiabatic				

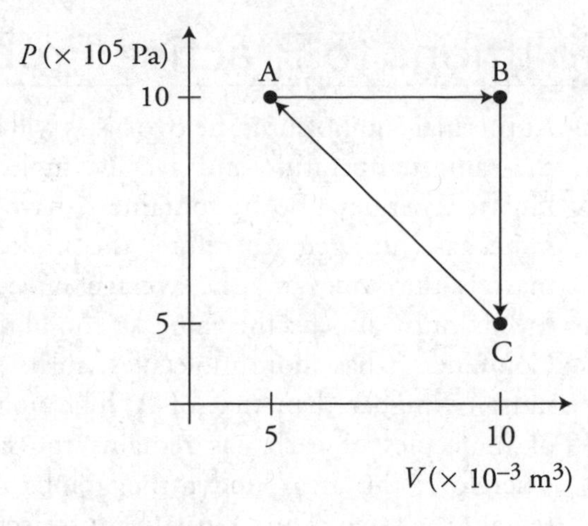

17. Two moles of a gas is taken through the thermodynamic process ABCA, as shown in the figure above.

(A) Rank the work in the steps of the process from most positive to most negative. Justify your reasoning.
(B) Rank the change in temperature of the gas for each step in the process from most positive to most negative. Justify your reasoning.
(C) Rank the thermal energies of the points A, B, and C from greatest to least. Justify your reasoning.
(D) Calculate the temperature of point B.
(E) Calculate the change in temperature for the entire cycle ABCA.
(F) Calculate the work done in process C→A.
(G) Calculate the heat flow for process C→A.

› Solutions to Practice Problems

1. (A) At thermal equilibrium, the two gases will have the same temperature and average molecular kinetic energies. Both containers have the same gas, nitrogen; therefore, the molecular mass is the same and the average molecular speeds must also be the same at equilibrium. Container B has more molecules, so its peak must be higher than that of A. The number of molecules of each gas remains the same. Therefore, the areas under the graphs must remain the same. Thus, curve A must decrease in height as it moves to a higher temperature. Curve B must increase in height as it moves to a lower temperature.

2. (C) Density $\rho = \frac{m}{V}$. The mass is given, molar mass times the number of moles, but we still need the volume. Using the ideal gas law ($PV = nRT$), we know that the pressure and the number of molecules are constant, which means that volume is proportional to temperature: $V \propto T$. Therefore, density is proportional to mass over temperature: $\rho \propto \frac{m}{v}$. So, taking the molar mass and dividing by the temperature will give the relative density of the three samples!

3. (A) Work equals the pressure times the change in volume. Paths 2, 3, and 4 have the same change in volume, but path 4 has the greatest pressure. Path 1 has no change in volume. Work also equals the area under the graph. Path 4 has the largest area. Path 1 has no area.

4. (D) The first law of thermodynamics ($\Delta U = Q + W_{done\ by\ the\ gas}$) can be rewritten as: $Q = \Delta U - W_{done\ by\ the\ gas}$. Remember that ΔU is directly related to the temperature of the gas. Thus the gas will absorb the most heat, Q, when its temperature goes up the most and when the gas does the most work. Both of these situations occur for path 4.

5. (B) and (C) Both processes end at the same PV value. Thus the temperature and thermal energy change for both processes are the same: $\Delta U = nR\Delta T$.

6. (B) The final pressure of the helium will equal the pressure of the atmosphere above the cylinder plus the additional pressure created by the combined weight of the mass and piston divided by the area of the piston.

$$P_{atmosphere} + \frac{\text{weight of mass and piston}}{\text{area of piston}} = P_{gas\ inside\ the\ piston}$$

$$100{,}000\ \text{Pa} + \frac{(60\ \text{kg} + 30\ \text{kg})\,(9.8\ \text{m/s}^2)}{\pi(0.1\ \text{m})^2} = 128{,}000\ \text{Pa}$$

7. (A) Use the first law of thermodynamics, $\Delta U = Q + W$. The candle adds heat to the gas, so Q is positive. Internal energy is directly related to temperature, so if ΔU is positive, then temperature goes up (and vice versa). Here we can be sure that ΔU is positive if the work done on the gas is either positive or zero. The only possible answer is A. In case B, the work done on the gas is negative because the gas expands. (Note that just because we add heat to a gas does NOT mean that temperature automatically goes up!) We have to take into account the work done on or by the gas.

8. (A) and (B) Both seats will absorb heat from the warm air in the car and after several hours would be in thermal equilibrium with the air. However, the driver's seat is in direct sunlight and will also absorb additional radiation from the sun, making it hotter. The reason it feels cooler to the driver is that cotton has a lower thermal conductivity than leather, so less heat per second transfers to the driver from the cotton-covered seat. Convection will indeed take heat away from the driver's seat but cannot cool it to a temperature below the air temperature inside the car. Therefore, it can't cool below the temperature of the passenger's seat.

9. (B) Gases with the same temperature will have the same average molecular kinetic energy: $\overline{K} = \frac{3}{2}k_B T$.

10. (A) $v_{rms} = \sqrt{\dfrac{3k_BT}{m_{\text{gas molecule}}}}$. Nitrogen is lighter than oxygen. Since the molecular velocity is inversely proportional to the mass of the gas molecules, the nitrogen in the air will have a higher overall speed.

11. (A)

(B) The temperature is greater right before the expansion. By definition, in an adiabatic process, no heat is added or removed. But because the gas expanded, work was done *by* the gas, meaning the W term in the first law of thermodynamics is negative. Since $\Delta U = Q + W$, $Q = 0$ and W is negative; ΔU is negative as well. Internal energy is directly related to temperature. Therefore, because internal energy decreases, so does temperature.

(C) In a full cycle, the gas begins and ends at the same state. So the total change in internal energy is *zero*. Now consider the total work done on or by the gas. When the gas expands in the first and second processes, there's more area under the graph than when the gas compresses in the second and third processes. So, the gas does more work expanding than compressing; the net work is thus done *by* the gas, and is *negative*. To get no change in internal energy, the Q term in the first law of thermodynamics must be positive; heat must be added to the gas.

12. (A) Gas atoms collide with and rebound off of a surface, imparting a tiny impulse on the surface. The sum of many billions of atoms constantly colliding with the surface imparts a constant force over the area of the surface. This creates a constant gas pressure on the surface.

(B) The motion of the atoms in a gas is random. Thus, the atoms collide with the surface in no preferred direction. In addition, when a gas atom bounces off a surface, only the atom's momentum that is perpendicular to the surface is changed. This means the impulse between the surface and the gas atom is also perpendicular to the surface. Thus, a gas always exerts a perpendicular force on the surface with which it is in contact.

13. (A) The nitrogen temperature decreases. The temperature of the carbon monoxide increases. The two gases reach thermal equilibrium at a temperature between 200 K and 400 K, but the final temperature of the combined gas will be closer to 200 K because there are more "cold" CO molecules than "hot" N_2 molecules.

(B) Thermal energy will flow from the hotter nitrogen gas to the colder carbon monoxide. On a macroscopic scale, heat always flows from hot to cold. This is due to the microscopic behavior of faster moving "hot" molecules colliding with, and transferring energy/momentum to, slower moving "cold" molecules.

(C) The carbon monoxide distribution will be shifted to the left of the nitrogen because it has a lower temperature and lower average kinetic energy. The CO peak must be higher than that of nitrogen because there are more molecules of CO, and all the molecules are bunched in a smaller speed distribution. The area under the CO curve should be twice as large as the area under the N_2 curve because there are twice as many molecules.

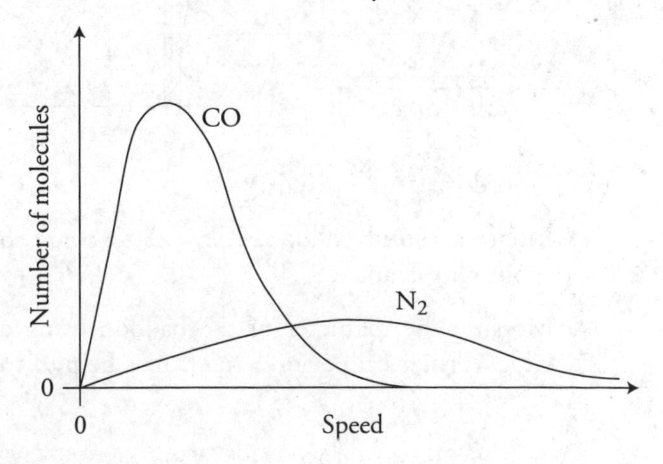

(D) The two gases have the same molecular mass; therefore, they will have the same shaped distribution with the same peak speed. The carbon monoxide distribution should have a peak lower than its initial condition because the gas is now spread out over a wider speed distribution. However, the CO peak must still be higher than the N_2 peak because there are more molecules of CO. The nitrogen speed distribution should be higher than its initial value. The area under the CO curve should be twice as large as the area under the N_2 curve because there are twice as many molecules.

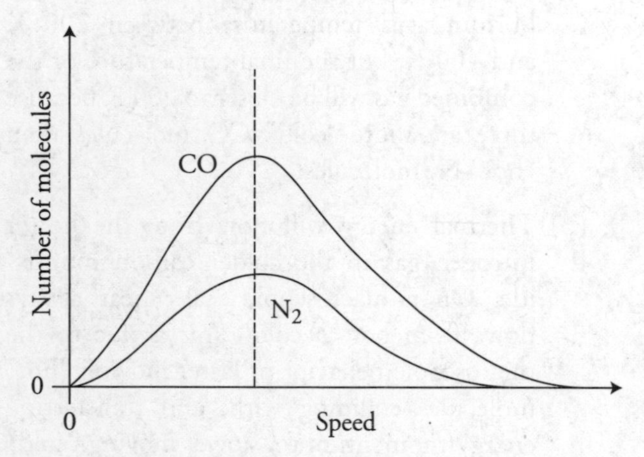

14. There are several different ways to perform this experiment. Here is one example.

(A) Equipment: balloon, ruler or large Vernier calipers, thermometer, incubator or oven, refrigerator, and freezer.

(B)

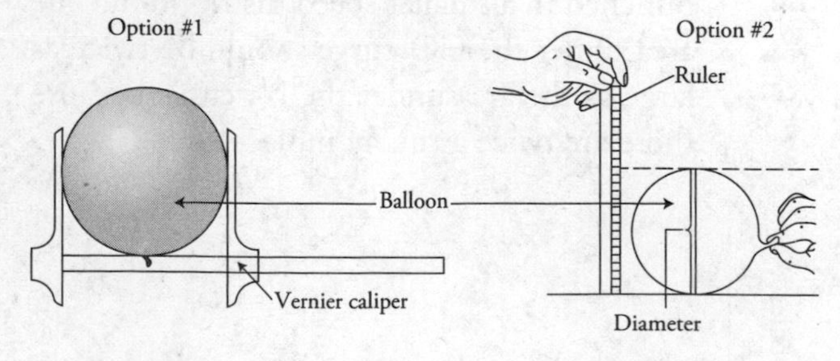

(C) Procedure:

1. Inflate a round balloon. Allow it to come to room temperature.
2. Measure the diameter of the balloon using a large Vernier caliper or a ruler. It is helpful to take a picture of the balloon next to the ruler and use the picture to determine the diameter of the balloon. Note: This measurement must be done quickly as the air temperature inside the balloon, and size of the balloon, change quickly when handled.
3. Repeat this process with hot air inside an incubator or oven and cold air inside a refrigerator/freezer. Be sure to allow time for the air in the balloon to come to thermal equilibrium with the external environment in each case.
4. Use the diameter measurement to estimate the volume of the balloon.

(D) Extrapolate the data to a zero volume. The temperature at V = 0 will be our experimental value for absolute zero.

(E) The graph should be linear. The *x*-intercept will be our experimental prediction for absolute zero.

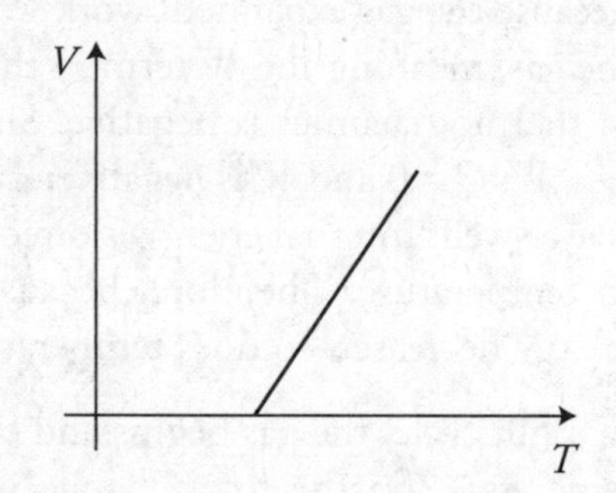

15. (A) Key features of the graph:

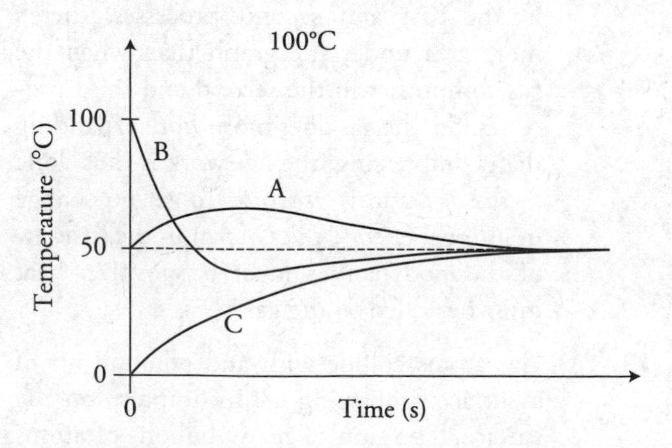

- All the blocks are identical and will reach equilibrium at the central temperature of 50°C.

- The temperature of block C continuously rises to equilibrium.
- Block A gains energy from block B, making block A initially rise above the equilibrium temperature of 50°C.
- Block B loses energy to both blocks, taking it below the temperature of block A. Block A then begins returning energy to block B. The sample solution shows that the temperature of block B drops below 50°C. This may or may not occur. The critical feature of the graph is that the temperature of block B drops below that of block A but remains above that of block C.

(B) The entropy rises as heat flows into block C. This expands the velocity and kinetic energy profile of the atoms. This, in turn, increases the overall disorder of the block as there are more thermal microstates for the atoms to occupy.

(C) This is an irreversible process. Therefore, the entropy of the system increases.

16.

Name of process	ΔT	ΔU	W	Q	Notes
1) Adiabatic	–	–	–	0	ΔU and W are the same value.
2) Isothermal	0	0	+	–	Q and W are the same magnitude.
3) Isochoric (Isovolumetric)	+	+	0	+	ΔU and Q are the same value.
4) Isobaric	–	–	+	–	Q has a larger magnitude than W.

ΔT is found by seeing how the path moves through PV values using the ideal gas law: $\Delta(PV) = nR\Delta T$.

ΔT & ΔU always have the same sign: $\Delta U_{\text{Thermal}} = N\left(\frac{3}{2}k_B\Delta T\right) = \frac{3}{2}nR\Delta T$.

Moving to the right is negative work. Moving to the left is positive work. Moving up or down the volume does not change and the work is zero: $W = -P\Delta V$.
After finding ΔU and W, use the First Law of Thermodynamics to calculate Q: $\Delta U = Q + W$.

17. (A) $W = -P\Delta V$ therefore, $W_{CA} > W_{BC} = 0 > W_{AB}$

(B) $\Delta(PV) = nR\Delta T$ therefore, $\Delta T_{AB} > \Delta T_{CA} = 0 > \Delta T_{BC}$

(C) $U_{\text{Thermal}} = N\left(\frac{3}{2}k_B T\right) = \frac{3}{2}nRT$ therefore, $U_B > U_A = U_C$

(D) $T = \frac{PV}{nR} = \frac{(10\times10^5\,\text{Pa})(10\times10^{-3}\,\text{m}^3)}{(2\text{ mol})\left(8.31\frac{\text{J}}{\text{mol K}}\right)} = 600\text{ K}$

(E) The process returns to the original state: $\Delta T = 0$.

(F) Work is the area under the path: 3750 J.

(G) Since $\Delta U_{CA} = 0$, $Q = -W = -3750$ J.

› Rapid Review

- Heat is a form of energy that can be transferred from one body to another. The internal energy of an object is the sum of the energies of all the molecules that make up that object. An object's temperature is directly related to average kinetic energy per molecule of the object.
- Heat is the transfer of thermal energy. This transfer occurs through conduction, convection, and electromagnetic radiation.
- The molecules inside an object are in constant random motion.
- Objects with the same temperature are in thermal equilibrium. The net heat transfer between them is zero.
- Two bodies with the same temperature do not necessarily have the same internal energy, and vice versa.
- When objects are heated, they expand.
- The pressure and volume of an ideal gas are related by the ideal gas law. The law can either be written as $PV = nRT$, or $PV = Nk_BT$.
- The first law of thermodynamics states that the change in a gas's internal energy equals the heat added to the gas plus the work done on the gas.
- *PV* diagrams illustrate the relationship between a gas's pressure and volume as it undergoes a process. Four types of processes a gas can undergo are (1) isothermal [temperature stays constant]; (2) adiabatic [no heat is transferred to or from the gas]; (3) isobaric [pressure stays constant]; and (4) isochoric [volume stays constant].
- Heat flows naturally from a hot object to a cold object through collisions of the molecules.
- Entropy is a measure of disorder. The entropy of a system cannot decrease unless work is done on that system.

Electric Force, Field, and Potential

IN THIS CHAPTER

Summary: Objects are made of atoms, which in turn are made of electrons, protons, and neutrons. An excess of electrons will cause an object to be negatively charged, while an excess of protons will create positively charged objects. This chapter focuses on how to deal with electric charges that aren't moving through circuit components, hence the name, electrostatics.

An electric field provides a force on a charged particle. Electric potential, also called voltage, provides energy to a charged particle. Once you know the force or energy experienced by a charged particle, Newtonian mechanics (i.e., kinematics, conservation of energy, etc.) can be applied to predict the particle's motion.

Key Ideas

- Electrons and protons are real particles that can move from place to place, or transfer from object to object by contact, but the net charge of the system always stays the same (conservation of charge).
- Most objects have the same number of electrons and protons. Only objects with excess protons or electrons have a net charge.
- Conductive objects that touch will transfer their charge so that they share any excess charge.
- One object can be charged even without touching another charged object, through a process called induction.

- A neutral object can have an excess of protons on one side and an excess of electrons on the other side. This charge separation is called charge polarization.
- The electric force on a charged particle is qE, regardless of what produces the electric field. The electric potential energy of a charged particle is qV.
- Positive charges are forced in the direction of an electric field; negative charges, experience a force in the opposite direction of the field.
- Positive charges are forced from high to low electric potential; negative charges are forced from low to high electric potential.
- Point charges produce non-uniform electric fields around themselves. Parallel plates produce a uniform electric field between the two oppositely charged plates.
- Electric field is a vector, and electric potential is a scalar.
- Capacitors store both electric charge and electric potential energy.

Relevant Equations

Electric force between two charged objects: $|\vec{F}_E| = \frac{1}{4\pi\varepsilon_0}\frac{|q_1q_2|}{r^2}$ where r is the distance between the centers of the charges.

The relationship between electric force and electric field: $\vec{E} = \frac{\vec{F}_E}{q}$

Electric field produced around a point charge: $|\vec{E}| = \frac{1}{4\pi\varepsilon_0}\frac{|q|}{r^2}$

Change in electric potential energy: $\Delta U_E = q\Delta V$

Electric potential produced by point charge: $V = \frac{1}{4\pi\varepsilon_0}\frac{q}{r}$

The relationship equation between electric field and electric potential difference: $|\vec{E}| = \left|\frac{\Delta V}{\Delta r}\right|$

Potential difference between two charged plates of a capacitor: $\Delta V = \frac{Q}{C}$

Capacitance of a parallel plate capacitor: $C = \kappa\varepsilon_0\frac{A}{d}$

Electric field strength between charged plates of a parallel plate capacitor: $E = \frac{Q}{\varepsilon_0 A}$

Electric potential energy stored in a charged capacitor: $U_C = \frac{1}{2}Q\Delta V = \frac{1}{2}C(\Delta V)^2$

Electric Charge

Electricity literally holds the world together. Sure, gravity is pretty important, too, but the primary reason that the molecules in your body stick together is because of electric forces. A world without electrostatics would be no world at all.

All matter is made up of three types of particles: protons, neutrons, and electrons. Protons have an intrinsic property called "positive charge." Neutrons don't contain any charge, and electrons have a property called "negative charge."

The unit of charge is the coulomb, abbreviated C. One proton has a charge of 1.6×10^{-19} coulombs. One electron has a charge of -1.6×10^{-19} C.

Most objects that we encounter in our daily lives are electrically neutral—things like couches, for instance, or trees, or bison. These objects contain as many positive charges as negative charges. In other words, they contain as many protons as electrons.

When an object has more protons than electrons, though, it is described as "positively charged"; and when it has more electrons than protons, it is described as "negatively charged." The reason that big objects like couches and trees and bison don't behave like charged particles is because they contain so many bazillions of protons and electrons that an extra few here or there won't really make much of a difference. So even though they might have a slight electric charge, that charge would be much too small, relatively speaking, to detect.

Tiny objects, like atoms, more commonly carry a measurable electric charge, because they have so few protons and electrons that an extra electron, for example, would make a big difference. Of course, you can have very large charged objects. When you walk across a carpeted floor in the winter, you pick up lots of extra charges and become a charged object yourself . . . until you touch a doorknob, at which point all the excess charge in your body travels through your finger and into the doorknob, causing you to feel a mild electric shock.

Electric charges follow a simple rule: *Like charges repel; opposite charges attract.* Two positively charged particles will try to get as far away from each other as possible, while a positively charged particle and a negatively charged particle will try to get as close as possible.

Let's take these ideas a step further.

Quanta of Charge, Conservation of Charge, and How Charge Moves Around

First, notice that charge comes in the smallest possible package, a quanta, of one proton or one electron, $\pm 1.6 \times 10^{-19}$ C. In our everyday world, you can't get a charge smaller than that. Every charged object comes in multiples of this quanta. You can have objects with a charge of $-3.2 \times 10^{-19}\ \text{C} = 2(-1.6 \times 10^{-19}\ \text{C}) = 2$ quanta of charge. But you can't have $-4.0 \times 10^{-19}\ \text{C} = 2.5(-1.6 \times 10^{-19}\ \text{C}) = 2.5$ quanta of charge, because there is no such thing as half of an electron. It's just like how money is quantized! You can't have half of a penny.

Second, atoms have a nuclear structure where the protons are buried deep inside with the electrons zipping around far away on the outside. Electrons are easy to remove or add to atoms. Moving protons in or out of an atom requires nuclear reactions! In the study of static electricity, we are not adding or removing protons from an object. (We discuss nuclear physics in Chapter 15.) So, when an object is negatively charged, it's because it has too many electrons placed on it. When positively charged, it has lost some electrons.

Third, we must obey the law of conservation of charge. Charge is carried by real things—electrons and protons don't just appear or disappear. Thus, charge can move around, but the initial charge of the system before will always equal the charge of the system after.

Fourth, how do we move charge around? Remember chemistry? Those atoms on the left side of the periodic table lose their outer electrons easily and can become positive ions. They are metals and good conductors of charge. Those atoms on the right side of the periodic table tend to hold on to their electrons more tightly. In fact, they may even steal electrons from others and become negative ions. These are nonmetals and insulators. Conductors allow charge to easily move through them. Insulators do not let charges move easily but hold them in place where they are.

There are really three ways for an object as a whole to become charged.

1. Charging by Friction

Charging by friction is one that you are probably familiar with. Place two materials into contact that have a different pull on their outer electrons, and electrons start jumping from one object to the other. Rub the objects together and the process speeds up. You have seen this when you comb your hair. Electrons jump from your hair to the comb. Your hair becomes positively charged and the comb negatively charged, but the net charge of the hair-comb system is still zero.

2. Charging by Contact or Conduction

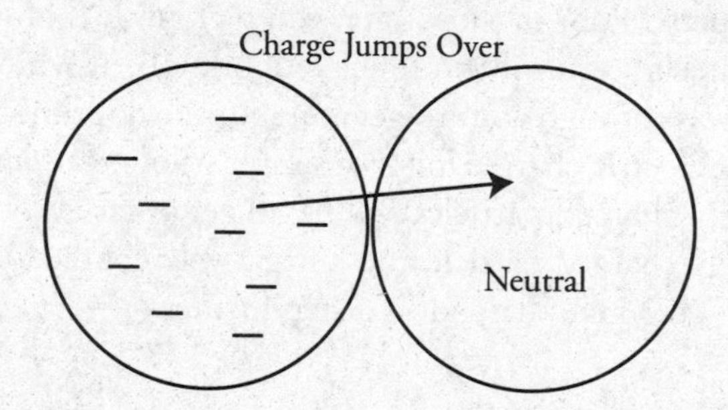

Let's say we have an object with extra electrons and we touch it to a neutral object. The extra electrons repel each other and some move onto the neutral object. They share the excess charge and become charged with the same sign. If they are the same size, they both get equal amounts of the charge. If one is bigger, it will end up with more charge than the smaller object because there is more room for the charges to spread out.

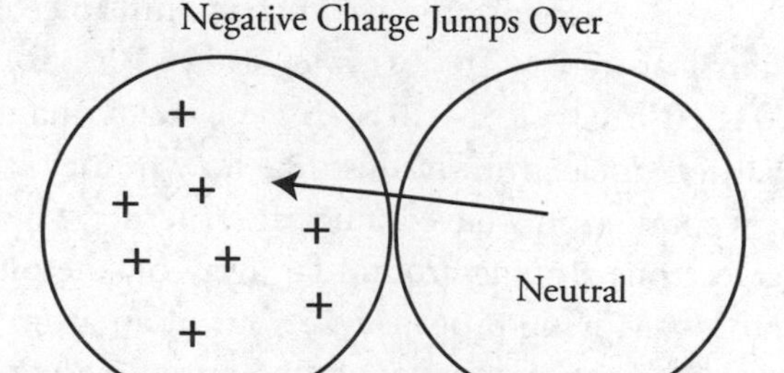

What if the charged object is positively charged? Positives can't jump to the neutral object because they are buried in the nucleus. In this case, negatives are attracted to the positive object, causing it to become less positive and the neutral object to become positive. It looks like positive charges moved to the right, but in reality negative charges moved to the left. So, sometimes we will say things like "positive charge moved from object A to object B," but in reality negative electrons moved in the reverse direction from object B to object A.

One last comment on contact: remember that insulators do not let charge move very easily. So, if you touch a charged insulator, you will share only the tiny amount of charge right where you touched it because the rest of the charges on the object are locked in place.

3. Induced Charge, Polarization, and Induction

You can also have something called "induced charge." An induced charge occurs when an electrically neutral object becomes polarized—when negative charges pile up in one part of the object and positive charges pile up in another part of the object. The drawing below illustrates how you can create an induced charge in an object.

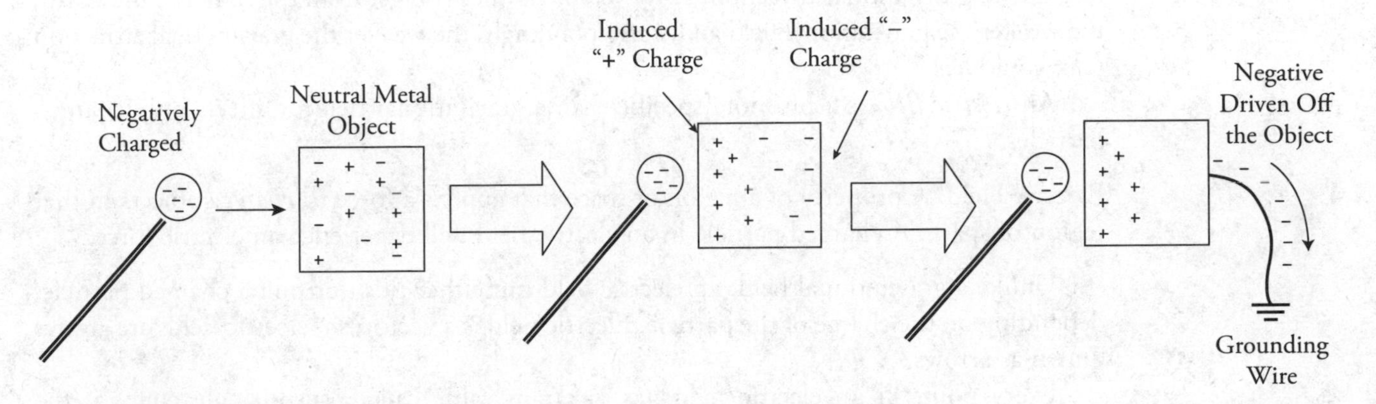

If we supply an escape route, like a grounding wire, for the induced charge piled up on the right, the negative charge will be driven completely off the object, leaving it with a net positive charge. Then disconnect the escape route, and like magic, we just gave the object a permanent positive charge by the process called induction. Note that the negatively charged sphere was brought close to, but did not touch, the neutral metal object! Make sure you understand this process of polarization and induction, as it is likely to show up on the exam.

Charge Distribution on Different Objects

You already know that like charges repel each other, but on an insulator they can't move. Therefore, a net charge will be stuck where it is. See the figure below. On conductors, any excess charge will force itself to the outside surface. For a uniformly shaped object, like a sphere, the excess charges are going to distribute themselves around the outside of the body evenly. But what if the body has an irregular shape? The excess charges will be forced to the farthest edges in an effort to get as far away from each other as possible. Charge will bunch up disproportionately on any pointy areas the body might have. The key idea to remember is, all the excess charge is on the outside surface of a conductor no matter what the shape is. There will be no excess charge inside a conductor.

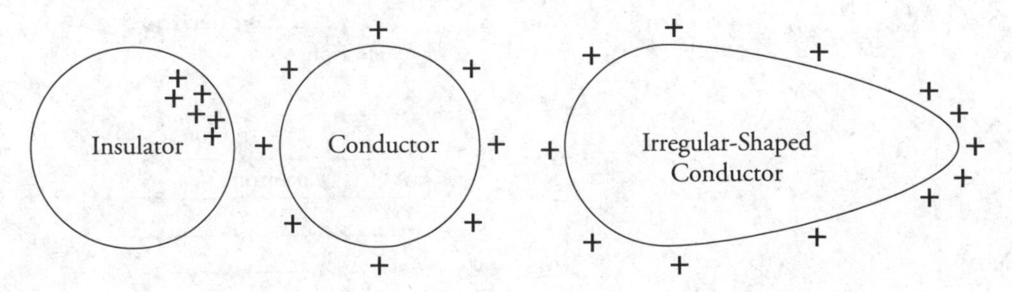

Electric Fields

Before talking about electric fields, I'll first define what a field, in general, is.

Field: A property of a region of space that can apply a force to objects found in that region of space.

A gravitational field is a property of the space that surrounds any massive object. There is a gravitational field that you are creating and which surrounds you, and this field extends infinitely into space. It is a weak field, though, which means that it doesn't affect other objects very much—you'd be surprised if everyday objects started flying toward each other because of gravitational attraction. The Earth, on the other hand, creates a strong gravitational field because of its tremendous mass. Objects are continually being pulled toward the Earth's surface due to gravitational attraction. However, the farther you get from the center of the Earth, the weaker the gravitational field and, correspondingly, the weaker the gravitational attraction you would feel.

An *electric field* is a bit more specific than a gravitational field: it affects only charged particles.

Electric Field: A property of a region of space that applies a force to *charged* objects in that region of space. A charged particle in an electric field will experience an electric force.

Unlike a gravitational field, an electric field can either push or pull a charged particle, depending on the charge of the particle. Electric field is a vector; so, electric fields are always drawn as arrows.

Every point in an electric field has a certain value called, surprisingly enough, the "electric field strength," or E, and this value tells you how strongly the electric field at that point would affect a charge. The units of E are newtons/coulomb, abbreviated N/C.

Force of an Electric Field

The force felt by a charged particle in an electric field is described by a simple equation:

$$\vec{F} = q\vec{E}$$

In other words, the force felt by a charged particle in an electric field is equal to the charge of the particle, q, multiplied by the electric field strength, E.

The direction of the force on a positive charge is in the same direction as the electric field; the direction of the force on a negative charge is opposite the electric field.

Let's try this equation on for size. Here's a sample problem:

An electron, a proton, and a neutron are each placed in a uniform electric field of magnitude 60 N/C, directed to the right. What is the magnitude and direction of the force exerted on each particle?

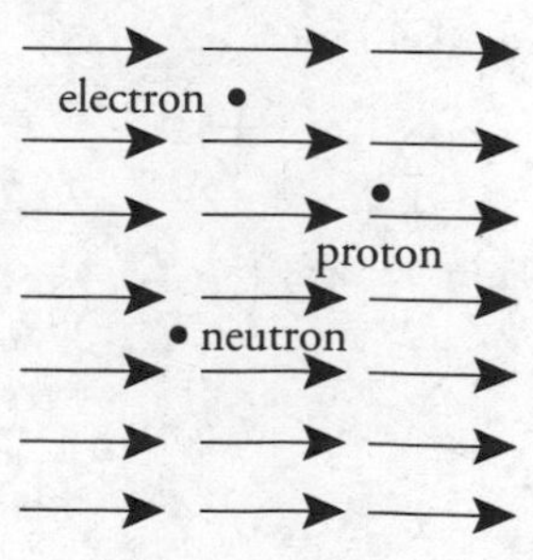

The solution here is nothing more than plug-and-chug into $\vec{F} = q\vec{E}$. Notice that we're dealing with a *uniform* electric field—the electric field vector lines are evenly spaced throughout the whole region AND all the electric field vectors are the same length. This means that, no matter where a particle is within the electric field, it always experiences an electric field of exactly 60 N/C.

Also note our problem-solving technique. To find the magnitude of the force, we plug in *just the magnitude* of the charge and the electric field—no negative signs allowed! To find the direction of the force, use the reasoning presented earlier (positive charges are forced in the direction of the E field, negative charges opposite the E field).

Let's start with the electron, which has a charge of 1.6×10^{-19} C (no need to memorize; you can look this up on the constant sheet):

$$\vec{F} = q\vec{E}$$

$$F = (1.6 \times 10^{-19}\ \text{C})(60\ \text{N/C})$$

$$F = 9.6 \times 10^{-18}\ \text{N to the LEFT}$$

Now the proton:

$$F = (1.6 \times 10^{-19}\ \text{C})(60\ \text{N/C})$$

$$F = 9.6 \times 10^{-18}\ \text{N to the RIGHT}$$

And finally the neutron:

$$F = (0\ \text{C})(60\ \text{N/C}) = 0\ \text{N}$$

Notice that the proton feels a force in the direction of the electric field, but the electron feels the same force in the opposite direction of the electric field.

Don't state a force with a negative sign. Signs just indicate the direction of a force, anyway. So, just plug in the values for q and E, then state the direction of the force in words.

Electric Field Vector Diagrams

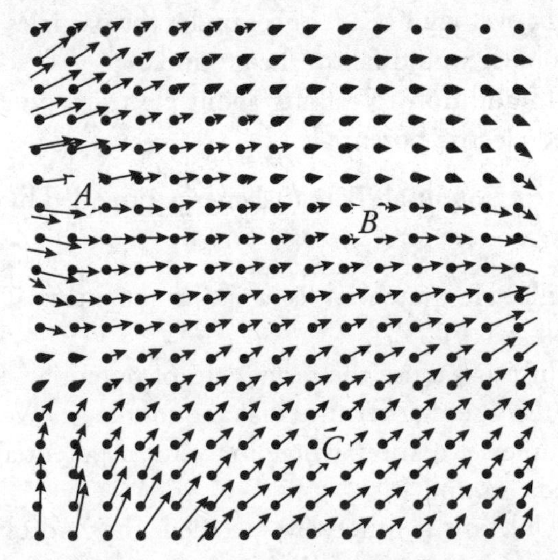

If we draw an electric field vector on a grid to represent both the direction and strength of the electric field at that point, we would get an electric field vector diagram. This is a visual picture that helps us see the force field as a whole. Place a proton at point A and it will receive a force to the right. Place the proton at point B and it will receive a smaller force to the right because, as we can see, the electric field is weaker. What would an electron experience if we place it at point *C*? A force down and to the left in the opposite direction of the E-Field because it is negative! Using the length of the electric field vectors as a guide, the magnitude of the electric field strength at these three points ranks in this order: (Greatest) $E_A > E_C > E_B$ (Weakest).

Look at the electric vector field in the next figure. What is going on at points *X* and *Y*? Electric field arrows are pointing inward toward *X* and are getting bigger as they get closer. *X* must be a negative charge location. *Y* must be a positive charge location. What direction is the electric field at *Z*? That's easy—to the left. What's the direction of the force on a charge placed at *Z*? Careful, it's a trick question! Is it a positive or negative charge? If it is negative, it gets pushed to the right. If it's positive, it gets forced to the left.

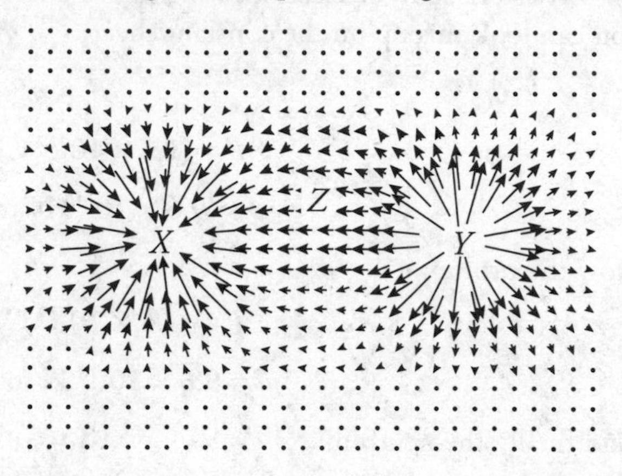

Electric Potential

When you hold an object up over your head, that object has gravitational potential energy. If you were to let it go, it would fall to the ground.

Similarly, a charged particle in an electric field can have electrical potential energy. For example, if you held a proton in your right hand and an electron in your left hand, those two particles would want to get to each other. Keeping them apart is like holding that object over your head; once you let the particles go, they'll travel toward each other just like the object would fall to the ground.

In addition to talking about electrical potential energy, we also talk about a concept called electric potential.

Electric Potential: Potential energy provided by an electric field per unit charge; also called *voltage*.

Electric potential is a scalar quantity. The units of electric potential are volts. 1 volt = 1 J/C.

Just as we use the term "zero of potential" in talking about gravitational potential, we can also use that term to talk about voltage. We cannot solve a problem that involves voltage unless we know where the zero of potential is. Often, the zero of electric potential is called "ground."

Unless it is otherwise specified, the zero of electric potential is assumed to be far, far away. This means that if you have two charged particles and you move them farther and farther from each another, ultimately, once they're infinitely far away from each other, they won't be able to feel each other's presence.

The electrical potential energy of a charged particle is given by this equation:

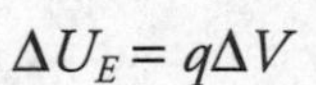

$$\Delta U_E = q\Delta V$$

Here, q is the charge on the particle, and ΔV is the difference in electric potential.

It is extremely important to note that electric potential and electric field are not the same thing. This example should clear things up:

> Three points, labeled *A*, *B*, and *C*, are found in a uniform electric field. At which point will a proton have the greatest electrical potential energy?

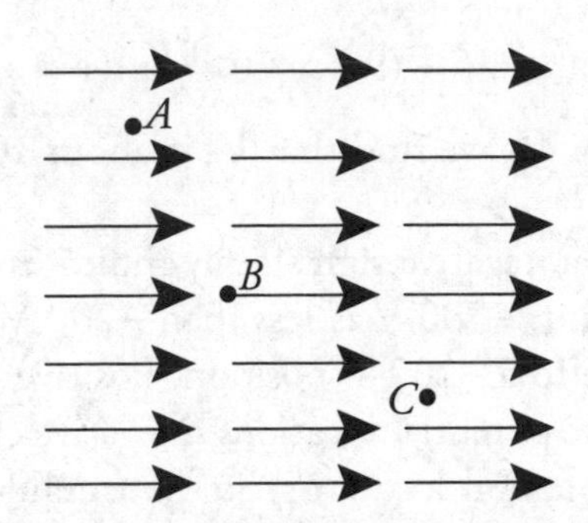

Electric field lines point in the direction that a positive charge will be forced, which means that our proton, when placed in this field, will be pushed from left to right. So, just as an object in Earth's gravitational field has greater potential energy when it is higher off the ground (think "*mgh*"), our proton will have the greatest electrical potential energy when it is farthest from where it wants to get to. The answer is *A*.

I hope you noticed that, even though the electric field was the same at all three points, the electric potential was different at each point.

How about another example?

> A positron (a positively charged version of an electron) is given an initial velocity of 6×10^6 m/s to the right. It travels into a uniform electric field, directed to the left. As the positron enters the field, its electric potential is zero. What will be the electric potential at the point where the positron has a speed of 1×10^6 m/s?

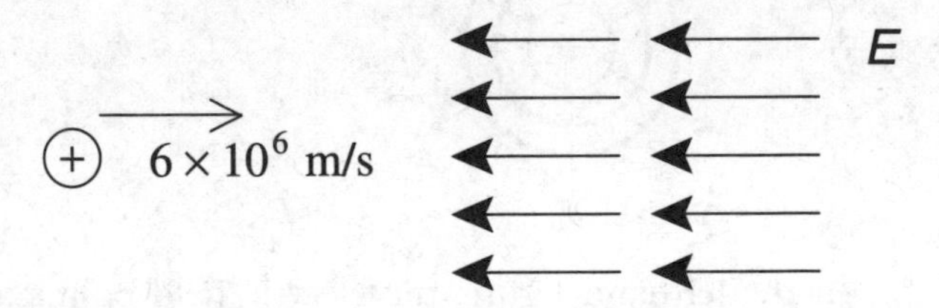

This is a rather simple conservation of energy problem, but it's dressed up to look like a really complicated electricity problem.

As with all conservation of energy problems, we'll start by writing our statement of conservation of energy.

$$KE_i + PE_i = KE_f + PE_f$$

Next, we'll fill in each term with the appropriate equations. Here the potential energy is not due to gravity (*mgh*), nor due to a spring ($\frac{1}{2}kx^2$). The potential energy is electric; so it should be written as qV.

$$\tfrac{1}{2}mv_i^2 + qV_i = \tfrac{1}{2}mv_f^2 + qV_f$$

Finally, we'll plug in the corresponding values. The mass of a positron is exactly the same as the mass of an electron, and the charge of a positron has the same magnitude as the charge of an electron, except a positron's charge is positive. Both the mass and the charge of an electron are given to you on the constants sheet. Also, the problem told us that the positron's initial potential V_i was zero.

$$\tfrac{1}{2}(9.1 \times 10^{-31}\text{ kg})(6 \times 10^{6}\text{ m/s})^2 + (1.6 \times 10^{-19}\text{ C})(0) =$$
$$\tfrac{1}{2}(9.1 \times 10^{-31}\text{ kg})(1 \times 10^{6}\text{ m/s})^2 + (1.6 \times 10^{-19}\text{ C})(V_f)$$

Solving for V_f, we find that V_f is about 100 V.

For *forces*, a negative sign simply indicates direction. For potentials, though, a negative sign is important. −300 V is less than −200 V, so a proton will seek out a −300 V position in preference to a −200 V position. Positive charges will naturally try to move to more negative electric potential locations. Negative charges will naturally try to move to more positive electric potential locations. So, be careful to use proper + and − signs when dealing with potential.

Equipotential Isolines

Just as you can draw electric field lines, you can also draw equipotential lines.

Equipotential Lines: Lines that illustrate every point at which a charged particle would experience the same potential.

The following figure shows a few examples of equipotential lines (shown with solid lines) and their relationship to electric field lines (shown with arrows):

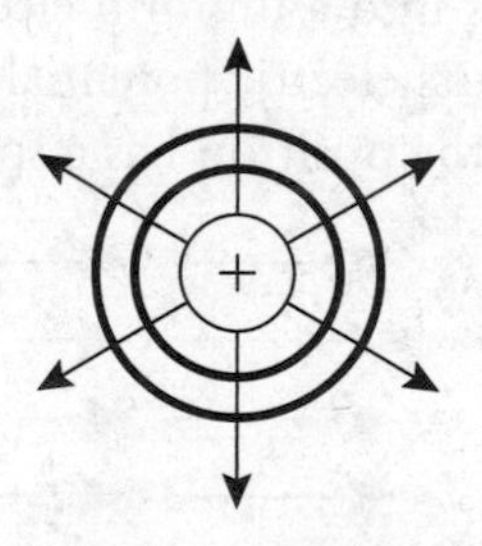

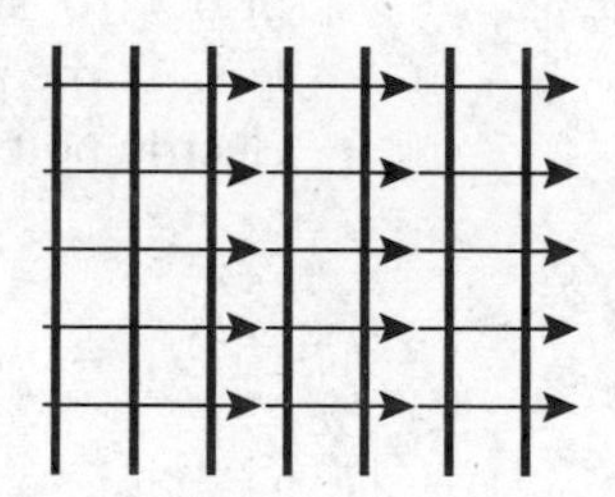

In the lefthand figure, the electric field points away from the positive charge. At any particular distance away from the positive charge, you would find an equipotential line that circles the charge—we've drawn two, but there are an infinite number of equipotential lines around the charge. If the potential of the outermost equipotential line that we drew was, say, 10 V, then a charged particle placed anywhere on that equipotential line would experience a potential of 10 V.

In the righthand figure, we have a uniform electric field. Notice how the equipotential lines are drawn perpendicular to the electric field lines. In fact, equipotential lines are always drawn perpendicular to electric field lines, but when the field lines aren't parallel (as in the drawing on the left), this fact is harder to see.

Moving a charge from one equipotential line to another takes energy. Just imagine that you had an electron and you placed it on the innermost equipotential line in the drawing on the left. If you then wanted to move it to the outer equipotential line, you'd have to push pretty hard, because your electron would be trying to move toward, and not away from, the positive charge in the middle.

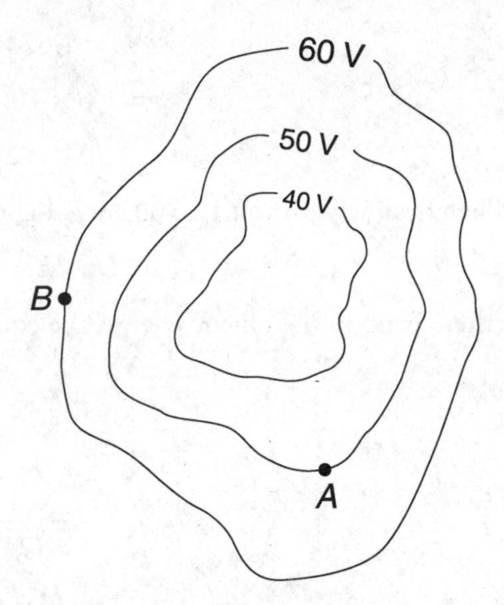

In the diagram above, point A and point B are separated by a distance of 30 cm. How much work must be done by an external force to move a proton from point A to point B?

The potential at point B is higher than at point A; so moving the positively charged proton from A to B requires work to increase the proton's potential energy. The question here really is asking how much more potential energy the proton has at point B.

Well, potential energy is equal to qV; here, q is 1.6×10^{-19} C, the charge of a proton. The potential energy at point A is $(1.6 \times 10^{-19}\ \text{C})(50\ \text{V}) = 8.0 \times 10^{-18}$ J; the potential energy at point B is $(1.6 \times 10^{-19}\ \text{C})(60\ \text{V}) = 9.6 \times 10^{-18}$ J. Thus, the proton's potential is 1.6×10^{-18} J higher at point B, so it takes 1.6×10^{-18} J of work to move the proton there.

Um, didn't the problem say that points A and B were 30 cm apart? Yes, but that's irrelevant. Since we can see the equipotential lines, we know the potential energy of the proton at each point; the distance separating the lines is irrelevant.

Look again at the figure. Notice how it looks like a topographical map that shows isolines of constant elevation. A plot of equipotential is just like that, except it shows isolines of equal potential. Positive charges will naturally be forced by the electric field to regions of more negative potential. Negative charges are forced by the electric field toward regions of higher or more positive potential. The flip side of this is that it takes work to "lift" a positive charge to a higher and more positive potential. It takes work to "lift" a negative charge to a more negative potential.

But, what happens if the charged object moves to a location of lower electric potential energy? Just like a falling ball picks up speed as it loses gravitational potential energy and gains kinetic energy, charged objects gain kinetic energy when they move to lower electric potential energy locations.

Example

A proton is accelerated from rest through an electric potential difference of 2000 V. Find the final speed of the proton.

A good idea is to list everything we know. We might not need all that data, but it is nice to have it all in one spot. Since it is a proton, we know its mass and charge:

$$m = 1.67 \times 10^{-27} \text{ kg}$$

$$q = 1.6 \times 10^{-19} \text{ C} = 1.0 \text{ e}$$

$$\Delta V = 2000 \text{ V}$$

When dealing with voltage, you are going to be solving an energy conservation problem:

$$U_i + K_i = U_f + K_f$$

Since there is no initial kinetic energy, the equation becomes:

$$U_i - U_f = \Delta U = K_f$$

$$q\Delta V = \frac{1}{2}mv^2$$

$$v = \sqrt{\frac{2q\Delta V}{m}}$$

Since you know that kinetic energy is in joules, your potential energy needs to be in joules. Your charge must be in coulombs and electric potential difference is in volts:

$$v = \sqrt{\frac{2(1.6 \times 10^{-19} \text{ C})(2000 \text{ V})}{1.67 \times 10^{-27} \text{ kg}}} = 620{,}000 \text{ m/s}$$

Practice

The figure below shows isolines of constant electric potential.

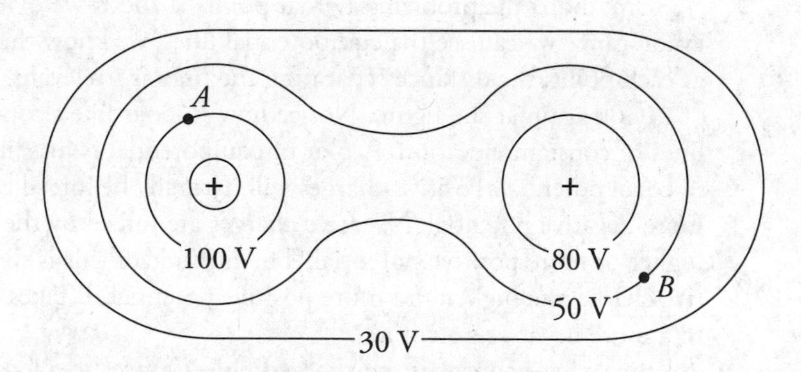

(a) Which way does the electric field vector point at *A*?
(b) If we place an electron at point *A*, in which direction will it receive a force?
(c) Does the electron gain or lose energy when it moves from point *A* to point *B*?

Answers to the Practice Problem:

(a) Remember that electric field vectors are always perpendicular to the isopotential lines, and directed away from positive charges. So, the electric field vector will be pointing mostly up and a little to the left.
(b) An electron is negative. It will experience a force in the opposite direction to the electric field, so mostly down and to the right—toward the positive charge.
(c) Keep in mind that positive charges want to move to lower electric potentials and negative charges want to move toward higher electric potentials. Since I am moving from

100 V to 50 V, I am moving a negative charge to a lower electric potential. That's the opposite of what the electron is going to naturally want, so I would have to add energy to the system. Left on its own, the electron would naturally "fall" toward one of the two positive charges. So we have to do work on the electron to move it from point *A* to point *B*. Notice how we cannot neglect the signs when considering the energy problem below.

$$\Delta U_E = q\Delta V = q(V_{final} - V_{initial})$$

$$\Delta U_E = (-1.6 \times 10^{-19}\ \text{C})(50\ \text{V} - 100\ \text{V})$$

$$\Delta U_E = +8 \times 10^{-18}\ \text{J}$$

Since the potential energy increased, there had to be work done on the electron.

Special Geometries for Electrostatics

There are two situations involving electric fields that are particularly nice because they can be described with some relatively easy formulas. Let's take a look.

Parallel Plates

If you take two metal plates, charge one positive and one negative, and then put them parallel to each other, you create a uniform electric field in the middle, as shown below.

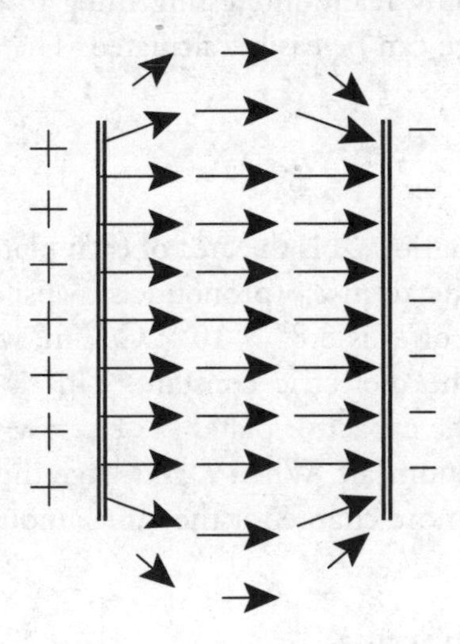

The electric field between the plates has a magnitude of

$$|E| = \frac{|\Delta V|}{|\Delta r|}$$

ΔV is the voltage difference between the plates, and r is the distance between the plates. Notice how the electric field is uniform in strength and direction near the center of the capacitor away from the edges. Near the edges of the capacitor, the electric field weakens and bends as shown in the figure.

Charged parallel plates can be used to make a *capacitor*, which is a charge-storage device. When a capacitor is made from charged parallel plates, it is called, logically enough, a "parallel-plate capacitor."

The battery in the following figure provides a voltage across the plates; once you've charged the capacitor, you disconnect the battery. The space between the plates prevents any charges from jumping from one plate to the other while the capacitor is charged. When you want to discharge the capacitor, you just connect the two plates with a wire.

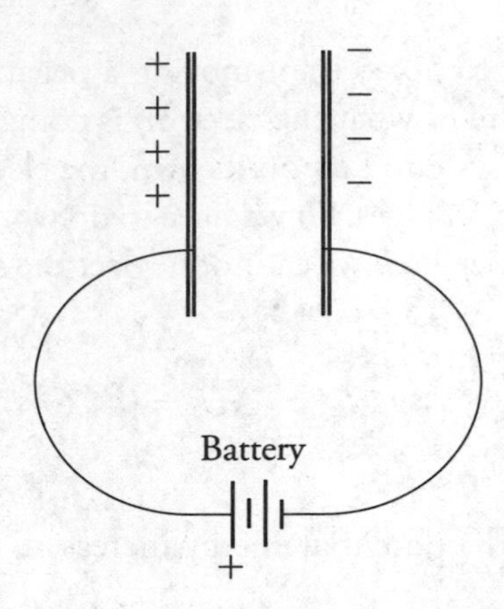

The amount of charge that each plate can hold is described by the following equation:

$$\Delta V = \frac{Q}{C}$$

Q is the charge on each plate, C is called the "capacitance," and ΔV is the voltage across the plates. The capacitance is a property of the capacitor you are working with, and it is determined primarily by the size of the plates and the distance between the plates. The units of capacitance are farads, abbreviated F; 1 coulomb/volt = 1 farad.

The only really interesting thing to know about parallel-plate capacitors is that their capacitance can be easily calculated. The equation is:

$$C = \kappa \varepsilon_0 \frac{A}{d}$$

In this equation, A is the area of each plate (in m^2), and d is the distance between the plates (in m). The term ε_0 (pronounced "epsilon-naught") is called the "vacuum permittivity." The value of ε_0 is 8.85×10^{-12} C/V·m, which is listed on the constants sheet.

κ is the dielectric constant. This is essentially how good of an insulator you have between the capacitor plates ($\kappa_{vacuum} = \kappa_{air} = 1.0$). Higher numbers mean a better insulator than a vacuum/air. When κ gets large the capacitance of the capacitor goes up, meaning it can store more charge for the same amount of potential difference.

Point Charges

As much as the writers of the AP exam like parallel plates, they *love* point charges. So you'll probably be using these next equations on the test.

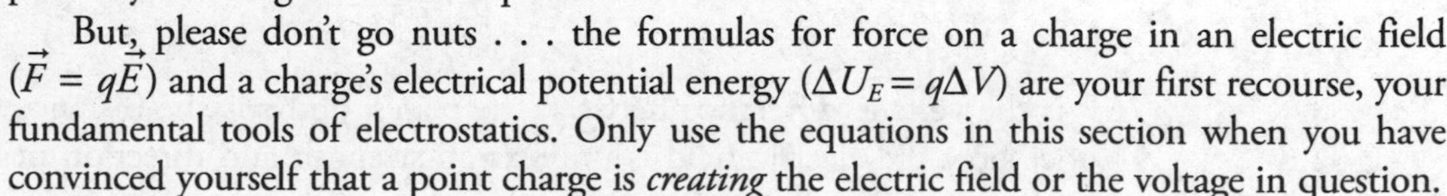

But, please don't go nuts . . . the formulas for force on a charge in an electric field ($\vec{F} = q\vec{E}$) and a charge's electrical potential energy ($\Delta U_E = q\Delta V$) are your first recourse, your fundamental tools of electrostatics. Only use the equations in this section when you have convinced yourself that a point charge is *creating* the electric field or the voltage in question.

First, the value of the electric field at some distance away from a point charge:

$$|\vec{E}| = \frac{1}{4\pi\varepsilon_0}\frac{|q|}{r}$$

q is the charge of your point charge, $\frac{1}{4\pi\varepsilon_0} = k$ is called the Coulomb's law constant ($\frac{1}{4\pi\varepsilon_0} = k = 9 \times 10^9$ N·m^2/C^2), and r is the distance away from the point charge. *The field produced by a positive charge points away from the charge; the field produced by a negative charge points toward the charge. When finding an electric field with this equation, do NOT plug in the sign of the charge or use negative signs at all.*

Second, the electric potential at some distance away from a point charge:

$$V = \frac{1}{4\pi\varepsilon_0}\frac{q}{r}$$

When using this equation, you must include a + or − sign on the charge creating the potential.

Electric field vectors point away from positive charges and toward negative charges. Equipotential "iso-lines" form circles around isolated point charges as seen in the figure below.

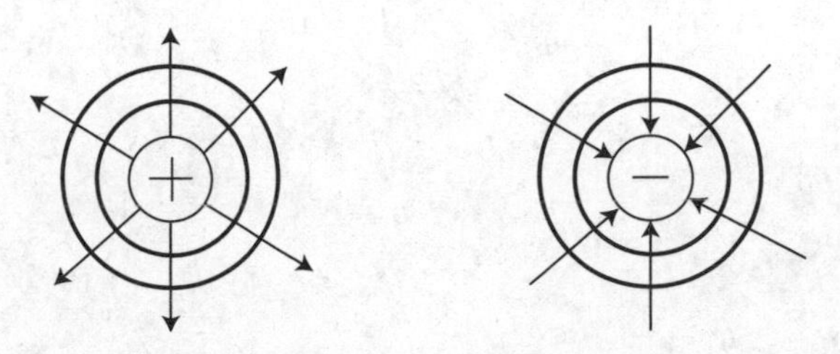

And third, the force that one point charge exerts on another point charge:

$$|\vec{F}_E| = \frac{1}{4\pi\varepsilon_0}\frac{|q_1 q_2|}{r^2}$$

In this equation, q_1 is the charge of one of the point charges, and q_2 is the charge on the other one. This equation is known as Coulomb's law.

Practice

To get comfortable with these three equations, here is a rather comprehensive problem.

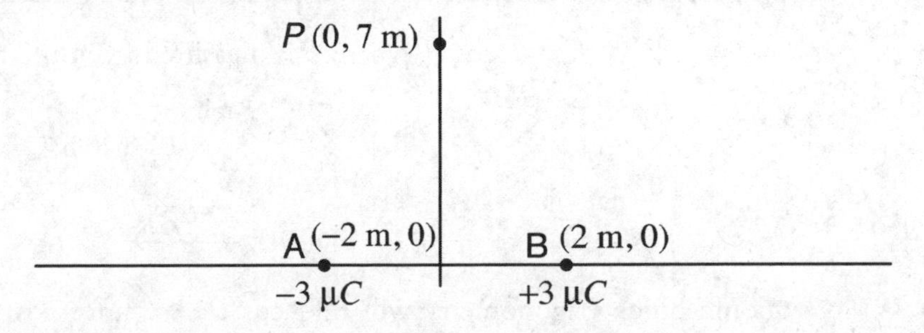

Two point charges, labeled "A" and "B", are located on the x-axis. "A" has a charge of −3 μC, and "B" has a charge of +3 μC. Initially, there is no charge at point P, which is located on the y-axis as shown in the diagram.

(1) What is the electric field at point P due to charges "A" and "B"?

(2) If an electron were placed at point P, what would be the magnitude and direction of the force exerted on the electron?

(3) What is the electric potential at point P due to charges "A" and "B"?

Yikes! This is a monster problem. But if we take it one part at a time, you'll see that it's really not too bad.

Part 1—Electric Field

Electric field is a vector quantity. So we'll first find the electric field at point *P* due to charge *A*, then we'll find the electric field due to charge *B*, and then we'll add these two vector quantities. One note before we get started: to find *r*, the distance between points *P* and *A* or between *P* and *B*, we'll have to use the Pythagorean theorem. We won't show you our work for that calculation, but you should if you were solving this on the AP exam.

$$E_{due\ to\ A} = \frac{(9\times10^{9})(3\times10^{-6}\ \text{C})}{(\sqrt{53}\ \text{m})^2} = 510\ \frac{\text{N}}{\text{C}},\ \text{pointing } \textit{toward} \text{ charge A}$$

$$E_{due\ to\ B} = \frac{(9\times10^{9})(3\times10^{-6}\ \text{C})}{(\sqrt{53}\ \text{m})^2} = 510\ \frac{\text{N}}{\text{C}},\ \text{pointing } \textit{away from} \text{ charge B}$$

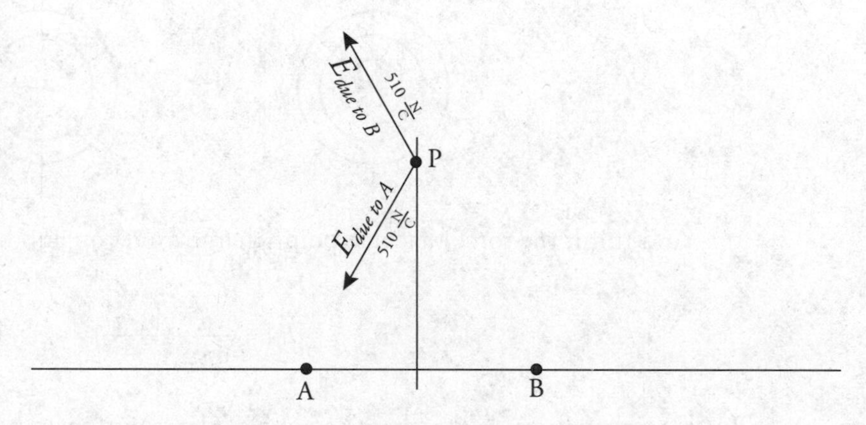

Note that we didn't plug in any negative signs! Rather, we calculated the magnitude of the electric field produced by each charge, and showed the direction on the diagram.

Now, to find the net electric field at point *P*, we must add the electric field vectors. This is made considerably simpler by the recognition that the *y*-components of the electric fields cancel . . . both of these vectors are pointed at the same angle, and both have the same magnitude. So, let's find just the *x*-component of one of the electric field vectors:

$E_x = E\cos\theta$, where θ is measured from the horizontal.

$E = 510$ N/C

θ

E_x (= 140 N/C)

Some quick trigonometry will find $\cos\theta$ . . . since $\cos\theta$ is defined as $\frac{\text{adjacent}}{\text{hypotenuse}}$, inspection of the diagram shows that $\cos\theta = \frac{2}{\sqrt{53}}$. So, the horizontal electric field $E_x = (510\ \text{m})\left(\frac{2}{\sqrt{53}}\right)$ . . . this gives 140 N/C.

And now finally, there are *TWO* of these horizontal electric fields adding together to the left—one due to charge *A* and one due to charge *B*. The total electric field at point *P*, then, is

280 N/C, to the left

Part 2—Force

The work that we put into Part 1 makes this part easy. Once we have an electric field, it doesn't matter what caused the E field—just use the basic equation $\vec{F} = q\vec{E}$ to solve for the force on the electron, where q is the charge of the electron. So,

$$F = (1.6 \times 10^{-19}\ \text{C})\ 280\ \text{N/C} = 4.5 \times 10^{-17}\ \text{N}$$

The direction of this force must be OPPOSITE the E field because the electron carries a negative charge; so, **to the right**.

Part 3—Potential

The nice thing about electric potential is that it is a scalar quantity, so we don't have to concern ourselves with vector components and other such headaches.

$$V_{due\ to\ A} = \frac{(9 \times 10^9)(-3 \times 10^{-6}\text{C})}{\sqrt{53}\ \text{m}} = -3700\ \text{V}$$

$$V_{due\ to\ B} = \frac{(9 \times 10^9)(+3 \times 10^{-6}\text{C})}{\sqrt{53}\ \text{m}} = +3700\ \text{V}$$

The potential at point P is just the sum of these two quantities. $V =$ zero!

Notice that when finding the electric potential due to point charges, you must include negative signs . . . negative potentials can cancel out positive potentials, as in this example.

Electric Fields Around a Point Charge and a Charged Conducting Sphere

To find the magnitude of the electric field outside a conducting sphere or point charge:

$$|\vec{E}| = \frac{1}{4\pi\varepsilon_0}\frac{|q|}{r^2}$$

Since the variables are all absolute valued, like Coulomb's law, and the electric field is a vector, the equation tells you only the magnitude of the electric field, you still need to do a vector diagram and solve for the resultant if there are multiple fields that overlap like in the practice problem above.

Inside a conducting sphere, things get interesting. First of all, any excess charge moves to the outside surface of the sphere. Inside the sphere, there is no electric field, because any charge placed inside the sphere is completely surrounded by the charge on the outside of the sphere. So the net force on any charge placed inside a conducting sphere is zero. If the electric field were graphed, it would look like this:

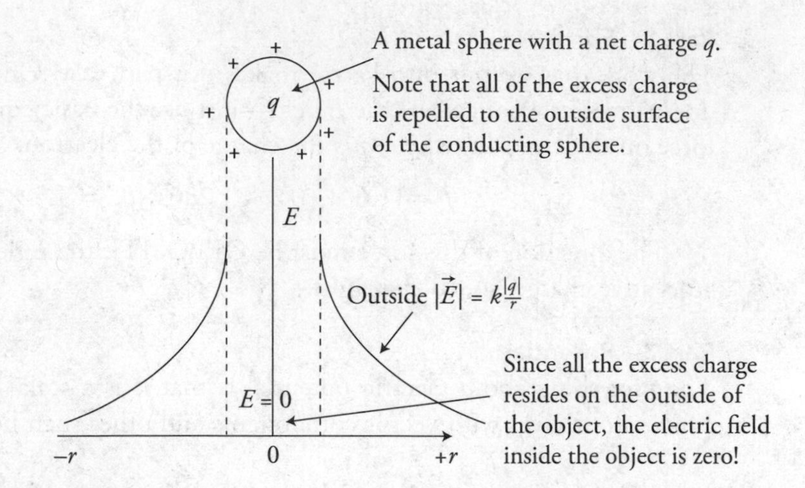

Remember: There will not be an electric field inside a charged *conducting object*! This is not true for *insulating objects*, because the excess charge cannot migrate to the outside surface.

Electric Potential Associated with a Point Charge and a Charged Conducting Sphere

To find electric potential outside a single point charge or charged sphere, the electric potential is:

$$V = \frac{1}{4\pi\varepsilon_0}\frac{q}{r}$$

If you need to find the potential due to a group of charges, don't panic—you are dealing with scalars. Simply solve for the potential from each charge individually and add them all up, like in the practice problem above. Once again, be careful and don't forget your signs; they are important!

What about inside a charged conducting sphere? We know there is no electric field inside the sphere, so there is no electrical potential difference from one point to another inside the sphere. To find the electrical potential inside the sphere:

$$V = \frac{1}{4\pi\varepsilon_0}\frac{q}{R}$$

where R is the radius of the sphere. As a graph, electrical potential looks like this:

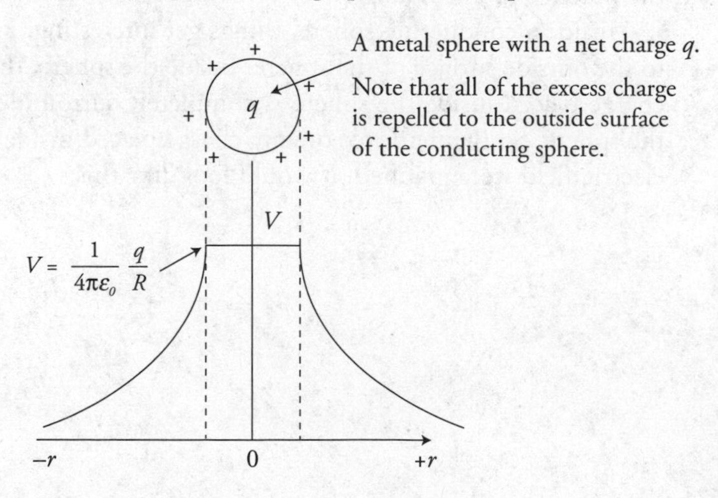

The Force Between Two Charges

Static electricity is nice to us because in many ways it mirrors things we have already learned in AP Physics 1, like gravity. Let's take a look at the force between two charged particles. The equation for this is Coulomb's law, and it is:

$$|\vec{F}_E| = \frac{1}{4\pi\varepsilon_0}\frac{|q_1 q_2|}{r^2}$$

where:

- ε_0 is a constant called vacuum permittivity, which is simply a measure of how easily an electric field passes through a vacuum. $\varepsilon_0 = 8.85 \times 10^{-12}\ C^2/N{\cdot}m^2$.
- $\frac{1}{4\pi\varepsilon_0}$ is the Coulomb's law electrostatic constant. $\frac{1}{4\pi\varepsilon} = k = 9 \times 10^9\ N{\cdot}m^2/C^2$.
- q_1 and q_2 are the two charges in coulombs.
- r is the distance between the centers of the bodies.

The AP exam will ask you to compare the electric force to gravity so make sure you know the similarities and differences. Notice how similar Coulomb's law is to Newton's law of universal gravitation:

$$\vec{F}_g = G\frac{m_1 m_2}{r^2}$$

Both are inversely related to the radius squared. But, gravity only attracts while the electric force can attract or repel. Also, the electric force only affects charged objects and is much, much stronger than gravity.

Mechanics and Charges

Many of the behaviors you learned in mechanics will be useful with charged objects.

Newton's Laws

- Newton's first law: If the charged object is at rest, then all the forces acting on the change must be canceling out: $\Sigma F = 0$.
- If there is a net force on a charge and the charge is free to move, they will follow Newton's second law: $\Sigma F = ma$.
- Newton's third law: the electric forces between the two charges will be equal but opposite in direction.
- Remember that force is a vector, so you must solve any problem involving multiple electric forces using a free-body diagram and vector analysis.

Charges in a Uniform Electric Field

A proton placed between two capacitor plates will accelerate toward the negative plate and away from the positive plate. But, if the proton is shot between the plates, it will experience parabolic trajectory motion just like a football in a gravitational field. In the following diagram, a proton accelerates to the right. The force on the charge is $F_E = qE = q\Delta V/d$. And the acceleration of the charge will be $a = \frac{F_E}{m_{proton}}$. Remember that electrons will accelerate in the opposite direction of the electric field!

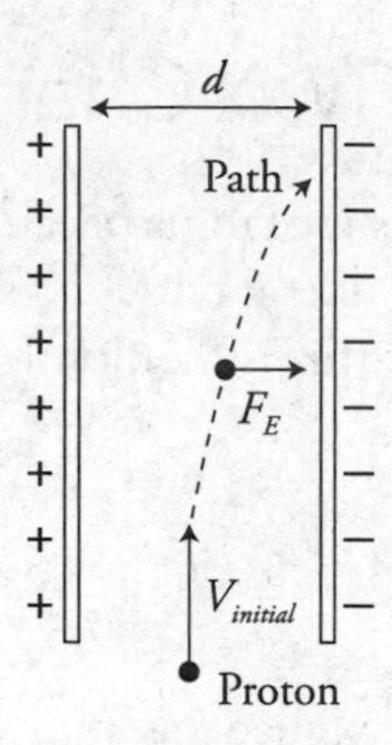

- An electric dipole, in a uniform electric field, will receive a torque that will cause it to rotate back and forth as it tries to align with the field. If the dipole has two opposite changes of equal size, it will not translate, because the two forces on each charge are equal and opposite in direction. See the following figure.

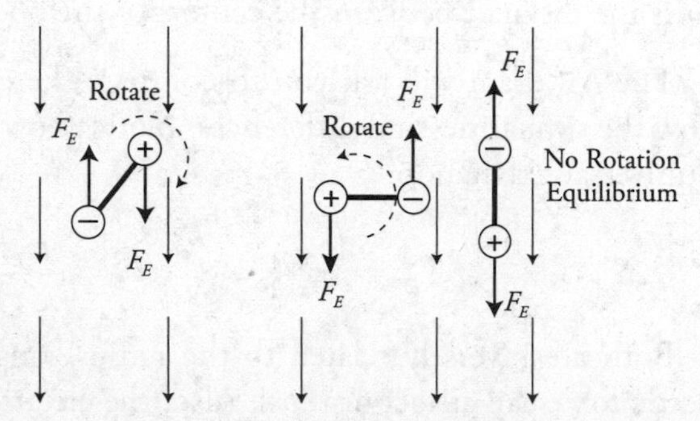

Conservation of Momentum Applied to Charges

- When charges attract, they can collide. Look at the next diagram. Charge A is negative and charge B is positive. Both charges start at rest.

(A)	(B)
$q = 3\ \mu C$	$q = +1\ \mu C$
$m = 5$ g	$m = 1$ g

(a) Which receives the larger force? Newton's third law—the force is the same.
(b) Which accelerates the greatest? B because of its smaller mass.
(c) What is the momentum of the system right before they collide? Conservation of momentum says it will be zero because they began with zero momentum.
(d) Where will they collide? They will collide closer to A. A has a larger mass and a smaller acceleration to the right. B has a smaller mass and a larger acceleration to the left. They will collide at the center of mass of the system.

- Charged particles can interact in a momentum collision without even touching. An electron shot at a stationary electron can "bounce off" due to the electrostatic repulsion and not even touch the other. Conservation of momentum is still in play. Both the x and y momentums before the interaction and after the interaction must be conserved.

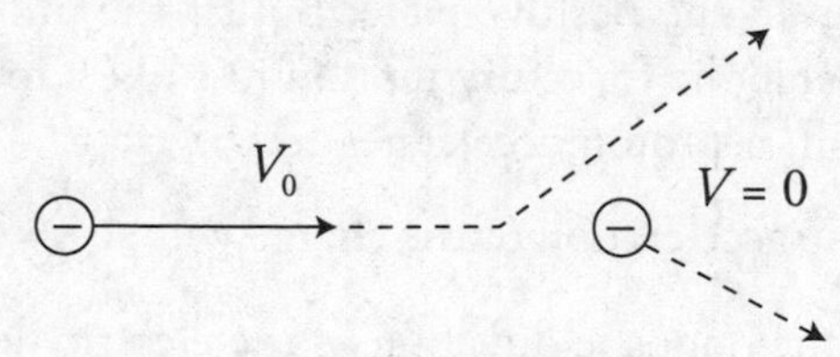

❯ Practice Problems

Multiple Choice

Questions 1 and 2

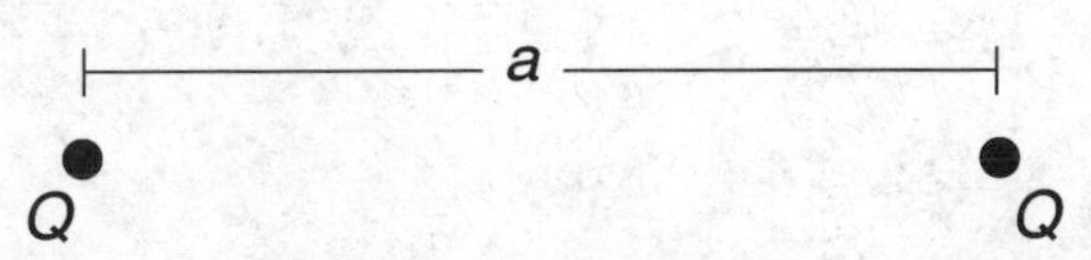

Two identical positive charges Q are separated by a distance a, as shown above.

1. What is the electric field at a point halfway between the two charges?

(A) $4\ kQ/a^2$
(B) $2\ kQ/a^2$
(C) zero
(D) $8\ kQ/a^2$

2. What is the electric potential at a point halfway between the two charges?

(A) $2\ kQ/a$
(B) zero
(C) $4\ kQ/a$
(D) $8\ kQ/a$

Questions 3 and 4

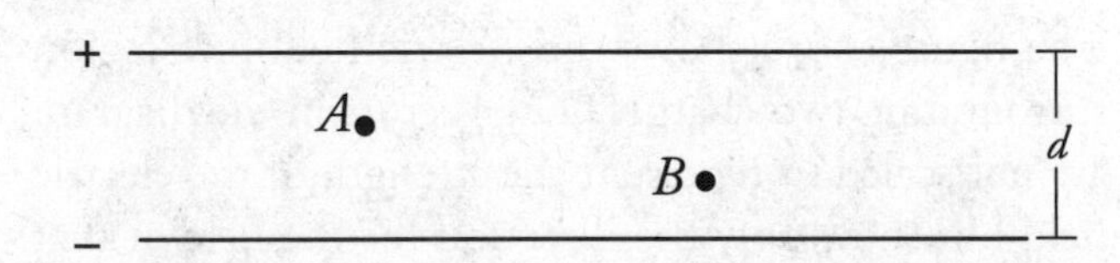

The diagram above shows two parallel metal plates that are separated by distance d. The potential difference between the plates is V. Point A is twice as far from the negative plate as is point B.

3. Which of the following statements about the electric potential between the plates is correct?

(A) The electric potential is the same at points A and B.
(B) The electric potential is two times larger at A than at B.
(C) The electric potential is two times larger at B than at A.
(D) The electric potential is four times larger at B than at A.

4. Which of the following statements about the electric field between the plates is correct?

(A) The electric field is the same at points A and B.
(B) The electric field is two times larger at A than at B.
(C) The electric field is two times larger at B than at A.
(D) The electric field is four times larger at B than at A.

5. Two identical neutral metal spheres are touching as shown in the figure. Which of the following locations of a positively charged insulating rod will create the largest positive charge in the sphere on the right?

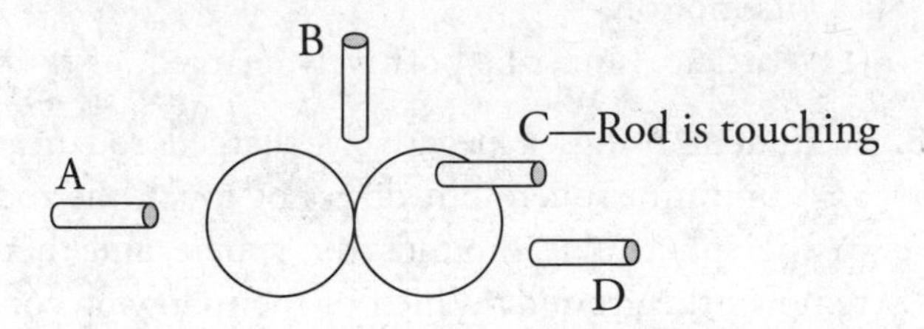

6. A student is comparing the gravitational field of a planet and the electric field of a positively charged metal sphere. Which of the following correctly describes the two fields?

(A) Both fields increase in magnitude as the size of the object creating the field increases.
(B) Both fields are proportional to 1/radius.
(C) Both fields are directed radially but in opposite directions.
(D) Both fields form concentric circles of decreasing strength at larger radii.

7. Three small droplets of oil with a density of ρ are situated between two parallel metal plates as shown. The bottom plate is charged positive and the top plate is charged negative. All the particles begin at rest. As time passes, particle 1 accelerates downward, particle 2 remains stationary, and particle 3 accelerates upward as shown. Which of the following statements is consistent with these observations?

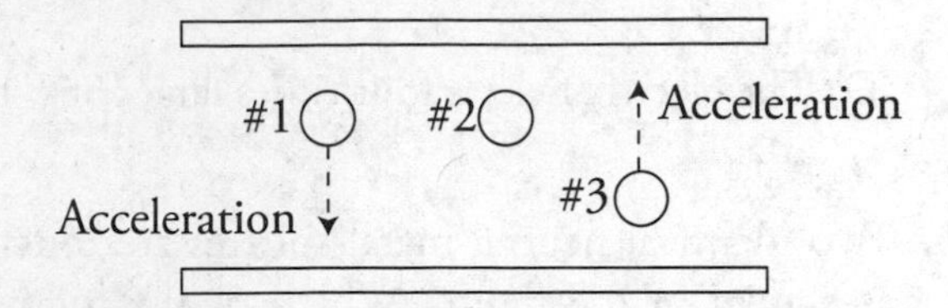

(A) Particle 1 must be negatively charged.
(B) Particle 2 must have no net charge.
(C) Particle 2 has a mass that is too small to affect its motion.
(D) Particle 3 must be positively charged.

8. A student brings a negatively charged rod near an aluminum sphere but does not touch the rod to the sphere. He grounds the sphere and then removes the ground. Which of the following correctly describes the force between the rod and sphere before and after the sphere is grounded?

	Before Grounding	After Grounding
(A)	attraction	attraction
(B)	attraction	repulsion
(C)	no force	attraction
(D)	no force	no force

Questions 9 and 10 refer to the following material.

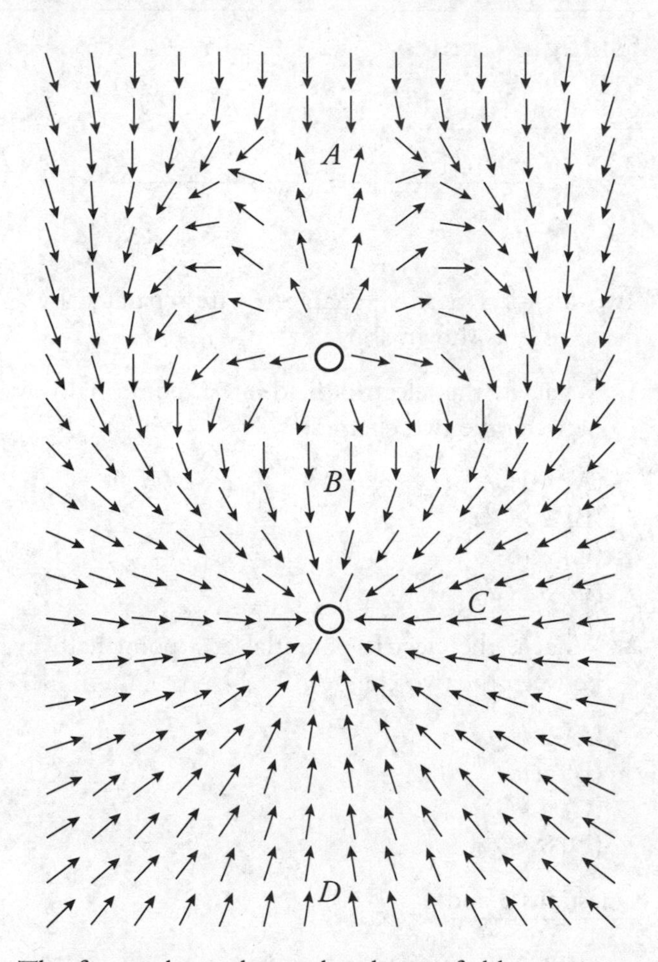

The figure above shows the electric field in a region surrounding two charges. The vectors in the diagram are not scaled to represent the strength of the electric field but show only the direction for the field at that point.

9. At which of the indicated points could you place a positive charge and have it receive the smallest force?

(A) *A*
(B) *B*
(C) *C*
(D) *D*

10. Which two points have the most similar electric potential?

(A) *A* and *B*
(B) *B* and *C*
(C) *C* and *D*
(D) *D* and *A*

11. A balloon rubbed with hair is suspended from the ceiling by a light thread. One at a time, a neutral wooden board and then a neutral steel plate of the same size are brought near to the balloon without touching. Which of the following correctly describes and explains the behavior of the balloon?

(A) The balloon is not attracted to the steel or the wood because both are neutral objects.
(B) The balloon is attracted to the steel because it is a conductor but not to the wood because it is an insulator.
(C) The balloon is attracted equally to both the steel and wood because both become polarized.
(D) The balloon is attracted to the steel more than it is attracted to the wood because steel is a conductor and the wood is an insulator.

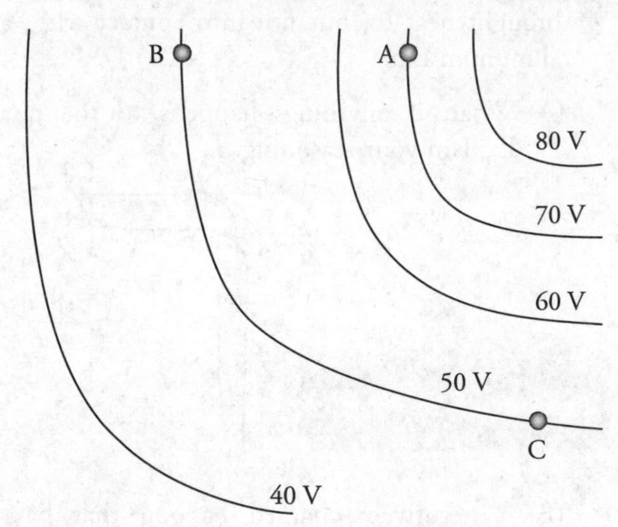

12. The figure shows isolines of electric potential in a region of space. Which of the following will produce the greatest increase in electric potential energy of the particle in the electric field?

(A) Moving an electron from point A to point C
(B) Moving an electron from point B to point A
(C) Moving a proton from point B to point C
(D) Moving a proton from point A to point C

Questions 13–16

The left figure shows a capacitor with a horizontal electric field. The distance between the plates is $4x$. The right figure shows two electrons, e_1 and e_2, and two protons, p_1 and p_2, which are placed between the plates at the locations shown.

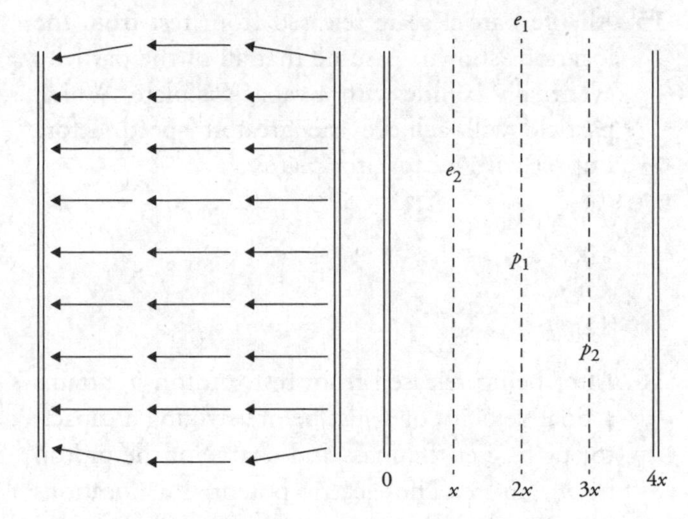

13. Which of the following correctly ranks the electric fields in between the capacitor plates at locations x, $2x$, and $3x$?

(A) $E_x > E_{2x} > E_{3x}$
(B) $E_x = E_{2x} = E_{3x}$
(C) $E_x = E_{3x} > E_{2x}$
(D) $E_{3x} > E_{2x} > E_x$

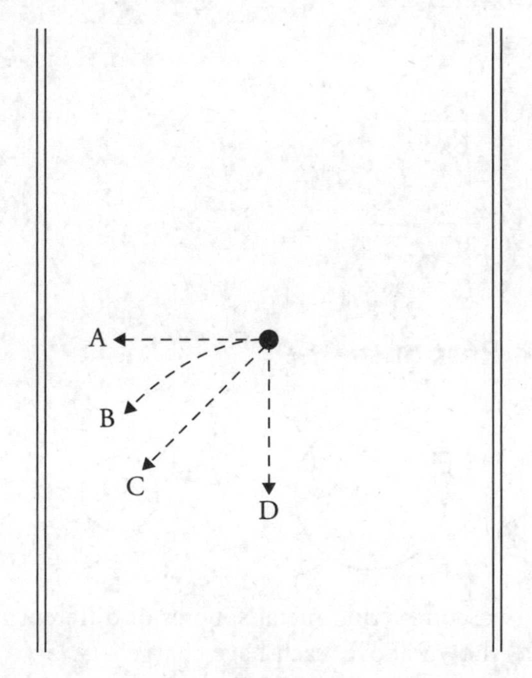

14. Charge p_1 is released from rest. Which of the trajectories shown in the figure above is a possible path of the released charge?

(A) A
(B) B
(C) C
(D) D

15. All the particles are released from rest from the locations shown. Assume that all of the particles eventually collide with a capacitor plate. Which particle will achieve the greatest speed before impact with a capacitor plate?

(A) e_1
(B) e_2
(C) p_1
(D) p_2

16. After being released from rest, proton p_2 attains a final velocity of v just before striking a capacitor plate. Let the mass and charge of the proton be m_p and e. The electric potential at locations 0, x, $2x$, $3x$, and $4x$ are V_0, V_x, V_{2x}, V_{3x}, and V_{4x}, respectively. What is the magnitude of the electric field between the plates? **(Select two answers.)**

(A) $\dfrac{V_{4x} - V_{3x}}{3x}$

(B) $\dfrac{V_0 - V_{3x}}{3x}$

(C) $\dfrac{m_p v^2}{6xe}$

(D) $\dfrac{e}{36\pi\varepsilon_0 x^2}$

Free Response

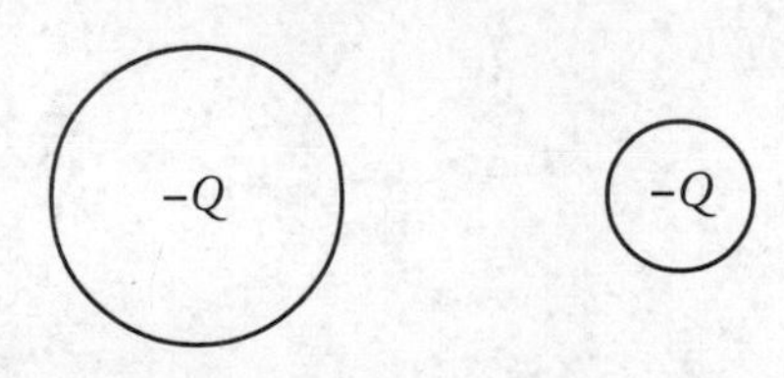

17. Two conducting metal spheres of different radii, as shown above, each have charge $-Q$.

(A) Consider one of the spheres. Is the charge on that sphere likely to clump together or to spread out? Explain briefly.
(B) Is the charge more likely to stay inside the metal spheres or on the surface of the metal spheres? Explain briefly.
(C) If the two spheres are connected by a metal wire, will charge flow from the big sphere to the little sphere, or from the little sphere to the big sphere? Explain briefly.
(D) Which of the following two statements are correct? Explain briefly.
  i. If the two spheres are connected by a metal wire, charge will stop flowing when the electric field at the surface of each sphere is the same.
  ii. If the two spheres are connected by a metal wire, charge will stop flowing when the electric potential at the surface of each sphere is the same.
(E) Explain how the correct statement you chose from part (D) is consistent with your answer to (C).

18. A negatively charged piece of clear sticky tape is brought near to, but not into contact with, an aluminum can.

(A) What, if anything, happens to the tape? Explain your reasoning.

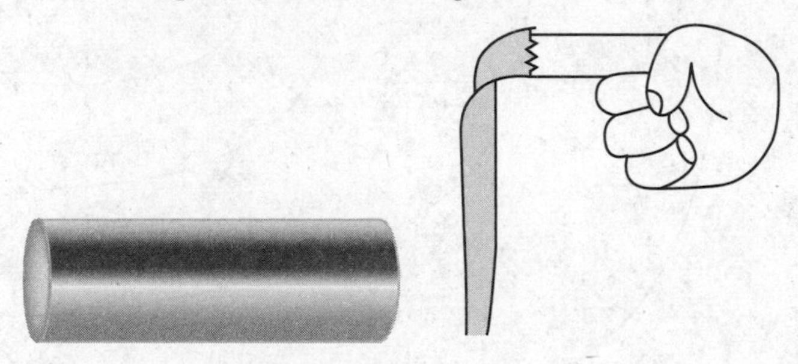

(B) A negatively charged balloon that has a charge much larger than that of the clear sticky tape, is brought near the aluminum can without touching. What, if anything, happens to the tape? Explain your reasoning.

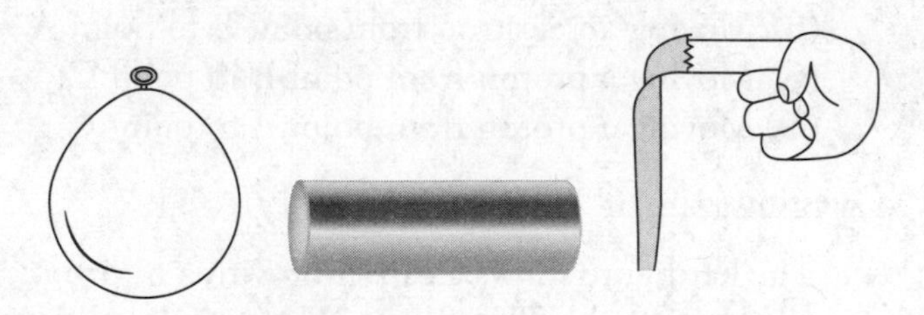

19. Your teacher gives you a charged metal sphere that rests on an insulating stand. The teacher asks you to determine if the charge on the object is positive or negative.

(A) List the items you would use to perform this investigation.
(B) Outline the experimental procedure you would use to make this determination. Indicate the measurements to be taken and how the measurements will be used to obtain the data needed. Make sure your outline contains sufficient detail so that another student could follow your procedure and duplicate your results.

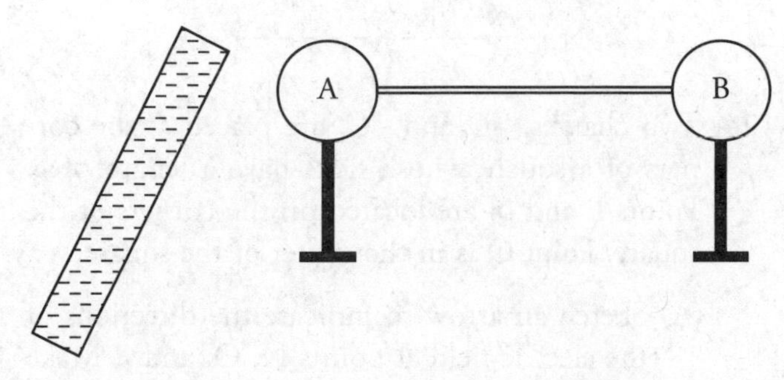

20. The figure above shows two conductive spheres (A and B) connected by a rod. Both spheres begin with no excess charge. A negatively charged rod is brought close to and held near sphere A as shown.

(A) If the connecting rod is made of wood, what is the net charge of the spheres while the rod is held in the position shown? Justify your answer.
(B) If the connecting rod is made of copper, what is the net charge of the spheres while the rod is held in the position shown? Justify your answer.
(C) The rod is now brought into contact with sphere A. How will this change the answers to the previous two questions? Explain.

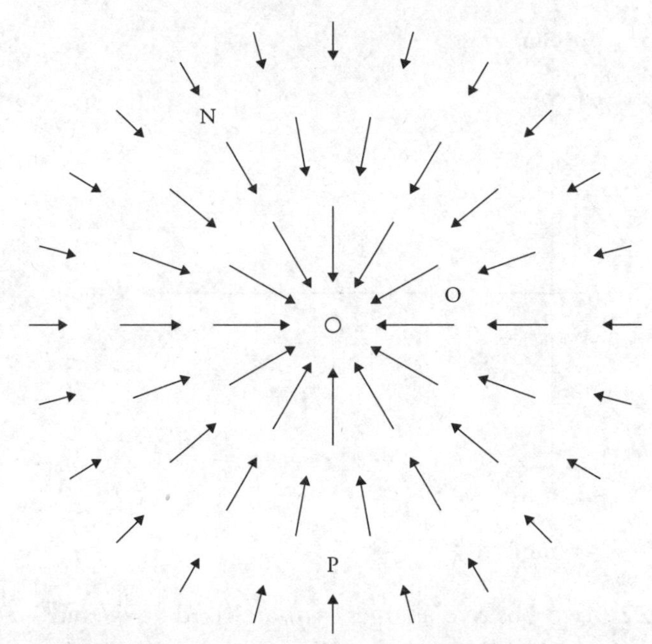

21. The electric field around a charged object is shown in the figure above.

(A) What aspects of the electric field indicate the sign on the charge?
(B) Rank the magnitudes of the electric field at points N, O, and P. Explain what aspects of the diagram indicate the strength of the electric field.
(C) A proton is placed at point P, and an electron is placed at point N. Both are released from rest at the same time. Compare and contrast the acceleration of the two particles at the instant they are released, and explain any differences.
(D) Describe the motions of the proton and the electron for a long time after they are released. Justify your claim.

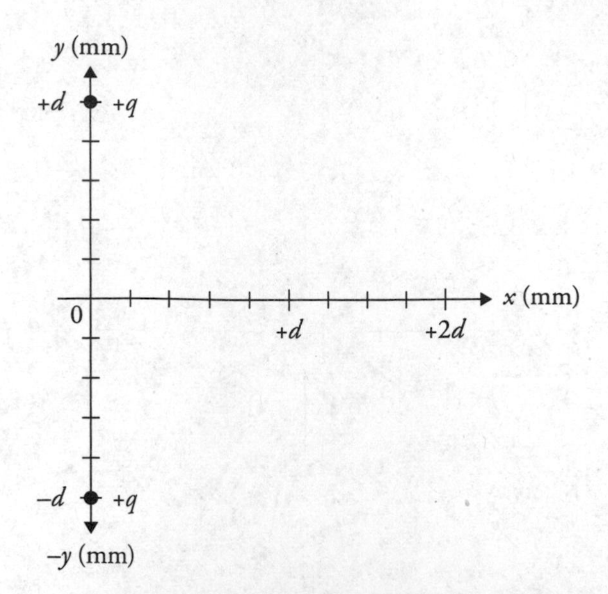

22. Two positive charges ($+q$) are fixed at $+d$ and $-d$ on the y-axis so they cannot move, as shown in the figure above.

(A) Calculate the force on a third charge, $-q$, placed at $+d$ on the x-axis. What direction is the force? Show all your work.

(B) If the charge $-q$ is moved to the origin, what will be the new force on the charge? Justify your response.

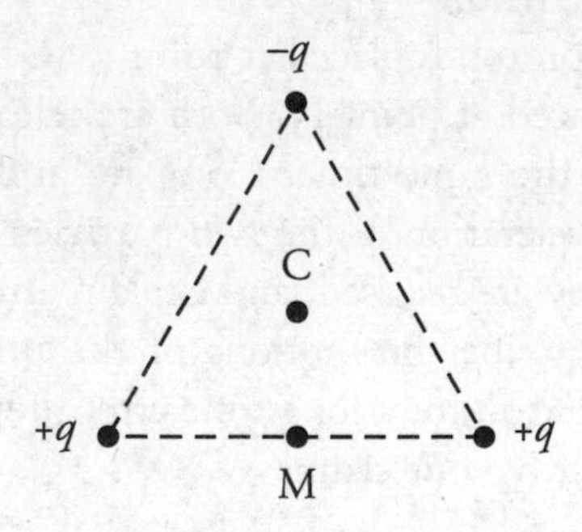

23. Three charges of magnitude q are placed at the corners of an equilateral triangle, as shown in the figure above.

(A) An electron is placed at C, the center of the triangle. Draw a force diagram of all the forces on the electron. All forces should be drawn proportionally. What is the direction of the net force on the electron?

(B) The electron is removed and a proton is placed at M, the midpoint of the bottom side of the triangle. Will the net force on the proton be greater than, less than, or the same as the net force on the electron from part (A) above? Justify your claim.

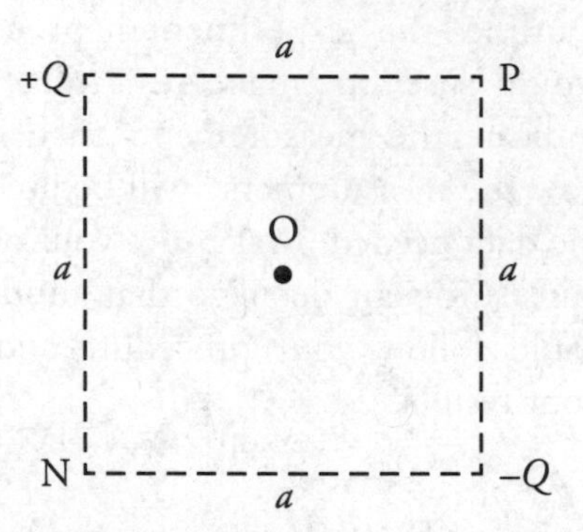

24. Two charges, $+Q$ and $-Q$, are placed at the corners of a square whose sides have a length of a. Points P and N are located on the corners of the square. Point O is in the center of the square.

(A) Sketch an arrow to indicate the directions of the electric field at points N, O, and P. Make sure the vectors are drawn to the correct proportion.

(B) Calculate the electric potentials at points N, O, and P.

(C) A proton is moved from point P to point O. How much total work is done by the electric field during this move? Explain.

(D) By moving only one of the charges, explain how the electric field at point O can be made to point directly to the right.

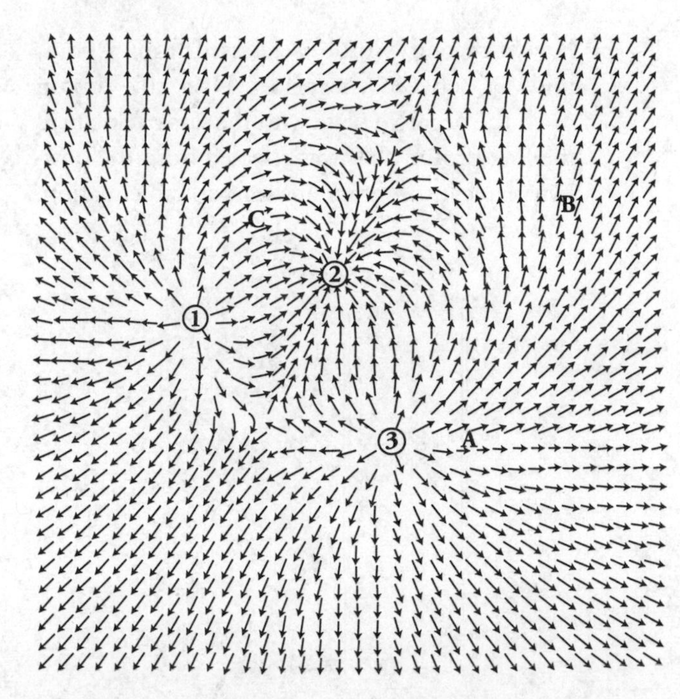

25. Electric field vectors around three charges 1, 2, and 3 are shown in the figure above.

(A) What are the signs of the three charges? Explain what aspects of the electric field indicate the sign of the charges.
(B) Draw the direction of the force on an electron placed at point C.
(C) Sketch two isoline lines of constant electric potential—one that passes through point A and another that passes through point B.
(D) Which isoline has a higher electric potential, the line that passes through point A or the one that passes through point B? Justify your answer.

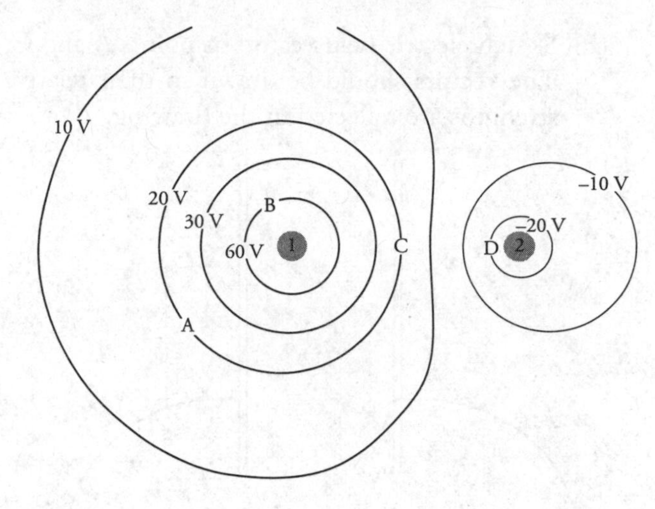

26. The figure shows isolines of electric potential. Circles 1 and 2 represent two spherical charges. Points A, B, C, and D represent locations on isolines of electric potential.

(A) What are the signs of the two charges, and how do their relative magnitudes compare? Explain how the isolines help you determine this.
(B) A proton is released from point C and moves through an electric potential difference of magnitude 40 V.
  i. On which isoline of electric potential will the proton end up?
  ii. The proton will have kinetic energy when it arrives at this new isoline. Where does this kinetic energy come from?
    a. For the system that includes the two charges and the proton, explain where this kinetic energy comes from.
    b. For the system that includes only the proton, explain where this kinetic energy comes from.
(C) An electron at point A is moved to point B. Has the electric potential energy of the electron-charges system increased or decreased? Justify your answer with an equation.
(D) The distance between points C and D is d. Derive a symbolic expression for the magnitude of the average electric field between the two points. Also, indicate the direction of the electric field.
(E) A particle with positive charge of Q is released from point C and gains kinetic energy on its path to point D. Derive a symbolic equation for the amount of work done by the electric field and the final kinetic energy of the proton.

(F) Sketch electric field vectors at points A and C. The vectors should be drawn so their relative strengths are reflected in the drawing.

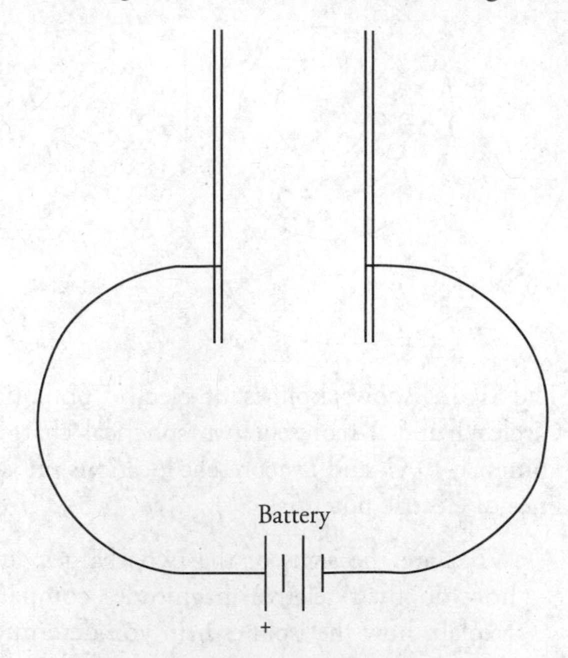

27. A battery of potential difference ΔV is connected to a parallel plate capacitor for a long time. The separation between the plates is d, and the area of one plate is A.

(A) Sketch the electric field between the plates of the capacitor.
(B) Sketch isolines of constant electric potential between the plates.
(C) Write an expression for the electric field strength between the plates.
(D) Write an expression for the charge on the left plate. Show your work.
(E) What is the net charge on both plates combined? Explain.
(F) A proton with a charge of $+e$ is released from the positive plate. Write an expression for the net force on the proton using known quantities. Do you need to include the force of gravity in your calculation? Justify your answer.
(G) Write an expression for the velocity of the proton when it reaches the negative plate. Derive this value using the concept of forces and the concept of energy.
(H) Now a second proton is released from a point midway between the plates. Does this proton reach the negative plate with the same velocity as the first proton that was released from the positive plate? Justify your answer with an equation.

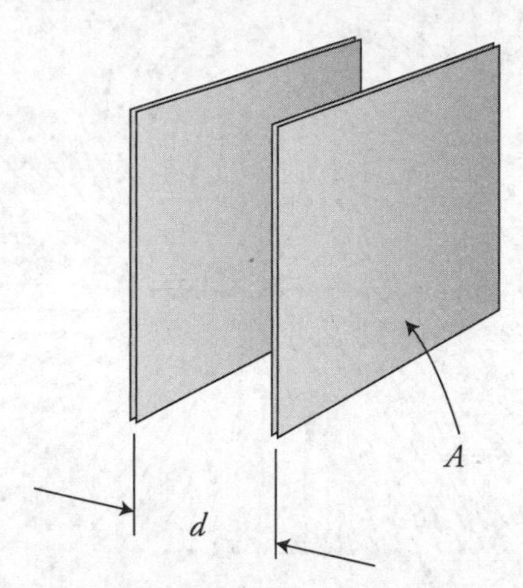

28. A parallel plate capacitor with a capacitance of C is shown in the figure above. The area of one plate is A, and the distance between the plates is d.

(A) If the area of both capacitor plates as well as the distance between them were doubled, what would be the effect on the capacitance of the capacitor? Explain.
(B) The capacitor is connected to a battery of potential difference ΔV. If the potential difference of the battery is doubled, what happens to the charge stored on the plates and the capacitance of the capacitor? Justify your answer.
(C) In an experiment, the area (A) of the capacitor plates is changed to investigate the effect on the capacitance (C) of the capacitor. Sketch the graph of the lab data you expect to see from this experiment.

(D) In another experiment, the distance between the plates (d) is changed to investigate the effect on the capacitance (C) of the capacitor. Sketch the graph of the lab data you expect to see from this experiment.

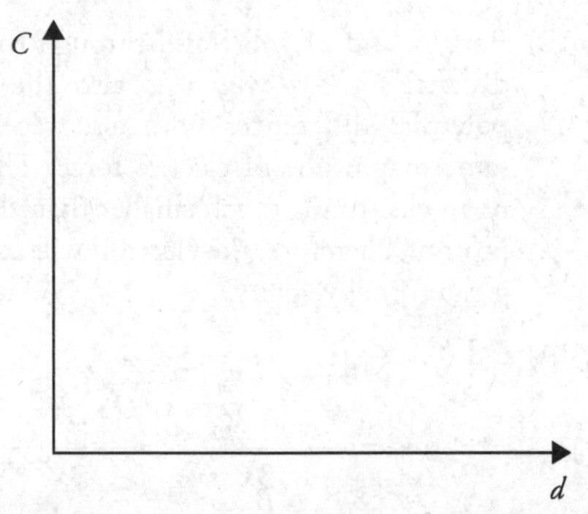

(E) You are going to use a capacitor to power a lightbulb. You need the bulb to shine for a long time. Describe the geometry of the capacitor you would choose to power the bulb. Explain your answer.

❯ Solutions to Practice Problems

1. (C) Electric field is a *vector*. Look at the field at the center due to each charge. The field due to the left-hand charge points away from the positive charge (i.e., to the right); the field due to the right-hand charge points to the left. Because the charges are equal and are the same distance from the center point, the fields due to each charge have equal magnitudes. So the electric field vectors cancel! $E = 0$.

2. (C) Electric potential is a *scalar*. Look at the potential at the center due to each charge: Each charge is distance $a/2$ from the center point, so the potential due to each is $kQ/(a/2)$, which works out to $2\ kQ/a$. The potentials due to both charges are positive, so add these potentials to get $4\ kQ/a$.

3. (B) If the potential difference between plates is, say, 100 V, then we could say that one plate is at +100 V and the other is at zero V. So, the potential must change at points in between the plates. The electric field is uniform and equal to V/d (d is the distance between plates). Thus, the potential increases linearly between the plates, and A must have twice the potential as B.

4. (A) The electric field by definition is *uniform* between parallel plates. This means the field must be the same everywhere inside the plates.

5. (A) Rods A, B, and D will each polarize the spheres, drawing negative charges toward themselves and leaving the opposite side positively charged. Thus A will cause the right sphere to be the most positive. Touching the spheres with the insulating rod will cause some of the polarized negative charge from the spheres to flow onto the rod. Since the rod is an insulator, this leaves the spheres with only a small excess positive charge that will be shared between both spheres.

6. (C) The magnitude of gravitational $\left(g = G\frac{m}{r^2}\right)$ and electric fields $\left(E = k\frac{q}{r^2}\right)$ depend on the mass and charge of the objects respectively. They are both proportional to $1/r^2$ and are directed along the radius from the center of mass or center of charge. Gravitational fields always point inward along the radius because it always attracts mass. Electric fields point inward for negative charges and outward for positive charges.

7. (D) All the droplets have mass ($m = \rho V$) and will experience a downward gravitational force. Particle 1 could be uncharged and simply falling due to the force of gravity. Particle 2 must have an electric force to cancel the gravity force. Particle 3 must be positive to receive an electric force upward larger than the force of gravity downward.

8. (A) Before grounding, the negatively charged rod polarized the sphere causing an attraction. After grounding, the sphere has been charged the opposite sign by the process of induction and the two will attract.

9. (A) The electric field vectors indicate that the electric field at location *A* is zero or very small.

10. (B) Isolines of constant potential are perpendicular to the electric field vectors. *B* and *C* appear to be on the same isoline that circles the bottom negative charge.

11. (D) The charged balloon will polarize both the wooden board and the steel plate. Therefore, it will be attracted to both. However, the polarization of the wood occurs on an atomic scale because it is an insulator, and its electrons do not move easily. The steel is a conductor that allows its electrons to migrate. This permits the electrons in the steel to move farther and create a larger charge separation in the process of polarization. This means the balloon will be attracted to the steel more strongly than to the wood.

12. (A) $\Delta U_E = q\Delta V$. To get the greatest increase in electric potential energy, we need the greatest change in electric potential times the charge. The charge of protons and electrons are the same magnitude. To increase the electric potential energy of a proton, we need to move the proton to higher potentials. To increase the electric potential energy of the electron, we need to move the electron to lower electric potentials.

13. (B) The electric field between the plates of a parallel plate capacitor is uniform and constant in strength as long as you are not too close to the edges of the capacitor.

14. (A) The electric force on a positive charge is in the direction of the electric field. The gravity force is much smaller than the electric force. All the other trajectories show gravity stronger than the electric force.

15. (B) Both e_2 and p_2 will travel through the same distance of $3x$, which is also the largest potential difference. Both also receive the same magnitude of electric force. The mass of an electron is much smaller than that of a proton. Therefore, the electron will achieve a greater final velocity.

16. (A) and (C)

$$E = \frac{\Delta V}{\Delta r} = \frac{V_{final} - V_{initial}}{3x} = \frac{V_0 - V_{3x}}{3x}$$

$$E = \frac{\Delta V}{\Delta r} = \left(\frac{1}{\Delta r}\right)\left(\frac{\Delta U}{q}\right) = \left(\frac{1}{3x}\right)\left(\frac{\Delta K}{e}\right)$$

$$= \left(\frac{1}{3x}\right)\left(\frac{\frac{1}{2}m_p v^2}{e}\right) = \frac{m_p v^2}{6xe}$$

17. (A) Like charges repel, so the charges are more likely to spread out from each other as far as possible.

(B) "Conducting spheres" mean that the charges are free to move anywhere within or onto the surface of the spheres. But because the charges try to get as far away from each other as possible, the charge will end up on the surface of the spheres. This is actually a property of conductors—charge will always reside on the surface of the conductor, not inside.

(C) Charge will flow from the smaller sphere to the larger sphere. Following the reasoning from parts (A) and (B), the charges try to get as far away from each other as possible. Because both spheres initially carry the same charge, the charge is more concentrated on the smaller sphere; so the charge will flow to the bigger sphere to spread out. (The explanation that negative charge flows from low to high potential, and that potential is less negative at the surface of the bigger sphere, is also acceptable here.)

(D) The charge will flow until the potential is equal on each sphere. By definition, negative charges flow from low to high potential. So, if the potentials of the spheres are equal, no more charge will flow.

(E) The potential at the surface of each sphere is $-kQ/r$, where r is the radius of the sphere. Thus, the potential at the surface of the smaller sphere is initially more negative, and the negative charge will initially flow from low-to-high potential onto the larger sphere.

18. (A) The clear sticky tape is attracted to the can due to charge polarization of the can.

(B) The tape is now repelled by the can because the stronger negative charge of the balloon will drive electrons toward the right side of the can, which will repel the tape.

19. There are several ways to accomplish this lab. Here is one example:

(A) A red balloon and a blue balloon, thread, kitchen plastic wrap, human hair.

(B) Procedure:

1. Blow up the balloons and tie a long thread to each.
2. Charge the red balloon positively by rubbing the kitchen plastic wrap all over its surface. Charge the blue balloon negatively by rubbing its surface on your hair.
3. Hold each balloon by the thread, and one at a time, bring them close to the charged metal sphere. Observe the results. One balloon should be attracted and the other repelled. The balloon that is repelled will be the same sign charge as the metal sphere.

20. (A) Both spheres will remain neutral because there is not a conductive pathway for charge to move onto or off of either sphere.

(B) Due to induction, the system that includes both spheres and the copper rod will become polarized. Sphere A will be positively charged, and sphere B will be negatively charged. The net charge of the system that includes both spheres and the copper rod is still zero because no charge has been added to the system.

(C) The answer to (A) is now: Sphere A becomes negative by contact, but sphere B remains neutral because wood is an insulator. The answer to (B) is now: Both spheres become negative by contact because there is a conductive pathway connecting both spheres.

21. (A) All the electric field vectors in the figure point inward toward the charge. Electric field vectors point in the direction of the force on positive charges; therefore, the charge in the figure is negative.

(B) $E_O > E_N = E_P$. The length of the electric field vector indicates the strength of the field.

(C) The electric field is the same magnitude at N and P. Both the electron and the proton have the same magnitude of charge. The electric force ($F_E = Eq$) for each is the same. However, the mass of the proton is larger than that of the electron. Therefore, the acceleration of the electron is greater. The electron accelerates in the opposite direction of the field. The proton accelerates in the same direction as the field.

(D) The proton accelerates from rest inward in the direction of the electric field. The acceleration of the proton increases as it moves into a larger electric field closer to the charge. The electron accelerates outward away from the charge. The acceleration decreases as it gets farther away from the charge where the field is weaker. The electron eventually reaches a constant velocity when it is very far away from the charge.

22. (A) See figure. Due to the symmetry of the arrangement of charges the force on $-q$ will be to the left along the x-axis. Only the x-component of the force

needs to be calculated. $F_E = k\dfrac{q_1 q_2}{r^2}$

The radius between the charges is: $r^2 = d^2 + d^2 = 2d^2$.

Therefore, the magnitude of the force between $+q$ and $-q$ is . $F_E = k\dfrac{q^2}{2d^2}$

The x-component of this force is: $F_{Ex} = F_E \cos\theta$.

Where, $\cos\theta$ is equal to:

$$\cos\theta = \frac{\text{adjacent side}}{\text{hypotenuse}} = \frac{d}{\sqrt{2d^2}} = \frac{1}{\sqrt{2}}$$

Therefore the x-component of this force becomes: $F_{Ex} = F_E \cos\theta = \frac{F_E}{\sqrt{2}}$

There are two charges with this force on $-q$. So the net force on $-q$ is $\sum F_{Ex} = 2\left(\frac{F_E}{\sqrt{2}}\right) = \sqrt{2}F_E = \sqrt{2}k\frac{q^2}{2d^2}$.

This net force is in the negative x-direction. See figure.

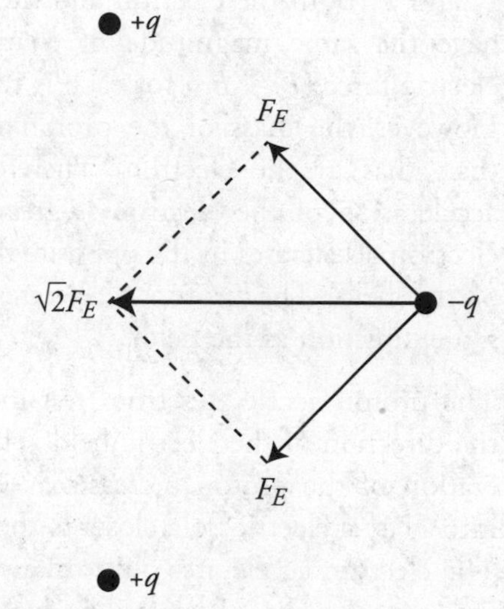

(B) The net force is zero because they are equal in size and opposite in direction. (Always look for any symmetries that will give you a simple answer!)

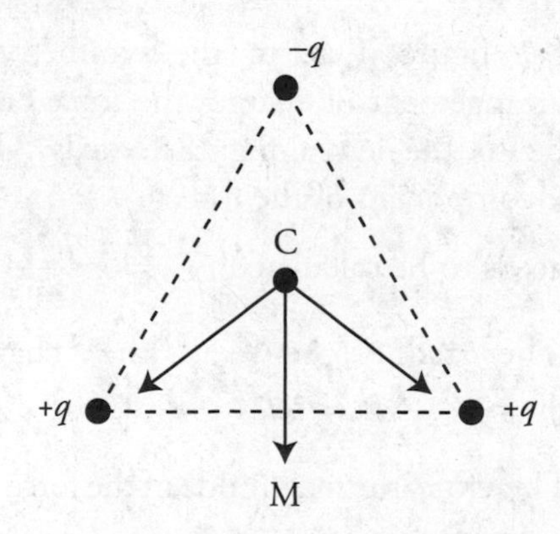

23. (A) See figure. Note: To get full credit, all the force vectors should be the same length. Two are diagonal toward the $+q$ charges, and one is directly down and away from the $-q$ charge.

(B) The force on the proton at M is less than the force on the electron at C. At point M, the forces from the two positive charges $+q$ are equal and opposite and cancel out. In addition, the force from the negative charge $-q$ is smaller in magnitude at point M than at point C.

24. (A) The electric fields at points N and P must be the same size. The electric field at O must be longer than the other two. See figure.

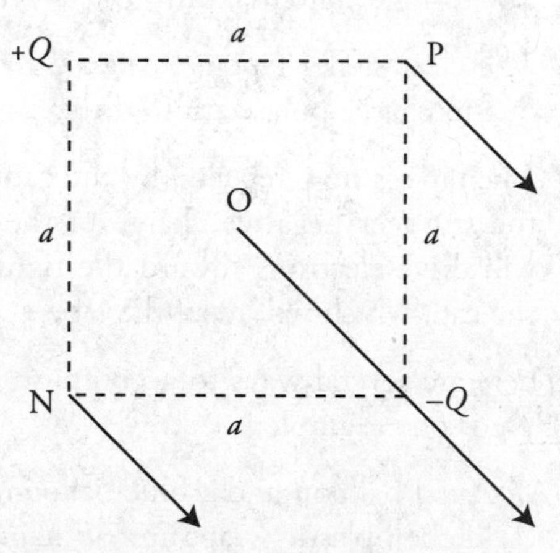

(B) Remember that electric potential is a scalar without direction. Simply sum up the individual potentials. $V_P = k\frac{(+Q)}{a} + k\frac{(-Q)}{a} = 0$. The electric potential will be zero at all three points because the charges are opposite in sign, and the radius is the same in each case.

(C) Work equals the negative change in electric potential energy. Since there is no change in electric potential moving from point P to point O, there is no change in the potential energy of the proton. Therefore, the work equals zero. In addition, the charge is moved perpendicular to the electric field; therefore, no work is done by the field.

(D) One way to accomplish this is to move $-Q$ to point P. The electric fields from the two charges will combine to create a net E-field to the right. See figure.

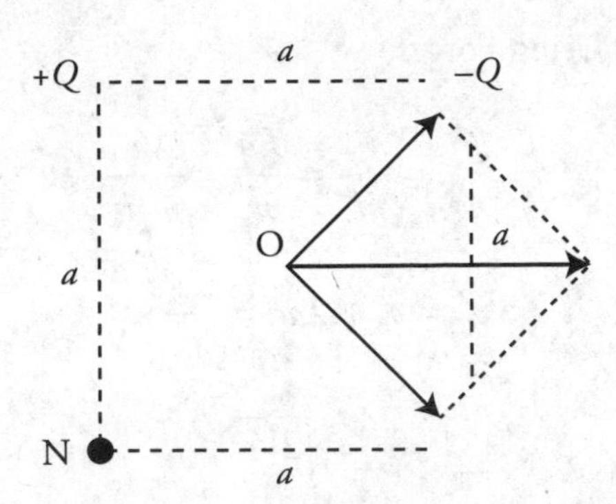

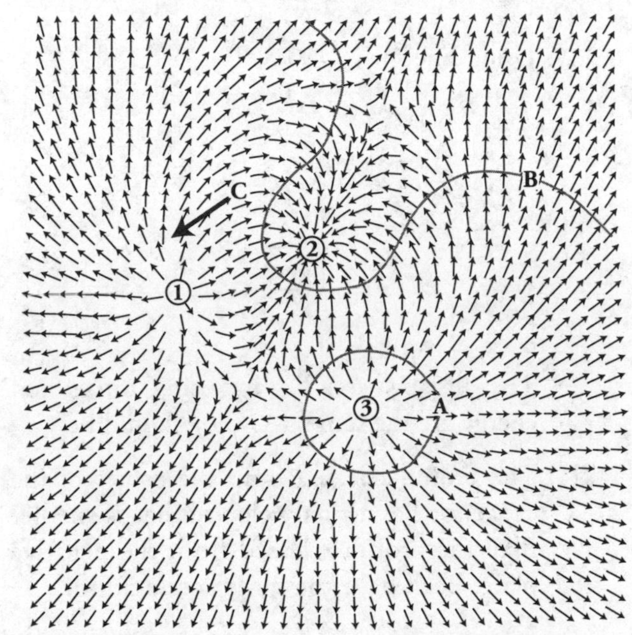

25. (A) Charges 1 and 3 are positive. Charge 2 is negative. The electric field vectors point toward negative charges and away from positive charges.

(B) The force will be in the opposite direction of the electric field. See figure.

(C) The isolines should be perpendicular to the electric field vectors. See figure for an example of the sketch.

(D) The isoline through point A will be at a higher electric potential because it is closer to the positive charge. In addition, electric field lines point toward lower electric potential.

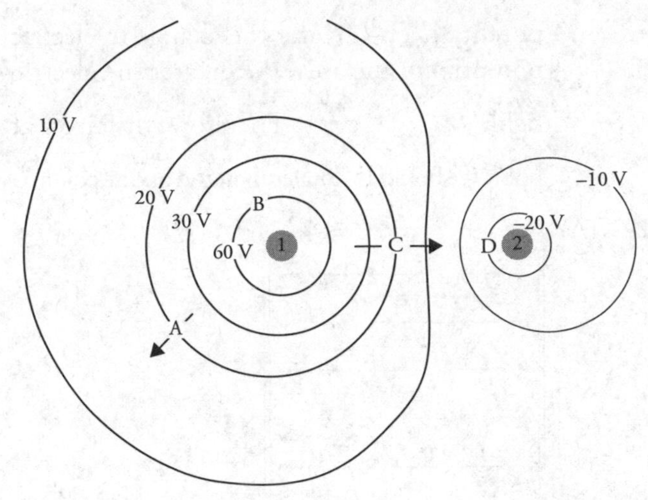

26. (A) The isolines closest to 1 are positive; therefore, 1 is positive. The isolines nearest 2 are more negative, so 2 is negative.

(B) i. Positive charges will naturally move toward more negative electric potential areas because the electric field will point in that direction. Therefore, the proton will end up 40 V lower in potential than it started; this will be the –20 V isoline.

ii. a. In the system that includes the two charges and the proton, the electric potential energy stored in the system decreases and converts to kinetic energy for the proton.

b. In the system that includes only the proton, the external electric field from the two charges does positive work on the proton, giving it kinetic energy.

(C) Decreased. $\Delta U_E = q\Delta V$ the change in electric potential is positive but the charge is negative. This gives us a negative change in electric potential energy.

(D) $E_{\text{average}} = \dfrac{\Delta V}{\Delta r} = \dfrac{\Delta V}{d}$. The electric field always points toward decreasing electric potential, which is to the right.

(E) $W = -\Delta U_E = -q\Delta V = -Q\Delta V$

Or: $W = Fd = Eqd = -\dfrac{\Delta V}{d}qd = -Q\Delta V$

(F) See the figure. The electric field vectors are always perpendicular to the isolines and point from more positive to less positive electric potential. The greater the change in electric potential in the area, the greater the electric field: $E_{average} = \frac{\Delta V}{\Delta r}$. Therefore, the arrow at point C should be longer than the one at point A.

27. (A)

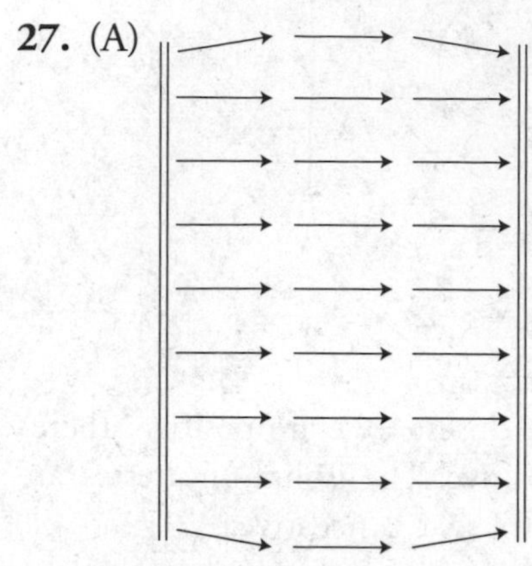

(B)

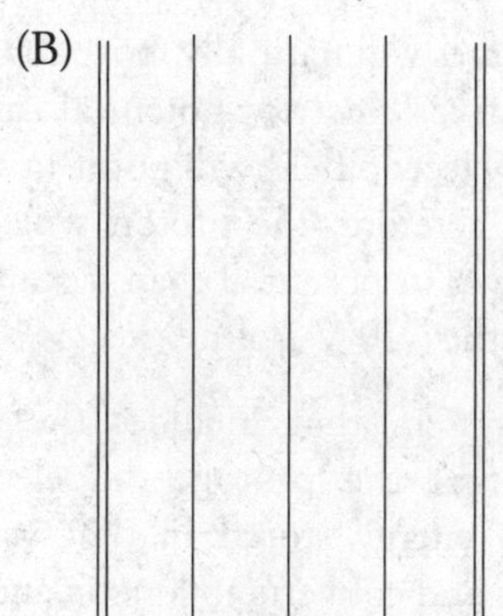

(C) $E = \frac{\Delta V}{\Delta r} = \frac{V}{d}$

(D) $Q = C\Delta V = \varepsilon_0 \frac{A}{d} V$

(E) The net charge is zero because the charge on the two plates are the same magnitude, but opposite in sign.

(F) $F = Eq = \frac{V}{d} e$. The gravitational force is usually much smaller than the electric force, and can be ignored in this case, because the magnitude of the proton mass is much smaller than the magnitude of the charge. The only time we need to worry about gravity is when the magnitude of the mass is much larger than the net charge of the object.

(G) Using Forces:

$$a = \frac{F_E}{m_p} = \frac{Eq}{m_p} = \frac{Ve}{m_p d}$$

$$v^2 = v_0^2 + 2ad = \frac{2Ve}{m_p}$$

$$v = \sqrt{\frac{2Ve}{m_p}}$$

Using Energy:

$$\Delta K = \Delta U_E$$

$$\frac{1}{2} m_p v^2 = \Delta V q = Ve$$

$$v = \sqrt{\frac{2Ve}{m_p}}$$

The answers you get by using forces and energy are the same. (As they should be!)

(H) The proton released from the center of the capacitor has a final velocity that is smaller than the proton released from the negative plate. Both have the same acceleration:

$$a = \frac{F_E}{m_p} = \frac{Eq}{m_p} = \frac{Ve}{m_p d}$$

But, the one released from the middle of the capacitor has less distance to accelerate: $v^2 = v_0^2 + 2ad$.

Or, we can just say that the proton released from the middle of the capacitor moves through a smaller electric potential, thus reducing its

final velocity: $v = \sqrt{\frac{2\Delta Ve}{m_p}}$.

28. (A) The two charges cancel each other out and the capacitance will remain the same:

$$C = \kappa\varepsilon_0 \frac{A}{d} = \kappa\varepsilon_0 \frac{2A}{2d}$$

(B) The capacitance remains the same as it is determined by the geometry of the capacitor itself:

$$C = \kappa\varepsilon_0 \frac{A}{d}$$

Since the capacitance stays the same, the charge on the plates will double: $2V = \frac{2Q}{C}$

(C) Capacitance is directly proportional to the plate area. So the graph is a line with a positive slope. See figure.

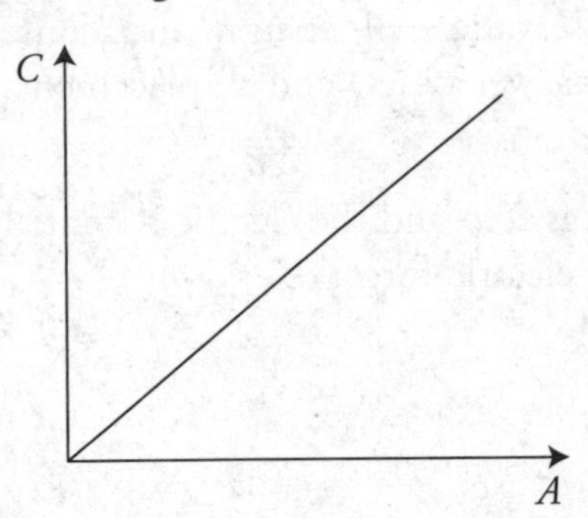

(D) Capacitance is inversely proportional to the distance between the plates. Therefore, the graph will be a hyperbola. See figure.

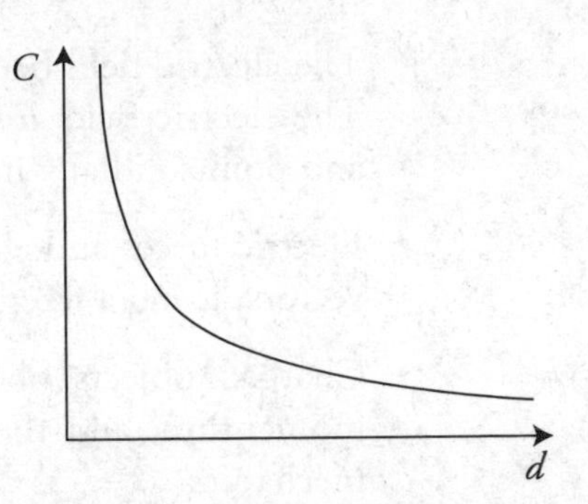

(E) We need the most stored energy in order to light the bulb for the longest possible time: $U_C = \frac{1}{2}C(\Delta V)^2$. We need the largest capacitance possible: $C = \kappa\varepsilon_0 \frac{A}{d}$. Therefore, we need the largest dielectric constant, largest plate area, and the smallest plate spacing we can get.

❯ Rapid Review

- Matter is made of protons, neutrons, and electrons. Protons are positively charged, neutrons have no charge, and electrons are negatively charged.
- Like charges repel; opposite charges attract.
- An induced charge can be created in an electrically neutral object by placing that object in an electric field.
- Objects can be permanently charged by contact, where the two objects touch and share the net charge.
- Objects can be charged by induction, where a charged object *A* is brought close to object *B* causing it to become polarized. An escape path is made such that the repelled charge is driven off of object *B*. Then the escape path is removed with the result that object *B* is permanently charged the opposite sign of *A*.
- When charge moves from object to object, the net charge of the system remains constant—conservation of charge.
- Charge is quantized, with the minimum size charge of an electron/proton: $\pm 1.6 \times 10^{-19}$ C.
- Protons are trapped on the nucleus. Electrons are easily moved from place to place.
- The electric force on an object depends on both the object's charge and the electric field it is in.
- Unless stated otherwise, the zero of electric potential is at infinity.
- Equipotential lines show all the points where a charged object would have the same electric potential. Equipotential isolines are always perpendicular to electric field vectors.

- The electric field between two charged parallel plates is constant except near the edges. The electric field around a charged particle depends on the distance from the particle and points radially inward for negative charges and outward for positive charges.
- Electric forces and electric fields are vector quantities that must be added like any other vectors, using a free-body vector diagram.
- Charged objects obey Newton's laws, conservation of energy, and conservation of momentum, and they can have accelerations, velocities, and displacements just as in mechanics.
- The electric field inside a charged conductor is zero and the electric potential inside is a constant. Any charge inside will not feel any electric force.

Electric Circuits

IN THIS CHAPTER

Summary: Electric charge flowing through a wire is called current. An electrical circuit is built to control the current. In this chapter, you will learn how to predict the effects of current flow. Besides discussing circuits in general, this chapter presents a problem-solving technique: the *V-I-R* chart, an incredibly effective way to organize a problem that involves circuits.

Key Ideas

- The physics concepts of conservation of charge and conservation of energy, as expressed in Kirchhoff's rules, explain the behavior of circuits.
- Current is the flow of charge through the circuit.
- A voltage source, such as batteries, generators, and solar cells, supply the energy required to create current.
- Resistance is a restriction to the flow of current.
- Resistance depends on the geometry of the resistor.
- Not all resistors and circuit elements are ohmic.
- Real batteries are not perfect and have internal resistance.
- The current through series resistors is the same through each, whereas the voltage across series resistors adds to the total voltage.
- The voltage across parallel resistors is the same across each, whereas the current through parallel resistors adds to the total current.
- The brightness of a lightbulb depends on the power dissipated by the bulb.

- A capacitor in a circuit blocks current and stores charge.
- Changing the components in a circuit, like opening or closing a switch, has effects on the circuit that can be predicted.

Relevant Equations

Definition of current: $I = \frac{\Delta Q}{\Delta t}$

Resistance of a wire in terms of its properties: $R = \rho\frac{L}{A}$

Ohm's law: $I = \frac{\Delta V}{R}$

Power in a circuit: $P = I\Delta V$

Equivalent parallel resistance: $\frac{1}{R_p} = \Sigma_i\frac{1}{R_i} = \frac{1}{R_1} + \frac{1}{R_2} + \frac{1}{R_3}$

Equivalent series resistance: $R_s = \Sigma_i R_i = R_1 + R_2 + R_3$

Capacitance of a capacitor based on its properties: $C = \kappa\varepsilon_0\frac{A}{d}$

Electric field between the plates of a capacitor: $E = \frac{Q}{\varepsilon_0 A}$

Charge stored on a capacitor: $\Delta V = \frac{Q}{C}$

Energy stored in a charged capacitor: $U_C = \frac{1}{2}Q\Delta V = \frac{1}{2}C(\Delta V)^2$

Equivalent series capacitance: $\frac{1}{C_s} = \Sigma_i\frac{1}{C_i} = \frac{1}{C_1} + \frac{1}{C_2} + \frac{1}{C_3}$

Equivalent parallel capacitance: $C_p = \Sigma_i C_i = C_1 + C_2 + C_3$

Current

The last chapter talked about situations where electric charges don't move around very much. Isolated point charges, for example, just sit there creating an electric field. But what happens when you get a lot of charges all moving together? That, at its essence, is what goes on in a circuit.

A circuit is simply any path that will allow charge to flow.

Current: The flow of positive electric charge. In a circuit, the current is the amount of charge passing a given point per unit time.

A current is defined as the flow of positive charge. We don't think this makes sense, because electrons—and not protons or positrons—are what flow in a circuit. But physicists have their rationale, and no matter how wacky, we won't argue with it—the AP exam always uses this definition.

In more mathematical terms, current is defined as follows:

$$I = \frac{\Delta Q}{\Delta t}$$

What this means is that the current, I, equals the amount of charge flowing past a certain point divided by the time interval during which you're making your measurement. This definition tells us that current is measured in coulombs/second. 1 C/s = 1 ampere, abbreviated as 1 A.

Resistance and Ohm's Law

You've probably noticed that just about every circuit drawn in your physics book contains a battery. The reason most circuits contain a battery is because batteries create a potential difference between one end of the circuit and the other. In other words, if you connect the terminals of a battery with a wire, the part of the wire attached to the "+" terminal will have a higher electric potential than the part of the wire attached to the "–" terminal. And positive charge flows from high potential to low potential. So, in order to create a current, you need a battery.

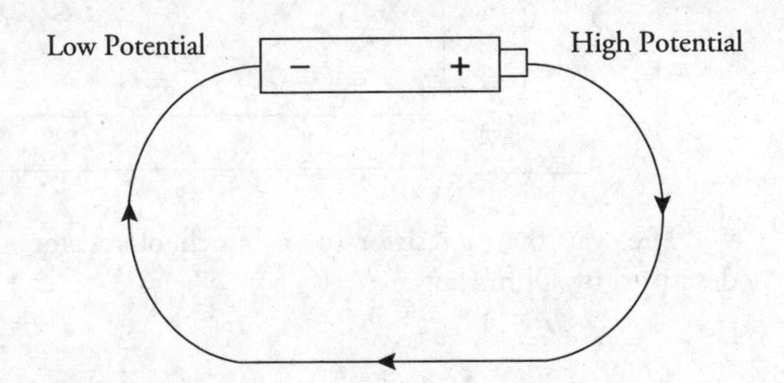

In general, the greater the potential difference between the terminals of the battery, the more current flows.

The amount of current that flows in a circuit is also determined by the resistance of the circuit.

Resistance: A property of a circuit that resists the flow of current.

Resistance is measured in ohms. One ohm is abbreviated as 1 Ω.

If we have some length of wire, then the resistance of that wire can be calculated. Three physical properties of the wire affect its resistance:

- The material the wire is made out of: the *resistivity*, ρ, of a material is an intrinsic property of that material. Good conducting materials, like gold, have low resistivities.[1]
- The length of the wire, L: the longer the wire, the more resistance it has.
- The cross-sectional area A of the wire: the wider the wire, the less resistance it has.

[1]Resistivity would be given on the AP exam if you need a value. Nothing here to memorize.

We put all of these properties together in the equation for resistance of a wire:

$$R = \rho\frac{L}{A}$$

Now, this equation is useful only when you need to calculate the resistance of a wire from scratch. Usually, on the AP exam or in the laboratory, you will be using resistors that have a pre-measured resistance.

Resistor: Something you put in a circuit to change the circuit's resistance.

Resistors are typically ceramic, a material that doesn't allow current to flow through it very easily. Another common type of resistor is the filament in a light bulb. When current flows into a light bulb, it gets held up in the filament. While it's hanging out in the filament, it makes the filament extremely hot, and the filament gives off light.

To understand resistance, an analogy is helpful. A circuit is like a network of pipes. The current is like the water that flows through the pipes, and the battery is like the pump that keeps the water flowing. If you wanted to impede the flow, you would add some narrow sections to your network of pipes. These narrow sections are your resistors.

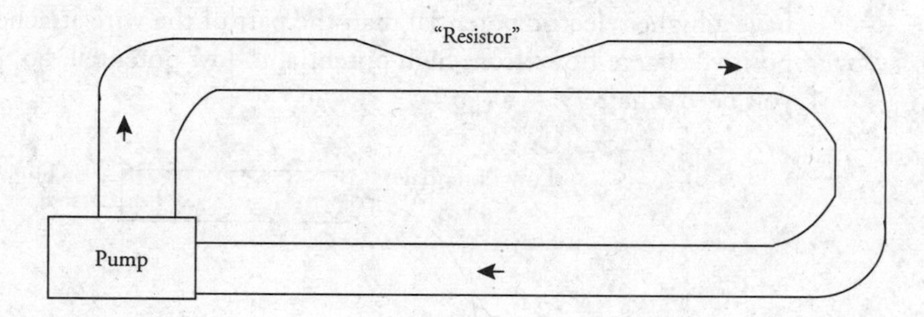

The way that a resistor (or a bunch of resistors) affects the current in a circuit is described by Ohm's law.

$$I = \frac{\Delta V}{R}$$

ΔV is the voltage across the part of the circuit you're looking at, I is the current flowing through that part of the circuit, and R is the resistance in that part of the circuit. Ohm's law is the most important equation when it comes to circuits, so make sure you know it well.

When current flows through a resistor, electrical energy is being converted into heat energy. The rate at which this conversion occurs is called the power dissipated by a resistor. This power can be found with the equation

$$P = I\Delta V$$

This equation says that the power, P, dissipated in part of a circuit, equals the current flowing through that part of the circuit multiplied by the voltage across that part of the circuit.

Using Ohm's law, it can be easily shown that $I\Delta V = I^2R = \frac{\Delta V^2}{R}$. It's only worth memorizing the first form of the equation, but any one of these could be useful.

Ohmic Versus Nonohmic

A circuit component that maintains the same resistance even when the voltage across it, or the current through it, are changed is said to be "ohmic." Ohmic simply means that the resistance is constant. The exam writers like to ask questions about this. Look at this table of data showing the voltage and current through two resistors. Are the resistors ohmic?

Voltage (ΔV)	Resistor #1: Current (mA)	Resistor #2: Current (mA)
1	100	50
2	190	100
3	270	150
4	330	200
5	360	250

An easy way to check is to rearrange Ohm's law to solve for resistance and calculate the resistance $R = \frac{\Delta V}{I}$.

Resistor #1:

First data point: $R = \frac{\Delta V}{I} = \frac{1\text{V}}{0.1\text{ A}} = 10\ \Omega$

Second data point: $R = \frac{\Delta V}{I} = \frac{2\text{V}}{0.19\text{ A}} = 10.5\ \Omega$. Close . . . let's check another point.

Third data point: $R = \frac{\Delta V}{I} = \frac{3\text{V}}{0.27\text{ A}} = 11.1\ \Omega$. Stop!

Resistor #1 is nonohmic, because the resistance does not stay constant.

When we repeat the same calculation for Resistor #2, we find that the resistance stays constant at 20 Ω. Resistor #2 is ohmic.

(a) Ohmic Material: Copper Wire

The current is directly proportional to the potential difference.

I

The resistance is $R = \frac{1}{\text{slope}}$

ΔV

(b) Nonohmic Material: Diode (Semiconductor Device)

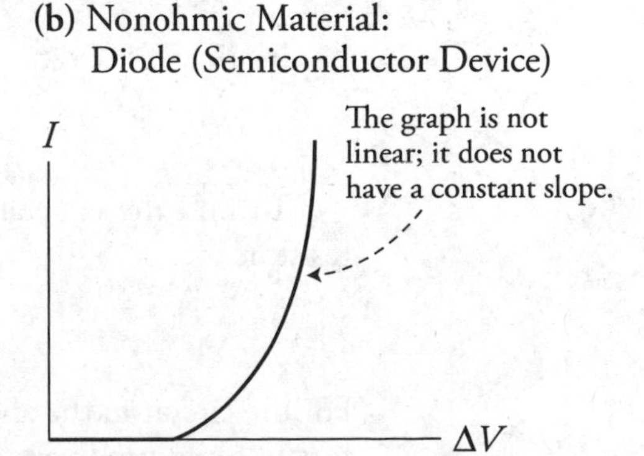

Another quick way to make this ohmic check is to graph the data. In a lab, you could vary the voltage applied to the resistor and measure the current. This would make ΔV the independent variable and I the dependent variable. Rearranging Ohm's law we would get: $I = \frac{1}{R}(\Delta V)$, which means that plotting ΔV on the x-axis and I on the y-axis will result in a slope of $1/R$. The slope will be constant (straight line) if the resistor is ohmic. In the above graph (a) is an ohmic material and graph (b) is a nonohmic material.

Resistors in Series and in Parallel

In a circuit, resistors can be arranged either in series with one another or parallel to one another. Before we take a look at each type of arrangement, though, we need first to familiarize ourselves with circuit symbols, shown next.

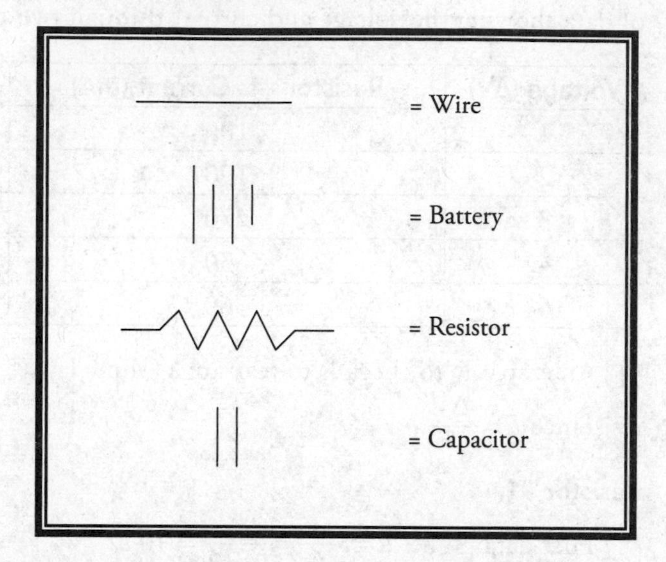

First, let's examine resistors in series. In this case, all the resistors are connected in a line, one after the other after the other. In other words, there is only one pathway for the current to travel through.

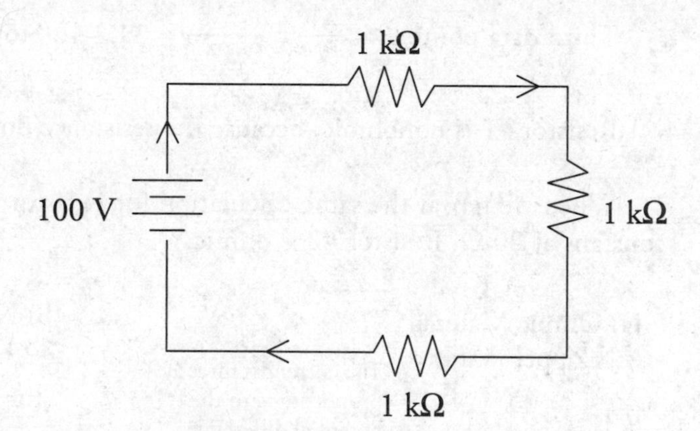

To find the equivalent resistance of series resistors, we just add up all the individual resistors.

$$R_s = \sum_i R_i$$

For the circuit in the above figure, $R_{eq} = 3000\ \Omega$. In other words, using three 1000-Ω resistors in series produces the same total resistance as using one 3000-Ω resistor.

Parallel resistors are connected in such a way that you create several paths through which current can flow. For the resistors to be truly in parallel, the current must split, go through only one resistor in each pathway, and then immediately come back together.

The equivalent resistance of parallel resistors is found by this formula:

$$\frac{1}{R_p} = \sum_i \frac{1}{R_i}$$

For the circuit in the next figure, the equivalent resistance is 333 Ω. So hooking up three 1000-Ω resistors in parallel produces the same total resistance as using one 333-Ω resistor. (Note that the equivalent resistance of parallel resistors is *less than* any individual resistor in the parallel combination.)

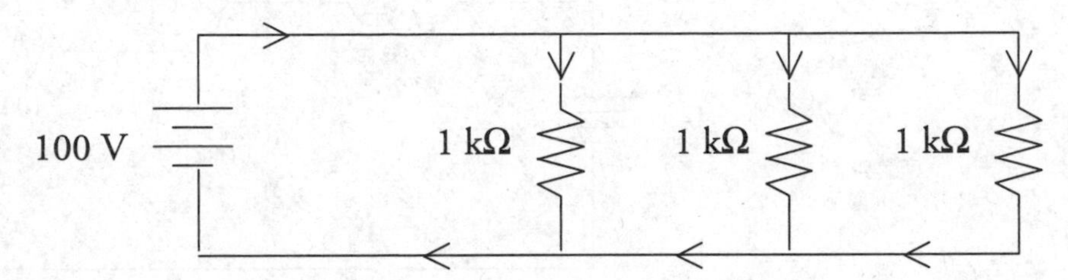

A Couple of Important Rules

Rule #1—When two resistors are connected in SERIES, the amount of current that flows through one resistor equals the amount of current that flows through the other resistor and is equal to the total current through both resistors.

Rule #2—When two resistors are connected in PARALLEL, the voltage across one resistor is the same as the voltage across the other resistor and is equal to the total voltage across both resistors.

The reason why the current is the same for everything connected in a series is because of conservation of charge. Remember that charge is carried by real things, protons and electrons, and they won't just disappear or be created in a circuit. Since there are no branches in a series pathway, all the charge that enters this single pathway must pass through every component in that pathway.

The reason why voltage is the same for resistors connected in parallel is due to conservation of energy. The voltage (energy per charge) used up in each parallel pathway must be the same. Think of this analogy: water in a river splits along parallel paths to flow around a large island in the middle of the stream. One path may go through rapids with rocks and drop off a waterfall, while the other path around the island may be calm and smooth. Even though the paths are different, the two paths begin and end together at the same height (the same gravitational potential). The same is true for current flowing through parallel paths in a circuit. No matter how different the paths may be, the current must begin and end at the same electric potential. Thus ΔV for parallel paths is always the same.

We will discuss conservation of charge and conservation of energy more a bit later in the section titled Kirchhoff's Rules. For now, let's discover a very useful tool: *V-I-R* charts.

The *V-I-R* Chart

Here it is—the trick that will make solving circuits a breeze. Use this method on your homework. Use this method on your quizzes and tests. But most of all, use this method on the AP exam. It works.

The easiest way to understand the *V-I-R* chart is to see it in action, so we'll go through a problem together, filling in the chart at each step along the way.

Find the voltage across each resistor in the circuit shown below.

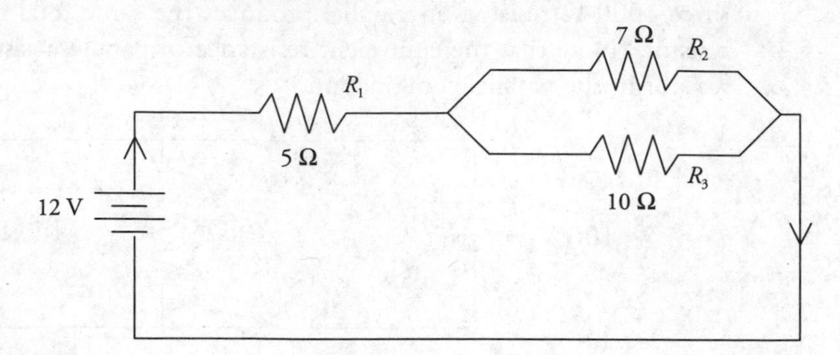

We start by drawing our *V-I-R* chart, and we fill in the known values. Right now, we know the resistance of each resistor, and we know the total voltage (it's written next to the battery).

	V	*I*	*R*
R_1			5 Ω
R_2			7 Ω
R_3			10 Ω
Total	12 V		

Next, we simplify the circuit. This means that we calculate the equivalent resistance and redraw the circuit accordingly. We'll first find the equivalent resistance of the parallel part of the circuit:

$$\frac{1}{R_{eq}} = \frac{1}{7\ \Omega} + \frac{1}{10\ \Omega}$$

Use your calculator to get $\frac{1}{R_{eq}} = 0.24\ \Omega$.

Taking the reciprocal, we get

$$R_{eq} = 4.1\ \Omega$$

So we can redraw our circuit like this:

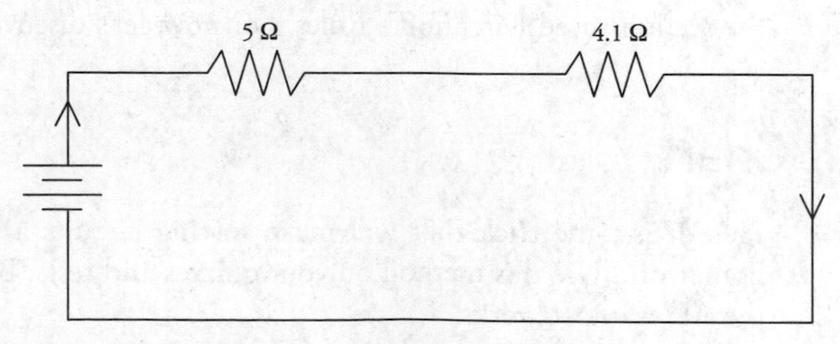

Next, we calculate the equivalent resistance of the entire circuit. Following our rule for resistors in series, we have

$$R_{eq} = 4.1\ \Omega + 5\ \Omega = 9.1\ \Omega$$

We can now fill this value into the *V-I-R* chart.

	V	I	R
R_1			5 Ω
R_2			7 Ω
R_3			10 Ω
Total	12 V		9.1 Ω

Notice that we now have two of the three values in the "Total" row. Using Ohm's law, we can calculate the third. That's the beauty of the *V-I-R* chart: *Ohm's law is valid whenever two of the three entries in a row are known.*

Then we need to put on our thinking caps. We know that all the current that flows through our circuit will also flow through R_1. (You may want to take a look back at the original drawing of our circuit to make sure you understand why this is so.) Therefore, the I value in the "R_1" row will be the same as the I in the "Total" row. We now have two of the three values in the "R_1" row, so we can solve for the third using Ohm's law.

	V	I	R
R_1	6.5 V	1.3 A	5 Ω
R_2			7 Ω
R_3			10 Ω
Total	12 V	1.3 A	9.1 Ω

Finally, we know that the voltage across R_2 equals the voltage across R_3, because these resistors are connected in parallel. The total voltage across the circuit is 12 V, and the voltage across R_1 is 6.5 V. So the voltage that occurs between R_1 and the end of the circuit is

$$12\text{ V} - 6.5\text{ V} = 5.5\text{ V}$$

Therefore, the voltage across R_2, which is the same as the voltage across R_3, is 5.5 V. We can fill this value into our table. Finally, we can use Ohm's law to calculate I for both R_2 and R_3. The finished *V-I-R* chart looks like this:

	V	I	R
R_1	6.5 V	1.3 A	5 Ω
R_2	5.5 V	0.79 A	7 Ω
R_3	5.5 V	0.55 A	10 Ω
Total	12 V	1.3 A	9.1 Ω

To answer the original question, which asked for the voltage across each resistor, we just read the values straight from the chart.

Now, you might be saying to yourself, "This seems like an awful lot of work to solve a relatively simple problem." You're right—it is.

However, there are several advantages to the *V-I-R* chart. The major advantage is that, by using it, you force yourself to approach every circuit problem exactly the same way.

So when you're under pressure—as you will be during the AP exam—you'll have a tried-and-true method to turn to.

Also, if there are a whole bunch of resistors, you'll find that the *V-I-R* chart is a great way to organize all your calculations. That way, if you want to check your work, it'll be very easy to do.

Tips for Solving Circuit Problems Using the *V-I-R* Chart

- First, enter all the given information into your chart. If resistors haven't already been given names (like "R_1"), you should name them for easy reference.
- Next, simplify the circuit to calculate R_{eq}, if possible.
- Once you have two values in a row, you can calculate the third using Ohm's law. *You CANNOT use Ohm's law unless you have two of the three values in a row.*
- Remember that if two resistors are in series, the current through one of them equals the current through the other. And if two resistors are in parallel, the voltage across one equals the voltage across the other.

Kirchhoff's Rules

Kirchhoff's rules are expressions of conservation of charge and conservation of energy. They are useful in any circuit but especially in complicated circuits such as circuits with multiple resistors and batteries. Kirchhoff's rules say:

1. Positive (+) currents entering a junction plus the negative (−) currents leaving a junction equal 0 ($\Sigma I_{junction} = 0$). Or, at any junction the current entering must equal the current leaving.
2. The sum of the voltages around a closed loop is zero ($\Sigma \Delta V_{loop} = 0$).

The first law is called the "junction rule," and the second is called the "loop rule." To illustrate the junction rule, we'll revisit the circuit from our first problem.

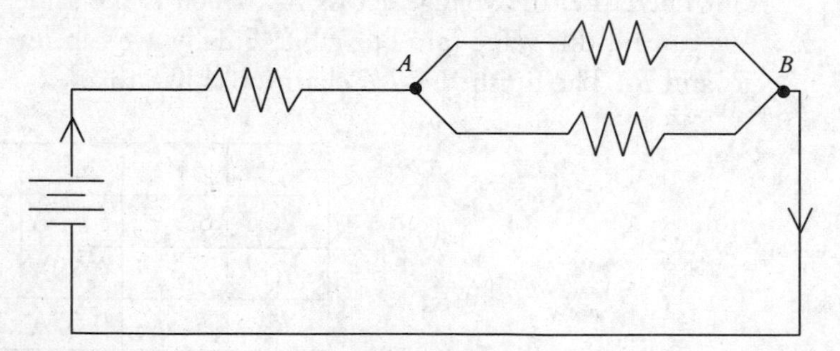

According to the junction rule, whatever current enters junction *A* must also leave junction *A*. So let's say that a current of 1.3 A enters junction *A* from the left, and then that current gets split between the two branches. If we measured the current in the top branch and the current in the bottom branch, we would find that the two currents still add up to a total of 1.3 A. And, in fact, when the two branches came back together at junction *B*, we would find that exactly 1.3 A was flowing out through junction *B* and through the rest of the circuit.

Kirchhoff's junction rule says that charge is conserved: you don't lose any current when the wire bends or branches. This seems remarkably obvious, but it's also remarkably essential to solving circuit problems.

Kirchhoff's loop rule is a bit less self-evident, but it's quite useful in sorting out difficult circuits.

As an example, I'll show you how to use Kirchhoff's loop rule to find the current through all the resistors in the circuit below.

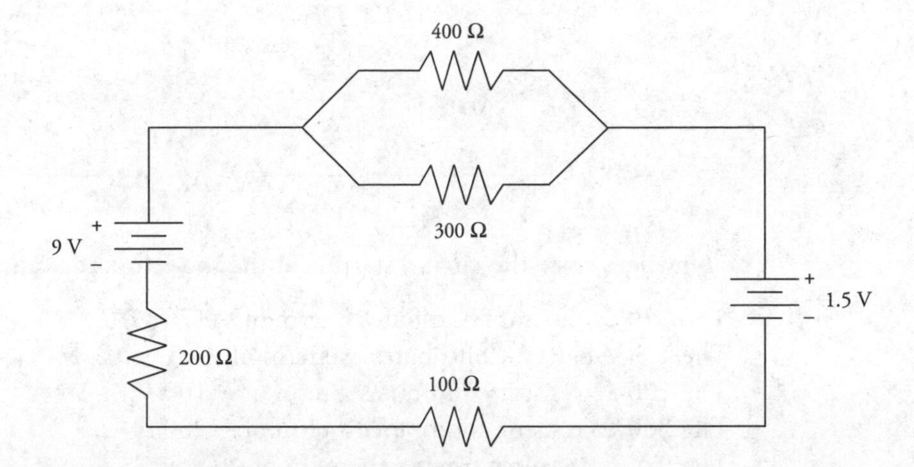

We will follow the steps for using Kirchhoff's loop rule:

- Arbitrarily choose a direction of current. Draw arrows on your circuit to indicate this direction.
- Follow the loop in the direction you chose. When you cross a resistor, the voltage is $-IR$, where R is the resistance, and I is the current flowing through the resistor. This is just an application of Ohm's law. (If you have to follow a loop *against* the current, though, the voltage across a resistor is written $+IR$.)
- When you cross a battery, if you trace from the – to the +, add the voltage of the battery; subtract the battery's voltage if you trace from + to –.
- Set the sum of your voltages equal to 0. Solve. If the current you calculate is negative, then the direction you chose was wrong—the current actually flows in the direction opposite to your arrows.

In the case of the following figure, we'll start by collapsing the two parallel resistors into a single equivalent resistor of 170 Ω. You don't *have* to do this, but it makes the mathematics much simpler.

Next, we'll choose a direction of current flow. But which way? In this particular case, you can probably guess that the 9 V battery will dominate the 1.5 V battery, and thus the current will be clockwise. But even if you aren't sure, just choose a direction and stick with it—if you get a negative current, you chose the wrong direction.

Here is the circuit redrawn with the parallel resistors collapsed and the assumed direction of current shown. Because there's now only one path for current to flow through, we have labeled that current I.

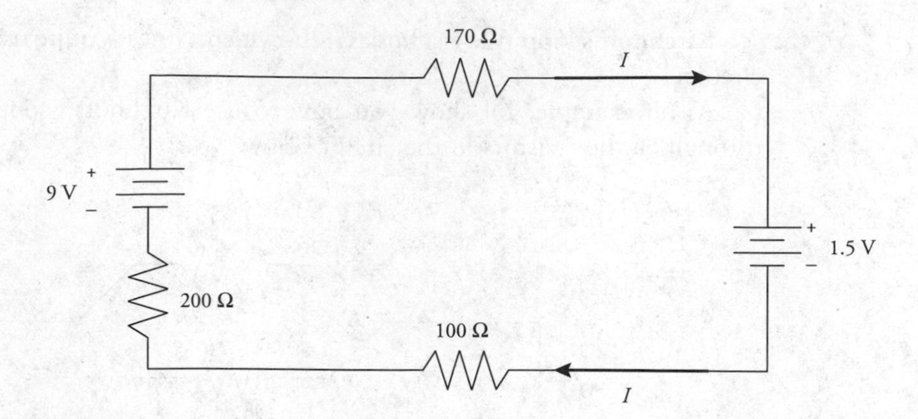

Now let's trace the circuit, starting at the top-left corner and working clockwise:

- The 170-Ω resistor contributes a term of −(170 Ω) I.
- The 1.5-V battery contributes the term of −1.5 volts.
- The 100-Ω resistor contributes a term of −(100 Ω) I.
- The 200-Ω resistor contributes a term of −(200 Ω) I.
- The 9-V battery contributes the term of +9 volts.

Combine all the individual terms, and set the result equal to zero. The units of each term are volts, but units are left off below for algebraic clarity:

$$0 = (-170)I + (-1.5) + (-100)I + (-200)I + (+9)$$

By solving for I, the current in the circuit is found to be 0.016 A; that is, 16 milliamps, a typical laboratory current.

The problem is not yet completely solved, though—16 milliamps go through the 100-Ω and 200-Ω resistors, but what about the 300-Ω and 400-Ω resistors? We can find that the voltage across the 170-Ω equivalent resistance is (0.016 A)(170 Ω) = 2.7 V. Because the voltage across parallel resistors is the same for each, the current through each is just 2.7 V divided by the resistance of the actual resistor: 2.7 V/300 Ω = 9 mA, and 2.7 V/400 Ω = 7 mA. Problem solved!

Oh, and you might notice that the 9 mA and 7 mA through each of the parallel branches adds to the total of 16 mA—as required by Kirchhoff's junction rule.

Since the AP exam has a lot of problems without numbers, let's look at Kirchhoff's rules symbolically. Look at the circuit in the next diagram. The battery has a voltage of ε; the resistors R_1, R_2, R_3, and R_4 are all the same size; and the currents are marked as I_1, I_2, and I_3.

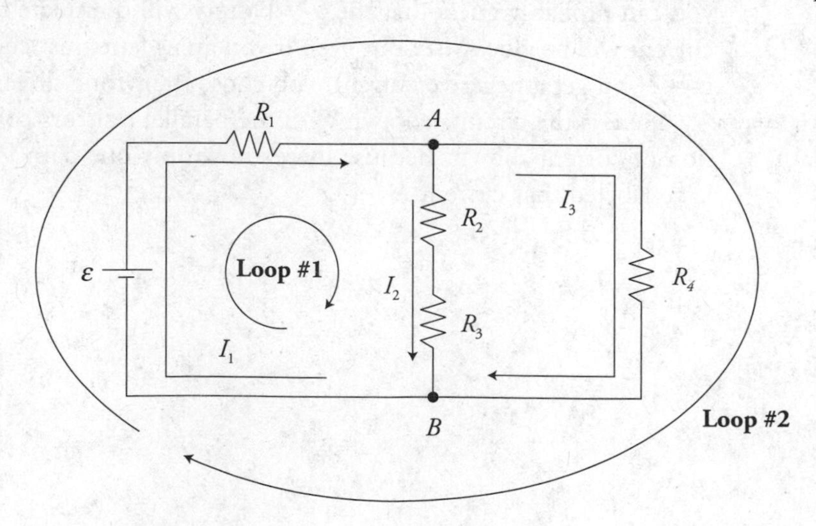

Let's apply the junction rule to junction A:

$$\Sigma I_{junction\ A} = I_1 - I_2 - I_3 = 0 \quad \text{or} \quad I_1 = I_2 + I_3$$

That makes sense. The current into the junction equals the current exiting the junction. We get the same equation for junction B.

Now let's work with the harder loop rule. We have several loops to choose from, but I've already marked loop #1 on the left and loop #2 that goes around the outside. Remember the procedures we have already learned!

$$\text{Loop \#1: } \Sigma\Delta V_{loop} = \varepsilon - I_1R_1 - I_2R_2 - I_2R_3 = 0 \text{ (Be careful of your subscripts!)}$$

If we rewrite the equation, we can see that the voltage supplied by the battery is completely consumed by the three resistors: $\varepsilon = I_1R_1 + I_2R_2 + I_2R_3$. This is exactly what conservation of energy says we should get. The voltage supplied to the loop is consumed by the components in the loop.

We can also make our math teacher proud and factor out I_2: $\varepsilon = I_1R_1 + I_2(R_2 + R_3)$. That tells us that the resistors $R_2 + R_3$ add in series, but you already saw that didn't you?

Now take a look at:

$$\text{Loop \#2: } \Sigma\Delta V_{loop} = \varepsilon - I_1R_1 - I_3R_4 = 0$$

Solving for ε:

$$\varepsilon = I_1R_1 + I_3R_4$$

Compare our two equations:

$$\varepsilon = I_1R_1 + I_2(R_2 + R_3) \text{ and } \varepsilon = I_1R_1 + I_3R_4$$

They both equal ε, which means we can set them equal to each other:

$$I_1R_1 + I_2(R_2 + R_3) = I_1R_1 + I_3R_4$$

Simplifying by canceling out the I_1R_1 term, we get:

$$I_2(R_2 + R_3) = I_3R_4$$

Remember that $\Delta V = IR$. This equation tells us that the voltage drop across R_2 and R_3 must equal the voltage drop across R_4. But we already knew that because the two combined resistors R_2 and R_3 are in parallel with R_4. That is the parallel rule!

Practice

Now for a good AP exam question:

Using this information and the previous figure:

(A) Rank the currents I_1, I_2, and I_3 from greatest to least.
(B) Rank the voltage drops across the resistors R_1, R_2, R_3, and R_4 from greatest to least.

(A) Think about it. I_1 has to be the greatest. I_2 and I_3 split off from I_1. But, there is more resistance in the I_2 path, so it must be the least.

Answer: $I_1 > I_3 > I_2$

(B) Remember that all the resistors are the same size. Since $\Delta V = IR$, the voltage drop will depend on the currents. More current = larger ΔV. (Notice that the current through resistors 2 and 3 must be the same because they are in the same pathway.)

Answer: $\Delta V_{R_1} > \Delta V_{R_4} > \Delta V_{R_2} = \Delta V_{R_3}$

Get comfortable using Kirchhoff's rules because it is common for the AP exam to ask you to write them and use them to justify a statement about the circuit.

Circuits from an Experimental Point of View

When a real circuit is set up in the laboratory, it usually consists of more than just resistors—lightbulbs and motors are common devices to hook to a battery, for example. For the purposes of computation, though, we can consider pretty much any electronic device to act like a resistor.

But what if your purpose is *not* computation? Often on the AP exam, as in the laboratory, you are asked about observational and measurable effects. A common question involves the brightness of lightbulbs and the measurement (not just computation) of current and voltage.

Brightness of a Bulb

The brightness of a bulb depends solely on the power dissipated by the bulb. (Remember, power is given by any of the equations I^2R, $I\Delta V$, or $(\Delta V)^2/R$.) You can remember that from your own experience—when you go to the store to buy a lightbulb, you don't ask for a "400-ohm" bulb, but for a "100-watt" bulb. And a 100-watt bulb is brighter than a 25-watt bulb. But be careful—a bulb's power can change depending on the current and voltage it's hooked up to. Consider this problem.

A lightbulb is rated at 100 W in the United States, where the standard wall outlet voltage is 120 V. If this bulb were plugged in in Europe, where the standard wall outlet voltage is 240 V, which of the following would be true?

(A) The bulb would be one-quarter as bright.
(B) The bulb would be one-half as bright.
(C) The bulb's brightness would be the same.
(D) The bulb would be twice as bright.
(E) The bulb would be four times as bright.

Your first instinct might be to say that because brightness depends on power, the bulb is exactly as bright. But that's not right! The power of a bulb can change.

The resistance of a lightbulb is a property of the bulb itself, and so will not change no matter what the bulb is hooked up to.

Since the resistance of the bulb stays the same while the voltage changes, by V^2/R, the power goes up, and the bulb will be brighter. How much brighter? Since the voltage in Europe is doubled, and because voltage is squared in the equation, the power is multiplied by 4—choice E.

Ammeters and Voltmeters

Ammeters measure current, and voltmeters measure voltage. This is pretty obvious, because current is measured in amps, voltage in volts. It is *not* necessarily obvious, though, how to connect these meters into a circuit.

Remind yourself of the properties of series and parallel resistors—voltage is the same for any resistors in parallel with each other. So if you're going to measure the voltage across a resistor, you must put the voltmeter in *parallel* with the resistor. In the next figure, the meter labeled V_2 measures the voltage across the 100-Ω resistor, while the meter labeled V_1 measures the potential difference between points A and B (which is also the voltage across R_1).

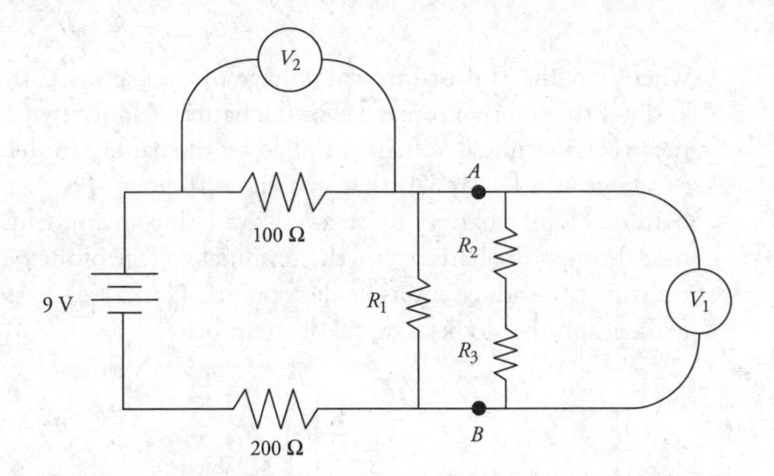

Current is the same for any resistors in *series* with one another. So, if you're going to measure the current through a resistor, the ammeter must be in series with that resistor. In the following figure, ammeter A_1 measures the current through resistor R_1, while ammeter A_2 measures the current through resistor R_2.

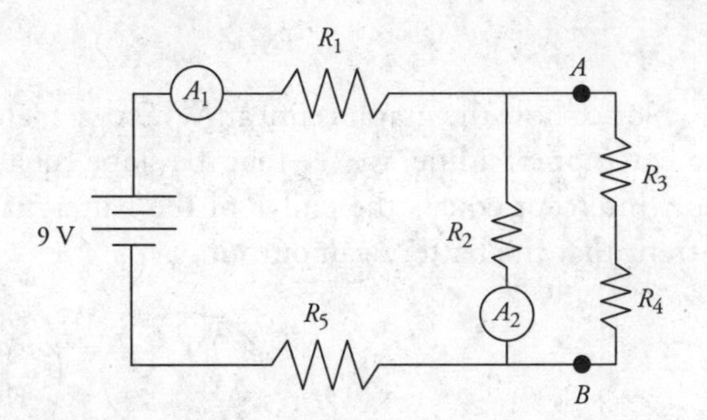

As an exercise, ask yourself, is there a way to figure out the current in the other three resistors based only on the readings in these two ammeters? Answer is in the footnote.[2]

Real Batteries and Internal Resistance

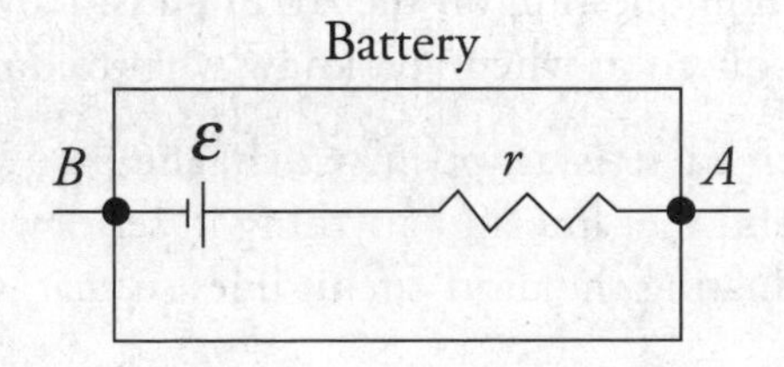

The batteries you use in your calculator or in the lab are not perfect. They have some internal resistance. This means that the voltage posted on the battery, the emf "ε," is not what actually comes out of the battery, because some of the voltage is lost before it ever gets out!

[2]The current through R_5 must be the same as through R_1, because both resistors carry whatever current came directly from the battery. The current through R_3 and R_4 can be determined from Kirchhoff's junction rule: subtract the current in R_2 from the current in R_1 and that's what's left over for the right-hand branch of the circuit.

The real voltage supplied by the battery is called the terminal voltage, $\Delta V_{terminal}$. Using the ideas we learned from Kirchhoff's rules, we can see that:

$$\Delta V_{terminal} = \varepsilon - Ir$$

where ε is the emf or internal voltage of the battery, I is the current through the battery, and r is the internal resistance of the battery. Notice that more current through the battery means less terminal voltage supplied by the battery to the external circuit.

Here is a handy lab that you should know—how to find the internal resistance of a battery. Hook a battery up to a resistor. Using an ammeter and voltmeter, measure the current through the battery and the terminal voltage of the battery. Repeat for several different resistors to produce multiple data points. Plot $\Delta V_{terminal}$ as a function of current I and you get a graph that looks like the diagram below.

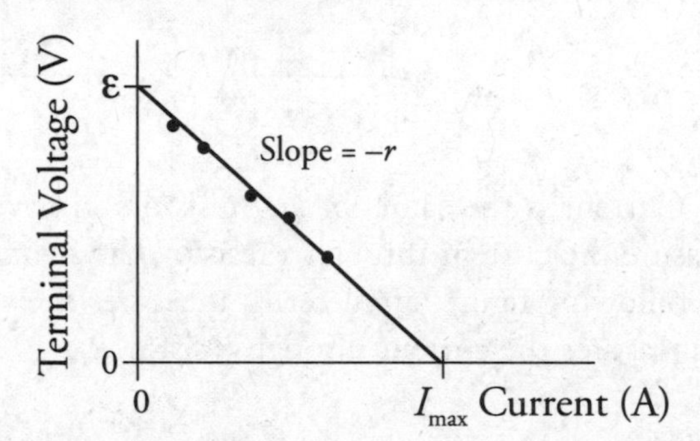

Notice how the graph is linear. When we match up the terminal voltage equations with the equation of a line, we see that the slope equals the negative of the internal resistance r, the y-intercept equals the emf ε of the battery, and the x-intercept will be the maximum current that the battery can output.

$$y = m\,x + b$$
$$\Delta V_{Terminal} = -r\,I + \varepsilon$$

Hint: This is one of those labs you might need to know about for the AP exam.

Changes in a Circuit—Switches

A common question on the AP exam is: How does closing a switch affect a circuit? This is an easy question when you know what to do.

- When there is an open switch, the part of the circuit that the switch is in goes dead. Pretend that line and anything in series with this line does not exist. Redraw the circuit eliminating the dead circuit lines so that you don't get confused.

Let's take a look at a circuit we used earlier, this time with a switch and let all the resistors be the same: $R_1 = R_2 = R_3 = R_4 = R$.

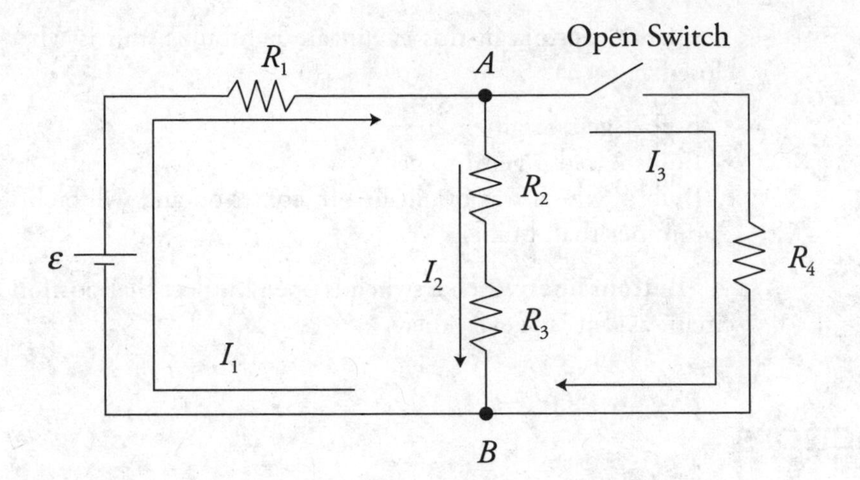

The circuit has an open switch. Therefore, $I_3 = 0$ and resistor R_4 receives no current. It is as if it isn't even there. So, what does that tell us about the circuit? With the switch open, all we have is a simple series circuit with three resistors where $I_1 = I_2$:

$$I_1 = I_2 = \frac{\Delta V}{R_{total}} = \frac{\varepsilon}{R_1 + R_2 + R_3} = \frac{\varepsilon}{3R}$$

Now close the switch. What happens to I_1 and I_2? With the switch closed, we now have a combination circuit with R_4 in parallel with R_2 and R_3. Remember that when resistors are added in parallel, the total resistance of the circuit is lowered. The resistance of the parallel set between points A and B is:

$$\frac{1}{R_p} = \frac{1}{R_2 + R_3} + \frac{1}{R_4} = \frac{1}{2R} + \frac{1}{R} = \frac{3}{2R}$$

Thus,

$$R_p = \frac{2R}{3}$$

The total resistance of the circuit is:

$$R_{total} = R_1 + \frac{2R}{3} = \frac{3}{3}R + \frac{2R}{3} = \frac{5R}{3}$$

To find the total current (I_1):

$$I_{total} = I_1 = \frac{\Delta V}{R_{total}} = \frac{\varepsilon}{\frac{5R}{3}} = \frac{3\varepsilon}{5R}$$

So the current I_1 has gone up from $\frac{\varepsilon}{3R}$ to $\frac{3\varepsilon}{5R}$.

What about I_2? Well, the current I_1 will split between I_2 and I_3, with I_2 only getting one-third of the current because it has twice the resistance in its pathway. One-third of $\frac{3\varepsilon}{5R}$ is $I_2 = \frac{\varepsilon}{5R}$, which is less than the original current of $\frac{\varepsilon}{3R}$ before the switch was closed.

If the resistors in this circuit are lightbulbs, this is what happens when the switch is closed:

- Bulb 1 gets brighter.
- Bulbs 2 and 3 get dimmer.
- Bulb 4, which was originally off, comes on and will be brighter than bulbs 2 and 3 but dimmer than bulb 1.

Bottom line: When a switch is open, neglect that portion of the circuit and analyze the circuit as if it isn't even there.

Capacitors

Capacitors were introduced in Chapter 11. Now let's look at them in more detail. Capacitors are really very simple devices. They have two metal plates, separated from each other by air or a material called a dielectric. A charge builds on the plates, one side positive and the other side negative, and energy is stored in the electric field between the plates. Capacitance is a way to describe how much charge a capacitor can hold for each volt of potential difference you hook it up to. The equation to find the capacitance of a capacitor is:

$$C = \kappa \varepsilon_0 \frac{A}{d}$$

where:

- C is the capacitance in farads.
- κ is the dielectric constant of the material between the plates. It has no units.
- ε_0 is the vacuum permittivity, the constant we have talked about already.
- A is the area of just one of the plates.
- d is the distance between the plates.

Diagram of a Capacitor

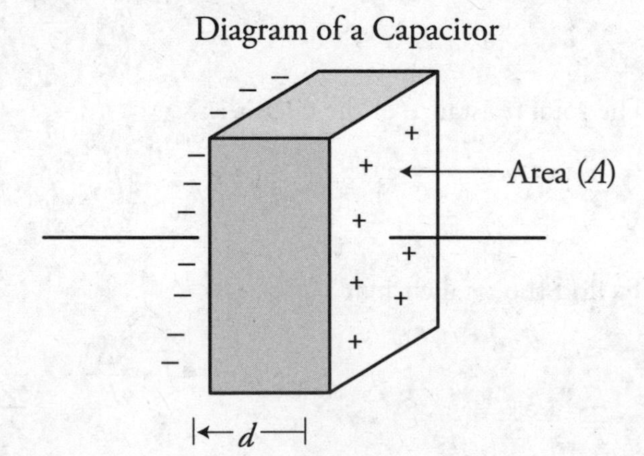

For any capacitor there is a direct relationship between the voltage across the plates and the charged stored on the plates:

$$\Delta V = \frac{Q}{C}$$

where:

- ΔV is the potential difference across the plates.
- Q is the charge that is stored on one plate. Be careful, if one plate has positive 2 C of charge on it and the other plate has –2 C of charge on it, the total charge is not 4 C or zero; it is 2 C.
- C is the capacitance of the capacitor in farads.

Example

A capacitor is made of two plates 10 cm by 10 cm separated by 0.1 mm with a 100-V battery across it and a switch. The switch is closed and the capacitor is allowed to charge.

1. Find the charge stored across the capacitor.
2. Find the energy stored in the capacitor.
3. Find the electric field strength between the plates.
4. The switch is now opened and the distance between the plates is increased to 0.2 mm. What happens to:

   (a) the capacitance
   (b) the voltage across the plates
   (c) the charge on the plates
   (d) the energy stored in the capacitor
   (e) the electric field strength between the plates

Answers

Part 1: First thing we need to do is find the area of the plates. Remember, everything must be in meters, so 0.1 m × 0.1 m = 0.01 m^2.

Now let's find the capacitance. Since again we must be in meters, distance between the plates is 1×10^{-4} m. Air is between the plates, so $\kappa = 1$:

$$C = \kappa\varepsilon_0 \frac{A}{d}$$

$$C = (1)(8.85 \times 10^{-12}\,\mathrm{C^2/N{\cdot}m^2})\frac{0.01\,\mathrm{m^2}}{1 \times 10^{-4}\,\mathrm{m}}$$

$$C = 8.85 \times 10^{-10}\,\mathrm{F}$$

To find the charge is now a snap!

$$\Delta V = \frac{Q}{C}$$

$$Q = C\Delta V = (8.85 \times 10^{-10}\,\mathrm{F})(100\,\mathrm{V}) = 8.85 \times 10^{-8}\,\mathrm{C}$$

Part 2: To find the energy stored, we could use either of these equations from the equation sheet. Both will give the same answer:

$$U_C = \tfrac{1}{2}Q\Delta V = \tfrac{1}{2}C(\Delta V)^2$$

$$U_C = \tfrac{1}{2}(8.85 \times 10^{-8}\,\mathrm{C})(100\,\mathrm{V}) = \tfrac{1}{2}(8.85 \times 10^{-10}\,\mathrm{F})(100\,\mathrm{V})^2 = 4.43 \times 10^{-6}\,\mathrm{J}$$

Part 3: To find electric field strength:

$$E = \frac{Q}{\varepsilon_0 A} = \frac{8.85 \times 10^{-8}\ \text{C}}{(8.85 \times 10^{-12}\ \text{C}^2/\text{N}\cdot\text{m}^2)(0.01\ \text{m}^2)} = 1 \times 10^6\ \frac{\text{N}}{\text{C}}$$

Now if we had been paying attention, we could have found electric field using $\frac{\Delta V}{\Delta r}$! That would have been a lot easier . . . we should pay better attention.

Part 4: Double the distance between the plates and what happens to everything else? This is a classic AP exam question. The first thing to do is decide what is changing and what is staying the same.

Changing: The distance d is doubled. New distance = $2d$.

Staying the same: Area of plates A and the charge on the plates Q because the switch was opened and it has nowhere to go. The charge is stuck on the plates. (Note that if the battery has stayed connected to the capacitor, the voltage would stay the same and the charge could change.)

(a) OK, now we have a place to start! Look at the capacitance equation:

$$\frac{C}{2} = \kappa \varepsilon_0 \frac{A}{2d}$$

When we double the distance, everything else stays the same, so capacitance is cut in half.

(b) Next, look at the voltage equation for a capacitor:

$$2\Delta V = \frac{Q}{\frac{C}{2}}$$

With the capacitance cut in half, the potential difference doubles.

(c) We have already decided that the charge stays the same because the switch was open and the charge could not move anywhere.

(d) Energy stored on the capacitor? Take your pick of equations:

$$2U_C = \frac{1}{2}Q(2\Delta V) = \frac{1}{2}\left(\frac{C}{2}\right)(2\Delta V)^2$$

Both equations tell us that the stored energy will double.

(e) OK, this time we are paying attention! Let's use the easy equation:

$$E = \frac{2\Delta V}{2\Delta r}$$

Hey! The electric field stays the same.

This is another one of those skills that the AP exam prizes—*semi-quantitative reasoning.* Make sure you understand how we worked part 4 of the problem, because it will be a skill that you will use a lot on the exam.

Parallel and Series Capacitors

Lucky for you, you already have a background in parallel and series, which will make your life much easier. Some hints to make solving problems easier:

1. Always draw a circuit diagram and label all the components clearly.
2. Make a chart to keep all your information clear and organized.
3. Keep in mind what is the same in the circuit—I can't stress enough the importance of this. I'll explain more as we go through each type of circuit.

Parallel Capacitors

Here is an example of a parallel capacitor circuit.

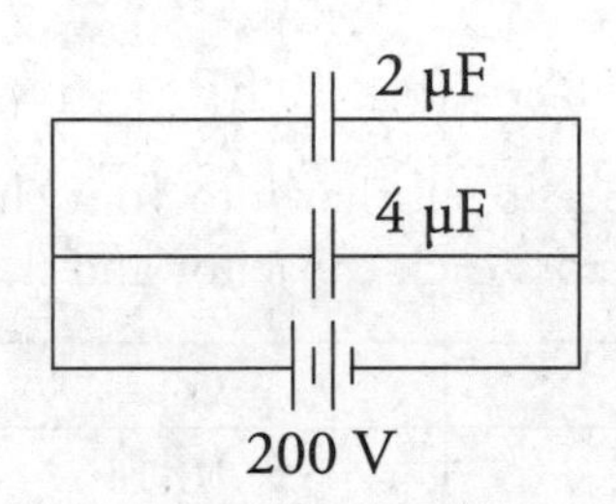

Remember that some things are the same as resistor circuits and some are different. For a parallel circuit:

- Electric potential difference is the same across each branch of a parallel circuit. This is the same for both capacitor circuits and resistor circuits:

$$V_P = V_1 = V_2 = V_3$$

- The total charge stored across the circuit equals the sum of the charges across each branch of the circuit.

$$Q_P = \sum_i Q_i = Q_1 + Q_2 + Q_3$$

- The total or equivalent capacitance of a parallel capacitor circuit equals the sum of the capacitance of each branch of the circuit. This is very different from a parallel resistance circuit, in which you add up reciprocals to find the total resistance:

$$C_P = \sum_i C_i = C_1 + C_2 + C_3$$

Example

Three capacitors, 3 μF, 6 μF, and 9 μF, are connected in parallel to a 300-V battery. What is the stored charge in the circuit and across each capacitor?

The first thing you want to do is create a *C*-Δ*V*-*Q* chart. Give each capacitor a name like C_1, C_2, C_3, and so on. Give the total or equivalent values a *T* as shown.

	C	ΔV	Q
1	3×10^{-6} F		
2	6×10^{-6} F		
3	9×10^{-6} F		
T		300 V	

Always ask yourself, "What is the same?" In the case of a parallel circuit, you know there are 300 volts across each leg of the circuit, so fill that into our chart.

	C	ΔV	Q
1	3×10^{-6} F	300 V	
2	6×10^{-6} F	300 V	
3	9×10^{-6} F	300 V	
T		300 V	

There are a couple of ways you can solve from this point. Let's find the total capacitance of the circuit:

$$C_P = \sum_i C_i = C_1 + C_2 + C_3$$

$$C_P = (3 \times 10^{-6}\text{ F}) + (6 \times 10^{-6}\text{ F}) + (9 \times 10^{-6}\text{ F}) = 18 \times 10^{-6}\text{ F}$$

Remember to fill that into your chart under the total capacitance. Now you can find the charge across each capacitor and the total charge using $Q = C\Delta V$.

	C	ΔV	Q
1	3×10^{-6} F	300 V	9×10^{-4} C
2	6×10^{-6} F	300 V	1.8×10^{-4} C
3	9×10^{-6} F	300 V	2.7×10^{-4} C
T	18×10^{-6} F	300 V	5.4×10^{-4} C

Notice that we could have simply added up the charges across C_1, C_2, and C_3 to find the total charge.

Series Capacitors

Just like in the parallel circuit, some things are the same as resistor circuits and some are different. For a series circuit:

- Electric potential difference adds up in a series circuit. This is the same for both capacitor circuits and resistor circuits:

$$V_S = \sum_i V_i = V_1 + V_2 + V_3$$

- The charge in each series capacitor is the same:

$$Q_S = Q_1 = Q_2 = Q_3$$

- The total or equivalent capacitance of a series capacitor circuit equals the sum of the reciprocal of each capacitor in the circuit. This is very different from a series resistance circuit, in which you simply add the resistors to find the total resistance:

$$\frac{1}{C_S} = \sum_i \frac{1}{C_i} = \frac{1}{C_1} + \frac{1}{C_2} + \frac{1}{C_3}$$

Example

Three capacitors—6 μF, 10 μF, and 15 μF—are connected in series to a 300-V battery. What is the stored charge in the circuit and the electric potential across each capacitor?

Let's make the chart C-ΔV-Q.

	C	ΔV	Q
1	6×10^{-6} F		
2	10×10^{-6} F		
3	15×10^{-6} F		
T		300 V	

Always ask yourself, "What is the same?" In the case of a series circuit, the charge is constant. In this problem the change is unknown, so the only thing we have enough information to determine is the total capacitance:

$$\frac{1}{C_S} = \sum_i \frac{1}{C_i} = \frac{1}{C_1} + \frac{1}{C_2} + \frac{1}{C_3}$$

$$\frac{1}{C_S} = \sum_i \frac{1}{C_i} = \frac{1}{6 \times 10^{-6}\text{ F}} + \frac{1}{10 \times 10^{-6}\text{ F}} + \frac{1}{15 \times 10^{-6}\text{ F}}$$

Be careful about your math at this point! Take each reciprocal separately, add them up, and don't forget to take a reciprocal at the end! You are trying to find C_S, not $\frac{1}{C_S}$!

$$\frac{1}{C_S} = (1.67 \times 10^5) + (1.0 \times 10^5) + (6.67 \times 10^6) = 3.34 \times 10^5$$

$$C_S = 3.0 \times 10^{-6}\text{ F}$$

	C	ΔV	Q
1	6×10^{-6} F		
2	10×10^{-6} F		
3	15×10^{-6} F		
T	3.0×10^{-6} F	300 V	

Now you can use $Q = C\Delta V$ to solve for the total charge, 3.0×10^{-6} F $\times$ 300 V = 9.0×10^{-4} C. Remember that charge is the same throughout a series circuit, so you can fill in the entire charge column.

	C	ΔV	Q
1	6×10^{-6} F		9.0×10^{-4} C
2	10×10^{-6} F		9.0×10^{-4} C
3	15×10^{-6} F		9.0×10^{-4} C
T	3.0×10^{-6} F	300 V	9.0×10^{-4} C

Now use $\Delta V = \frac{Q}{C}$ to find the electric potential across C_1, C_2, and C_3. You know that in a series circuit, the potentials add up to the total, so it is easy to check your answers.

	C	ΔV	Q
1	6×10^{-6} F	150 V	9.0×10^{-4} C
2	10×10^{-6} F	90 V	9.0×10^{-4} C
3	15×10^{-6} F	60 V	9.0×10^{-4} C
T	3.0×10^{-6} F	300 V	9.0×10^{-4} C

Combination Parallel Series Circuits

What if the capacitors are in both parallel and series? Don't fret; just find the equivalent capacitance using the rules you have already learned.

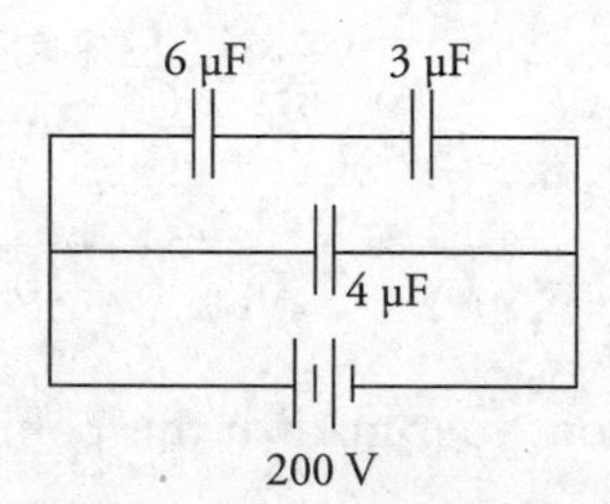

What is the equivalent capacitance of this combination? First, we can see that the top two capacitors are in series. Their equivalent resistance is 2 μF. This 2-μF capacitor is in parallel with the 4-μF capacitor for a grand total of 6 μF. Easy peasy lemon squeezy.

The AP exam probably won't ask you anything more complicated than that. But what the exam will ask you is how capacitors affect a circuit with resistors in it. So let's move on to that.

RC Circuits

Sometimes you will be asked to solve a problem with both resistors and capacitors in the circuits. These are called RC circuits. You will only be asked about RC circuits in two different states:

1. When you first connect a capacitor to a circuit, no charge has built up across the plates, so the capacitor freely allows charge to flow. Treat the capacitor like wire with no electrical potential across it.
2. After a long time, steady state will be achieved. The capacitor has built up its maximum charge, and no more current will flow onto the device. The capacitor is "full." In steady state, the potential difference across the capacitor plates equals the voltage of whatever devices are connected in parallel with it. In steady state, treat the capacitor as an open switch. Every circuit component connected in a series to a capacitor in steady state will receive no current.

Example 1

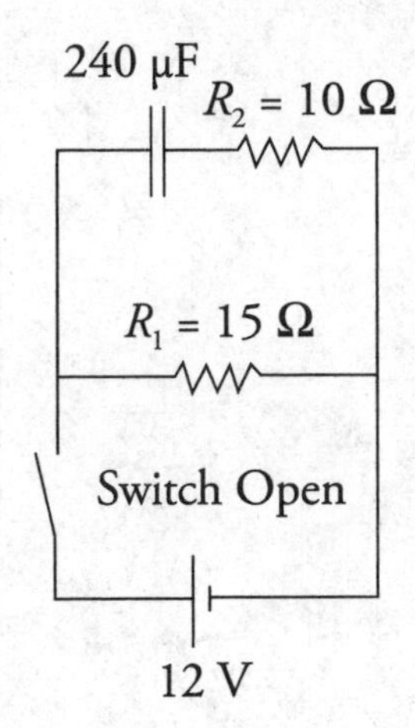

In the circuit in the diagram above, the capacitor is initially uncharged and the switch is opened.

1. Find the currents in the circuit when the switch is first closed.

2. Find the currents in the circuit and the charge on the capacitor after a long time.

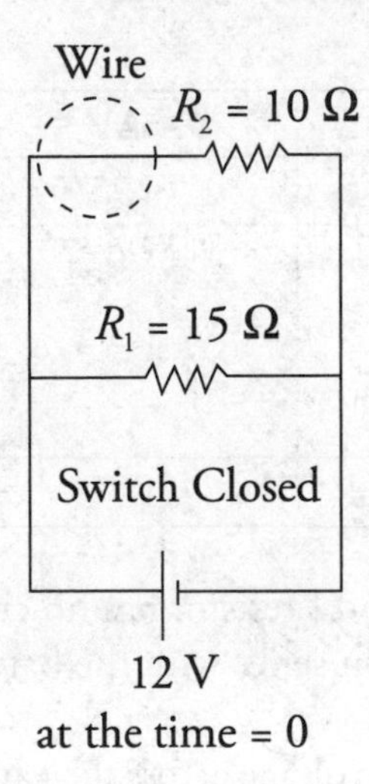

When the switch is first closed, the uncharged capacitor acts like a wire. It is like the capacitor isn't there, and there is just a parallel circuit as pictured in the diagram above. We can fill in the shaded part of our *V-I-R* chart just like we did earlier in the chapter.

	ΔV	I	R
R_1	12 V	1.2 A	10 Ω
R_2	12 V	0.8 A	15 Ω
Total	12 V	2 A	6 Ω

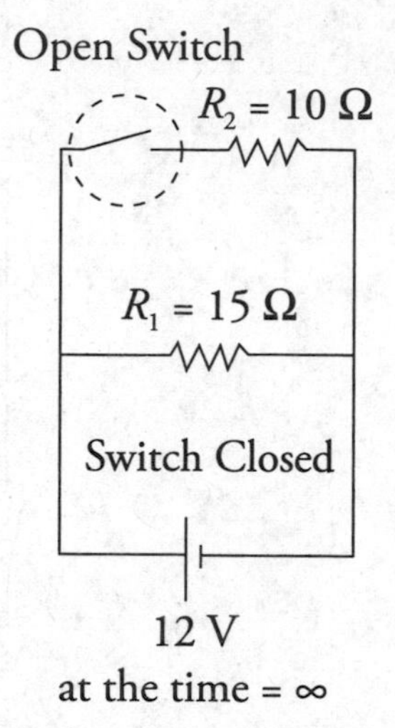

When the switch has been closed for a long time (time approaches infinity), the capacitor becomes charged and it no longer allows current to flow. It becomes like an open switch, as seen in the diagram above. This means the 10-Ω resistor will no longer carry any current and the potential difference across it must be zero. It's as if the top part of the circuit has been disconnected from the circuit by a switch! Filling in the shaded area of our *V-I-R* chart, we get:

	ΔV	***I***	***R***
R_1	12 V	1.2 A	10 Ω
R_2	0	0	15 Ω (This resistor does nothing in the circuit!)
Total	12 V	1.2 A	10 Ω

Since the 15-Ω resistor has no current passing through it, the effective circuit is simply a 10-Ω resistor in series with the battery. Therefore, the total resistance of the circuit is just 10 Ω.

What is the voltage across the capacitor? It is connected in parallel to R_1 and the battery. So, 12 volts. The charge on the capacitor would be:

$$\Delta V = \frac{Q}{C}$$

$$12\text{ V} = \frac{Q}{240\ \mu\text{F}}$$

$$Q = 0.00288\text{ C}$$

Example 2

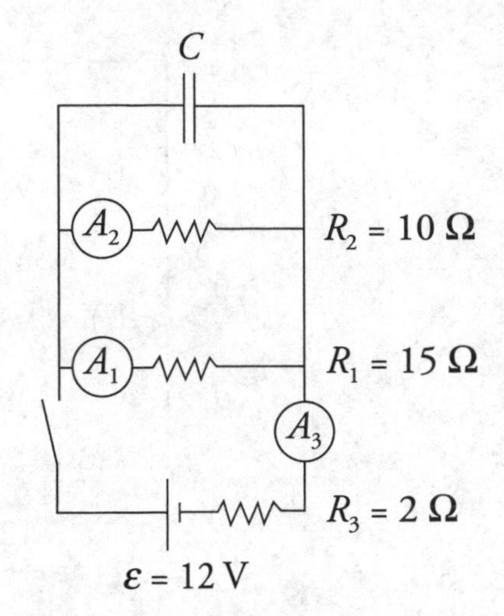

In the circuit in the diagram above, we have moved the initially uncharged capacitor to the top in parallel with the resistor R_2 and added another resistor R_3, and ammeters A_1, A_2, and A_3 to measure the currents.

Here is a great AP exam question: Rank the currents I_1, I_2, and I_3:

(A) immediately after the switch is closed

(B) after the switch has been closed a long time

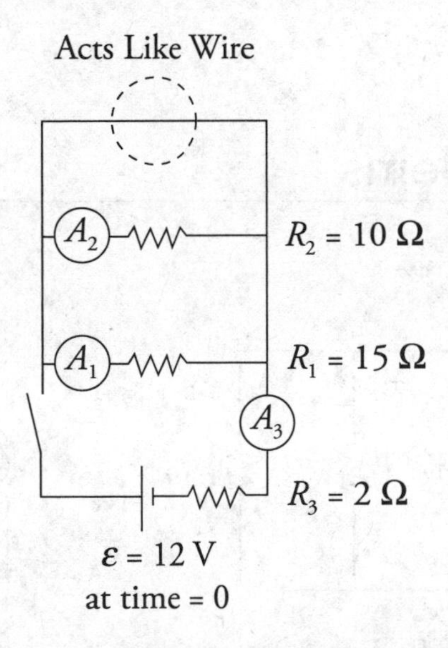

(A) When the switch is first closed, the uncharged capacitor acts like a wire, and we get the circuit pictured in the diagram above. Notice that the capacitor is acting like a short-circuit wire with no resistance at all. This means all current from the battery will bypass resistors R_1 and R_2. I_1 and I_2 will read zero.

Answer: $I_3 > I_1 = I_2$

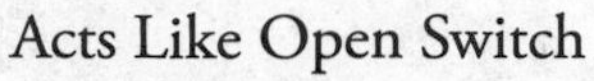

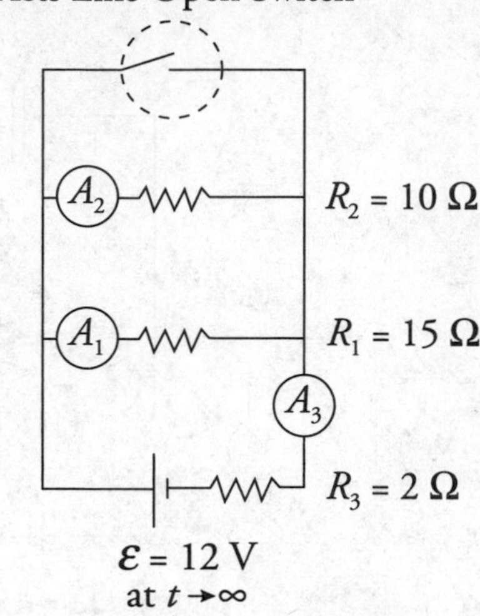

(B) After time goes by and the capacitor is full of charge, it acts like an open switch. The circuit is a simple combination circuit. Using what we have learned with Kirchhoff's rules, we can find the answer.

Answer: $I_3 > I_2 > I_1$

Note that the capacitor is in parallel with R_1 and R_2. To find the charge stored on the capacitor, we will need to find the potential difference across R_1 and R_2.

› Practice Problems

Multiple Choice

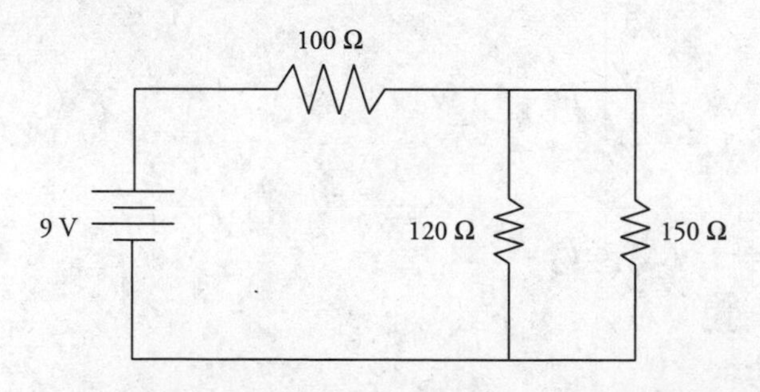

1. 100-Ω, 120-Ω, and 150-Ω resistors are connected to a 9-V battery in the circuit shown above. Which of the three resistors dissipates the most power?

(A) the 100-Ω resistor
(B) the 120-Ω resistor
(C) the 150-Ω resistor
(D) both the 120-Ω and 150-Ω resistors

2. A 1.0-F capacitor is connected to a 12-V power supply until it is fully charged. The capacitor is then disconnected from the power supply, and used to power a toy car. The average drag force on this car is 2 N. About how far will the car go?

(A) 36 m
(B) 72 m
(C) 144 m
(D) 24 m

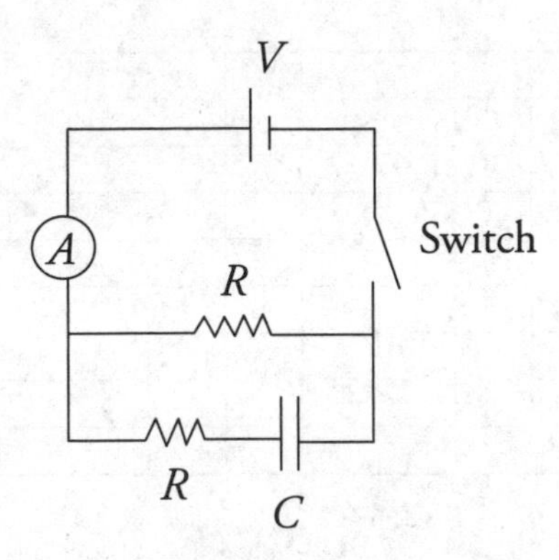

3. The circuit shown in the figure above has two resistors, an uncharged capacitor, a battery, an ammeter, and a switch initially in the open position. After the switch is closed, what will happen to the current measured in the ammeter *A*?

(A) It will increase to a constant value.
(B) It will remain constant.
(C) It will decrease to a constant value.
(D) It will decrease to zero.

Questions 4 and 5 refer to the following material.

A student is investigating what effect wire diameter *D* has on a simple circuit. The student has five wires of various diameters made of the same material and length. She connects each wire to a power supply and measures the current *I* passing through the wire with an ammeter. The data from the investigation is given in the chart.

Output voltage of the power supply: 1.5 V
Length of wires: 1.0 m
Data table:

Trial	Diameter of Wire (mm)	Current Through Wire (mA)
1	0.50	1.0
2	1.0	4.0
3	1.5	9.0
4	2.0	16
5	3.0	36

4. What can be concluded from this data?

(A) $I \propto D^2$
(B) $I \propto D$
(C) $I \propto D^{\frac{1}{2}}$
(D) There does not seem to be a relationship between the diameter of the wire and the current.

5. Using the information already gathered, what additional steps would need to be taken to extend this investigation to see if there is a relationship between wire diameter and the resistance of the wire?

(A) Determine what material the wires are made of and look up the resistivity of the wire.
(B) Multiply the current through the wire by the voltage of the power supply.
(C) Divide the current through the wire by the voltage of the power supply.
(D) Divide the voltage of the power supply by the current through the wire.

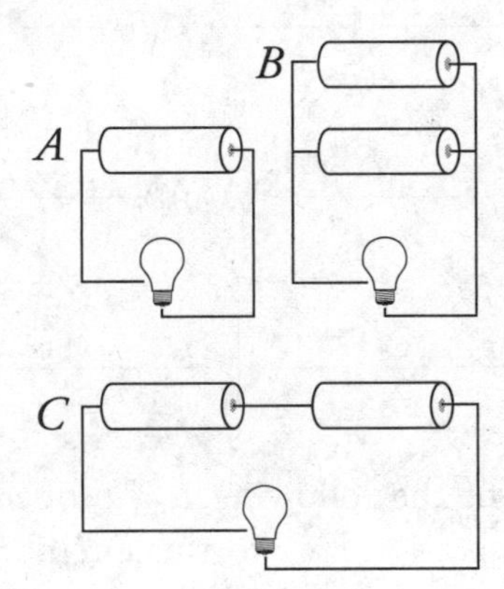

6. Identical lightbulbs are connected to batteries in different arrangements, as shown in the figure above. Which of the following correctly ranks the brightness of the bulbs?

(A) $C > B > A$
(B) $C > B = A$
(C) $C = B > A$
(D) $B > C > A$

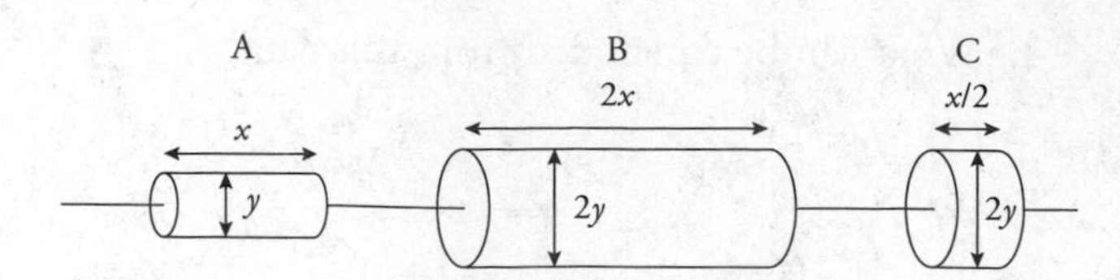

7. Three cylindrical resistors made of the same material but different dimensions are connected, as shown in the figure above. A battery is connected across the resistors to produce current. Which is the correct ranking of the currents for the resistors?

(A) $I_A = I_B = I_C$
(B) $I_A > I_B > I_C$
(C) $I_C > I_A = I_B$
(D) $I_C > I_B > I_A$

Questions 8 and 9

Two batteries and two resistors are connected in a circuit as shown in the figure below. The currents through R_1, R_2, and $\mathcal{E}_2$ are shown.

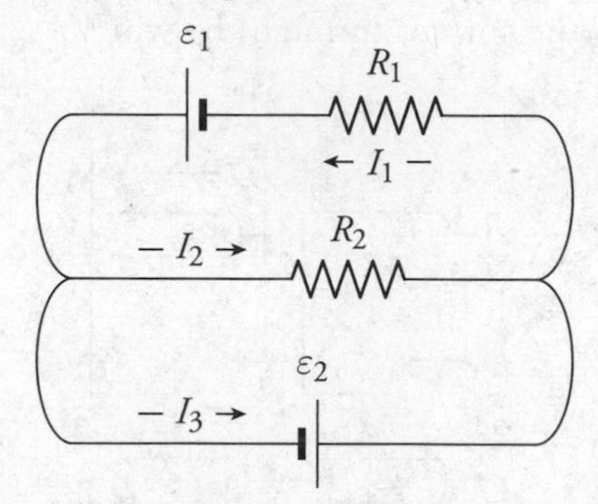

8. Which of the following is a proper application of conservation laws to this circuit? **(Select two answers.)**

(A) $\mathcal{E}_1 - I_1R_1 - I_2R_2 = 0$
(B) $\mathcal{E}_1 - \mathcal{E}_2 - I_1R_1 = 0$
(C) $I_1 + I_2 - I_3 = 0$
(D) $I_2 + I_3 - I_1 = 0$

9. The resistors R_1 and R_2 have the same resistance. If the potential differences of the batteries are $\mathcal{E}_1 =$ 9 V and $\mathcal{E}_2 =$ 6 V, which resistor will have the most current passing through it?

(A) R_1
(B) R_2
(C) R_1 and R_2 have the same current
(D) It is not possible to determine the currents through the resistors without more information

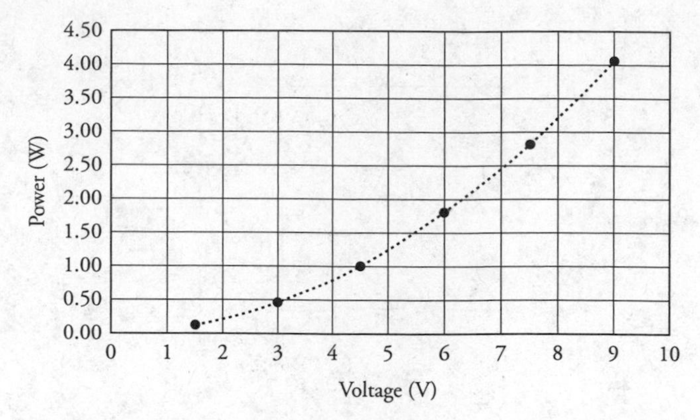

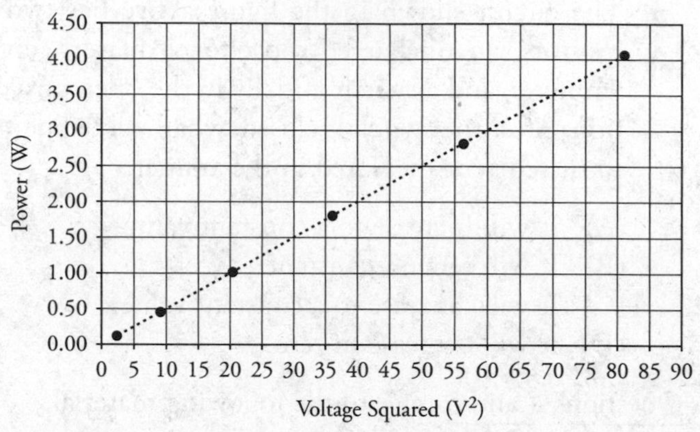

10. A single resistor is connected to a voltage source that consists of batteries with the same voltage connected in series. The power dissipated by the resistor for various voltages is shown in the two graphs. Which of the following can be deduced from the graphs? **(Select two answers.)**

(A) The batteries have a potential difference of 1.5 V.
(B) The batteries have internal resistance.
(C) The resistor is non-ohmic.
(D) The power dissipated by the resistor is proportional to the voltage squared.

Free Response

11.

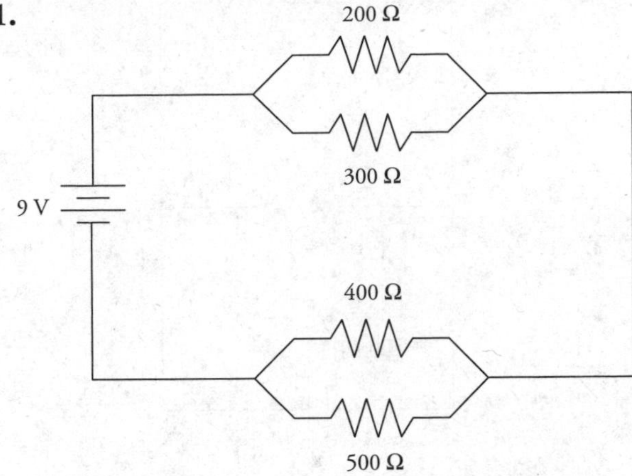

(A) Simplify the above circuit so that it consists of one equivalent resistor and the battery.

(B) What is the total current through this circuit?

(C) Find the voltage across each resistor. Record your answers in the spaces below.

Voltage across 200-Ω resistor: ______

Voltage across 300-Ω resistor: ______

Voltage across 400-Ω resistor: ______

Voltage across 500-Ω resistor: ______

(D) Find the current through each resistor. Record your answers in the spaces below.

Current through 200-Ω resistor: ______

Current through 300-Ω resistor: ______

Current through 400-Ω resistor: ______

Current through 500-Ω resistor: ______

(E) The 500-Ω resistor is now removed from the circuit. State whether the current through the 200-Ω resistor would increase, decrease, or remain the same. Justify your answer.

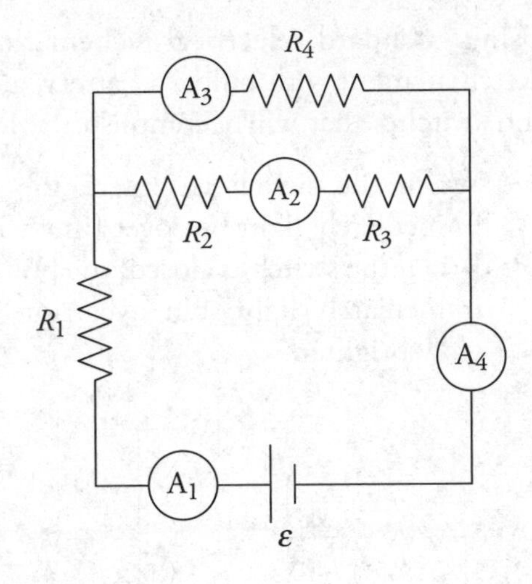

12. A circuit with four identical resistors (R_1, R_2, R_3, and R_4), a battery of potential difference (ε), and four ammeters (A_1, A_2, A_3, and A_4) is shown in the figure above.

(A) Use Kirchhoff's junction rule to prove that currents I_1 and I_4 are the same.

(B) Write Kirchhoff's loop rule for the loop that contains the battery and resistor R_3.

(C) Rank the currents (I_1, I_2, I_3, and I_4) from greatest to least. Justify your answer.

(D) Write an expression for the equivalent resistance of the circuit in terms of known quantities.

13. Two lightbulbs have power ratings of 40 W and 100 W when connected to a potential difference of 120 V.

(A) Calculate the resistance of both bulbs. Show your work.

(B) Which bulb glows brightest when connected in parallel? Justify your answer.

(C) Which bulb glows brightest when connected in series? Justify your answer.

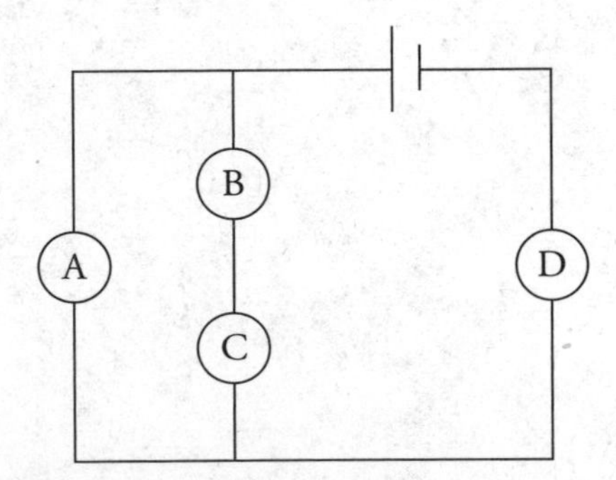

14. Four identical bulbs are attached in a circuit, as shown above. Rank the brightness of the bulbs. Justify your prediction in terms of power.

15. Using standard electrical schematic figures, sketch a circuit with bulbs, a battery, a capacitor, and switches that will accomplish the following.

(A) When the switch is closed, the bulb will immediately light but over time will go out.
(B) When the switch is closed, the bulbs will not immediately light, but over time they will glow brighter.

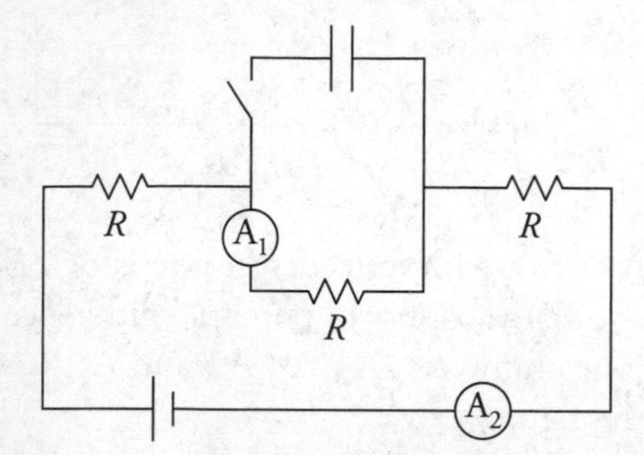

16. The circuit shown in the figure consists of three identical resistors, two ammeters, a battery, a capacitor, and a switch. The capacitor is initially uncharged, and the switch is open. Explain what happens to the readings of the two ammeters from the instant the switch is closed until a long time has passed.

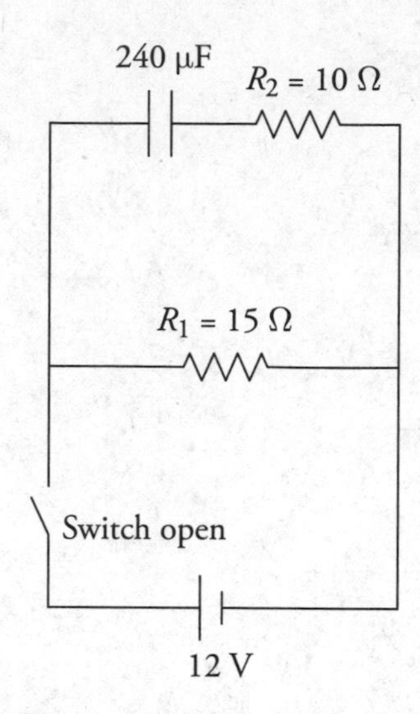

17. The figure shows a circuit with two resistors, a battery, a capacitor, and a switch. Originally, the switch is open, and the capacitor is uncharged.

(A) Complete the voltage-current-resistance-power chart for the circuit immediately after the switch is closed.

Location	***V***	***I***	***R***	***P***
1			15 Ω	
2			10 Ω	
Total for circuit	12 V			

(B) Complete the voltage-current-resistance-power chart for the circuit after the switch is closed for a long time.

Location	***V***	***I***	***R***	***P***
1			15 Ω	
2			10 Ω	
Total for circuit	12 V			

(C) What is the energy stored in the capacitor after the switch has been closed a long time?

› Solutions to Practice Problems

1. (A) On one hand, you could use a *V-I-R* chart to calculate the voltage or current for each resistor, then use $P = IV$, I^2R, or $\frac{V^2}{R}$ to find power. On the other hand, there's a quick way to reason through this one. Voltage changes across the 100-Ω resistor, then again across the parallel combination. Because the 100-Ω resistor has a bigger resistance than the parallel combination, the voltage across it is larger as well. Now consider each resistor individually.

By power $= \frac{V^2}{R}$, the 100-Ω resistor has both the biggest voltage and the smallest resistance, giving it the most power.

2. (A) The energy stored by a capacitor is $\frac{1}{2}C(\Delta V)^2$. By powering a car, this electrical energy is converted into mechanical work, equal to the force times the displacement. Solve for displacement, you get 36 m.

3. (C) When the switch is initially closed, the uncharged capacitor acts as a "wire" or "closed switch" with no resistance. Thus the initial circuit "looks" like a parallel circuit. As the capacitor charges, the current through the bottom resistor in series with the capacitor drops to zero, as the capacitor acts as a "broken wire" or "open switch" with infinite resistance. Thus, after a long time the circuit becomes a series circuit with current passing only through a single top resistor. As the circuit transitions from this "parallel to series," the equivalent resistance of the circuit increases. This produces a current through the ammeter that drops from its maximum starting value to a steady-state lower value.

4. (A) As the diameter increases, the current also increases, indicating a direct relationship of some kind. Comparing trials 1 and 2, the diameter doubles and the current quadruples. This suggests a quadratic relationship. To confirm this, compare trials 1 and 3—the diameter is tripled and the current increases by a factor of 9.

5. (D) To investigate the relationship of diameter and resistance, the resistance of the wire needs to be calculated. We already know the output voltage of the power supply and the current through the wire. We just need to calculate the resistance for each entry in the table: $R = \frac{\Delta V}{I}$.

6. (B) Using Kirchhoff's loop rule, we can see that the batteries in arrangement *C* add up to double the voltage. Connecting batteries as in arrangement *B* does not add any extra voltage to the bulb, and it will not be any brighter than arrangement *A*.

7. (A) There is only one pathway. The current is the same.

8. (A) and (D) Applying Kirchhoff's loop rule to the top loop, we get answer choice A. Applying Kirchhoff's junction rule to the right junction, we get answer choice D.

9. (A) The emfs of both batteries point in the same direction for the outer loop, for a combined potential difference of 15 V. The emfs point in the opposite direction for the line that contains R_2, for a combined potential difference of 3 V.

10. (A) and (D) The batteries are connected in series; therefore, their electric potentials add up. The difference between the voltages in the left graph is 1.5 V. Thus, the batteries must be 1.5 V each. The right graph shows power proportional to the voltage squared because the graph is linear.

11. (A) Combine each of the sets of parallel resistors first. You get 120 Ω for the first set, 222 Ω for the second set, as shown in the diagram below. These two equivalent resistances add as series resistors to get a total resistance of 342 Ω.

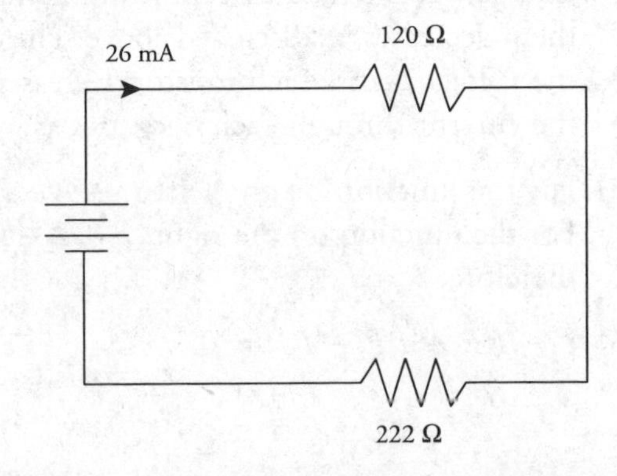

(B) Now that we've found the total resistance and we were given the total voltage, just use Ohm's law to find the total current to be 0.026 A (also known as 26 mA).

(C) and (D) should be solved together using a *V-I-R* chart. Start by going back one step to when we began to simplify the circuit: 9-V battery, a 120-Ω combination, and a 222-Ω combination, shown above. The 26-mA current flows through each of these . . . so use $V = IR$ to get the voltage of each: 3.1 V and 5.8 V, respectively.

Now go back to the original circuit. We know that voltage is the same across parallel resistors. So both the 200-Ω and 300-Ω resistors have a 3.1-V voltage across them. Use Ohm's law to find that 16 mA goes through the 200-Ω resistor, and 10 mA through the 300 Ω. Similarly, both the 400-Ω and 500-Ω resistors must have 5.8 V across them. We get 15 mA and 12 mA, respectively.

Checking these answers for reasonability: the total voltage adds to 8.9 V, or close enough to 9.0 V with rounding. The current through each set of parallel resistors adds to just about 26 mA, as we expect.

(E) Start by looking at the circuit as a whole. When we remove the 500-Ω resistor, we actually *increase* the overall resistance of the circuit because we have made it more difficult for current to flow by removing a parallel path. The total voltage of the circuit is provided by the battery, which provides 9.0 V no matter what it's hooked up to. So by Ohm's law, if total voltage stays the same while total resistance increases, total current must *decrease* from 26 mA.

Okay, now look at the first set of parallel resistors. Their equivalent resistance doesn't change, yet the total current running through them decreases, as discussed above. Therefore, the voltage across each resistor decreases, and the current through each decreases as well.

12. (A) For the junction on the left: $I_1 - I_2 - I_3 = 0$. For the junction on the right: $I_3 + I_2 - I_4 = 0$, therefore, $I_1 = I_4$.

(B) $\varepsilon_1 - I_1R_1 - I_2R_2 - I_2R_3 = 0$

(C) $I_1 = I_4 > I_3 > I_2$. Currents 1 and 4 are the same and equal to the combination of currents 2 and 3. The currents 2 and 3 have the same potential but different resistances. The resistance is less for current 3; therefore, it is larger than current 2.

(D) $R_{eq} = R_1 + \dfrac{R_4(R_2 + R_3)}{R_2 + R_3 + R_4}$

13. (A) $P = \dfrac{V^2}{R}$, $R_{40\text{ W}} = 360\,\Omega$, $R_{100\text{ W}} = 144\Omega$

(B) $P = \dfrac{V^2}{R}$. When connected in parallel, the bulbs receive the same electric potential, as shown by Kirchhoff's loop rule. Therefore, the bulb with the smallest resistance (100 W) is brighter.

(C) $P = I^2R$. When connected in series there is only one pathway for the current. The bulbs receive the same current. Therefore, the bulb with the largest resistance (40 W) is brightest.

14. D > A > C = B, $P = I^2R$. All the bulbs have the same resistance. Therefore, the bulb that receives the greatest current will have the greatest power consumption and be brightest. Bulb D is in the main current pathway and receives the greatest current. The main current splits between the two pathways. Bulb A sits in the pathway with the least resistance, so it will receive more current than bulbs C and B. Bulbs C and B are in the same pathway and receive the smallest current.

15. There is more than one way to draw each of these, but here is an example.

(A)

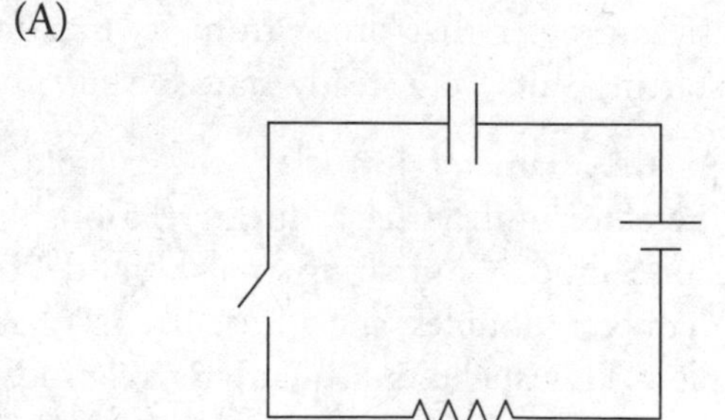

(B)

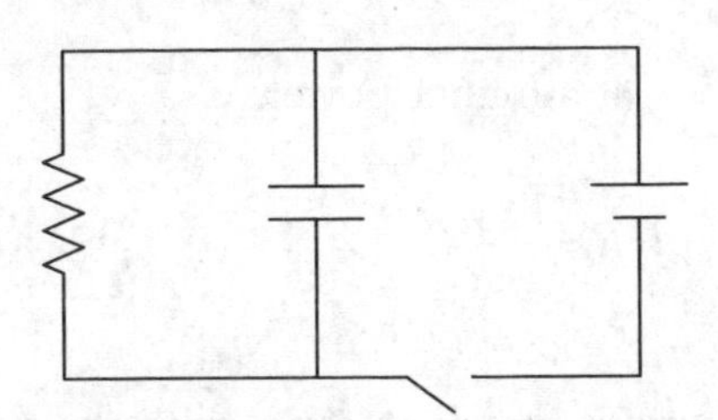

16. When the switch is first closed, the capacitor behaves like a wire. This creates a short circuit around ammeter 1, and it will read zero. All the current is flowing through only two resistors in series and ammeter 2. When the capacitor becomes fully charged, it behaves like an open switch in the circuit. The current is now passing through all three resistors in series. This increases the current in ammeter 1 but decreases the current in ammeter 2 because the total resistance of the circuit has increased.

17. (A)

Location	***V***	***I***	***R***	***P***
1	12 V	0.80 A	15 Ω	9.6 W
2	12 V	1.2 A	10 Ω	14 W
Total for circuit	12 V	2.0 A	6.0 Ω	24 W

(B) All numbers are rounded to two significant digits.

Location	***V***	***I***	***R***	***P***
1	12 V	0.80 A	15 Ω	9.6 W
2	0	0	10 Ω	0
Total for circuit	12 V	0.80 A	15 Ω	9.6 W

(C)

$$U = \frac{1}{2}C(\Delta V)^2 = \frac{1}{2}(240\ \mu\text{F})(12\ V)^2 = 0.017\,\text{J}$$

❯ Rapid Review

- Current is the flow of positive charge. It is measured in amperes.
- Resistance is a property that impedes the flow of charge. Resistance in a circuit comes from the internal resistance of the wires and from special elements inserted into circuits known as "resistors."
- Resistance is related to current and voltage by Ohm's law: $\Delta V = IR$.
- Resistors that have a constant resistance no matter what the current through them or voltage across them are said to be ohmic. If the resistance changes, then these resistors are nonohmic.
- When resistors are connected in series, the total resistance equals the sum of the individual resistances. And the current through one resistor equals the current through any other resistor in series with it.
- When resistors are connected in parallel, the inverse of the total resistance equals the sum of the inverses of the individual resistances. The voltage across one resistor equals the voltage across any other resistor connected parallel to it.
- The *V-I-R* chart is a convenient way to organize any circuit problem.
- Kirchhoff's junction rule says that the exact same amount of current coming into a junction will leave the junction. This is a statement of conservation of charge. Kirchhoff's loop rule says that the sum of the voltages across a closed loop equals zero. This rule is helpful especially when solving problems with circuits that contain more than one battery.
- Ammeters measure current, and are connected in series; voltmeters measure voltage, and are connected in parallel.

- Real batteries have internal resistance that cuts the amount of voltage the battery supplies to the circuit.
- Bulbs are brighter when they are operating at a higher power.

$$P = I\Delta V = I^2 R = \frac{(\Delta V)^2}{R}$$

- When a switch is open, the part of the circuit that is in series with the switch does not receive any current and is "dead."
- When capacitors are connected in series, the inverse of the total capacitance equals the sum of the inverses of the individual capacitances. When capacitors are connected in parallel, the total capacitance just equals the sum of the individual capacitances.
- A capacitor's purpose in a circuit is to store charge and energy. After it has been connected to a circuit for a long time, the capacitor becomes fully charged and prevents the flow of current.
- An uncharged capacitor behaves like a wire in a circuit, but once it is charged, it behaves like an open switch.

Magnetism and Electromagnetic Induction

IN THIS CHAPTER

Summary: Magnetic fields produce forces on moving charges; moving charges, such as current-carrying wires, can create magnetic fields. This chapter discusses the production and the effects of magnetic fields.

Key Ideas

- Magnetic poles always come in pairs (dipoles); you will never find a monopole (north or south by itself).
- Magnetic fields, just like electric fields, extend to infinity.
- The study of magnetic behavior is inherently three-dimensional (3D) and we use "right-hand rules" to help us visualize its behavior.
- The force on moving charges due to the magnetic field is qvB.
- The direction of the magnetic force on a moving charge is given by a right-hand rule and is *not* in the direction of the magnetic field.
- The magnetic force is always perpendicular to the moving charges' velocity. This causes a moving charge in magnetic field to arc into circular paths.
- Current-carrying wires produce magnetic fields.
- When the magnetic flux through a wire loop changes, a voltage is induced.
- Materials placed in a magnetic field will exhibit one of three behaviors: ferromagnetism, paramagnetism, or diamagnetism.

Relevant Equations

Force on a charged particle moving in a magnetic field: $\vec{F}_M = q\vec{v} \times \vec{B}$ $\quad |\vec{F}_M| = |q\vec{v}||\sin\theta||\vec{B}|$

Force on a current-carrying wire: $\vec{F}_M = I\vec{\ell} \times \vec{B}$ $\quad |\vec{F}_M| = |I\vec{\ell}||\sin\theta||\vec{B}|$

Magnetic field due to a long, straight, current-carrying wire: $B = \dfrac{\mu_0}{2\pi}\dfrac{I}{r}$

Magnetic flux: $\Phi_B = \vec{B}\cdot\vec{A}$ $\quad \Phi_B = |\vec{B}|\cos\theta|\vec{A}|$

Induced emf: $\varepsilon = -N\dfrac{\Delta\Phi_B}{\Delta t}$

Induced emf for a wire moving through a magnetic field: $\varepsilon = B\ell v$

Magnetic Fields

Some people might think of a screwdriver that holds on to a screw or magnets on a refrigerator when they think of magnetism, or a horseshoe-shaped piece of metal doing amazing things in a cartoon. But physicists know that magnetism goes far beyond picking up a bunch of paper clips. Magnetism is different from any other subject you've learned before because much of magnetic behavior is three-dimensional. There are many right-hand rules that are indispensable to understanding magnetism, as well as concepts that are completely different from anything you've learned before. But hang in there; I'm going to help you get a pile of points on your magnetism questions on the AP exam.

All magnets have north and south poles at their ends. There is no such thing as a monopole or single pole magnet.

You can never create a monopole magnet with just a north pole or just a south pole. If you took the magnet in the next figure,

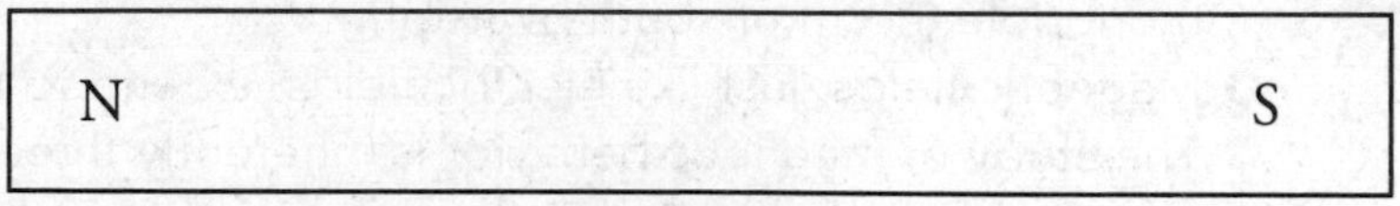

N S

Bar magnet

and cut it down the middle, you would not separate the poles. Instead, you would create two magnets like those shown in the following figure.

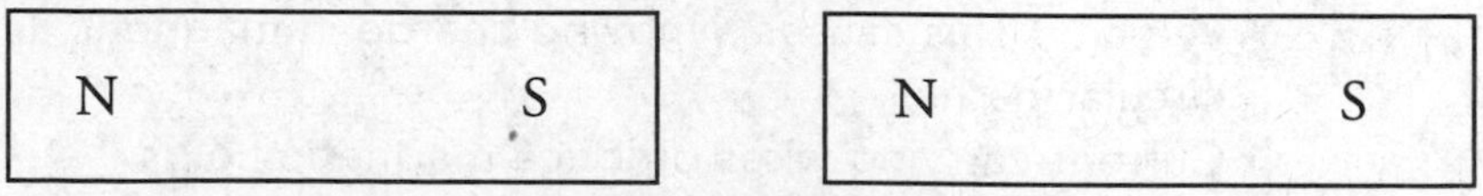

N S N S

Cutting the bar magnet in half just gives you two smaller bar magnets. You can never get an isolated north or south pole.

Be careful; do not confuse magnetism with static electricity. It's an easy thing to do because there are some things that are very similar.

Similarities:

- Electric charges exert forces on each other. Opposite charges attract and like charges repel. Magnets exert forces on each other. Opposite poles attract and like poles repel.
- Charges produce electric fields that extend out into space to infinity and get weaker with distance. Magnets produce magnetic fields that extend out into space to infinity and get weaker with distance as well.

***Huge* Difference:**

- Electric fields exert forces on positive charges in the direction of the field. Negative charges get pushed in the opposite direction of the electric field. Magnetic fields are not that simple. Magnetic fields affect only moving charges, *and* the force is perpendicular to both the velocity of the charge and the direction of the magnetic field!

But, we are getting ahead of ourselves. Let's get back to magnetic fields. We will discuss all the unique behaviors of magnetism as we go along.

A magnet creates a magnetic field (see the next figure). Unlike electric field lines, which either emanate from a positive charge dead end on a negative charge, or extend infinitely into space, magnetic field lines always form loops. These loops extend infinitely into space. The loops point away from the north end of a magnet, and inward toward the south end. Near the magnet, the lines point nearly straight into or out of the poles.

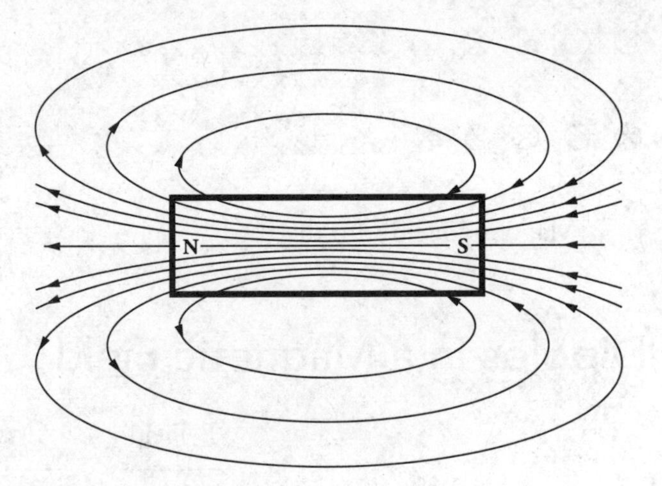

The figure below shows how iron filings sprinkled onto a bar magnet gather on the magnetic field lines, making them visible.

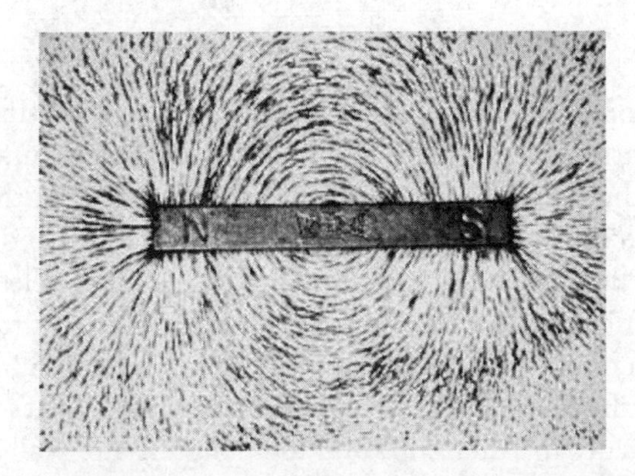

The next diagram shows the magnetic vector field representation. Each individual arrow represents the magnetic field vector at that point. Longer arrows mean a larger field strength.

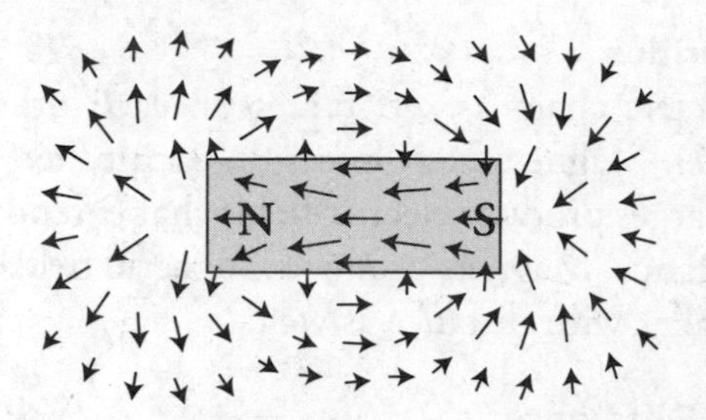

The space around a magnet contains a magnetic field. All magnetic fields are created by moving charges. In a permanent magnet, like the ones you stick on your refrigerator, the moving charges are the electrons spinning around the nucleus of the atoms. Remember that currents are moving charges in a wire, so they also create magnetic fields. We will talk about that in a minute.

Remember that I said we would be dealing with three dimensions? A piece of paper is two-dimensional, so physicists came up with some nice ways to draw magnetic fields that are going in or out of the page. For a magnetic field going into the page, a bunch of X's will be drawn. For a magnetic field coming out of the page, a bunch of dots will be drawn. In the next figure you will see two different variations of how each of these is drawn.

Magnetic Field into the Page

Magnetic Field out of the Page

Magnetic Dipoles in a Magnetic Field

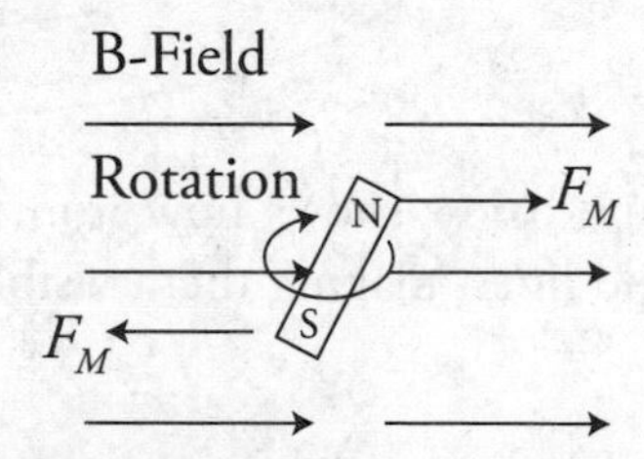

I hope you played with a compass as a kid. (If not, you probably have the app on your smartphone.) A compass is a tiny magnet mounted on a rotation point so that it can turn in the Earth's magnetic field. It works like this: the north end of the magnet is forced in the direction of the magnetic field and the south end is forced in the opposite direction. This is just like the behavior of electric dipoles in an electric field. This is why a magnetic compass points north. Hey, wait a minute. Isn't the north end of a magnet attracted to the south end of another magnet? Why yes, yes it is. That means that the *geographic North Pole* of the Earth is actually the *magnetic south pole*! (That's why we can't find Santa. His workshop is actually in Antarctica and the penguins are really elves. It all makes sense. . . .)

Magnetic Field Around a Straight Current-Carrying Wire

Maybe as a kid, you wrapped wire around a nail, connected it to a battery, and made an electromagnet. To start things out, we are going to keep things simple with a single straight piece of wire carrying a current. The magnetic field forms circles in the plane perpendicular to the wire. I know that sounds confusing, but picture a wire going through the center of a bunch of washers stacked one on top of the other. The washers are all circles in planes perpendicular to the wire. You are going to have to be able to find the direction of the magnetic field going around the wire, and we're going to use something called the right-hand rule to help us out.

The right-hand rule will help us find the direction of the magnetic field by a long, straight, current-carrying wire. Call this the right-hand "curly fingers" rule. You're going to use your pen or pencil to help out. Take your pencil and hold it in the direction the wire is oriented, with the tip pointing in the direction of the current. Next, grasp the pencil with your right hand and your thumb pointing in the direction of the current, the same direction the tip of your pencil is pointing. Your fingers will curl around the pencil in the same direction the magnetic field curls about the current.

Magnetic Field Around a Wire

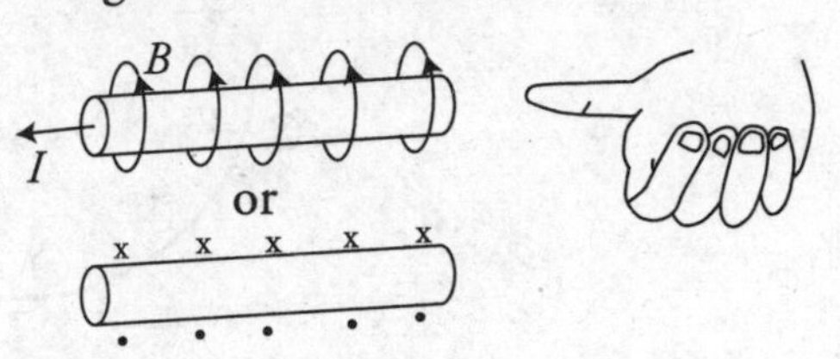

With the fingers of your right hand still holding the pencil, point the tip of the pencil directly at your face. See how your thumb points toward your eyes and your fingers circle around the pencil in a counterclockwise direction? When the current is directed upward, out of the page and toward your face, the magnetic field will form counterclockwise circles around the wire, as shown in the figure below.

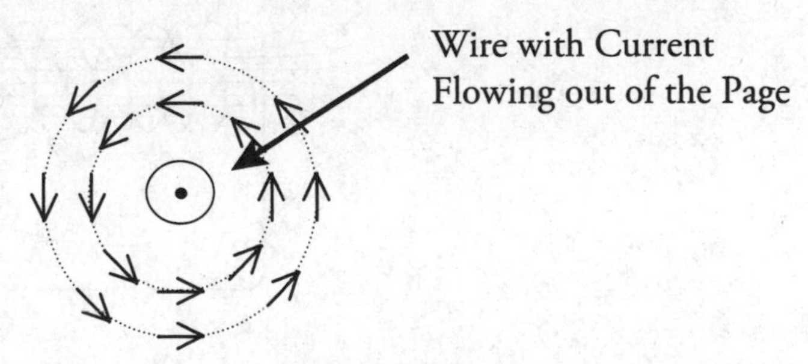

Here's another example. What does the magnetic field look like around a wire in the plane of the page with current directed upward toward the top of the page? We won't walk you through this one; just use the right-hand rule, and you'll be fine. The answer is shown in the next figure. (Notice how the magnetic field points into the page on the right-hand side of the current and out of the paper on the left-hand side of the current. Just like the fingers of your right-hand curl around the current when you point your right-hand thumb in the direction of the current.)

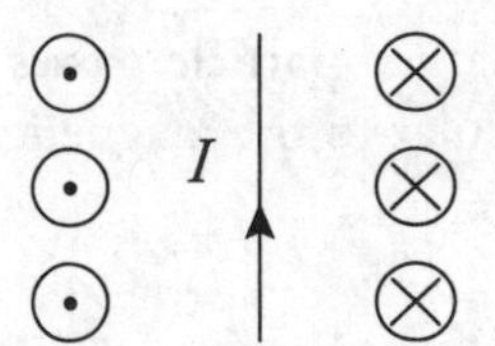

The formula to find the strength of a magnetic field outside a long straight wire is:

$$B = \frac{\mu_0}{2\pi}\frac{I}{r}$$

where:

- B is the magnetic field strength measured in teslas.
- μ_0 is the vacuum permeability. (Don't sweat it; μ_0 is a constant. $\mu_0 = 4\pi \times 10^{-7}\ \text{T} \cdot \text{m/A}$.)
- I is the current going to the wire in amperes.
- r is the distance from the center of the wire to the point where you're trying to find the magnetic field strength.

Question: What would the magnetic field look like in the space around a loop of wire with a counterclockwise current in it? Remember to use your right-hand rule!

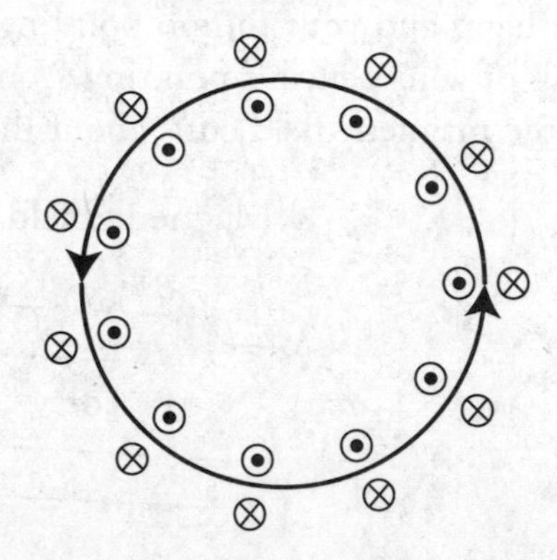

Answer: It would point out of the paper, in the middle of the loop, and into the paper on the outside of the loop.

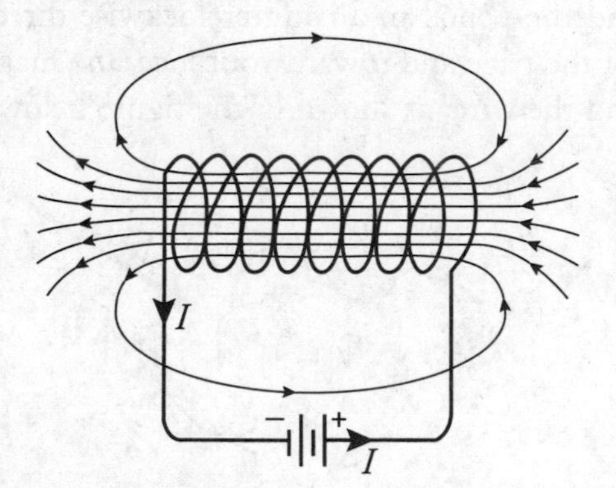

Now loop the wire around many times to create a coil called a *solenoid.* Hook it up to a battery and you get a dipole magnetic field that looks like a bar magnet, just like the one you got when you wrapped wire around that nail to produce an electromagnet as a kid. By the way, which end of the figure is the north end of the electromagnet? (Choose the left side!)

Force on a Moving Charged Particle

Whenever a charged particle passes through a magnetic field, if a component of its velocity is perpendicular to the magnetic field, it will experience a force. The magnitude of that force is:

$$\vec{F}_M = q\vec{v} \times \vec{B}$$

Now you might be asking yourself, why is there a "×" between the velocity and the magnetic field strength? Well, you're multiplying two vectors together, and that is called a "cross product." Cross products act a little weirdly. First of all, the angle between the velocity vector and magnetic field vector has an effect on the magnitude of the force. If they're at a right angle, you get a maximum force. If they're parallel to each other, the force is zero. Another way of writing a cross product is:

$$|\vec{F}_M| = |q\vec{v}||\sin\theta||\vec{B}|$$

where:

- $\vec{F}_M$ is the force acting on the moving charge in Newtons.
- q is the charge of the particle moving to the field in coulombs.
- $\vec{v}$ is the velocity of the particle in meters per second.
- $\vec{B}$ is the magnetic field strength in teslas.
- θ is the angle between the velocity and the magnetic field vectors.

Don't worry too much about the cross-product equation. I know it's on the reference table, but the equation with "sin θ" will serve you fine.

Here comes the really strange thing about a cross product: the force acting on the charge is at a right angle to both the magnetic field and the velocity. Of course, there is a hand rule to help you out. (If the right-hand rule that follows is not the one that your teacher taught you, don't worry! There are several different variations of this right-hand rule. The point is for you to find one that works for you and stick with it.)

Here's the right-hand rule for a moving positive charge in a magnetic field. (You might want to call this the right-hand "flat finger" rule.) Hold your hand flat with your thumb at a right angle to your fingers.

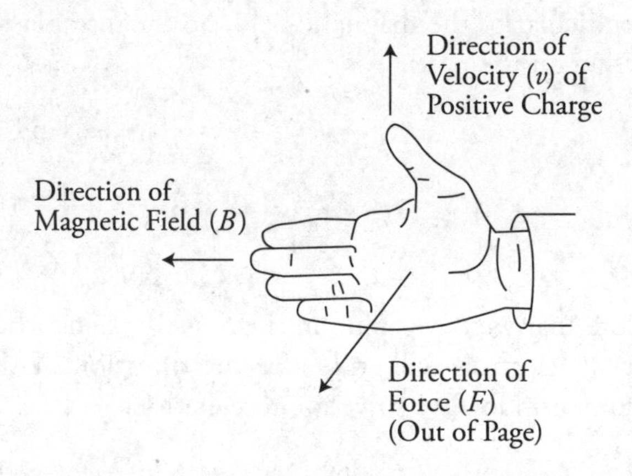

- Your fingers should point in the direction of the magnetic field. There are usually lots of magnetic field lines and you have lots of fingers.
- Your thumb will point in the direction of the velocity of the particle.
- Perpendicular to the palm of your hand is the direction of the force on the particle. A nice way to think of this is that the direction of the force on the particle is the same direction you would push on something with the palm of your hand.

Notice that everything is at right angles. If the velocity is not perpendicular to the magnetic field, you can bring your thumb a little closer or farther away (if you can) from your fingers. The force is always going to be perpendicular to the two of them and will be in the direction your palm is facing.

What if there is a negative particle moving through the field? Everything is reversed for the negative particle, so simply use your left hand and follow the same rules. So remember:

- Right hand for positive particles
- Left hand for negative particles

Now is a good time for "body art." Take a pen and write a *B* on the tip of each of your fingers. Write *v* on your thumb and put a big *F* and a positive sign "+" on your right palm. Repeat this for your left hand except put a big *F* and a negative sign "–" on your left palm. Now you have a useful physics tattoo to help you with these problems. (Remember, you can't walk into the AP exam with anything written on your hands. But once the exam starts, you can tattoo yourself if you need to!)

The key to this right-hand rule is to remember the sign of your particle. This next problem illustrates how important the sign can be.

> An electron travels through a magnetic field, as shown below. The particle's initial velocity is 5×10^6 m/s, and the magnitude of the magnetic field is 0.4 T. What are the magnitude and direction of the particle's acceleration?

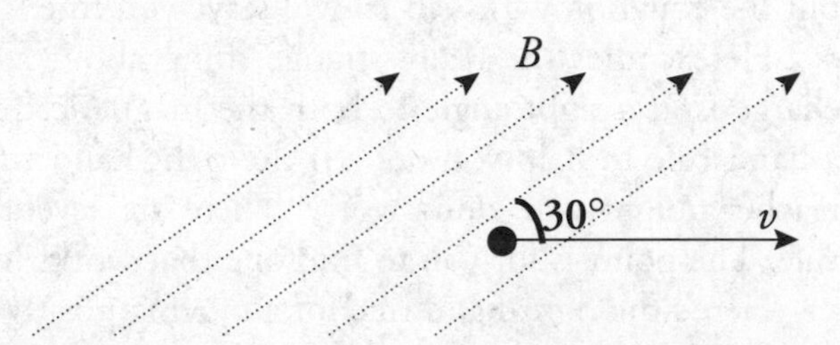

This is one of those problems where you're told that the particle is *not* moving perpendicular to the magnetic field. So the formula we use to find the magnitude of the force acting on the particle is

$$|\vec{F}_M| = |q\vec{v}||\sin\theta||\vec{B}|$$

$$F_M = (1.6 \times 10^{-19}\text{ C})(5 \times 10^6\text{ m/s})(\sin 30°)(0.4\text{ T})$$

$$F_M = 1.6 \times 10^{-13}\text{ N}$$

Note that we never plug in the negative signs when calculating force. The negative charge on an electron will influence the direction of the force, which we will determine in a moment. Now we solve for acceleration:

$$F_{net} = ma$$

$$a = 1.8 \times 10^{17}\text{ m/s}^2$$

Wow, you say . . . a bigger acceleration than anything we've ever dealt with. Is this unreasonable? After all, in less than a second the particle would be moving faster than the speed of light, right? The answer is still reasonable. In this case, the acceleration is perpendicular to the velocity. This means the acceleration is *centripetal,* which means the force is toward the center and the particle must move in a circle at constant speed. But even if the particle were speeding up at this rate, either the acceleration wouldn't act for very long, or relativistic effects would prevent the particle from traveling faster than light.

Now that you found the magnitude of the force, you want to find the direction. Be careful; it's an electron, so you need to use your left hand. To do this you're going to hold your left hand with your fingers pointing in the direction of the magnetic field and your thumb pointing in the direction of the velocity. Notice this leaves your palm pointing straight into the page, so the force acting on the particle is directly into the page. Some questions will tell you that toward the right of the page is the positive x-axis, the top of the page is the positive y-axis, and directly out a page, toward your face, is the positive z-axis. So this force would be in the negative z direction ($-z$).

Quick question: If the electron in the previous problem was traveling parallel to the magnetic field, what would happen? *Nothing*. The angle θ between the field and the velocity would be 0°, or maybe 180°. In either case the $\sin 0° = \sin 180° = 0$. Since the magnetic force is zero, the electron would just keep moving like the magnetic field wasn't even there. Can you hear Newton's ghost saying, "An object in motion stays in motion unless acted upon by a net outside force"?

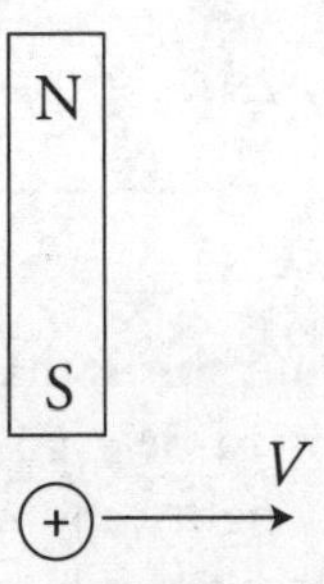

Another quick question: A positive charge, moving to the right, passes the south end of a magnet. What is the direction of the force on the magnet? That's not a typo. What is the force on the magnet? Remember Newton's third law—for every force there is another one that acts in the opposite direction on the other object. Just figure out the direction of the force on the charge and the force on the magnet will be in the opposite direction. We already know that magnetic fields enter the south end of the magnet, which means upward on the paper. Velocity is to the right. Charge is positive—use your right tattooed hand to find the force on the charge will be out of the paper. Therefore, the force on the south end of magnet will be into the paper.

Magnetic Force on a Wire

Before we get into this, let's first think about two bar magnets side by side. See the diagram below. If the two magnets have opposite poles near each other, they're going to attract each other, and if they have like poles near each other, they're going to repel each other. But let's think about how the magnetic fields interact.

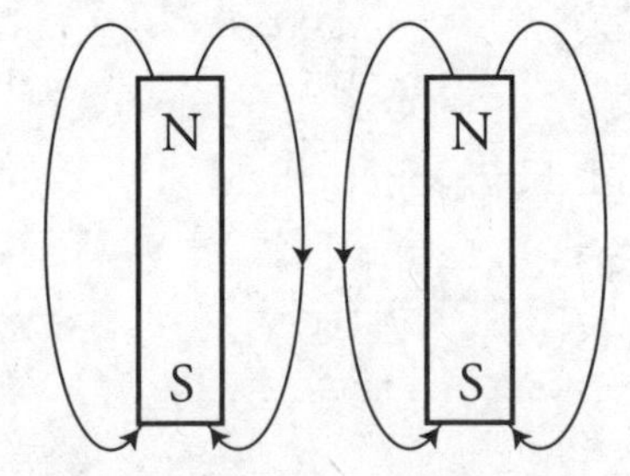

Magnetic fields in the same direction repel each other.

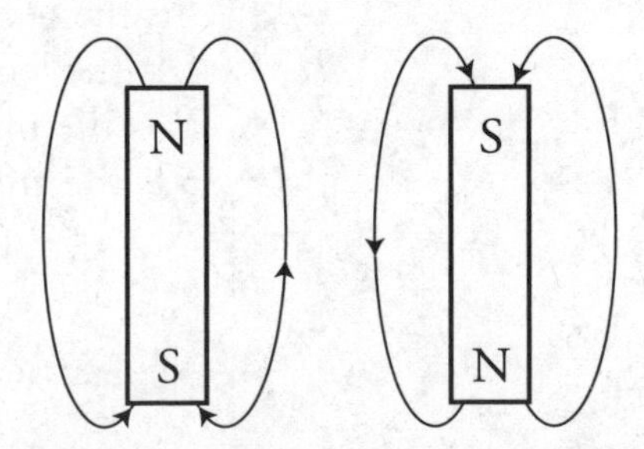

Magnetic fields in the opposite direction attract each other.

Keep this in mind as we think about a wire in a magnetic field. Looking at the next figure, there's a wire with current coming out of the page sitting in a magnetic field going toward the right side of the page. The magnetic field around the wire is going counterclockwise as indicated by our right hand "curly fingers" rule. Above the wire the magnetic field from the wire is in the opposite direction of the magnetic field from the outside magnets, so those magnetic fields cause attraction pulling the wire up and pushing the magnets down. Below the wire, the magnetic fields are in the same direction, so they repel each other pushing the wire up and the magnets down. So both of these interactions are trying to push the wire up toward the top of the page.

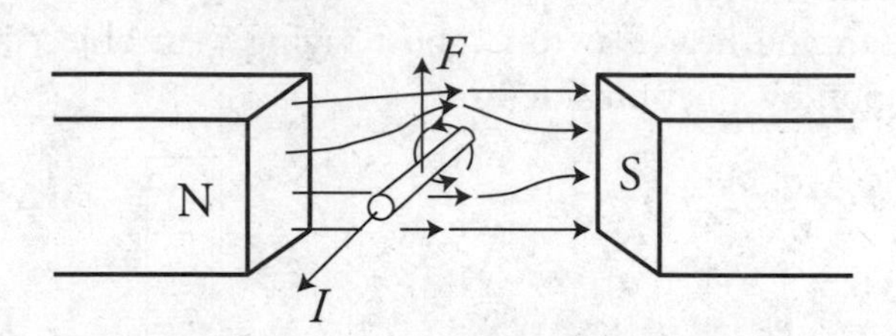

What's nice about this is that we can use the same hand rule we just used. Remember, we consider current to be a positive charge flow, so for a current-carrying wire, we will always be using our right hand, never the left hand.

To use the right-hand rule for a current-carrying wire in a magnetic field, once again, you are going to hold your hand flat with your thumb perpendicular to your fingers:

- Your fingers point in the direction of the external magnetic field (to the right).
- Your thumb points in the direction of the current flowing in the wire (out of the page).
- Your palm aims in the direction of the force (upward toward the top of the page).

Practice this with the diagram above and prove that it works.

The equation for force on a current-carrying wire in a magnetic field is:

$$\vec{F}_M = I\ell \times \vec{B}$$

This is a cross product just like before, so we can write it like this:

$$|\vec{F}_M| = |I\ell||\sin \theta||\vec{B}|$$

where:

- $\vec{F}_M$ is the force acting on the wire in newtons.
- I is the current going to the wire in amperes.
- ℓ is the length of the wire.
- $\vec{B}$ is the magnetic field strength in teslas.
- θ is the angle between the current and the magnetic field.

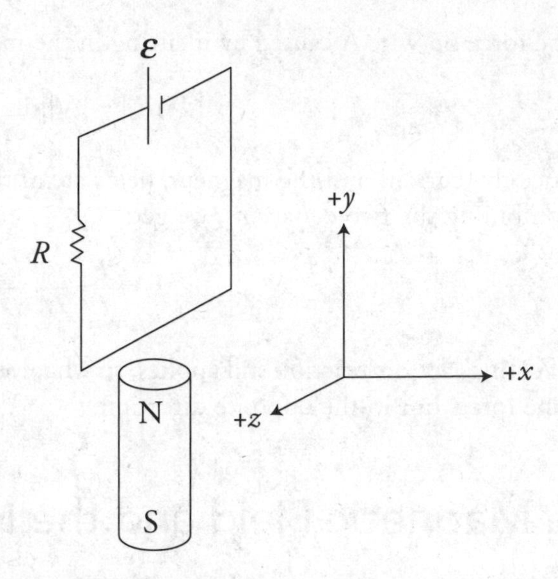

Question: The lower part of a circuit sits just above a magnet as shown in the diagram above. In what direction does the bottom wire of the circuit receive a force?

Answer: The north end of the magnet will produce a magnetic field upward in the $+y$ direction. The battery produces a current that is counterclockwise or in the $-z$ direction for the bottom wire. Using our right-hand "flat fingers" rule, we see that the force is in the $+x$ direction.

Force Between Two Parallel Wires

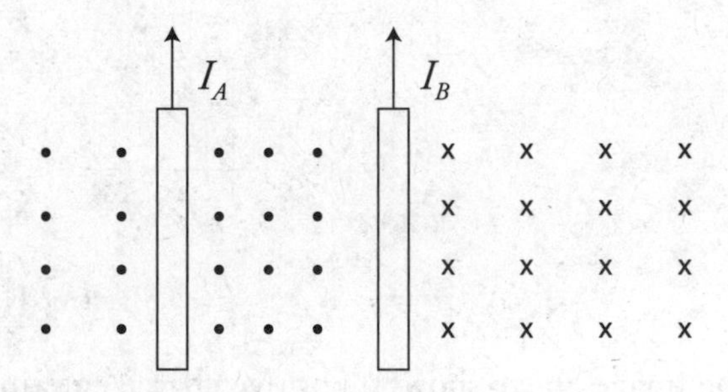

Finding the force between two parallel wires is easier than you would think. Just find the magnetic field around one wire and see how it affects the other wire. Let's start with wire B in the figure above. What will the magnetic field look like surrounding the B wire? Using our right-hand "curly fingers" rule, we find that the magnetic field will point into the paper on the right of the wire and out of the paper on the left of the wire as seen in the figure. Notice that wire A is located in the magnetic field pointing out of the paper. So, just use the right-hand "flat fingers" rule to find that the force on the left wire will be to the right. What happens to wire B? What would Newton say? Newton's third law says the force on wire B will be exactly the same size and to the left.

Finding the magnitude of the force is a little trickier but not bad. Remember that wire A is sitting in the magnetic field caused by wire B. If the wires are a distance r apart from each other, the strength of the magnetic field from wire B is:

$$B_B = \frac{\mu_0}{2\pi}\frac{I_B}{r}$$

The force on wire A caused by it sitting in the magnetic field from wire B is:

$$|\vec{F}_A| = |I_A \ell||\sin\theta||\vec{B}_B|$$

Since the current and the magnetic fields are at the right angle, $\theta = 1$, we can drop it out. Combining the two equations, we get:

$$\frac{F}{\ell} = \frac{\mu_0}{2\pi}\frac{I_A I_B}{r}$$

Of course, action/reaction still applies, so whatever force wire A experiences, wire B gets the same force, but in the opposite direction.

Charges in a Magnetic Field and the Mass Spectrometer

We have learned that magnetic forces are directed perpendicular to the velocity vector of the charge. Does that sound vaguely familiar to something you learned last year in AP Physics 1? Like the gravitational force from the Sun acting perpendicular to the velocity of Earth causing the Earth to move in a circular orbit?

Since the magnetic force is always perpendicular to the velocity of the charge, it does no work on the charge. It only changes its direction and gives the charge a centripetal acceleration. The force causes the direction of the charge's velocity to change, but not its magnitude. So we get a nice circular path.

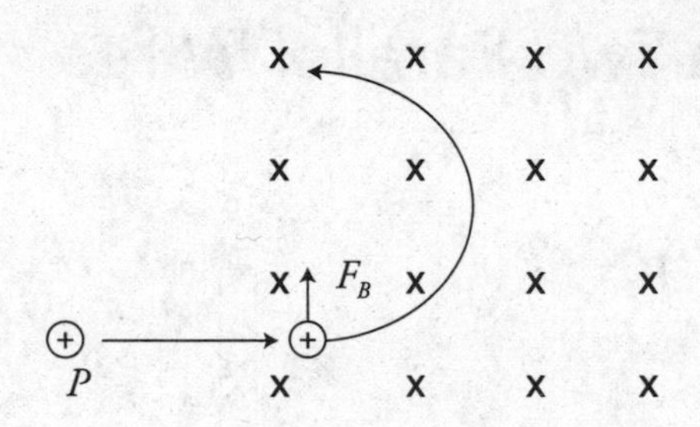

Magnetic Field into the Page

The image above shows a positive charge entering a magnetic field directed into the page. The magnetic force is always perpendicular to the velocity, causing the charge to move in a circular path.

You could be asked to find the radius of the path or the diameter of the circle. The magnetic force on the charge acts as a centripetal force:

$$F_C = F_B$$

$$\frac{mv^2}{r} = qvB$$

Don't worry about θ in this example because the velocity and magnetic fields are at a right angle to each other:

$$r = \frac{mv}{qB}$$

Notice how the radius depends on the momentum (mv) of the particle. Higher momentum = larger radius. The radius of the path is inversely related to the charge of the particle and the magnetic field strength. A larger charge and/or larger magnetic field give us a smaller radius. This equation for the radius of a charged particle's path moving through a magnetic field is not on the official equation sheet you will receive. Make sure you know how to derive it!

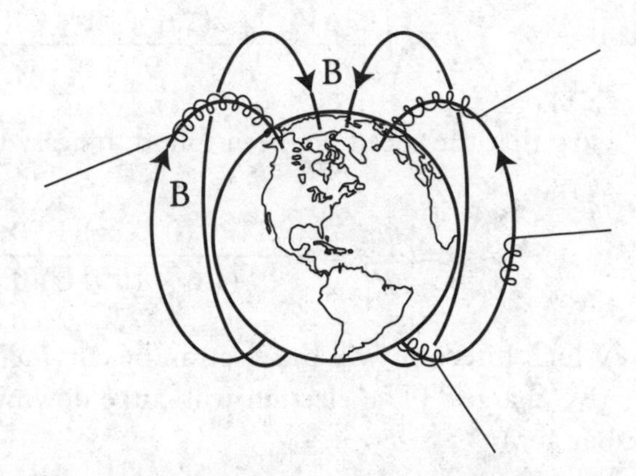

If the particle enters the magnetic field at an angle with part of the velocity parallel to the field, the charge will take a helical path. (A helical path is like a stretched out Slinky. See the diagram above.) Remember that velocities parallel to the field do not create magnetic forces. Only velocities perpendicular to the field create a force. This is what causes Earth's northern and southern lights. Charged particles from space hit the Earth's magnetic field at an angle and helix along the field lines until they enter the atmosphere near the North and South Poles.

A device called a mass spectrometer uses arcing charges in a magnetic field to determine the charge-to-mass ratio *q/m* of an unknown particle by measuring the radius of the path the particle takes in the magnetic field.

Example

Diagram of a Mass Spectrometer

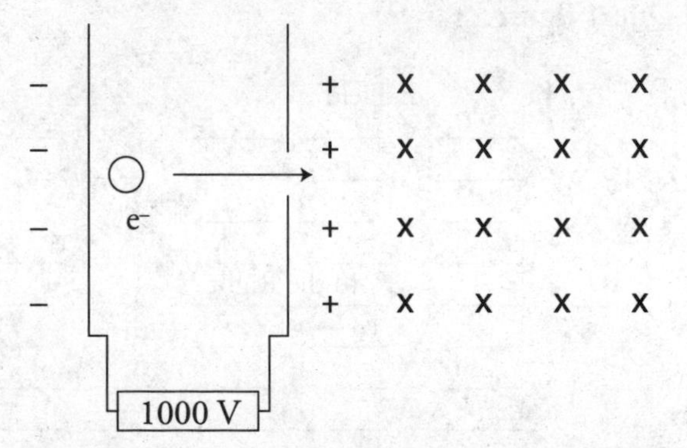

An electron is accelerated between two plates with the potential difference of 1000 V. It passes through a hole in one of the plates and enters a magnetic field of 0.1 T directed into the page. Find the radius of the path the electron makes in the magnetic field.

Looking at the figure above, we know that the electric field between the plates goes toward the left, but electrons experience a force in the opposite direction of the field. Therefore, the electron is accelerated toward the right. To find the speed of the electron as it enters the magnetic field, you can use conservation of energy:

$$\Delta U_E = K$$

$$q\Delta V = \frac{1}{2}mv^2$$

$$v = \sqrt{\frac{2q\Delta V}{m}} = \sqrt{\frac{2(1.6 \times 10^{-19}\text{ C})(1000\text{ V})}{9.11 \times 10^{-31}\text{ kg}}} = 1.9 \times 10^7\text{ m/s}$$

Now that the speed has been found, it's easy to find the radius:

$$r = \frac{mv}{qB} = \frac{(9.11 \times 10^{-31}\text{ kg})(1.9 \times 10^7\text{ m/s})}{(1.6 \times 10^{-19}\text{ C})(0.1\text{ T})} = 1.1 \times 10^{-3}\text{ m}$$

Which direction will the electron be curving? Remember to use the left-hand rule for negative charges. (The electron will curve downward toward the bottom of the page in a circular arc.)

What happens to the electron if it leaves the magnetic field? Remember Newton's first law. (The electron will travel off in a straight line along the final direction it had when it leaves the magnetic field.)

Practice sketching several possible paths of the figure above because the AP exam likes to ask conceptual questions like this. If you find that your sketch looks "bad," just write a note to the exam reader. Something simple and to the point like: "The path in the field is circular. Once it leaves the field it will travel in a straight line." That is all you need to do! Remember that this is not an art exam. The exam readers are looking for your physics understanding, and a quick explanation will secure you points.

Particles Moving Through Both Magnetic and Electric Fields

Let's review the differences between electric and magnetic fields. Look at the four fields in the next figure.

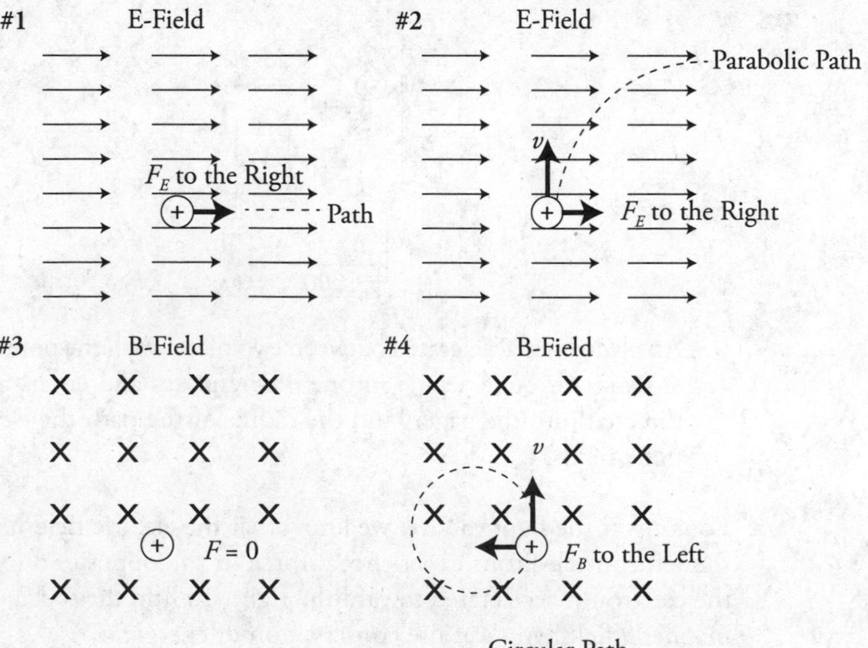

- **Electric Fields:** A positive charge in an electric field receives a force to the right. It does not matter if the particle is moving or not. Starting at rest (see #1), the particle will accelerate to the right. Starting with a velocity upward (see #2), the particle receives a force to the right and will travel along a parabolic path just like a baseball in Earth's gravitational field.
- **Magnetic Fields:** A positive charge will not experience a force unless it is moving (see #3). Starting with a velocity upward (see #4), the particle will experience a centripetal force that causes it to arc into a circular path.
- **Remember:** If the charge is negative, the forces will be reversed in direction.

Make sure you keep these two different behaviors in mind because the AP exam likes to test you on them. For example: A proton is moving at 5×10^4 m/s toward the right between two charged parallel plates. The electric field between the plates is 2×10^5 N/C. Find the magnitude and direction of the magnetic field between the plates needed so the electron will continue in a straight line.

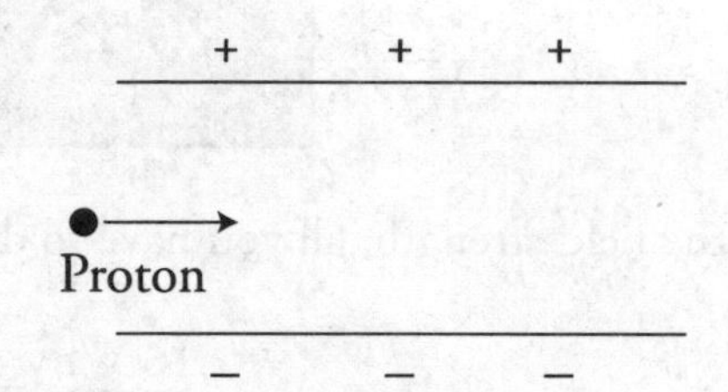

Let's first work out the direction of the electric field. You know that the electric field between the plates points in the direction from high to low electric potential. Since the upper plate is positive and the lower plate is negative, the electric field is toward the bottom of the page. The proton is positive so it experiences an electric force toward the bottom of the page. But remember, we want the proton to move in a straight line, which means the force from the magnetic field has to balance out the force from the electric field and point toward the top of the page. Now it's time to break out our right-hand "flat finger" rule. The magnetic force must be toward the top of the page so your palm needs to point toward the top of the page. The velocity is toward the right side of the page, so your thumb will point toward the right. If you set your hand up correctly, your fingers will point into the page, telling you that the magnetic field is aimed into the page.

We've already mentioned that the forces from the magnetic and electric fields must cancel each other out:

$$\vec{F}_E = \vec{F}_M$$

$$q\vec{E} = qv\vec{B}$$

What's nice is that the charge drops out, and since the velocity and magnetic field are at right angles to each other, we don't need any trigonometry functions:

$$\vec{B} = \frac{\vec{E}}{\vec{v}} = \frac{2 \times 10^5 \text{ N/C}}{5 \times 10^4 \text{ m/s}} = 4.0 \text{ T}$$

Magnetic Flux

Magnetic flux is a measure of the total magnetic field passing through an area. It is measured in webers. The connection between magnetic field strength (which is sometimes called magnetic flux density) and magnetic flux can be thought of like this: picture a square area 2 meters on each side with a total magnetic flux of 16 webers passing through that area. You can draw that with 16 X's inside of that area, as shown in the next figure.

To find the field strength, all you have to do is divide the flux by the area:

$$\frac{16\text{ Wb}}{4\text{ m}^2} = 4\frac{\text{Wb}}{\text{m}^2} = 4\text{T}$$

The equation for magnetic flux is:

$$\Phi_B = \vec{B} \cdot \vec{A}$$

Notice the "dot" in the equation. This is called a dot product. When using a dot product, you are multiplying the magnitudes of the vectors and ending up with a scalar answer. Another way we can look at a dot product is:

$$\Phi_B = |\vec{B}|\cos\theta|\vec{A}|$$

Notice that we are using cosine. How can this relationship be applied? Picture a magnetic field going toward the right. You are holding a piece of paper in front of you with the magnetic field passing through the paper. As you turn the paper, at one point there is the most flux possible passing through the paper. (See the left side of the following figure.) If the paper is then turned 90°, none of the magnetic flux will pass through the piece of paper. (See the right side of the figure.)

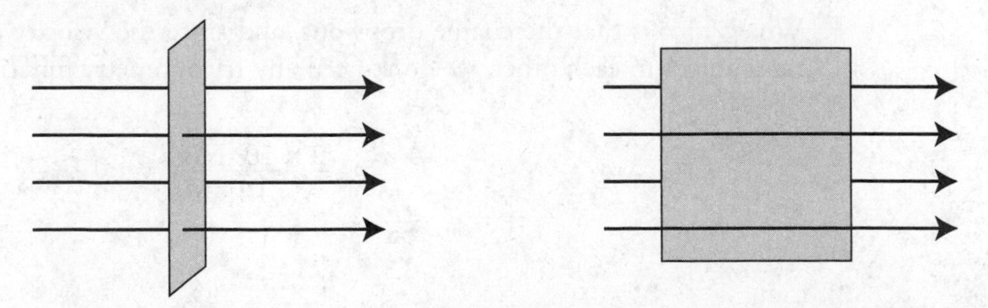

Just remember: There's maximum flux when the field passes directly through the area and no flux when the field passes by, but not through, the area.

Electromagnetic Induction

Of all the technological innovations that have been implemented, generating electricity may be the most useful. Most people don't really have any idea how the majority of our electricity is actually generated, but you, as an AP Physics ace, do. So let's take this step by step.

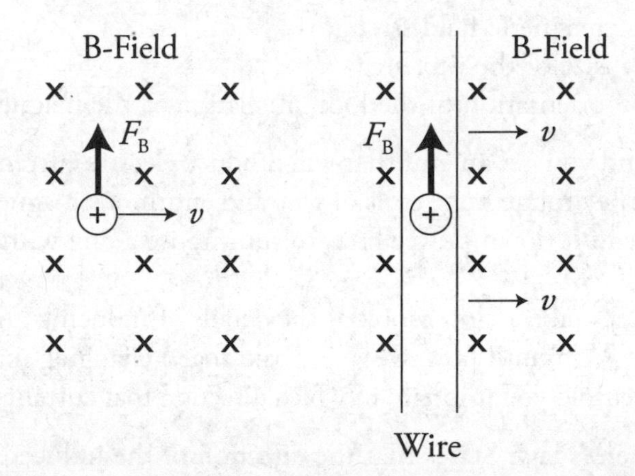

To understand how electromagnetic induction works, picture a proton moving toward the right in a magnetic field aiming into the page. Using the right-hand rule, the force acting on the proton will be toward the top of the page. Now let's take a look at a wire being dragged toward the right as shown in the second image. Conventional current is considered to be the flow of positive charge, so if you picture a positive charge in that wire, it will get a force toward the top of the page as the wire is moved to the right. In fact, every positive charge in the wire will get that force, which will produce a current moving toward the top of the wire.

This is called an electromotive force, emf, or ε, measured in volts. To find the electromotive force, or emf, of a wire moving through a magnetic field:

$$\varepsilon = B\ell v$$

where:

- ε is the electromotive force in volts.
- B is the magnetic field strength in teslas.
- ℓ is the length of the wire embedded within the magnetic field in meters.
- v is the velocity of the wire in meters per second.

Another way we can use electromagnetic induction to produce currents is to have a loop of wire that has a changing magnetic field through the inside of that loop. This can be done by:

- Changing the magnetic field strength
- Changing the flux area of the loop
- Changing the direction the loop is facing in the field (Turning the loop so it's parallel to the field and then turning it so that it's perpendicular to the field will change the amount of magnetic flux in the loop.)

To find the emf:

$$\varepsilon = -N\frac{\Delta\Phi_B}{\Delta t}$$

where:

- N is the number of wraps of wire in the loop.

Remember that an emf is like a battery. That means, if you have a loop of wire in a magnetic field, all you have to do is change one of these three things:

1. magnetic field strength
2. size of the flux area
3. orientation of the loop in relation to the magnetic field

and you get an emf that will produce electric current! The electric company uses option #3. They rotate huge coils of wire and enormous magnets past each other to induce current and send it down power lines to your house along with a monthly electric bill.

Let's take a closer look at the details of inducing an emf.

Up until now, we've just said that a changing magnetic flux creates a current. We haven't yet told you, though, in which direction that current flows. To do this, we'll turn to Lenz's law.

Lenz's law: States that the direction of the induced current opposes any change in flux.

When a current flows through a loop, that current creates its own magnetic field. So what Lenz said is that the current that is induced will flow in such a way that the magnetic field it creates points opposite to the direction in which the already existing magnetic flux is changing.

Sound confusing?[1] It'll help if we draw some good illustrations. So here is Lenz's law in pictures.

We'll start with a loop of wire that is next to a region containing a magnetic field. Initially, the magnetic flux through the loop is zero.

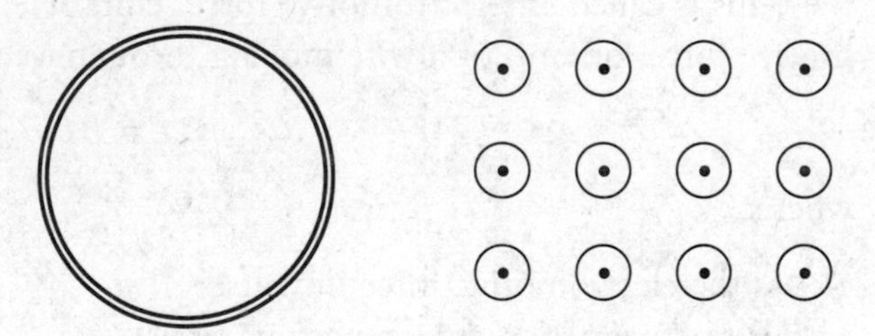

Now, we will move the wire into the magnetic field. When we move the loop toward the right, the magnetic flux will increase as more and more field lines begin to pass through the loop. The magnetic flux is increasing out of the page—at first, there was no flux out of the page, but now there is some flux out of the page. Lenz's law says that the induced current will create a magnetic field that opposes this increase in flux. So the induced current will create a magnetic field into the page to oppose the increasing flux out of the page. By the right-hand rule, the current will flow clockwise. This situation is shown in the next illustration.

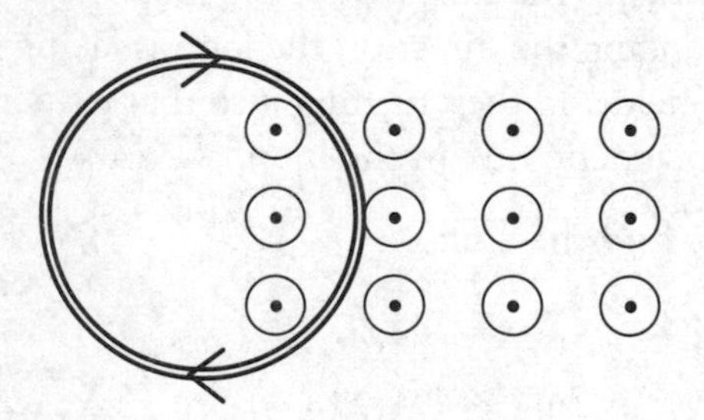

[1] "Yes!"

After a while, the loop will be entirely in the region containing the magnetic field. Once it enters this region, there will no longer be a changing flux, because no matter where it is within the region, the same number of field lines will always be passing through the loop. Without a changing flux, there will be no induced emf, so the current will stop. This is shown in the next illustration.

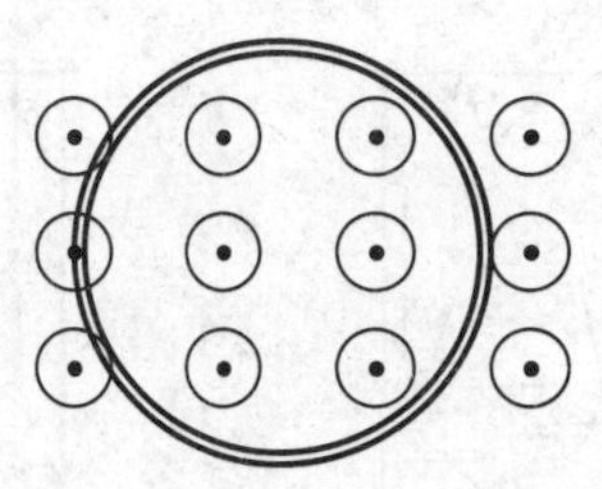

To solve a problem that involves Lenz's law, use this method:

- Point your right thumb in the initial direction of the magnetic field.
- Ask yourself, "Is the flux increasing or decreasing?"
- If the flux is decreasing, Lenz says we need to cancel out this decrease in flux by adding magnetic field in the direction your thumb is pointing. Just curl your fingers (with your thumb still pointed in the direction of the magnetic field). Your fingers show the direction of the induced current.
- If flux is increasing in the direction you're pointing, then Lenz says we need to cancel this increase. So, point your thumb in the opposite direction of the magnetic field, and curl your fingers. Your fingers show the direction of the induced current.

Induced emf in a Rectangular Wire

Consider the example in the earlier figures with the circular wire being pulled through the uniform magnetic field. It can be shown that if instead we pull a *rectangular* wire into or out of a uniform field B at constant speed v, then the induced emf in the wire is found by

$$\varepsilon = B\ell v$$

Here, ℓ represents the length of the side of the rectangle that is cutting through the magnetic field lines, as shown below.

X X X X
B
X X X X
X X X X
X X X X
$\ell \longrightarrow v$

Notice how the rectangle is being pulled from the field. That means the magnetic flux is decreasing. Using Lenz's law rules, we find that the induced current in the rectangle of wire will be clockwise.

We use electromagnetic induction:

- to generate electricity
- in microphones to turn our voices into electrical currents and then convert those currents back into sound with speakers
- to run motors
- in MRIs and other medical devices
- on those magnetic strips on the back of credit cards
- and the list goes on and on

It is truly a powerful tool.

Magnetic Behavior of Materials

You may have noticed not every material seems to react to a magnet. There are actually three types of magnetic behavior that materials exhibit.

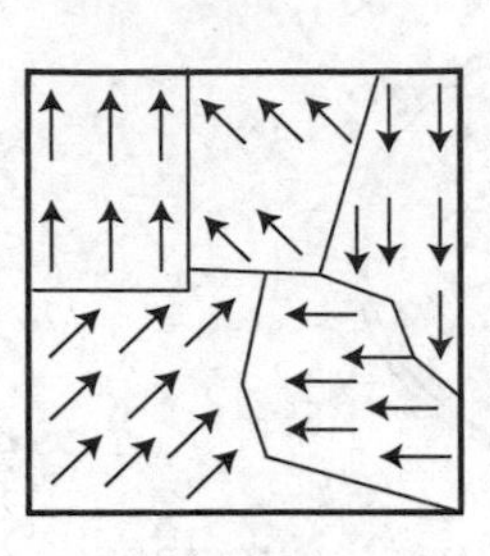
No Field

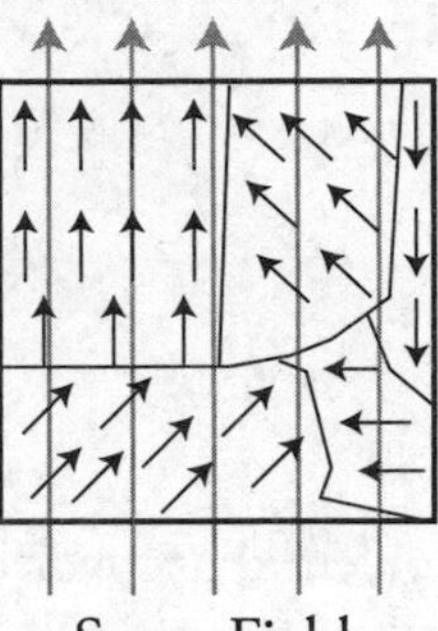
Some Field

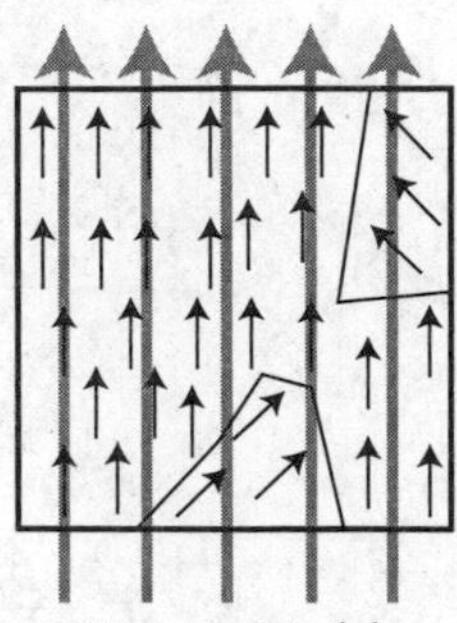
Strong Field

1. **Ferromagnetism:** Materials like iron, nickel, and cobalt have multiple unpaired electrons that tend to create localized regions of magnetic fields inside the material called domains. When these materials are placed in an external magnetic field, the domains align with the external magnetic field, amplifying it many times. In a strong enough external field, the domains can grow and merge, creating a permanent magnet. (See the figure above.) This is why a current-carrying wire wrapped around an iron nail will produce such a strong magnet. Ferromagnetic materials are strongly attracted by magnets.
2. **Paramagnetism:** The magnetic properties of the material tend to align with an external magnetic field, but the result is weak and does not enhance the magnetic field very much. These materials will not produce a permanent magnet. Paramagnetic materials are very weakly attracted to magnets.
3. **Diamagnetism:** When placed in a magnetic field, the internal magnetic properties of the material align opposite to the external field, cancelling out part of the field. Water and graphite behave this way. Diamagnetic materials are very weakly repelled by magnets.

› Practice Problems

Multiple Choice

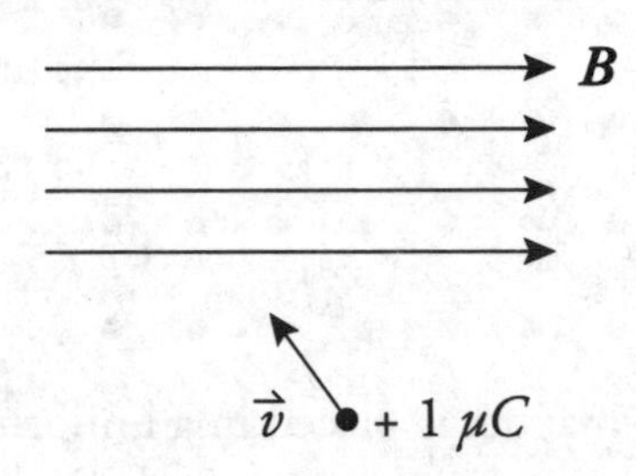

1. A point charge of +1 μC moves with velocity v into a uniform magnetic field B directed to the right, as shown above. What is the direction of the magnetic force on the charge?

(A) to the right and up the page
(B) directly out of the page
(C) directly into the page
(D) to the right and into the page

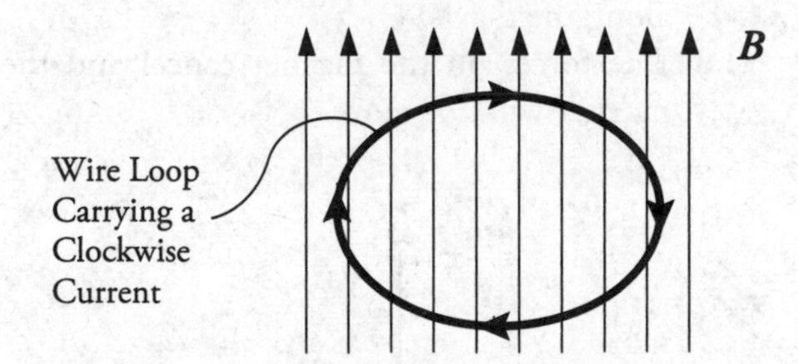

2. A uniform magnetic field B points up the page, as shown above. A loop of wire carrying a clockwise current is placed at rest in this field, as shown above, and then let go. Which of the following describes the motion of the wire immediately after it is let go?

(A) The wire will expand slightly in all directions.
(B) The wire will contract slightly in all directions.
(C) The wire will rotate, with the top part coming out of the page.
(D) The wire will rotate, with the left part coming out of the page.

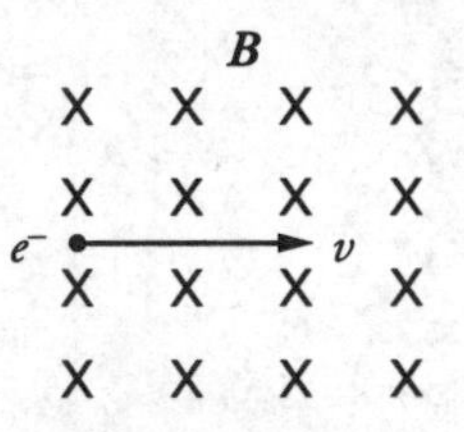

3. An electron moves to the right in a uniform magnetic field that points into the page. What is the direction of the electric field that could be used to cause the electron to travel in a straight line?

(A) down toward the bottom of the page
(B) up toward the top of the page
(C) into the page
(D) out of the page

4. The vector force field surrounding a bar magnet has a shape similar to the force field surrounding which of the following?

(A) two stars of the same mass in a binary star system
(B) a proton and electron in a hydrogen atom
(C) two equally charged Van de Graaff generators placed close to each other on a lab table
(D) an alpha particle

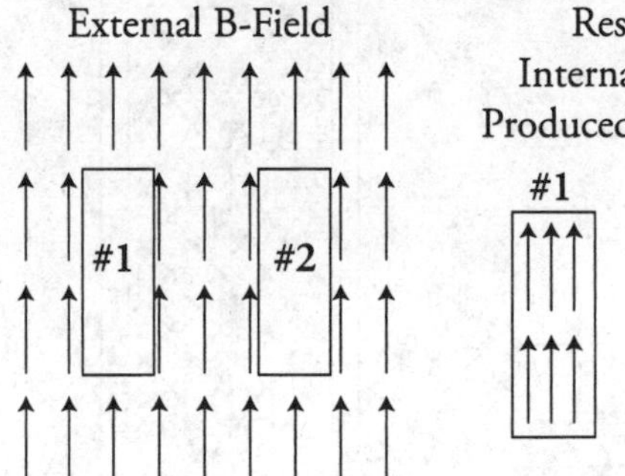

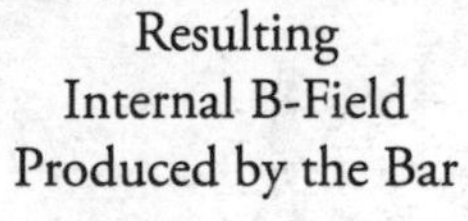

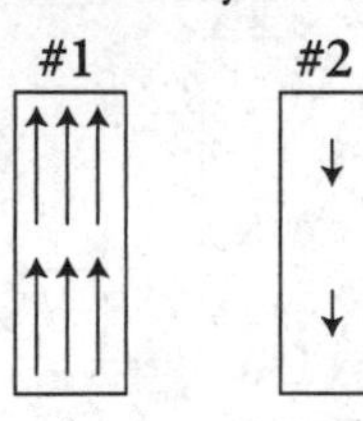

5. Two bars of differing materials are placed in an external magnetic field as shown above. The resulting internal magnetic field in each bar is depicted in the second figure. Which of the following correctly describes the two materials?

(A) Bar 1 is a ferromagnetic material; bar 2 is a paramagnetic material.
(B) Bar 1 is a ferromagnetic material; bar 2 is a diamagnetic material.
(C) Bar 1 is a non-magnetic material; bar 2 is a paramagnetic material.
(D) Bar 1 is a non-magnetic material; bar 2 is a diamagnetic material.

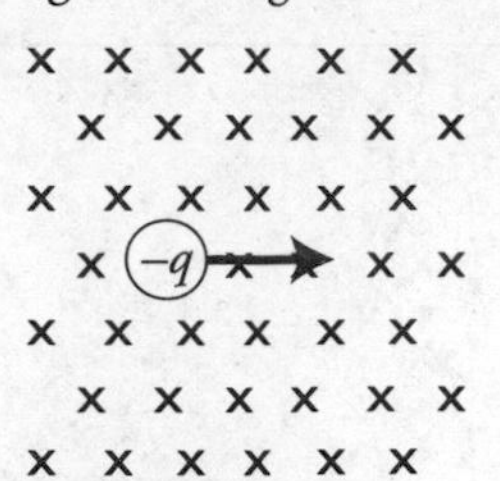

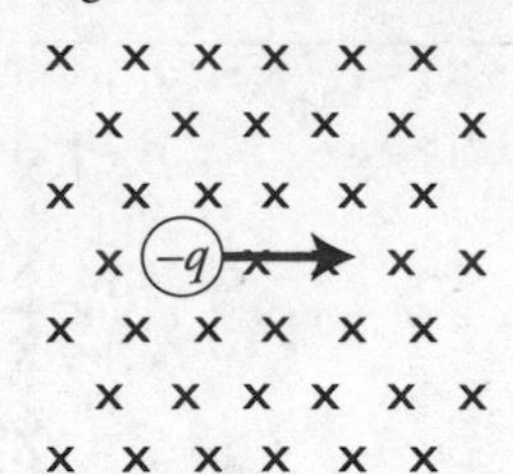

6. A scientist sends two identical negative charges $-q$ moving to the right. One is in a magnetic field and one is in an electric field as shown in the figures above. The scientist observes the motion of both charges. Which of the following correctly indicates the direction each charge will turn?

	Magnetic Field	Electric Field
(A)	toward the top of the page	into the page
(B)	toward the top of the page	out of the page
(C)	toward the bottom of the page	out of the page
(D)	toward the bottom of the page	The charge does not change direction

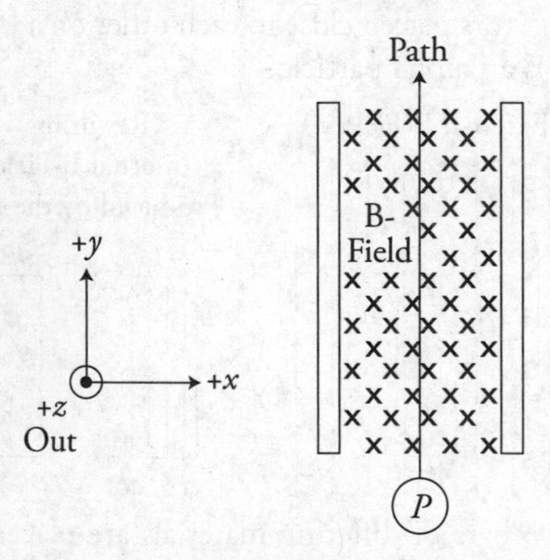

7. A magnetic field directed into the page in the $-z$ direction is placed between the plates of a charged capacitor as shown in the figure. The magnetic and electric fields are adjusted so that a particle of charge $+1e$ moving at a velocity of v will pass straight through the fields in the $+y$ direction. Which direction is the electric field in the region between the plates?

(A) $+x$ direction
(B) $-x$ direction
(C) $+z$ direction
(D) $-z$ direction

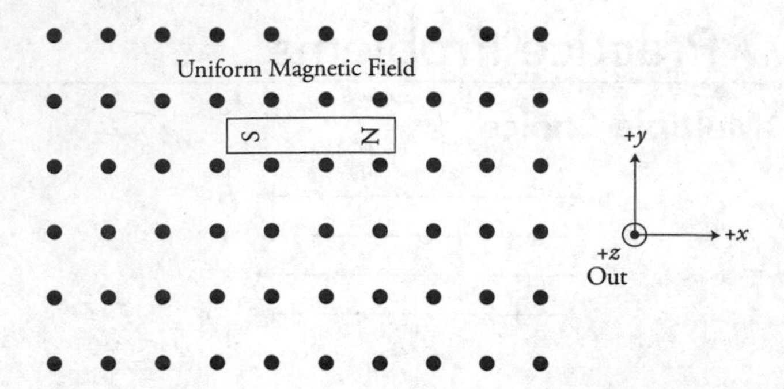

8. A small magnet is placed in a uniform magnetic field that is pointing up out of the paper in the $+z$ direction. If the magnet is free to move in the field, which of the following is a correct statement about its motion?

(A) There is a torque that rotates the magnet about the y-axis.
(B) There is a torque that rotates the magnet about the z-axis.
(C) There is a force that moves the magnet along the z-axis.
(D) The forces on the magnet cancel and the magnet will not move.

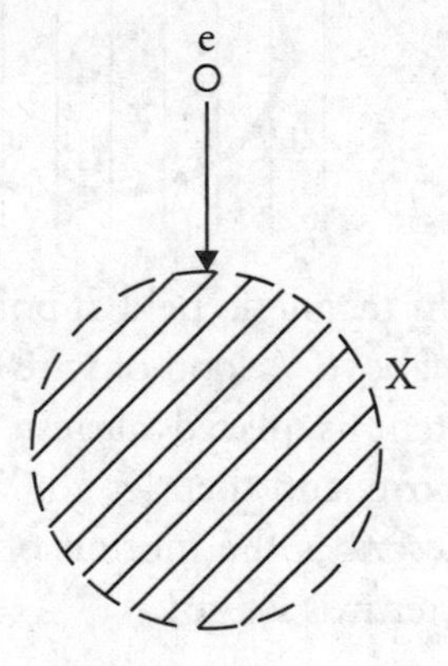

9. An electron is moving downward toward the bottom of the page when it passes through a region of magnetic field, as shown in the figure above by the shaded area. The electron travels along a path that takes it through the spot marked X. The gravitational force on the electron is very small. What is the direction of the magnetic field?

(A) Toward the bottom of the page
(B) Toward the top of the page
(C) Out of the page
(D) Into the page

Questions 10–11

A proton is moving at a velocity of 6.0×10^4 m/s to the right, in the plane of the page, when it encounters a region of magnetic field with a magnitude 0.12 T perpendicular to the page, as shown in the figure below.

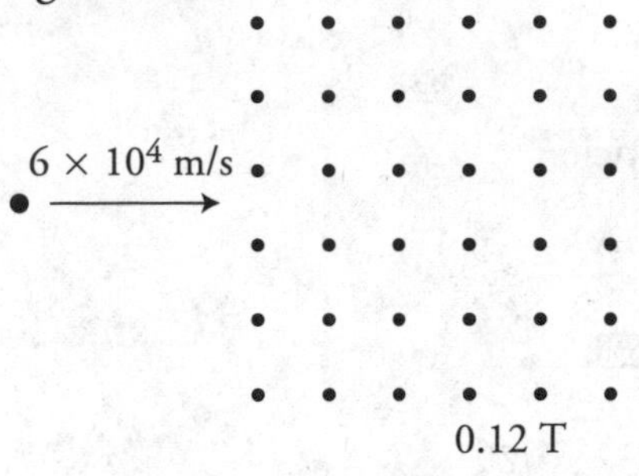

10. Which of the following is the radius of curvature of the path of the proton?

(A) 5×10^1 m
(B) 5×10^{-3} m
(C) 5×10^{-5} m
(D) 5×10^{-7} m

11. The proton is replaced with an electron moving in the same direction and at the same speed. Which of the following best describes the deflection direction and the radius of curvature of the electron in the magnetic field?

Deflection direction	Radius of curvature
(A) Same as proton	Larger than proton's
(B) Same as proton	Smaller than proton's
(C) Opposite of proton	Larger than proton's
(D) Opposite of proton	Smaller than proton's

Questions 12–13

An electron is traveling at a constant speed of v parallel to a wire carrying a current of I, as shown in the figure below. The electron is a distance of d from the wire.

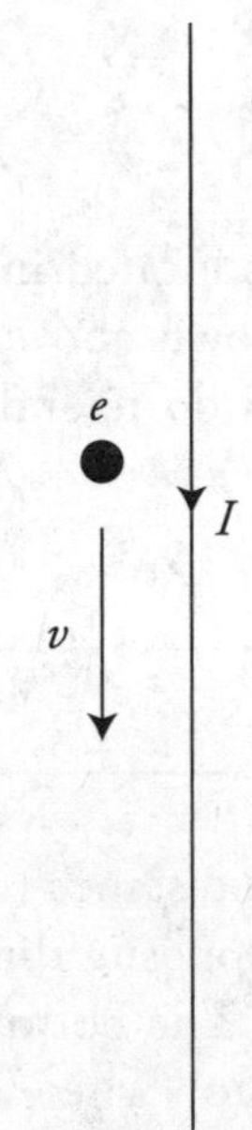

12. Which of the following is true concerning the force on the current-carrying wire due to the electron?

(A) The force is directed toward the right.
(B) The force is directed toward the left.
(C) The force is directed into the page.
(D) There is no force on the current-carrying wire due to the electron.

13. The force on the electron from the current is F. Which of the following will increase the force to $2F$? **(Select two answers.)**

(A) Halve the distance of the electron to the wire.
(B) Halve the velocity of the electron.
(C) Double the current in the wire.
(D) Double the current in the wire and halve the distance of the electron to the wire.

Free Response

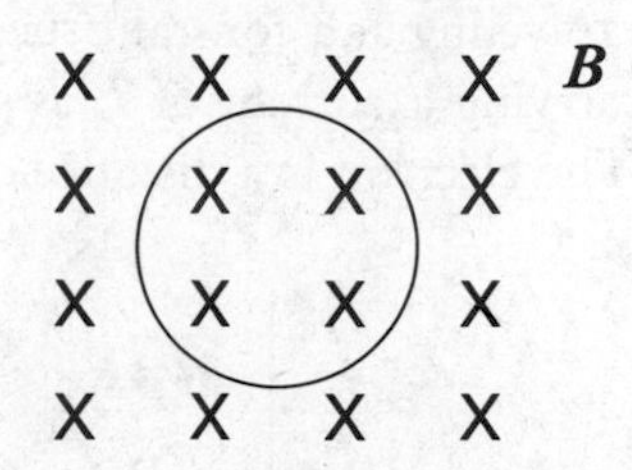

14. A loop of wire is located inside a uniform magnetic field, as shown above. Name at least four things you could do to induce a current in the loop.

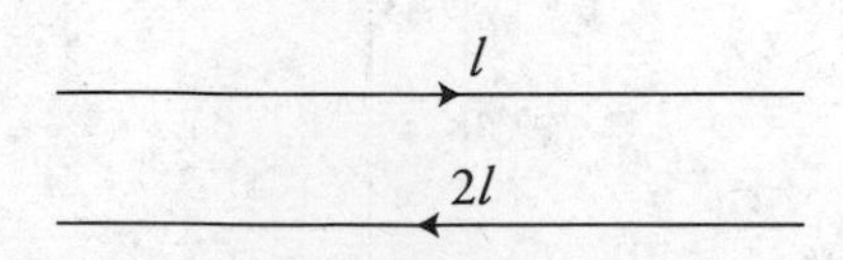

15. Two long wires, a distance (d) apart, carry different currents in opposite directions, as shown in the figure above. The bottom wire has a current twice that of the top wire.

(A) What is the direction of the magnetic force on the top wire?
(B) How does the magnetic force on the bottom wire compare in strength and direction to the force on the top wire? Justify your claim.
(C) Which way is the net magnetic field directed at the midpoint between the wires? Explain your reasoning.
(D) Derive an algebraic expression for the force per unit length on the top wire by the lower wire.

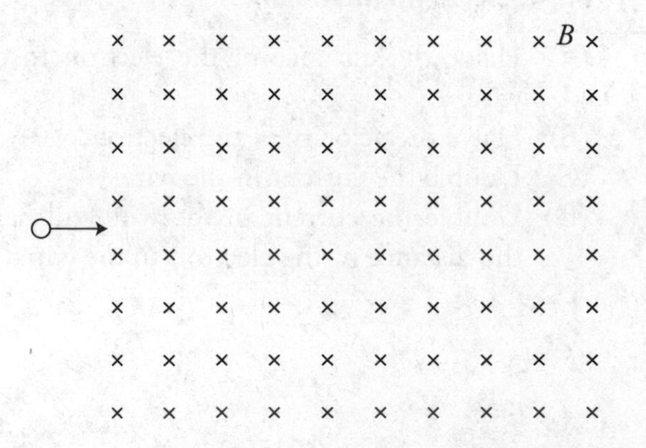

16. A particle is moving toward a magnetic field directed into the page, as shown in the figure above. Sketch and label the path taken by each of the following particles. Draw all of the pathways in proportion to the paths taken by all other particles. All particles enter the magnetic field with the same initial velocity and direction.

(A) Neutron
(B) Proton
(C) Electron
(D) Positron
(E) Alpha particle

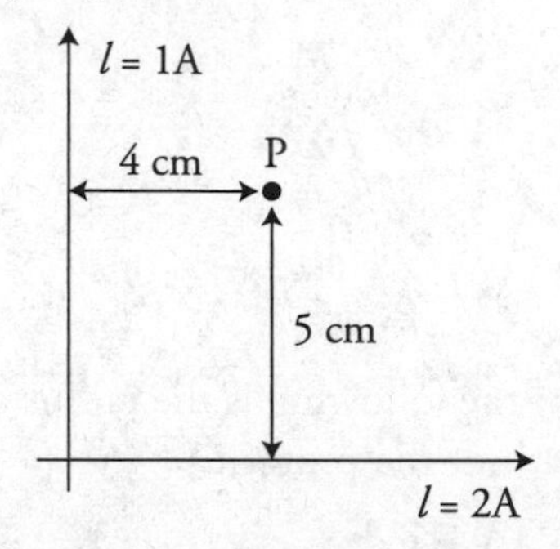

17. Two perpendicular wires carry currents of 1.0 A and 2.0 A, as shown in the figure above. Point P is 4 cm from the 1.0 A wire and 5 cm from the 2.0 A wire. Show all your work.

(A) Calculate the net magnitude and direction of the magnetic field at point P.
(B) Calculate the magnitude and direction of the force on a proton placed at point P.
(C) Calculate the magnitude and direction of a force on an electron at point P, moving at 3.5×10^6 m/s toward the right.

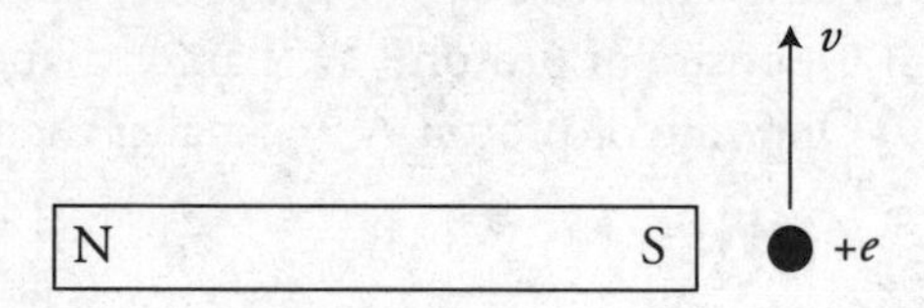

18. A proton traveling toward the top of the page at a velocity of v passes by a magnet, as shown in the figure above.

(A) What is the direction of the force on the proton?
(B) What is the direction of the force on the magnet? Explain your answer.
(C) What happens to the force on the proton if the velocity is doubled to $2v$?

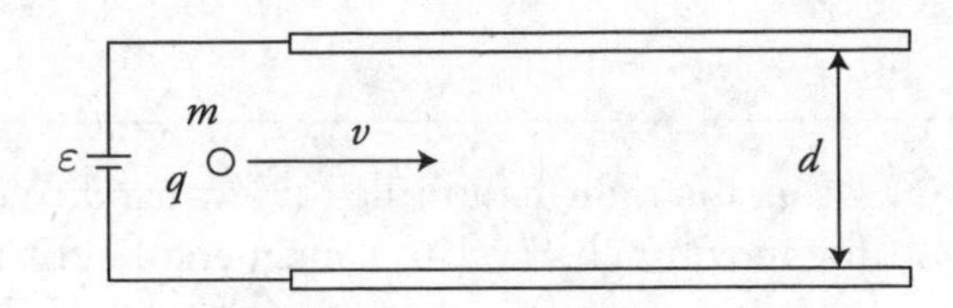

19. A positively charged particle (q) of mass m travels horizontally, with a velocity of v, through the center of two capacitor plates. The plates are separated by a distance of d and connected to a battery of potential difference (ε), as shown in the figure above.

(A) Sketch the electric field between the plates.
(B) Derive an algebraic expression for the electric field between the plates in terms of given quantities. Show all your work.
(C) Describe the motion of the particle as it passes through the capacitor plates. What shape is the path? Which direction is the acceleration? Explain your reasoning.
(D) What direction of magnetic field is needed to make the particle travel horizontally straight through the capacitor plates? Justify your answer.
(E) Derive an algebraic expression for the magnitude of the magnetic field needed to cause this straight, horizontal motion between the plates in terms of given quantities. Show all your work.
(F) The crossed electric and magnetic fields are adjusted to cause positively charged particles with a velocity of v to travel straight. What happens to a particle traveling at $2v$? Will it travel straight, or will it curve? Justify your answer.
(G) The crossed electric and magnetic fields are tuned to cause positively charged particles with a velocity of v to travel straight. What happens to a *negatively* charged particle traveling at v? Will it travel straight, or will it curve? Justify your answer.

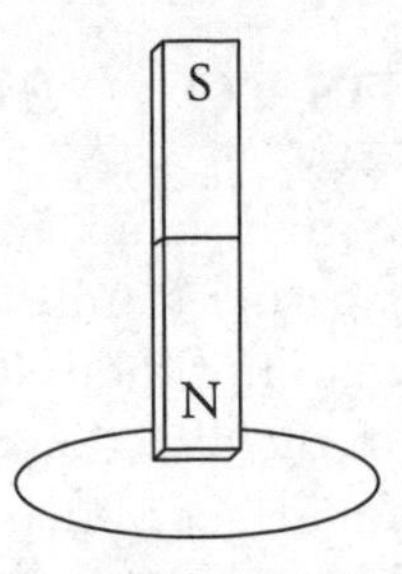

20. A loop of wire is positioned perpendicular to the end of a magnet, as shown in the figure above.

(A) Describe at least two ways to move the wire loop that will produce current in the wire. Explain why each method produces current.
(B) Describe at least two ways to move the magnet so current is produced in the wire. Explain why each method produces current.
(C) Describe at least two methods of how the magnet and/or wire loop can be moved that will not produce current. Explain why neither of these methods produce current.

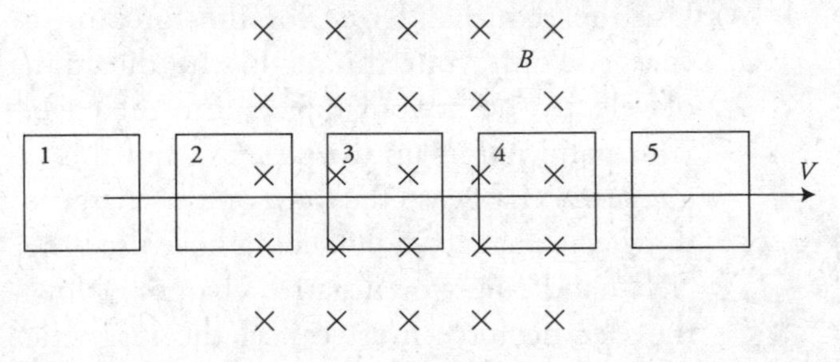

21. A rectangular loop of wire travels to the right and passes through a region of magnetic field into the page, as shown in the figure above.

(A) At which locations will current be induced in the wire? Explain your answer.
(B) At the locations where current is induced, what direction is the current in the wire loop—clockwise or counterclockwise? Explain your answer.

› Solutions to Practice Problems

1. (C) Use the right-hand rule for the force on charged particles. You point your thumb in the direction of the velocity, and point your fingers in the direction of the magnetic field. This should get your palm facing into the page. Because this is a positive charge, no need to switch the direction of the force.

2. (C) Use the right-hand rule for the force on a wire. Look at each part of this wire. At the leftmost and rightmost points, the current is along the magnetic field lines. Thus, these parts of the wire experience no force. The topmost part of the wire experiences a force out of the page (point your thumb to the right, fingers point up the page, the palm points out of the page). The bottommost part of the wire experiences a force *into* the page. So, the wire will rotate.

3. (A) Use the right-hand rule for the force on a charge. Point your thumb in the direction of velocity, point your fingers into the page, your palm points up the page . . . but this is a *negative* charge, so the force on the charge is down the page. (Or, you could have used the "left-hand rule" for negative charges!) Now, the electric force must cancel the magnetic force for the charge to move in a straight line, so the electric force should be up the page. (E and B *fields* cannot cancel, but forces sure can.) The direction of an *electric* force on a negative charge is opposite the field; so the field should point down, toward the bottom of the page.

4. (B) The two opposite poles of a magnet will have similar field characteristics to the two opposite sign charges in a hydrogen atom.

5. (B) When placed in an external magnetic field, magnetic domains inside a ferromagnetic material will strongly align with the external field, which intensifies the strength of the field. Diamagnetic materials, on the other hand, will generate an internal magnetic field in the opposite direction, weakening the strength of the field.

6. (C) Using the right-hand rule (or left-hand rule!) for moving charges in a magnetic field, the negative charge receives a force toward the bottom of the page. (Remember that this is a negative charge!) Electric forces are always parallel to the field, but opposite in direction for negative charges.

7. (A) Using the right-hand rule for magnetic forces on moving charges, we see that the magnetic force will be directed to the left. Therefore, we need an electric force to the right. Since the particle is positively charged, we need an electric field to the right to accomplish this.

8. (A) The north pole receives a force out of the page, and the south pole receives a force in the opposite direction into the page. This will create a torque on the magnet, rotating it about the y-axis.

9. (C) The original velocity of the electron is toward the bottom of the page. The force on the electron is to the right. Therefore, by the right-hand rule, the magnetic field is out of the page. Remember that the electron is negative and receives a force opposite to what positive receives!

10. (B) $F_C = F_M = qvB = ma_C = m\frac{v^2}{r}$

$$r = \frac{mv}{qB} = 2 \times 10^{-2}\ \text{m}$$

11. (D) $r = \frac{mv}{qB}$. The mass of the electron is smaller; therefore, the radius is smaller as well. The charge is opposite, so the electric force is in the opposite direction.

12. (A) The magnetic field, due to the current in the wire near the electron, is into the page. By the right-hand rule, the negative electron will receive a force to the left due to the current. By Newton's third law, the force on the wire will be equal and opposite to the right.

13. (A) and (C) Combining the magnetic force on a moving charge equation with the magnetic field around a wire equation we get: $F_M = qvB = (qv)\frac{\mu_0}{2\pi}\frac{I}{r}$. From this we can see that cutting the distance in half and doubling the current will cause the force to increase by a factor of two.

14. The question might as well be restated, "name four things you could do to change the flux through the loop," because only a changing magnetic flux induces an emf.

(A) Rotate the wire about an axis in the plane of the page. This will change the θ term in the expression for magnetic flux, $BA \cos \theta$.

(B) Pull the wire out of the field. This will change the area term, because the magnetic field lines will intersect a smaller area of the loop.

(C) Shrink or expand the loop. This also changes the area term in the equation for magnetic flux.

(D) Increase or decrease the strength of the magnetic field. This changes the B term in the flux equation.

15. (A) The bottom wire produces a magnetic field into the page, in the region of the top wire, which causes a force directed upward toward the top of the page and away from the lower wire.

(B) According to Newton's third law, the force is equal in strength but opposite in direction.

(C) Using the right-hand rule for wires we see that the magnetic field produces by the top wire points into the page between the wires. The magnetic field from the bottom wire also points into the page between the wires.

(D) The magnetic force on the top current-carrying wire is due to the magnetic field of the lower wire:

$$F_M = I_{top} l B_{bottom} = I_{top} l\left(\frac{\mu_0}{2\pi}\frac{I_{bottom}}{d}\right) = Il\left(\frac{\mu_0}{2\pi}\frac{2I}{d}\right)$$

Therefore, the force per unit length of wire is

$$\frac{F_M}{l} = \left(\frac{\mu_0}{\pi}\frac{I^2}{d}\right).$$

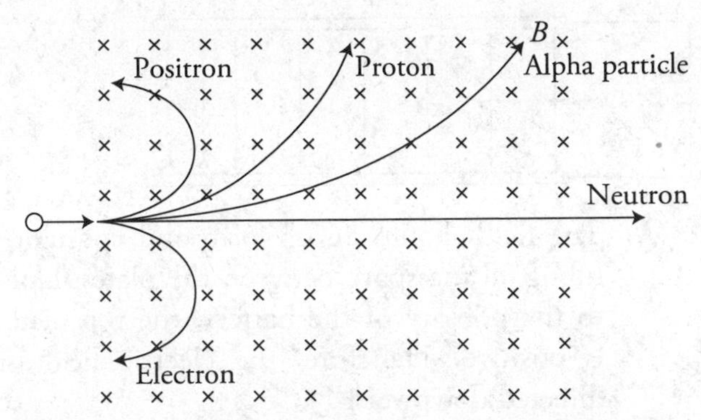

16. Note that the neutron travels along a straight path. All positive particles curve toward the top of the page. The proton, electron, and positron all have the same charge, velocity, and magnetic force acting on them. Therefore, the lightest particles will curve with a tighter radius. The positron and electron have the same radius of curvature. The alpha particle has twice the charge of the proton, which means it will receive twice the force. However, it is also four times the mass of the proton. Therefore, it will have a greater momentum per magnetic force and will curve less.

17. (A) Using the right-hand rule for magnetic fields produced around wires we see that the fields are in opposite directions at point P. Remember to convert your distance to meters.

$$B_{1A} = \frac{\mu_0}{2\pi}\frac{I}{r} = 5\times10^{-6} \text{ T into the page}$$

$$B_{2A} = \frac{\mu_0}{2\pi}\frac{I}{r} = 8\times10^{-6} \text{ T out of the page}$$

$$B_{net} = 3\times10^{-6} \text{ T out of the page}$$

(B) The force is zero because the proton is not moving!

(C) $F_M = qv(\sin\theta)B = evB = 1.7\times10^{-18}$ N toward the top of the page.

18. (A) The magnetic field is to the left. By the right-hand rule, the force is out of the page.

(B) Into the page, by Newton's third law.

(C) The force is proportional to the velocity; therefore, the force is doubled. The direction is the same.

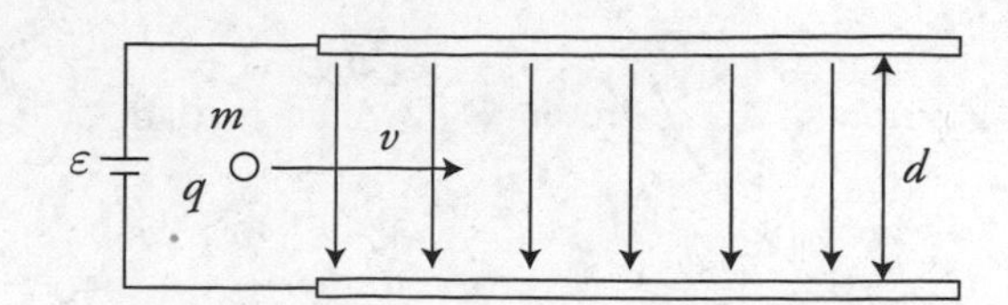

19. (A) The field should be uniform and constant, filling all the space between the plates. Due to the polarity of the battery, the top plate is positive. Therefore, the electric field is directed downward.

(B) $E = \frac{\Delta V}{\Delta r} = \frac{\mathcal{E}}{d}$

(C) The particle will accelerate downward toward the bottom plate in a parabolic path.

(D) We need a magnetic force upward to cancel the electric force downward on the positively charged particle. Using the right-hand rule, the magnetic field must point into the page.

(E) The electric and magnetic forces must cancel out: $F_E = F_M$

$$Eq = \frac{\mathcal{E}q}{d} = qvB$$

$$B = \frac{\mathcal{E}}{dv}$$

(F) The electric force remains the same, but the upward magnetic force increases because it is proportional to the velocity. Therefore, the particle will curve upward toward the top plate. $Eq < q(2v)B$.

(G) Because the charge cancels out of the equation, the negative particle will travel straight through without curving: $Eq = qvB$.

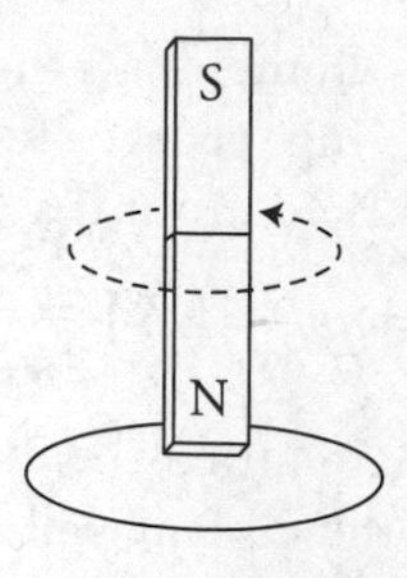

20. (A) There are many answers to this one. All we need to do is change one of the variables in the magnetic flux equation: $\Phi_M = BA\cos\theta$. Here are two examples: move the loop to the left, out of the magnetic field, or rotate the loop about a diameter.

(B) There are many answers. Here are two: move the magnet away from the loop, or drop the magnet through the loop.

(C) Here are two examples: rotate the magnet about its long axis as shown in the figure, or move both the magnet and the loop in the same direction at the same time, keeping the same distance between them, so there is no relative motion between them.

21. (A) To induce a current in the rectangular loop of wire, we need a change in magnetic flux. There is a change in magnetic flux only at locations 2 and 4.

(B) The flux is increasing into the page at location 2. Therefore, the current is counterclockwise to produce an out-of-the-page field to oppose the increase. At location 4, the flux is decreasing. Therefore, the current is clockwise to produce an into-the-page field to oppose the decrease in the flux.

❯ Rapid Review

- Don't confuse magnetic and electric fields! Charged particles behave differently in each field.
- Magnetic fields can be drawn as loops going from the north pole of a magnet to the south pole.
- A long, straight, current-carrying wire creates a magnetic field that wraps around the wire in concentric circles. The direction of the magnetic field is found by a right-hand "curly fingers" rule.
- Similarly, loops of wire that carry current create magnetic fields. The direction of the magnetic field is, again, found by a right-hand rule.
- A magnetic field exerts a force on a charged particle if that particle is moving perpendicular to the magnetic field.
- When a charged particle moves perpendicular to a magnetic field, it ends up going in circles. This phenomenon is the basis behind mass spectrometry.
- A changing magnetic flux creates an induced emf, which causes current to flow in a wire.
- Lenz's law says that when a changing magnetic flux induces a current, the direction of that current will be such that the magnetic field it induces is pointed in the opposite direction of the original change in magnetic flux.
- Domains inside ferromagnetic materials align with external magnetic fields, greatly enhancing the field. Paramagnetic materials weakly enhance external magnetic fields. Diamagnetic materials create magnetic fields that cancel the effect of external magnetic fields.

Geometric and Physics Optics

IN THIS CHAPTER

Summary: This chapter reviews some of what you learned about mechanical waves in AP Physics 1 and then extends your knowledge to include electromagnetic waves. We'll talk about interference, diffraction, reflection, refraction, and optics.

Key Ideas

- Waves are a disturbance in a medium that transports energy.
- Light is an electromagnetic wave that can travel through a vacuum.
- Electromagnetic waves are created by oscillating charges.
- The speed of a wave is constant within a given material. When a wave moves from one material to another, its speed and its wavelength change, but its frequency stays the same.
- Interference patterns can be observed when a wave goes through two closely spaced slits or even a single slit.
- Interference patterns can also be observed when light reflects off of a thin film.
- Waves can bend around corners and spread out when passing through an opening. This is called diffraction and is a natural behavior of all waves.
- The point source model, also known as Huygens' principle, explains many of the unique behaviors of waves.

- When waves encounter a boundary, they can transmit, reflect, and be absorbed.
- Transverse waves can be polarized.
- When light hits an interface between materials, the light reflects and refracts. The angle of refraction is given by Snell's law.
- Total internal reflection can occur when light strikes a boundary going from a higher-index material to a lower-index material at a large incident angle.
- Concave mirrors and convex lenses are optical instruments that converge light to a focal point. Concave lenses and convex mirrors are optical instruments that cause light to diverge.
- Virtual images are formed right-side up; real images are upside down and can be projected on a screen.

Relevant Equations

Mathematical equation of a wave:	$x = A\cos(\omega t) = A\cos(2\pi ft)$
Relationship between frequency and time period:	$f = \frac{1}{T}$
Wavelength:	$\lambda = \frac{v}{f}$
Path length difference in an interference pattern:	$\Delta L = m\lambda$
Angle between bright spots (constructive interference locations) in an interference pattern:	$d\sin\theta = m\lambda$
Index of refraction:	$n = \frac{c}{v}$
Snell's law:	$n_1 \sin\theta_1 = n_2 \sin\theta_2$
The lensmaker's equation/mirror equation:	$\frac{1}{s_i} + \frac{1}{s_o} = \frac{1}{f}$
Magnification equation:	$\lvert M\rvert = \left\lvert\frac{h_i}{h_o}\right\rvert = \left\lvert\frac{s_i}{s_o}\right\rvert$

Transverse and Longitudinal Waves

Waves come in two forms: transverse and longitudinal.

Wave: A rhythmic oscillation that transfers energy from one place to another.

A *transverse wave* occurs when the particles in the wave move perpendicular to the direction of the wave's motion. When you jiggle a string up and down, you create a transverse wave. A transverse wave is shown next.

Longitudinal waves occur when particles move parallel to the direction of the wave's motion. Sound waves are examples of longitudinal waves. A good way to visualize how a sound wave propagates is to imagine one of those "telephones" you might have made when you were younger by connecting two cans with a piece of string. When you talk into one of the cans, your vocal cords cause air molecules to vibrate back and forth, and those vibrating air molecules hit the bottom of the can, which transfers that back-and-forth vibration to the string. Molecules in the string get squished together or pulled apart, depending on whether the bottom of the can is moving back or forth. So the inside of the string would look like the next figure, in which dark areas represent regions where the molecules are squished together, and light areas represent regions where the molecules are pulled apart.

The terms we use to describe waves can be applied to both transverse and longitudinal waves, but they're easiest to illustrate with transverse waves. Take a look at the next figure.

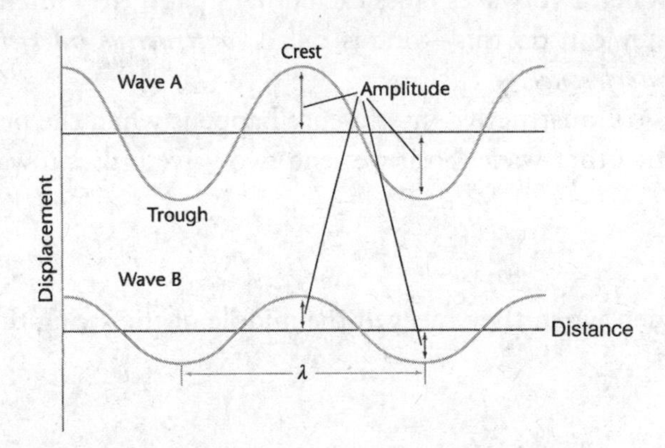

A *crest* (or a peak) is a high point on a wave, and a *trough* is a low point on a wave. The distance from peak-to-peak or from trough-to-trough is called the *wavelength*. This distance is abbreviated with the Greek letter λ (lambda). The distance that a peak or a trough is from the horizontal axis is called the *amplitude*, abbreviated with the letter A.

The time necessary for one complete wavelength to pass a given point is the period, abbreviated T. The number of wavelengths that pass a given point in 1 second is the *frequency*, abbreviated f. The period and frequency of a wave are related with a simple equation:

$$f = \frac{1}{T}$$

The Wave Equation

On the AP exam, you might be expected to look at a graph of a wave and produce the mathematical equation of the wave. It's actually quite easy to do. Look at the equation of a wave on the equation sheet:

$$x = A\cos(\omega t) = A\cos(2\pi f t)$$

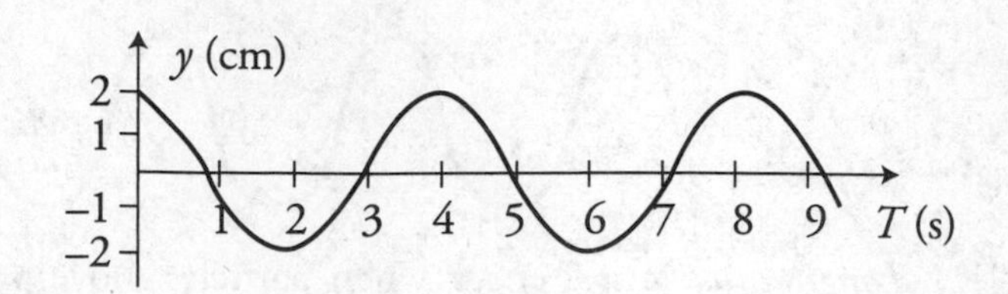

It looks kind of scary but no worries. All you have to do is fill in the blanks. Look at this graph of a wave in the figure above. What can we find on the graph?

1. amplitude: $A = 2$ cm
2. time period: $T = 4$ s (which means $f = ¼$ Hz)

Plug these into the wave equation: (2 cm) cos $[2\pi(0.25 \text{ Hz})t]$ and we are done! OK, you might want to convert units and this is really not a cosine wave. It looks like a sine wave, but remember the only difference between sine and cosine is a phase shift. Your math teacher can tell you about phase shifts. The point is, don't make it hard. Remember this is a timed exam. Answer the question and move on.

Interference

When two waves cross each other's path, they interact with each other. There are two ways they can do this—one is called *constructive interference*, and the other is called *destructive interference*.

Constructive interference happens when the peaks of one wave align with the peaks of the other wave. So if we send two wave pulses toward each other, as shown below,

then when they meet in the middle of the string they will interfere constructively.

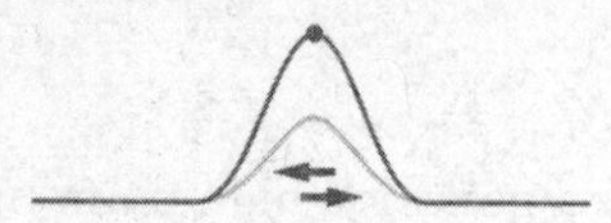

The two waves will then continue on their ways.

However, if the peaks of one wave align with the troughs of the other wave, we get destructive interference. For example, if we send the two wave pulses in the figure below toward each other,

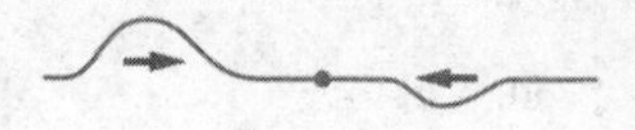

they will interfere destructively,

and then they will continue along their ways.

Electromagnetic Waves

When radio waves are beamed through space, or when X-rays are used to look at your bones, or when visible light travels from a lightbulb to your eye, electromagnetic waves are at work. All these types of radiation fall in the electromagnetic spectrum, shown below.

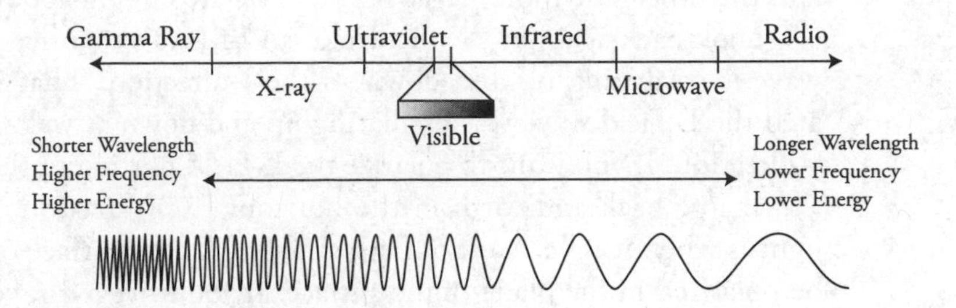

The unique characteristic about electromagnetic waves is that all of them travel at exactly the same speed through a vacuum—3×10^8 m/s. The more famous name for "3×10^8 m/s" is "the speed of light," or "*c*."

What makes one form of electromagnetic radiation different from another form is simply the frequency of the wave. AM radio waves have a very low frequency and a very long wavelength; whereas, gamma rays have an extremely high frequency and an exceptionally short wavelength. But they're all just varying forms of light waves.

Remember that charges create an electric field around themselves that extends infinitely far away. When charges oscillate, they create a disturbance in the electric field they are producing that moves outward through the field. This is a wave!

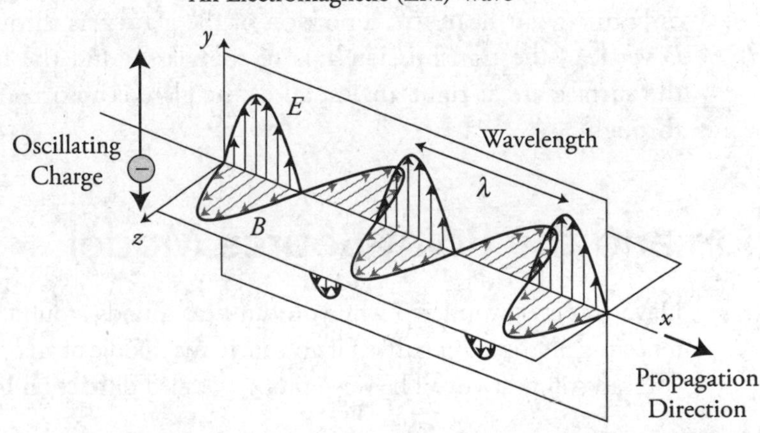

The really interesting part is that the oscillating electric field wave induces an oscillating magnetic field wave at a right angle to itself. This re-creates the electric field wave, and the whole thing keeps oscillating back and forth and never stops. This electromagnetic, or EM, wave is self-propagating and can travel through a vacuum at the speed of light. This is a major departure from mechanical waves, like sound, that require a physical medium to move through. This wave is produced by any oscillating charge. If you charge up a balloon and shake it around, you are producing an EM wave. Remember from thermodynamics we learned that atoms are in constant, random, *oscillating* motion. That means anything with a temperature above absolute zero—which is everything—will be radiating EM waves all the time. You are emitting infrared radiation right now. If we heat you up to about 1000°C, you will glow red!

Look back at the diagram of the EM wave. Notice that EM waves are transverse. This will be important in the next section.

Polarization

One of the interesting differences between transverse and longitudinal waves is polarization. First off, this is not the same polarization that we talked about in static electricity. That was charge polarization where a neutral object can be induced to have one side more positive and the other side more negative. Here we are talking about wave polarization.

Look back again at the picture of an EM wave. Notice how the electric portion of the wave is oscillating up and down in the *y*-direction? That is because the charge that created the E-Field wave was oscillating up and down as well. It is said to be polarized in the *y*-direction. If you want to polarize the E-Field portion of the wave horizontally, just wiggle the charge back and forth in the horizontal (*z*) direction. Another way we get polarized light is when it reflects off of flat surfaces, like the surface of a lake or a mirror. It tends to be polarized in the plane of the surface. If you drive a car, you know about road glare. The bright reflection off the road tends to be horizontally polarized.

Without going into any details, there are substances that will allow EM waves polarized in the *y*-direction to pass through them, but will absorb waves polarized in the *z*-direction. We use these in "polarized" sunglasses to block glare from the road and in 3D movies so that each eye sees only one of the two images projected onto the screen. Longitudinal waves cannot be polarized, because they vibrate forward and backward.

Here is a quick experiment you can perform: On a sunny day, put on polarizing sunglasses and look at the horizontal surface of a road or lake. Observe how the glasses block the horizontally polarized light reflecting from the surface. Now tilt your head sideways 90°. What will you see? Why do you now see so much glare from the surface? When you tilt your head 90° you align the transmission axis of the lenses with the axis of the polarized light reflecting from the surface and all the glare goes right through the sunglasses. If you only turn your hear 45°, a portion of the glare gets through the glasses. When your head is vertical, the transmission axis of the glasses and the horizontally polarized light from the surface are at right angles, all of the glare is absorbed by the glasses and no glare gets through.

Diffraction and the Point-Source Model

Have you ever wondered why you can hear sounds around corners? It's not just due to waves reflecting off something. Even if you are in the middle of an open field, someone standing behind you can still hear you. The wave property called diffraction helps explain this phenomenon.

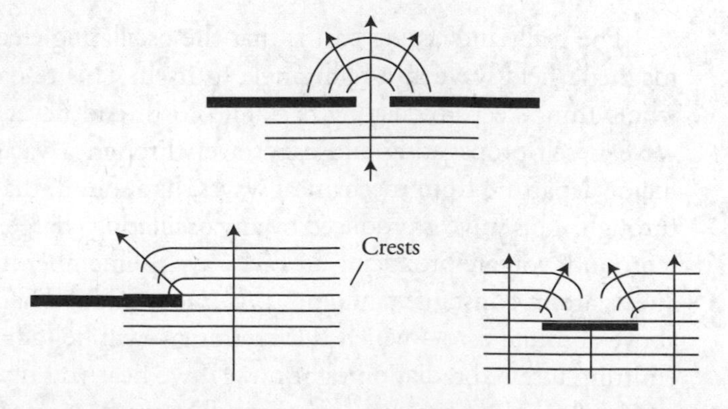

Look at the wave fronts traveling upward toward the top of the page in the diagram above. Notice how they all bend around the obstruction. Why does this occur? Huygens explained it this way: each point on a wave is the starting spot for a new wave. This is called the point-source model. So, each point of the wave that goes past a boundary is the new starting spot for a wave. These waves don't just travel forward, they also travel outward to the sides as seen in this diagram below.

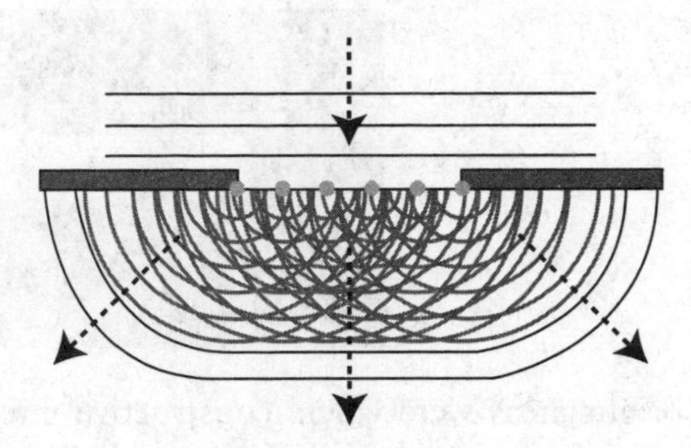

As a result, we can hear things around corners. If you are in the band, you know that the mouth of wind instruments curves outward to enhance diffraction, helping the sound to bend outward and fill the entire auditorium.

So we can hear things around corners but can't see things around corners. Does this mean that light is not a wave? For years that was the argument, but it turns out that the smaller the wave is, compared to the boundary it is passing by, the less we see the effects of diffraction. In general, the wavelength of the wave needs to be about the same size as or larger than the obstacle, or we won't get much diffraction. Visible light has a wavelength between 400 nm and 700 nm. That's very small. So we would need an opening that is about that size or smaller before we would start to notice diffraction effects in visible light.

Single and Double Slits

The way that physicists showed that light behaves like a wave was through slit experiments. Consider light shining through two very small slits, located very close together—slits separated by tenths or hundredths of millimeters. The light shone through each slit and then hit a screen. But here's the kicker: rather than seeing two bright patches on the screen (which would be expected if light was made of particles), the physicists saw lots of bright patches. The only way to explain this phenomenon was to conclude that light behaves like a wave.

Look at the next figure. When the light waves go through each slit, they were diffracted. As a result, the waves that came through the top slit overlap with the waves that came through the bottom slit—everywhere that peaks or troughs crossed paths, either constructive or destructive interference occurred.

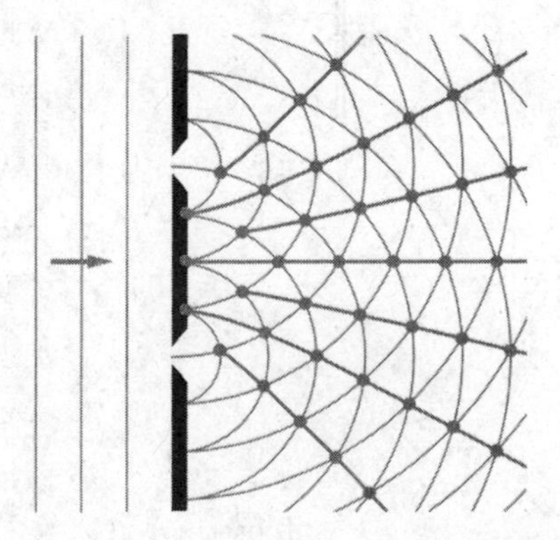

So when the light waves hit the screen, at some places they constructively interfered with one another and in other places they destructively interfered with one another. That explains why the screen looked like the following image.

The bright areas were where constructive interference occurred, and the dark areas were where destructive interference occurred. Particles can't interfere with one another—only waves can—so this experiment proved that light behaves like a wave.

When light passes through slits to reach a screen, the equation to find the location of bright spots is as follows:

$$d \sin \theta = m\lambda$$

Here, d is the distance between slits, λ is the wavelength of the light, and m is the "order" of the bright spot; we discuss m below. θ is the angle at which an observer has to look to see the bright spot.

The variable m represents the "order" of the bright or dark spot, measured from the central maximum as shown in the next figure. Bright spots get integer values of m; dark spots get half-integer values of m. The central maximum represents $m = 0$. The first constructive interference location to the side of the central maxima is called the first-order maxima and would be represented by $m = 1$. The second-order maxima would be $m = 2$, etc. The first destructive interference location at $m = ½$, would be the first-order minima.

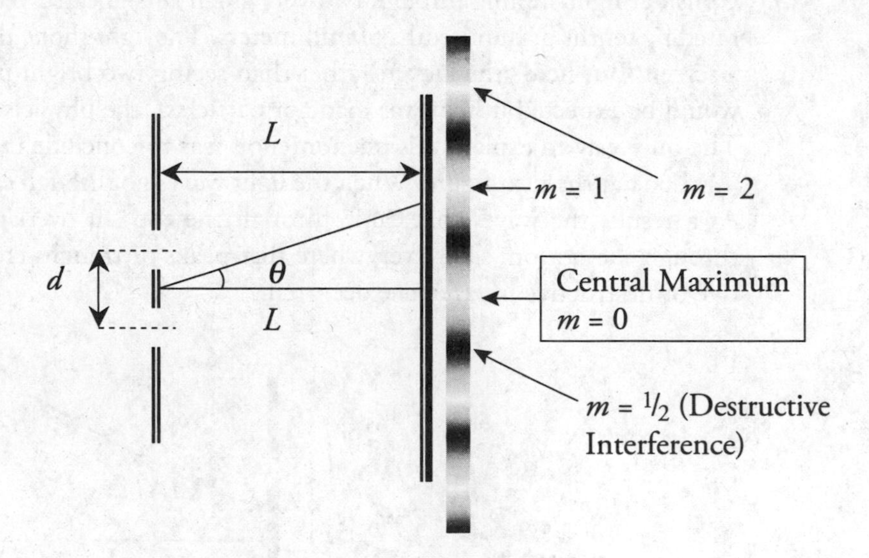

So, for example, if you wanted to find how far from the center of the pattern the first bright spot labeled $m = 1$ is, you would plug in "1" for m, If you wanted to find the dark region closest to the center of the screen, you would plug in "½" for m.

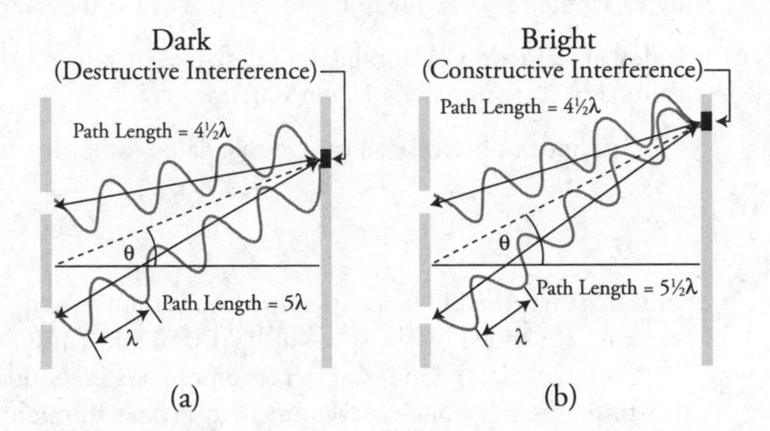

To better understand why waves produce this interference pattern but particles do not, look at the above diagram where the waves are shown oscillating toward the screen on the right.

Count the number of wavelengths from the top opening to the screen in figure (a): 4½ wavelengths. This is called the path length to the screen. Now count the path length from the bottom opening to the screen: 5 wavelengths. That means the waves will arrive ½ a wavelength off. Or, a crest of one wave is meeting a trough for the other wave, creating destructive interference, or a dark spot on the screen. Now look at figure (b). One wave travels 4½ wavelengths, while the other travels 5½ wavelengths. They are in phase and will have constructive interference, producing a bright spot on the screen.

The path length difference equation, $\Delta L = m\lambda$, shows us that whenever the difference in the length to the screen ΔL is a whole multiple ($m = 1, 2, 3, 4, 5$, etc.) of the wavelength, we will get constructive interference. When the waves are off by a ½ wavelength (m = ½, ³⁄₂, ⁵⁄₂, etc.), there will be destructive interference.

Single Slits and Diffraction Gratings

Once you understand the double-slit experiment, single slits and diffraction gratings are simple.

A diffraction grating consists of a large number of slits, not just two slits. The locations of bright and dark spots on the screen are the same as for a double slit, but the bright spots produced by a diffraction grating are very sharp dots.

A single slit produces interference patterns as well because the light that bends around each side of the slit interferes upon hitting the screen. For a single slit, the central maximum is bright and very wide; the other bright spots are regularly spaced, but dim relative to the central maximum.

Index of Refraction

Light also undergoes interference when it reflects off of thin films of transparent material. Before studying this effect quantitatively, though, we have to examine how light behaves when it passes through different materials.

Light—or any electromagnetic wave—travels at the speed c, or 3×10^8 m/s. But it only travels at this speed through a vacuum, when there aren't any pesky molecules to get in the way. When it travels through anything other than a vacuum, light slows down. The amount by which light slows down in a material is called the material's *index of refraction.*

Index of refraction: A number that describes by how much light slows down when it passes through a certain material, abbreviated n.

The index of refraction can be calculated using this equation.

$$n = \frac{c}{v}$$

This says that the index of refraction of a certain material, n, equals the speed of light in a vacuum, c, divided by the speed of light through that material, v.

For example, the index of refraction of glass is about 1.5. This means that light travels 1.5 times faster through a vacuum than it does through glass. The index of refraction of air is approximately 1. Light travels through air at just about the same speed as it travels through a vacuum.

Another thing that happens to light as it passes through a material is that its wavelength changes. When light waves go from a medium with a low index of refraction to one with a high index of refraction, they get squished together. So, if light waves with a wavelength of 500 nm travel through air ($n_{air} = 1$), enter water ($n_{water} = 1.33$), and then emerge back into air again, it would look like the figure below.

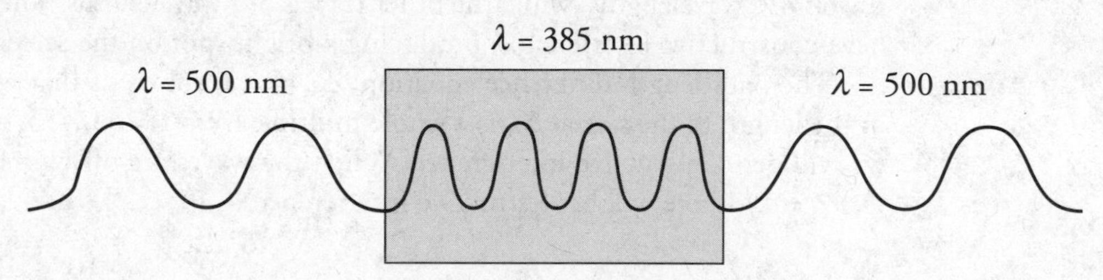

The equation that goes along with this situation is the following:

$$\lambda_n = \frac{\lambda}{n}$$

In this equation, λ_n is the wavelength of the light traveling through the transparent medium (like water, in the figure above), λ is the wavelength in a vacuum, and n is the index of refraction of the transparent medium.

It is important to note that, even though the wavelength of light changes as it goes from one material to another, its frequency remains constant. The frequency of light is a property of the photons that comprise it (more about that in Chapter 15), and the frequency doesn't change when light slows down or speeds up.

Thin Films

When light hits a thin film of some sort, the interference properties of the light waves are readily apparent. You have likely seen this effect if you've ever noticed a puddle in a parking lot. If a bit of oil happens to drop on the puddle, the oil forms a very thin film on top of the water. White light (say, from the sun) reflecting off of the oil undergoes interference, and you see some of the component colors of the light.

Consider a situation where monochromatic light (meaning "light that is all of the same wavelength") hits a thin film, as shown in the next figure. At the top surface, some light will be reflected, and some will penetrate the film. The same thing will happen at the bottom surface: some light will be reflected back up through the film, and some will keep on traveling out through the bottom surface. Notice that the two reflected light waves overlap; the wave that reflected off the top surface and the wave that traveled through the film and reflected off the bottom surface will interfere.

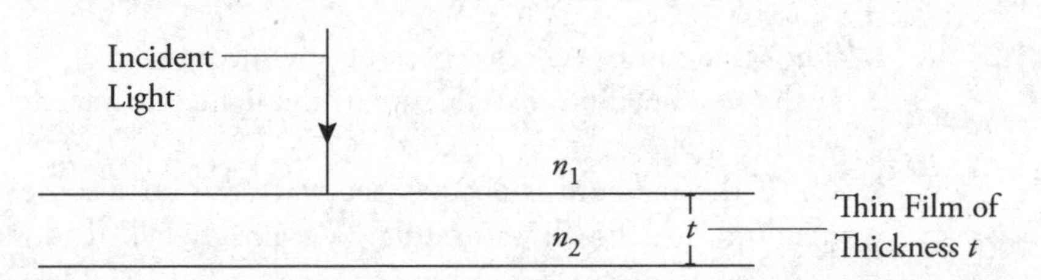

The important thing to know here is whether the interference is constructive or destructive. The wave that goes through the film travels a distance of $2t$ before interfering, where t is the thickness of the film. If this extra distance is precisely equal to a wavelength, then the interference is constructive. You also get constructive interference if this extra distance is precisely equal to two, or three, or any whole number of wavelengths.

But be careful what wavelength you use . . . because this extra distance occurs inside the film, we're talking about the wavelength *in the film*, which is the wavelength in a vacuum divided by the index of refraction of the film.

The equation for constructive interference turns out to be

$$2t = m\lambda_n$$

where m is any whole number, representing how many extra wavelengths the light inside the film went.

So, when does *destructive* interference occur? When the extra distance in the film precisely equals ½ wavelength . . . or 1½ wavelengths, or 2½ wavelengths . . . so for destructive interferences, plug in a half-integer for m.

There's one more complication. If light reflects off of a surface while going from low to high index of refraction, the light "changes phase." For example, if light in air reflects off oil (n ~1.2), the light changes phase. If light in water reflects off oil, though, the light does not change phase. For our purposes, a phase change simply means that the conditions for constructive and destructive interference are reversed.

Summary: For thin film problems, go through these steps.

1. Count the phase changes. A phase change occurs for every reflection from low to high index of refraction.
2. The extra distance traveled by the wave in the film is twice the thickness of the film.
3. The wavelength in the film is $\lambda_n = \frac{\lambda}{n}$, where n is the index of refraction of the film's material.
4. Use the equation $2t = m\lambda_n$. If the light undergoes zero or two phase changes, then plugging in whole numbers for m gives conditions for constructive interference. (Plugging in half-integers for m gives destructive interference.) If the light undergoes one phase change, conditions are reversed—whole numbers give destructive interference, half-integers, constructive.

Finally, why do you see a rainbow when there's oil on top of a puddle? White light from the sun, consisting of *all* wavelengths, hits the oil. The thickness of the oil at each point allows only one wavelength to interfere constructively. So, at one point on the puddle, you see just a certain shade of red. At another point, you see just a certain shade of orange, and so on. The result, over the area of the entire puddle, is a brilliant, swirling rainbow.

Wave Behavior at Boundaries

When a light wave encounters a boundary, several things can occur:

1. The wave can be reflected back off the medium.
2. If the new medium is transparent, the light can transmit through, carrying its energy with it.
3. Or, if the medium is opaque, the wave can be absorbed, giving up its energy to the medium, which will warm it up. (We already talked about this in Chapter 10.)

In reality, more than one of these can happen at the same time. Some of the wave can be absorbed, some transmitted, and some reflected. The key is that all of the wave energy goes somewhere. Nothing is gained or lost (conservation of energy).

The remainder of this chapter will concentrate on how light reflects and transmits when it encounters a new medium.

Reflection and Mirrors

Okay, time to draw some pictures. Let's start with plane (flat) mirrors.

The key to solving problems that involve plane mirrors is that the angle at which a ray of light hits the mirror equals the angle at which it bounces off, as shown in the next figure.

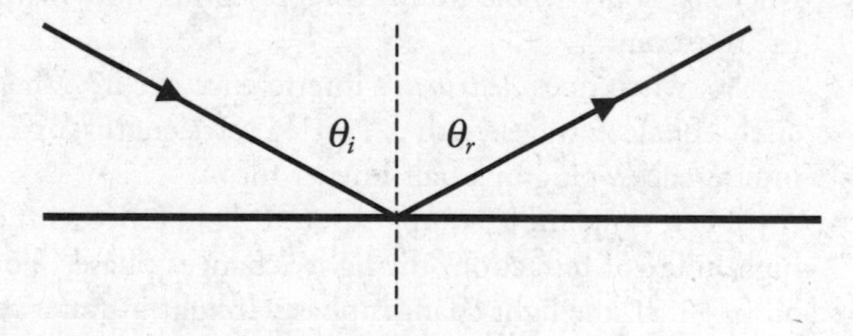

In other words—or, more accurately, "in other symbols"—$\theta_i = \theta_r$, where θ_i is the incident[1] angle, and θ_r is the reflected angle.

So, let's say you had an arrow, and you wanted to look at its reflection in a plane mirror. We'll draw what you would see in the following figure.

The image of the arrow that you would see is drawn in dotted lines. To draw this image, we first drew the rays of light that reflect from the top and bottom of the arrow to your eye. Then we extended the reflected rays through the mirror.

Whenever you are working with a plane mirror, follow these rules:

- The image is upright. Another term for an upright image is a *virtual image.*
- The image is the same size as the original object. That is, the magnification, *m*, is equal to 1.
- The image distance, s_i, equals the object distance, s_o.

[1]The "incident" angle simply means "the angle that the initial ray happened to make."

A more challenging type of mirror to work with is called a spherical mirror. Before we draw our arrow as it looks when reflected in a spherical mirror, let's first review some terminology (this terminology is illustrated in the next figure).

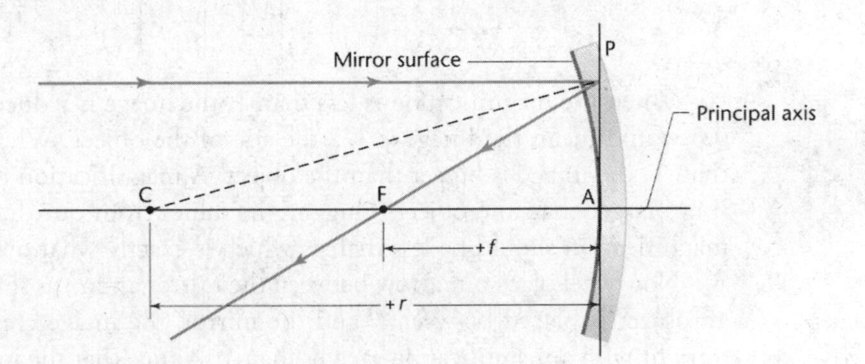

A spherical mirror is a curved mirror—like a spoon—that has a constant radius of curvature, r. The imaginary line running through the middle of the mirror is called the *principal axis*. The point labeled C, which is the center of the sphere, lies on the principal axis and is located a distance r from the middle of the mirror. The point labeled F is the focal point, and it is located a distance f, where $f = (r/2)$, from the middle of the mirror. The focal point is also on the principal axis. The line labeled P is perpendicular to the principal axis.

There are several rules to follow when working with spherical mirrors. Memorize these.

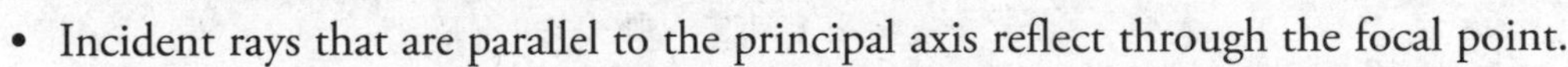

- Incident rays that are parallel to the principal axis reflect through the focal point.
- Incident rays that go through the focal point reflect parallel to the principal axis.
- Any points that lie on the same side of the mirror as the object are a *positive* distance from the mirror. Any points that lie on the other side of the mirror are a *negative* distance from the mirror. Notice how the figure above shows a positive focal length and radius.
- $\dfrac{1}{f} = \dfrac{1}{s_o} + \dfrac{1}{s_i}$

That last rule is called the "mirror equation." (You'll find this equation to be identical to the "lensmaker's equation" later.)

To demonstrate these rules, we'll draw three different ways to position our arrow with respect to the mirror. In the first scenario, we'll place our arrow on the principal axis, beyond point *C*, as shown in the following figure.

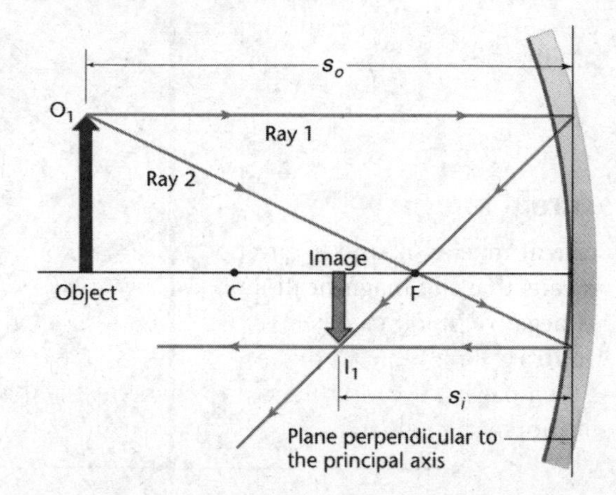

Notice that the image here is upside down. Whenever an image is upside down, it is called a real image. A real image can be projected onto a screen, whereas a virtual image cannot.[2]

The magnification, *m*, is found by

$$|M| = \left|\frac{h_i}{h_o}\right| = \left|\frac{s_i}{s_o}\right|$$

When the magnification is less than 1, the image is reduced in size. A magnification of 0.5 would mean the image is 1/2 the size of the object. When the magnification is larger than 1, the image is bigger than the object. A magnification of 2.5 means the image is 2.5 times larger than the object. Plugging in values from our drawing above, we see that our magnification should be less than 1, which is exactly what our figure shows. Good for us!

Now we'll place our arrow between the mirror and *F*, as shown in the next figure. When an object is placed between *F* and the mirror, the image created is a virtual image—it is upright with a magnification greater than 1. Notice that the image distance will be negative because it is on the other side or opposite side of the mirror from the object.

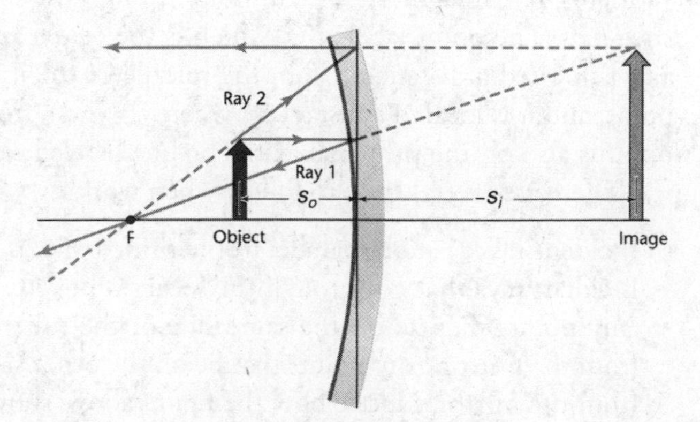

[2]TRY THIS! Light a candle; reflect the candle in a concave mirror; and put a piece of paper where the image forms. . . . You will actually see a picture of the upside-down, flickering candle!

Finally, we will place our object on the other side of the mirror, as shown in the next figure. Now we have a convex mirror, which is a *diverging* mirror—parallel rays tend to spread away from the focal point. In this situation, the image is again virtual with a magnification less than 1.

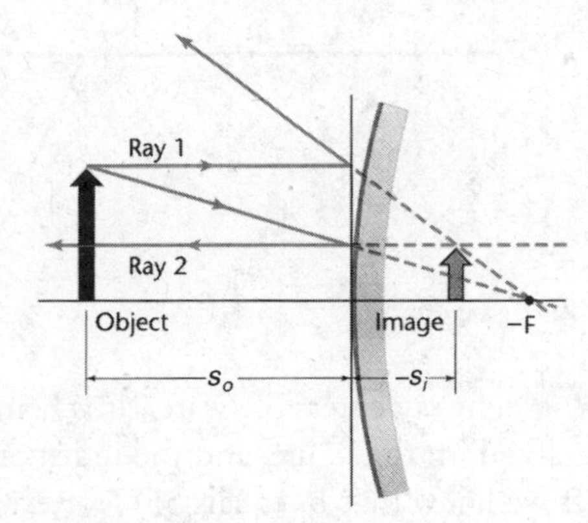

When we use the mirror equation here, we have to be especially careful: s_o is positive, but both s_i and f are negative, because they are not on the same side of the mirror as the object.

Note that the convex (diverging) mirror *cannot* produce a real image. Give it a try—you can't do it!

Snell's Law and the Critical Angle

Two things before we begin:

1. Remember that light can partially transmit into a new material, as well as partially reflect off the surface at the same time. You know this because you can see through a window and see your reflection in the window at the same time.
2. When light transmits into a new material, the index of refraction of the new medium determines the new velocity of the light, $n = \frac{c}{v}$, and the new wavelength of the light, $\lambda_n = \frac{\lambda}{n}$. The frequency of the light stays the same.

In addition to changing its speed and its wavelength, light can also change its direction when it travels from one medium to another. The way in which light changes its direction is described by Snell's law.

$$n_1 \sin \theta_1 = n_2 \sin \theta_2$$

To understand Snell's law, it's easiest to see it in action. The next figure should help.

In the following figure, a ray of light is going from air into water. The dotted line perpendicular to the surface is called the normal.[3] This line is not real; rather it is a reference line for use in Snell's law. *In optics, ALL ANGLES ARE MEASURED FROM THE NORMAL, NOT FROM A SURFACE!*

[3]Remember that in physics, "normal" means "perpendicular"—the normal is always perpendicular to a surface.

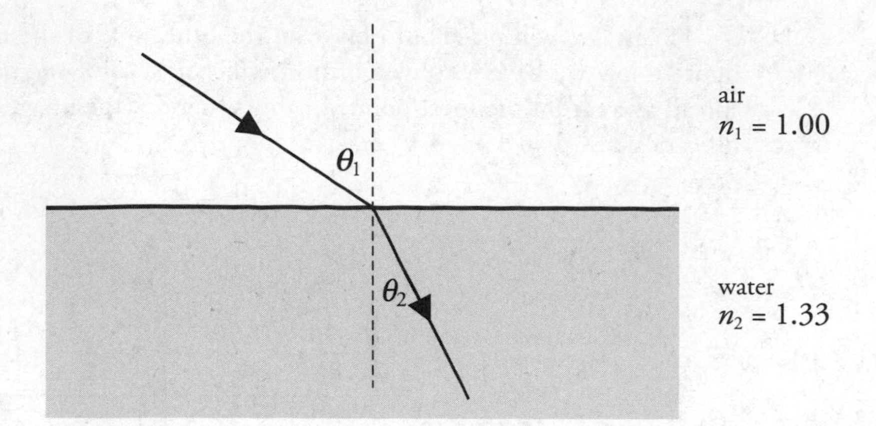

As the light ray enters the water, it is being bent toward the normal. The angles θ_1 and θ_2 are marked on the figure, and the index of refraction of each material, n_1 and n_2, is also noted. If we knew that θ_1 equals 55°, for example, we could solve for θ_2 using Snell's law.

$$(1.00)(\sin 55°) = (1.33)(\sin \theta_2)$$

$$\theta_2 = 38°$$

Whenever light goes from a medium with a low index of refraction to one with a high index of refraction—as in our drawing—the ray is bent toward the normal. Whenever light goes in the opposite direction from a larger index of refraction to a smaller index of refraction—say, from water into air—the ray is bent away from the normal.

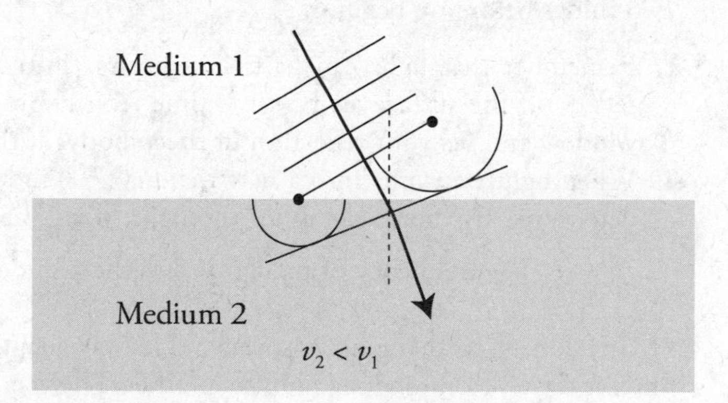

But why, you may be asking, does light change directions just because it moves into a new material? The point source model of waves (Huygens' principle) tells us why. The waves moving through medium 1 are traveling faster than the waves traveling in medium 2. Using the endpoints of the wave front as point sources, we can draw the two waves that are produced. Notice how the wave in medium 1 is farther out because it is moving faster. When we connect these two point sources, the new wave front has changed direction. It's heading in a new direction closer to the normal line. This happens because the wave slows down, resulting in an angle of refraction that is smaller than the angle of incidence.

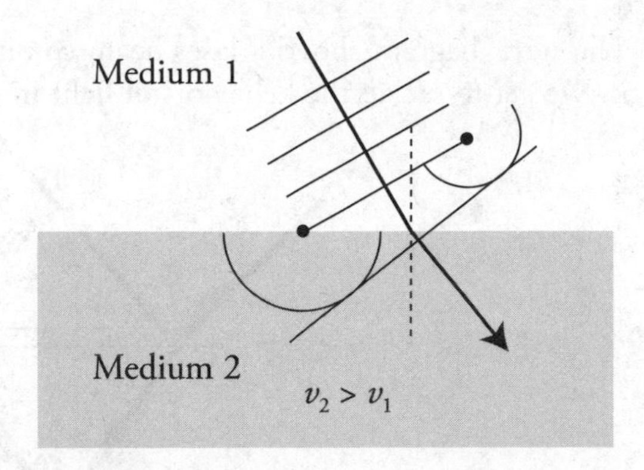

Now let's reverse the situation. Medium 1 is the slow material and medium 2 is the fast medium. The effect is reversed. When the wave speeds up, it will turn away from the normal. It has to.

Here is another way to remember the "bending" rule. Bigger velocity of light means a bigger angle with the normal. Smaller velocity of light means a smaller angle with the normal.

There is only one exception to this rule! If light hits the new medium head-on (angle of incidence of 0°), it won't change direction. It just plows straight into the new medium with no turning. You can prove that to yourself using the point-source model.

If you have a laser pointer, try shining it into some slightly milky water . . . you'll see the beam bend into the water. But you'll also see a little bit of the light reflect off the surface, at an angle equal to the initial angle. (Be careful the reflected light doesn't get into your eye!) In fact, at a surface, if light is refracted into the second material, some light must be reflected.

Sometimes, though, when light goes from a medium with a high index of refraction to one with a low index of refraction, we could get *total internal reflection*. For total internal reflection to occur, the light ray must be directed at or beyond the *critical angle*.

Critical Angle: The angle past which rays cannot be transmitted from one material to another, abbreviated θ_c.

$$\sin \theta_c = \frac{n_1}{n_2}$$

Again, pictures help, so let's take a look at the next figure.

In this figure, a ray of light shines up through a glass block. The critical angle for light going from glass to air is 42°; however, the angle of the incident ray is greater than the critical angle. Therefore, the light cannot be transmitted into the air. Instead, all of it reflects inside the glass. Total internal reflection occurs anytime light cannot leave a material.

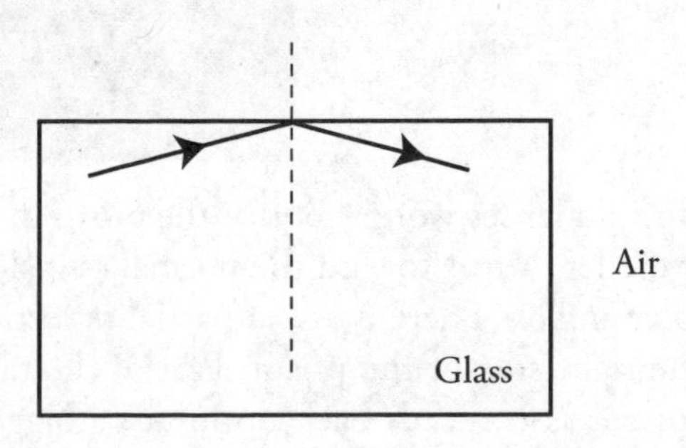

The next diagram shows a laser beam coming from the upper left shining on a glass block. We get to see all the behaviors of light in one image.

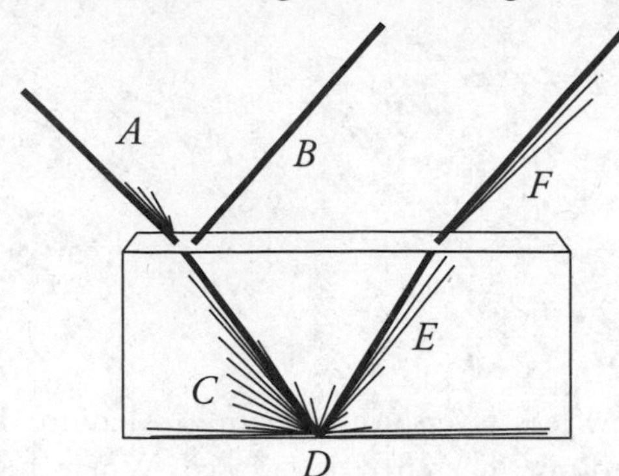

- First, the light strikes the surface at point A. Part of the light reflects, B, and part transmits, C. The reflected light takes off at the same angle equal to the angle of incidence.
- The transmitted wave, C, refracts toward the normal because it is slowing down.
- None of the wave exits the bottom of the glass block. The angle of incidence must be larger than the critical angle. Total internal reflection occurs at point D.
- After reflecting off the bottom of the glass, the wave, E, strikes the top of the glass and refracts away from the normal, F, as the wave speeds back up when it reenters the air.

This could easily be the basis for a question on the AP exam.

Lenses

We have two types of lenses to play with: convex and concave. A *convex lens*, also known as a "converging lens," is shown below.

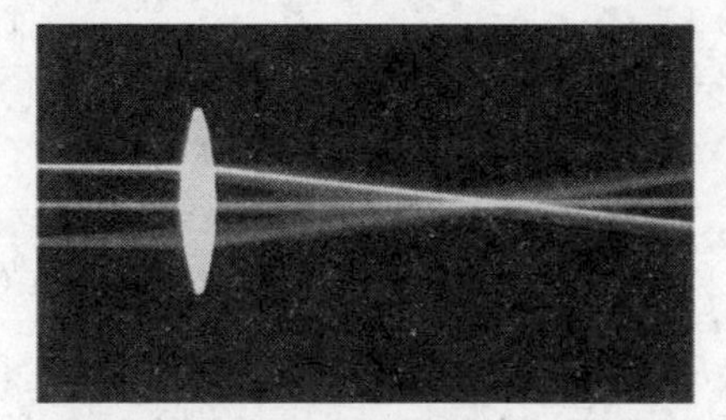

And a *concave lens*, or "diverging lens," is shown in the next figure.

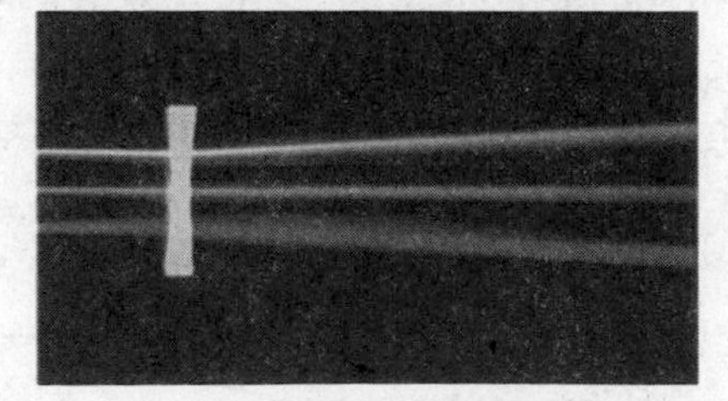

How do lenses work? Look at the prism in the following diagram. The ray coming from air on the left, turns toward the normal as it slows down in the glass: a smaller speed means a smaller angle. (There is also a partial reflection as well.) Once inside the prism, the light hits the right side of the prism. Part of the ray reflects again. The rest refracts away from the normal as it speeds back up in air: a bigger speed means a bigger angle. After leaving the prism, the ray has a new downward direction.

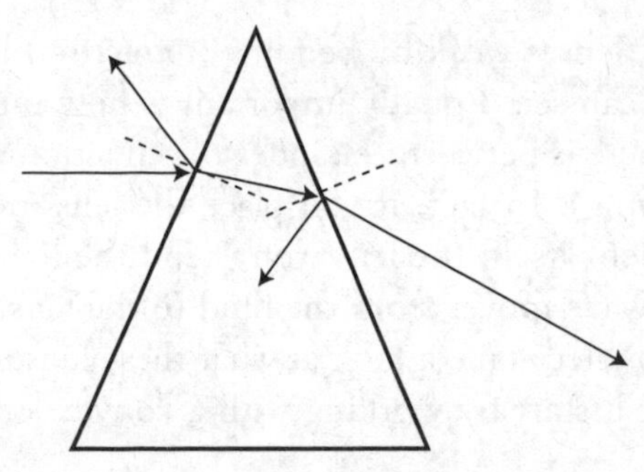

Let's concentrate only on the rays that transmit through the prism and make some lenses! Place two prisms together flat side to flat side, smooth it out, and *shazam poof*, you have a converging or convex lens.

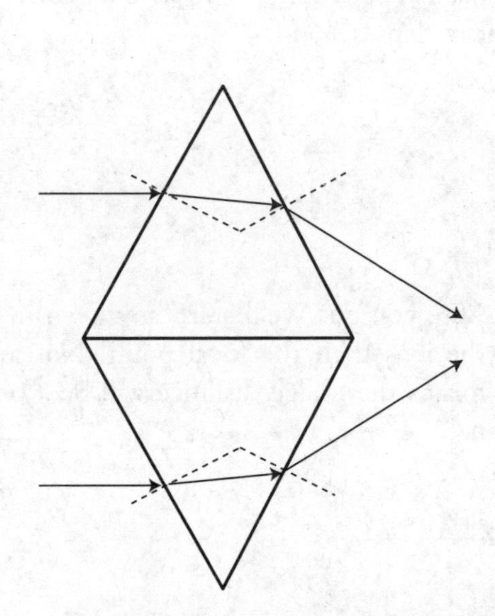

Refraction by a Converging Lens

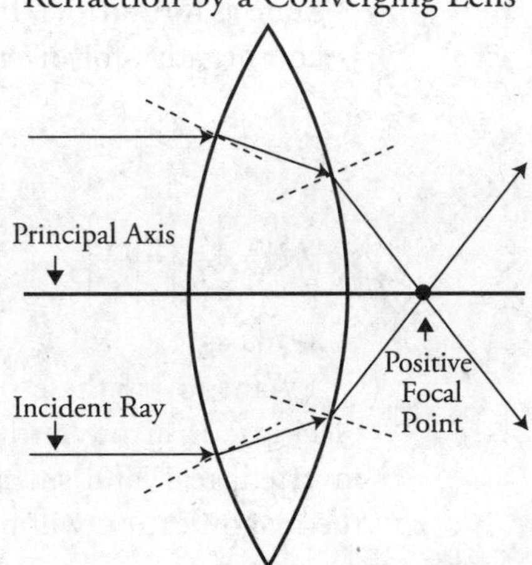

Incident rays that travel parallel to the principal axis will refract through the lens and converge to the focal point. The focal length will be positive for converging lenses.

Place two prisms point to point, smooth them out, and you have a diverging or concave lens.

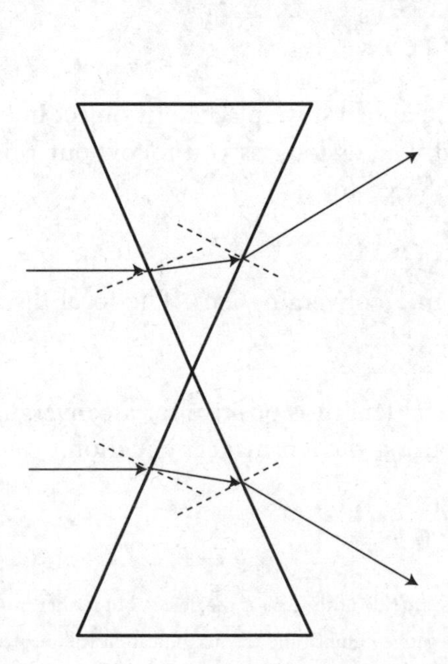

Refraction by a Diverging Lens

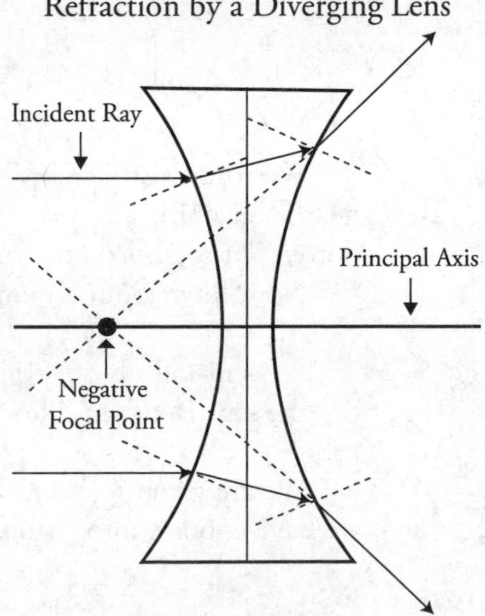

Incident rays that travel parallel to the principal axis will refract through the lens and diverge as if coming from the focal point, never intersecting. The focal length will be negative for diverging lenses.

So, lenses work by bending (refracting) light. The shape of the lens is very important, as you can see. Equally important is how much the light changes speed. The more difference there is between the index of refraction of the lens and the air, the greater the bending (refracting). In fact, if you place the lens in a fluid that has the same index of refraction as the lens itself, the lens won't bend the light at all, because there is no change in velocity as the wave moves from the fluid to the lens. No change in velocity equals no refraction.

Now let's take a look at what these lenses can do for us.

We'll start by working with a convex lens. The rules to follow with convex lenses are these:

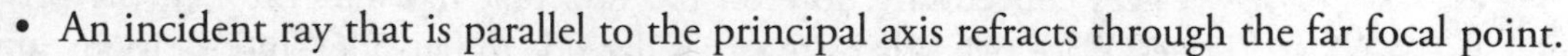

- An incident ray that is parallel to the principal axis refracts through the far focal point.
- An incident ray that goes through near focal point refracts parallel to the principal axis.
- The lensmaker's equation and the equation to find magnification, shown below, are the same as for mirrors. In the lensmaker's equation, f is positive for converging/convex lenses and negative for diverging/concave lenses.
- $\frac{1}{f} = \frac{1}{s_o} + \frac{1}{s_i}$
- $|M| = \left|\frac{h_i}{h_o}\right| = \left|\frac{s_i}{s_o}\right|$

Want to try these rules out? Sure you do. We'll start, as shown in the next figure, by placing our arrow farther from the lens than the focal point. Notice how the image is inverted, real, and smaller. This means the image distance will be a positive number and the magnification will be less than 1.

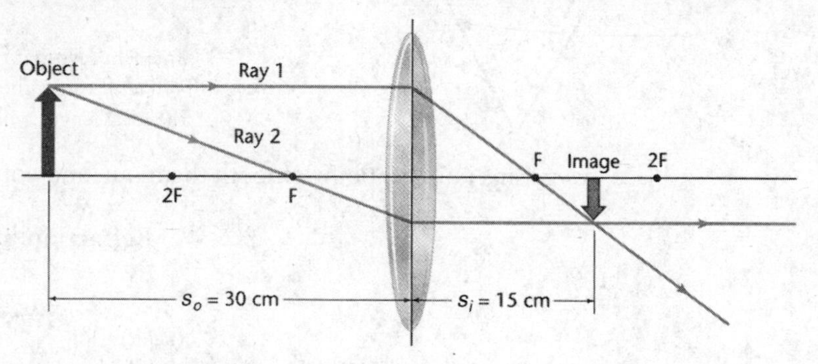

We could also demonstrate what would happen if we placed our object in between the near focal point and the lens. But so could you, as long as you follow our rules. And we don't want to stifle your artistic expression. So go for it.[4]

Now, how about a numerical problem?

A 3-cm-tall object is placed 20 cm from a converging lens. The focal distance of the lens is 10 cm. How tall will the image be?

We are given s_o and f. (Note that the focal length is positive for a converging lens.) So we have enough information to solve for s_i using the lensmaker's equation.

$$\frac{1}{0.1} = \frac{1}{0.2} + \frac{1}{s_i}$$

[4]Answer: The image is a virtual image, located on the same side of the lens as the object, but farther from the lens than the object. This means that the image distance will be a negative number and the magnification will be greater than 1.

Solving, we have $s_i = 20$ cm. Since the image distance is positive, we know that the image will appear on the far side of the lens as a real/inverted image. Now we can use the magnification equation.

$$m = \left| -\frac{0.2}{0.2} \right| = 1$$

Our answers tell us that the image is exactly the same size as the object. So our answer is that the image is 3 cm tall, that it is inverted and real.

When working with diverging lenses, follow these rules:

- An incident ray parallel to the principal axis will refract as if it came from the near focal point.
- An incident ray toward the far focal point refracts parallel to the principal axis.
- The lensmaker's equation and the magnification equation still hold true. With diverging lenses, though, f is negative.

We'll illustrate these rules by showing what happens when an object is placed farther from a concave lens than the focal point. This is shown in the next figure.

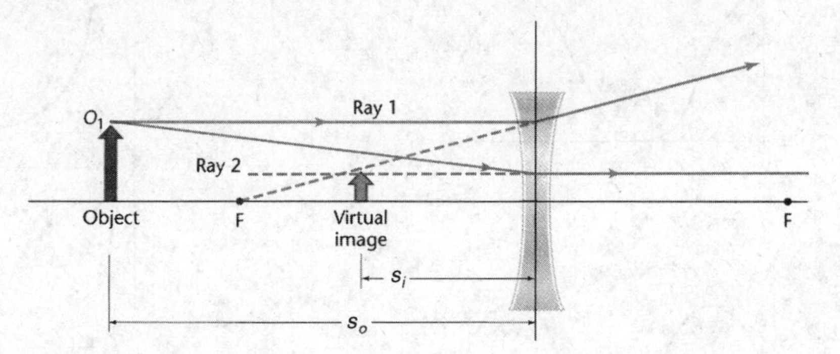

The image is upright, so we know that it is virtual.

Now go off and play with lenses. And spoons. And take out that box of crayons that has been collecting dust in your cupboard and draw a picture. Let your inner artist go wild. (Oh, and do the practice problems, too!)

› Practice Problems

Multiple Choice

1. Monochromatic light passed through a double slit produces an interference pattern on a screen a distance 2.0 m away. The third-order maximum is located 1.5 cm away from the central maximum. Which of the following adjustments would cause the third-order maximum instead to be located 3.0 cm from the central maximum?

(A) Moving light source farther from the slits
(B) Decreasing the wavelength of the light
(C) Using a screen closer to the slits
(D) Decreasing the distance between slits

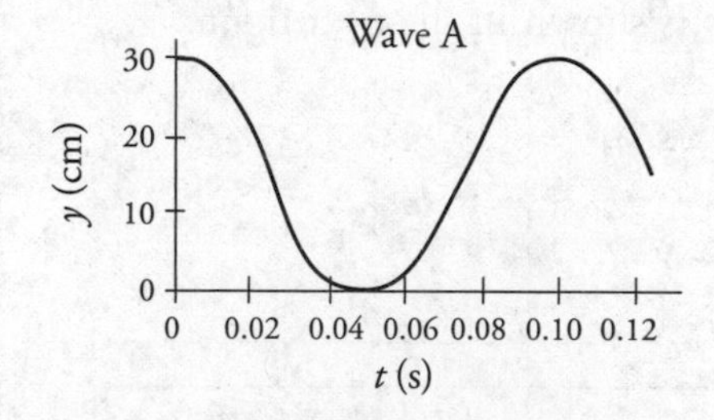

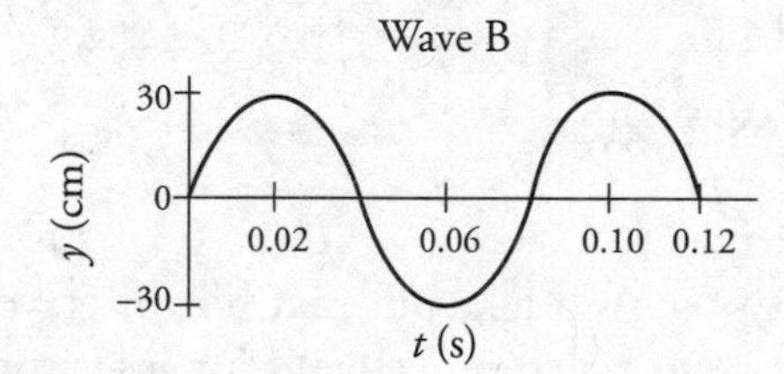

2. Two waves travel through different materials. The graph of the vertical position y of a point in the medium as a function of time t for each wave is shown in the figure. Which of the following statements can be verified using the data presented in the figure?

(A) Wave A has a high frequency than wave B.
(B) Both waves have the same amplitude.
(C) Both waves have the same time period of motion.
(D) The equation of motion for wave B is $30 \sin\left(\frac{2\pi}{0.08}t\right)$.

3. An object is placed in front of a spherical convex mirror. Which of the following object distances s_o will produce an inverted image?

(A) $s_o < f$
(B) $2f > s_o > f$
(C) $s_o > 2f$
(D) There is not a location s_o that produces an inverted image.

4. Using a convex lens, students take measurements of both the distance of the object from the lens s_o and the image distance from the lens s_i and then graph the data. The focal length F is shown each axis. Which of the following figures best depicts the results of the graphed data?

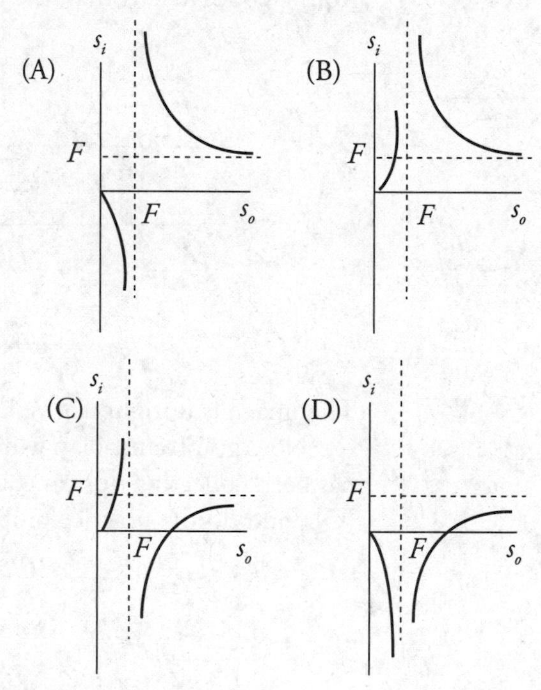

5. Light travels from air into a plastic prism. Which of the following ray diagrams correctly shows the path of the ray that travels through the prism?

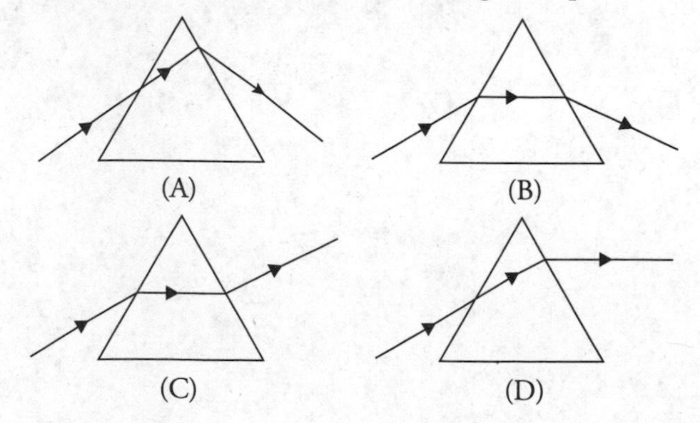

6. In an aquarium, light traveling through water (n = 1.3) is incident upon the glass container (n = 1.5) at an angle of 36° from the normal. What is the angle of transmission in the glass?

(A) The light will not enter the glass because of total internal reflection.
(B) 31°
(C) 36°
(D) 41°

7. Light waves traveling through air strike the surface of water at an angle. Which of the following statements about the light's wave properties upon entering the water is correct?

(A) The light's speed, frequency, and wavelength all change.
(B) The light's speed and frequency change, but the wavelength stays the same.
(C) The light's wavelength and frequency change, but the light's speed stays the same.
(D) The light's wavelength and speed change, but the frequency stays the same.

8. An object is placed at the center of a concave spherical mirror. What kind of image is formed, and where is that image located?

(A) A real image is formed at the focal point of the mirror.
(B) A real image is formed at the center of the mirror.
(C) A real image is formed one focal length beyond the center of the mirror.
(D) A virtual image is formed one radius behind the mirror.

9. In a laboratory experiment, you shine a green laser past a strand of hair. This produces a light and dark pattern on a screen. You notice that the lab group next to you has produced a similar pattern on a screen, but the light and dark areas are spread farther apart. Which of the following could cause the light and dark pattern to spread? **(Select two answers.)**

(A) The second group used thinner hair.
(B) The second group is using a red laser.
(C) The second group had the screen closer to the hair.
(D) The second group held the laser farther from the hair.

10. In a human eye, the distance from the lens to the retina, on which the image is focused, is 20 mm. A book is held 30 cm from the eye, and the focal length of the eye is 16 mm. How far from the retina does the image form, and what lens should be used to place the image directly on the retina?

Distance of image from retina	Corrective lens
(A) 3.1 mm in front of the retina	Concave lens
(B) 3.1 mm in front of the retina	Convex lens
(C) 14 mm behind the retina	Concave lens
(D) 14 mm behind the retina	Convex lens

Free Response

11. Laser light is passed through a diffraction grating with 7000 lines per centimeter. Light is projected onto a screen far away. An observer by the diffraction grating observes the first-order maximum 25° away from the central maximum.

(A) What is the wavelength of the laser?
(B) If the first-order maximum is 40 cm away from the central maximum on the screen, how far away is the screen from the diffraction grating?
(C) How far, measured along the screen, from the central maximum will the second-order maximum be?

12. Your eye has a lens that forms an image on the retina.

(A) What kind of lens does your eye have? Justify your answer.
(B) You develop an eye disorder where your eye forms the image in front of your retina. What kind of corrective lens is needed so that you can see clearly? Justify your answer.

13. Which of the following optical instruments can produce a virtual image with magnification 0.5? Justify your answer.

(A) Convex mirror
(B) Concave mirror
(C) Convex lens
(D) Concave lens

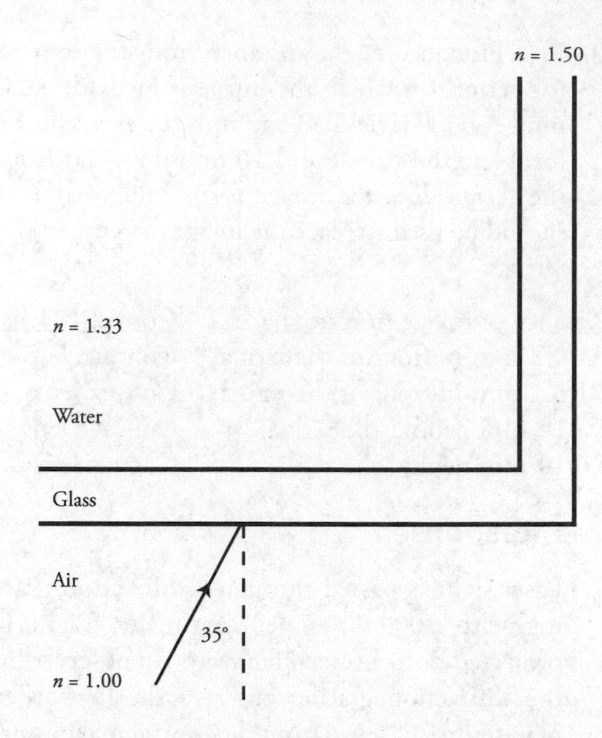

14. Light traveling through air encounters a glass aquarium filled with water. The light is incident on the glass from the front at an angle of 35°.

(A) At what angle does the light enter the glass?
(B) At what angle does the light enter the water?
(C) On the diagram above, sketch the path of the light as it travels from air to water. Include all reflected and refracted rays; label all angles of reflection and refraction.

After entering the water, the light encounters the side of the aquarium, hence traveling back from water to glass. The side of the tank is perpendicular to the front.

(D) At what angle does light enter the glass on the side of the aquarium?
(E) Does the light travel out of the glass and into the air, or does total internal reflection occur? Justify your answer.

15. (A) Write the appropriate wave equation for the following electromagnetic wave representation.

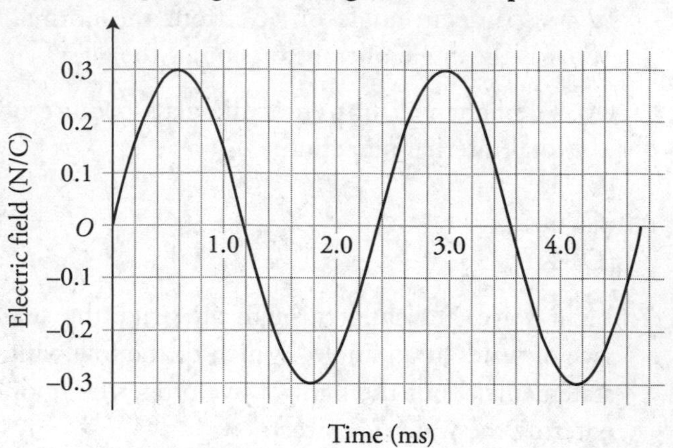

(B) Produce a sketch of this wave equation on the axis: $B = 4.0\cos\left(\frac{\pi x}{2}\right)$, where B is in units of T, and x is in units of m.

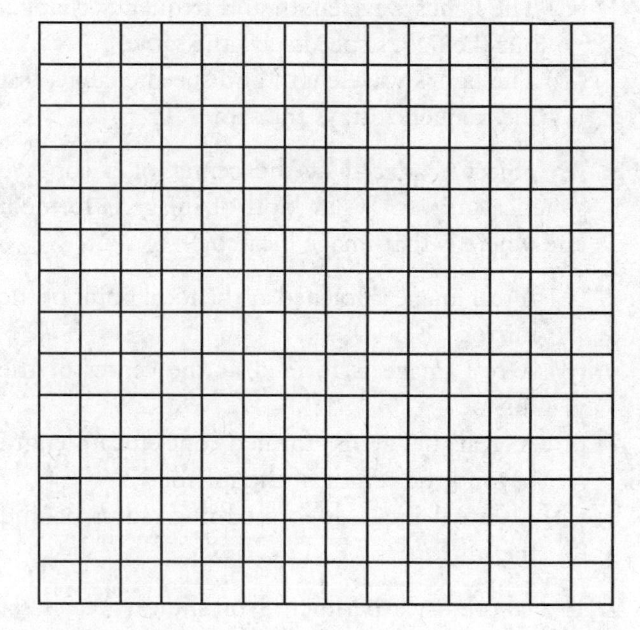

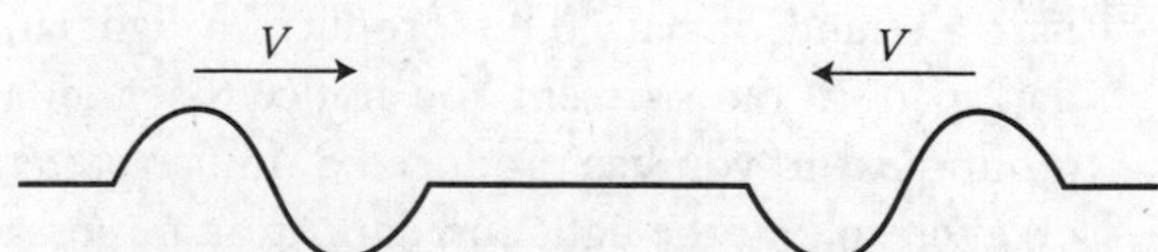

16. Two waves travel toward each other, as shown in the figure. Sketch at least three unique interference patterns that will be seen as the waves pass each other.

17. Use the point source model of wave propagation to describe and explain the behaviors of the waves listed here. Sketch a representation to assist in your explanation.

(A) Diffraction of a plane wave front as it passes by a boundary (see figure).

(B) Sound waves can be heard around corners, but light waves do not seem to bend around corners. Why is this?

(C) Diffraction of a plane wave front as it passes through a small opening comparable in size to the wavelength (see figure).

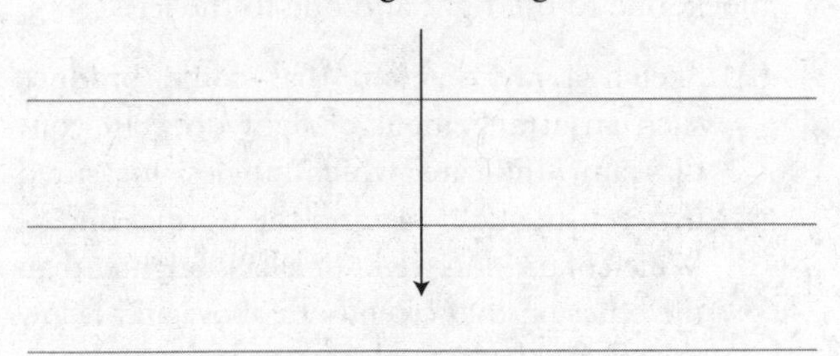

(D) Diffraction of a plane wave front as it passes through an opening that is larger than the wavelength (see figure).

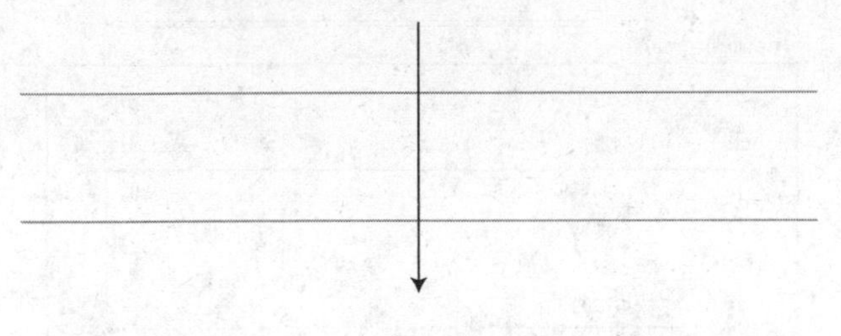

(E) Double-slit interference pattern that appears when a plane wave front passes through two small openings (see figure).

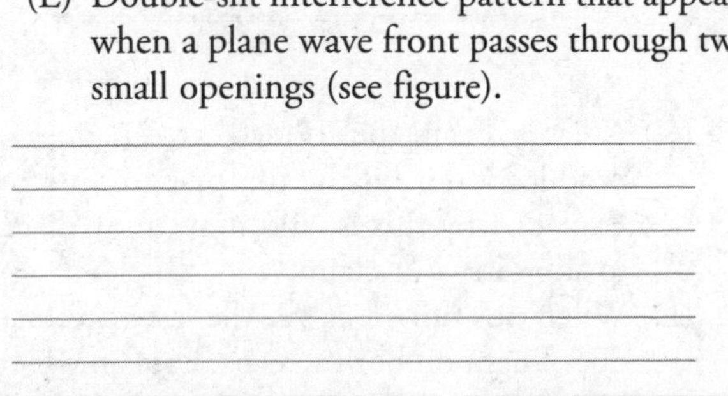

(F) Refraction of light as light passes from air straight into a block of glass (see figure). Which way will the light turn and why?

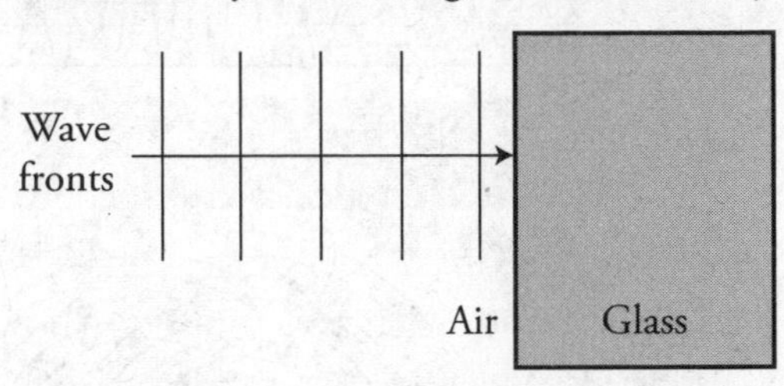

(G) Refraction of light as light passes at an angle from a faster speed medium into a slower speed medium (see figure). Which way will the light turn and why?

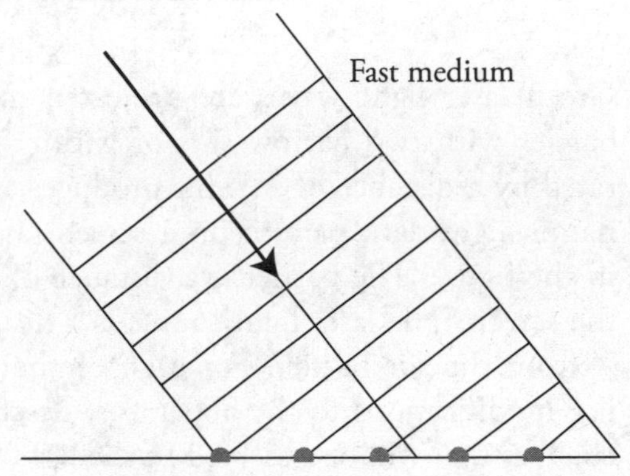

(H) A plane wave front of light reflecting off a flat mirror (see figure).

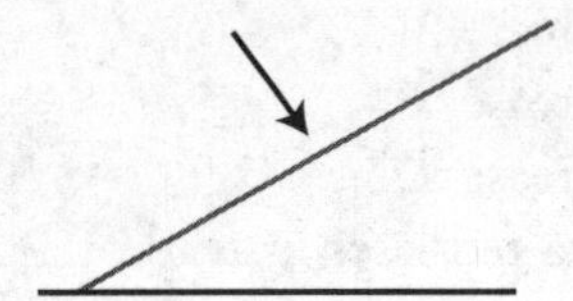

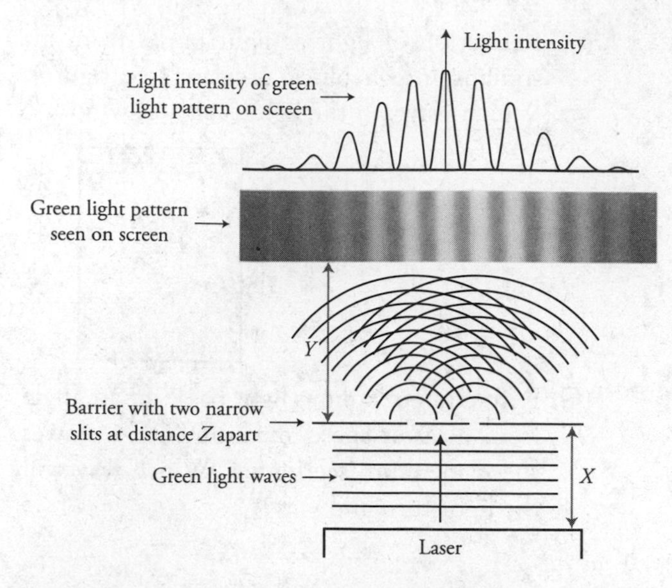

18. Green laser light waves are projected toward a barrier with two narrow slits of width W separated by a distance of Z. This produces an alternating light-dark pattern on a screen, as shown in the figure. The barrier is a distance of Y from the screen. The laser light source is a distance of X from the slit barrier. For each of the following modifications to the apparatus, describe the changes that will be observed in the light pattern seen on the screen, and sketch the new pattern. Justify each claim with an equation.

(A) Decrease W.
(B) Decrease X.
(C) Decrease Y.
(D) Decrease Z.
(E) Use a red laser instead of a green one.
(F) Use a violet laser instead of a green one.
(G) Replace the double-slit barrier with a multi-slit diffraction grating with the same slit spacing of Z.

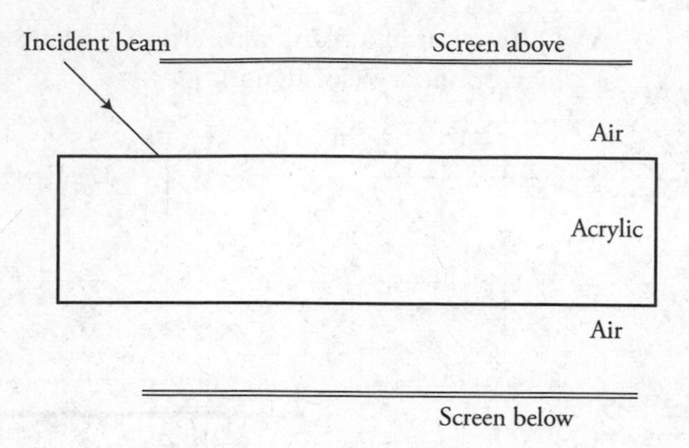

19. Students shine a laser beam at a rectangular block of acrylic plastic, as shown in the figure. Two dots of light appear on a screen above the acrylic block, one to the right and one to the left. Likewise, two dots of light appear below the block, one to the right and one to the left.

(A) Sketch a ray diagram that could produce such an arrangement of light dots. In your diagram, indicate which angles measured between the light ray and the normal line.
(B) Which of the dots (right or left) is brighter than the other on the screen both above and below the acrylic block? Explain your reasoning.

The original acrylic block is replaced by a new block with a rectangular air gap in the middle, as shown in the figure.

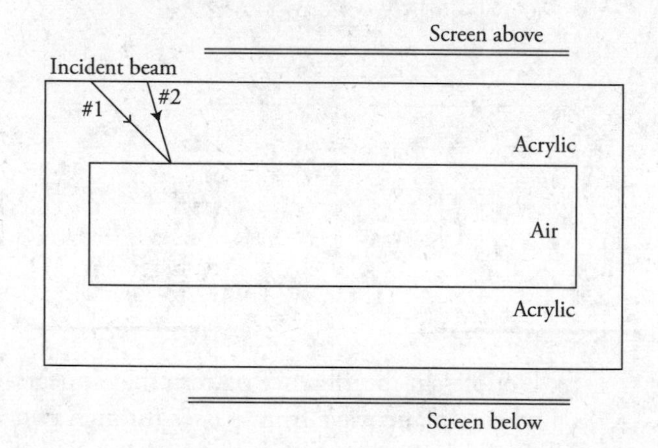

(C) A laser beam positioned at #1 does not produce a dot on the top or bottom screen. Explain why this is, and draw a ray diagram to support your claim.
(D) When positioned at #2, the laser produces a dot of light on both screens. Explain why this happens, and draw a ray diagram to support your claim.

20. Your physics teacher instructs you to determine the index of refraction of a glass prism.

(A) List the items you would use to perform this investigation.

(B) Sketch a simple diagram of your investigation. Make sure to label all items, and label the measurements you would make.

(C) Outline the experimental procedure you would use to gather the necessary data. Indicate the measurements to be taken and how the measurements will be used to obtain the data needed. Make sure your outline contains sufficient detail so that another student could follow your procedure and duplicate your results.

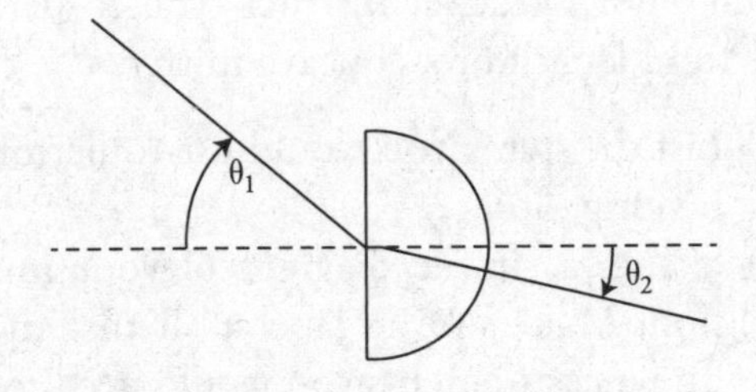

Incidence angle (degree)	Refraction angle (degree)	$\sin \theta_i$	$\sin \theta_r$
10	6	0.17	0.10
20	11	0.34	0.19
30	16	0.50	0.28
40	21	0.64	0.36
50	26	0.77	0.43
60	29	0.87	0.49
70	32	0.94	0.53
80	34	0.98	0.56

21. A group of students are given a semicircular sapphire prism through which they shine a beam of light, as shown in the figure. They measure the incident and refracted angle of the light and produce the table shown.

(A) Use the data to determine the index of refraction of the prism.

(B) Graph the refraction angle as a function of incidence angle. Are the two variables directly related? Justify your claim.

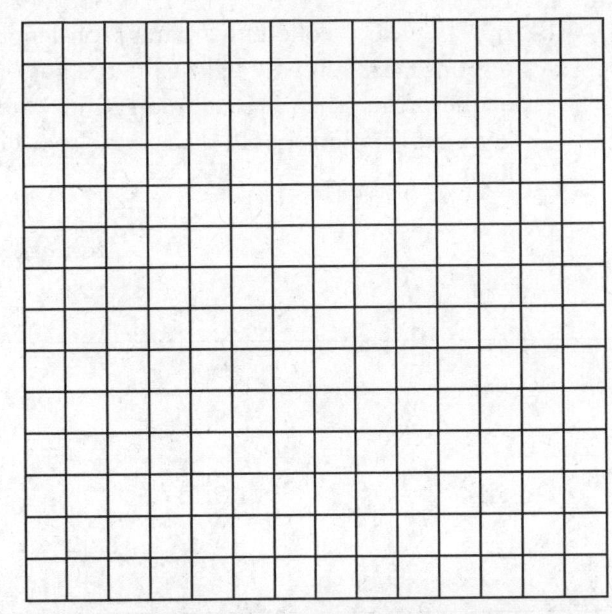

(C) Graph the sine of the refraction angle as a function of the sine of the incidence angle. Use the slope of the graph to calculate the index of refraction of the prism. Show your work.

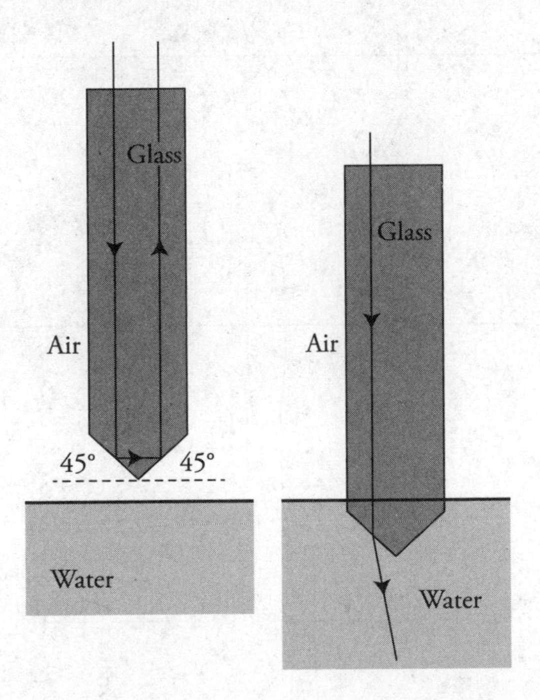

22. A glass prism has a point at the bottom made up of 45° angled surfaces, as shown in the figure. A light beam directed downward through the top of the prism is completely reflected off the bottom surfaces and exits back through the top of the prism. When submerged in water, the light beam exits the bottom of the prism.

(A) Calculate the minimum index of refraction of the glass.

(B) In a clear, coherent, paragraph-length response, explain why light exits the bottom of the prism when it is submerged in water. Your explanation should discuss the speed of light.

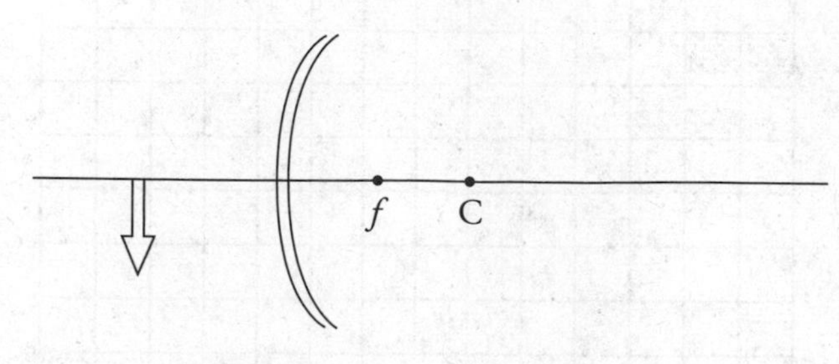

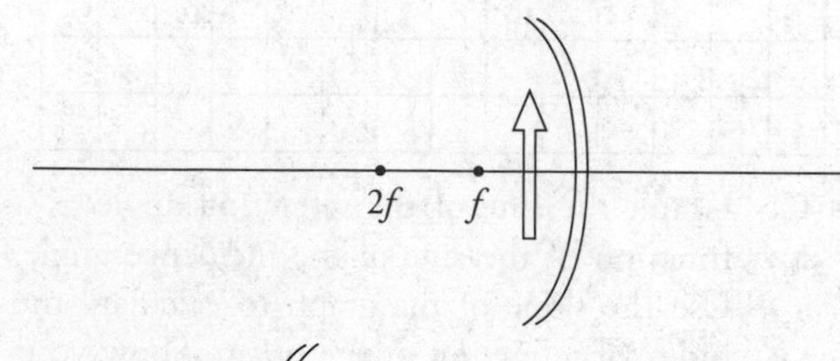

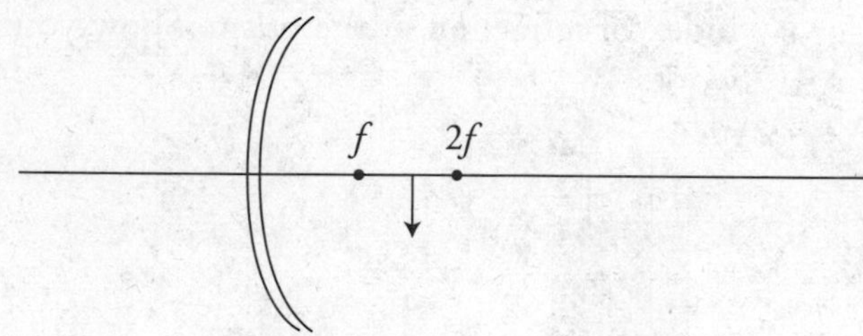

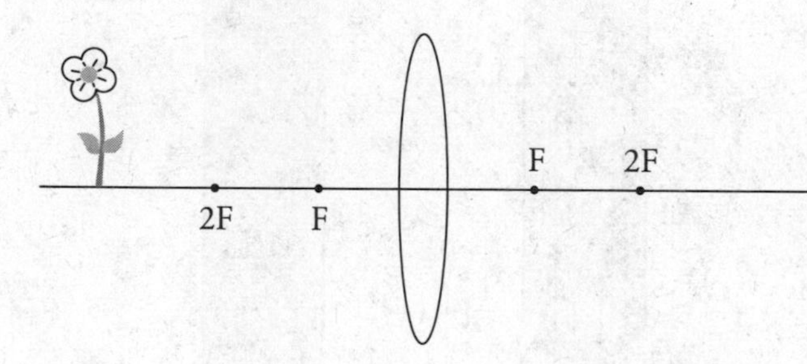

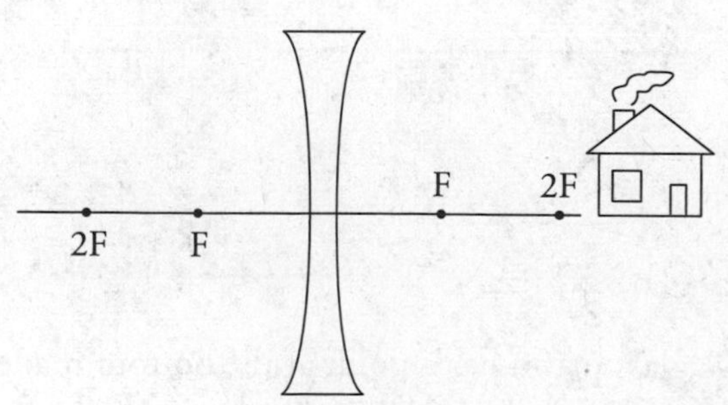

23. An object is placed in front of each of the lenses/mirrors shown in each figure below.

(A) Sketch a ray diagram with at least two rays to locate the position of the image. Sketch the image.

(B) Next to each image, indicate whether the image is real or virtual, inverted or upright, and enlarged or reduced in size.

(C) Sketch a stick figure on each diagram to indicate where a person would have to be standing and in which direction the person needs to look to see the image formed by the mirror.

(D) What distinguishes a real image from a virtual image in these ray diagrams? Explain.

(E) What would happen to the image of the flower if the bottom half of the lens were covered by a piece of cardboard? Justify your claim by making reference to the ray diagram.

24. Your physics teacher instructs you to determine the focal length of a concave mirror.

(A) List the items you would use to perform this investigation.

(B) Sketch a simple diagram of your investigation. Make sure to label all items, and indicate measurements you would need to make.

(C) Outline the experimental procedure you would use to gather the necessary data. Indicate the measurements to be taken and how the measurements will be used to obtain the data needed. Make sure your outline contains sufficient detail so that another student could follow your procedure and duplicate your results.

(D) Could this procedure be used to find the focal length of a convex mirror? Justify your response.

25. A lens lab produces these data and the graph.

s_o	s_i	$1/s_o$	$1/s_i$
50	12.5	0.020	0.080
30	15.0	0.033	0.067
25	16.7	0.040	0.060
20	20.0	0.050	0.050
15	30.0	0.067	0.033
12	60.0	0.083	0.017

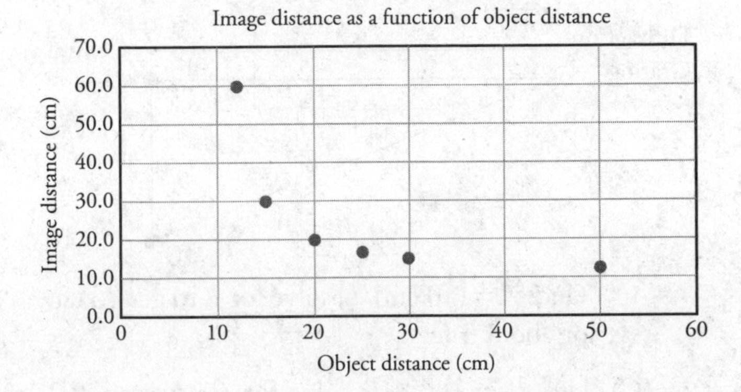

(A) What data would you plot to produce a straight line? Plot the data.

(B) What information from your straight line plot will allow you to determine the focal length of the lens? Justify your claim with an equation. Calculate the focal length of the lens using your straight line graph.

› Solutions to Practice Problems

1. (D) Using the equation, $d \sin \theta = m\lambda$, we want to increase θ while keeping $m = 3$. We could do this by decreasing d or increasing λ. Or we could keep θ the same but just move the slits farther from the screen, which will enlarge the pattern by spreading it out with added distance. Moving the light source farther from the slits won't change anything.

2. (D) The two waves do not have the same time period. Wave B has a higher frequency. Wave A has an amplitude of only 15 cm. The equation for a wave is $x = A \sin(2\pi ft)$. Wave B has an amplitude of 30 cm and a time period of 0.08 seconds and, therefore, a frequency of 1/0.08 Hz.

3. (D) A convex/diverging mirror produces only virtual/upright images of reduced size.

4. (A) Convex lenses produce real images (positive s_i) when the object distance is beyond the focal point, and virtual images (negative s_i) when the object distance is inside the focal length.

5. (B) A and D do not refract the light correctly at the first surface. Light should be bending toward the normal as it is slowing down. Exiting the prism, light will speed back up and should bend away from the normal. So C is right out. That leaves answer choice B.

6. (B) If you had a calculator, you could use Snell's law, calling the water medium "1" and the glass medium "2": $1.3 \sin 36° = 1.5 \sin \theta_2$. You would find that the angle of transmission is 31°. But, you don't need a calculator . . . so look at the choices. The light must bend *toward* the normal when traveling into a material with higher index of refraction, and choice B is the only angle smaller than the angle of incidence. Choice A is silly because total internal reflection can occur only when light goes from high to low index of refraction.

7. (D) The speed of light (or any wave) depends upon the material through which the wave travels; by moving into the water, the light's speed slows down. But the frequency of a wave does not change, even when the wave changes material. This is why tree leaves still look green under water—color is determined by frequency, and the frequency of light under water is the same as in air. So, if speed changes and frequency stays the same, by $v = \lambda f$, the wavelength must also change.

8. (B) You could approximate the answer by making a ray diagram, but the mirror equation works, too:

$$\frac{1}{f} = \frac{1}{s_o} + \frac{1}{s_i}$$

Because the radius of a spherical mirror is twice the focal length, and we have placed the object at the center, the object distance is equal to $2f$. Solve the mirror equation for s_i by finding a common denominator:

$$s_i = \frac{fs_o}{s_o - f}$$

9. (A) and (B) $\sin\theta = \dfrac{m\lambda}{d}$. Increasing the wavelength of the laser (l) and/or decreasing the hair width (d) will both increase the angle of the pattern.

10. (A) *Be careful of units*! Convert the object distance from 30 cm to 300 mm.

$$\frac{1}{f} = \frac{1}{s_i} + \frac{1}{s_o}$$

$$\frac{1}{16\,\text{mm}} = \frac{1}{s_i} + \frac{1}{300\,\text{mm}}$$

$$s_i = 16.9\,\text{mm}$$

This means the image is 3.1 mm in front of the retina, which is at a distance of 20 mm. We need a diverging lens to move the image back to the retina. Diverging lenses are concave.

11. (A) Use $d \sin \theta = m\lambda$. Here d is not 7000! d represents the distance between slits. Because there are 7000 lines per centimeter, there's 1/7000 centimeter per line; thus, the distance between lines is 1.4×10^{-4} cm, or 1.4×10^{-6} m. θ is 25° for the first-order maximum, where $m = 1$. Plugging in, you get a wavelength of just about 6×10^{-7} m, also known as 600 nm.

(B) This is a geometry problem.

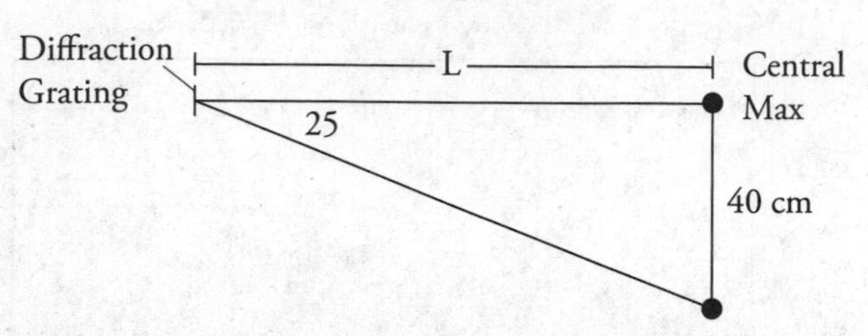

$\tan 25° = (40\text{ cm})/L$; solve for L to get 86 cm, or about 3 feet.

(C) Use $d \sin \theta = m\lambda$; solve for θ using m = 2, and convert everything to meters. We get $\sin \theta = 2(6.0 \times 10^{-7}\text{ m})/(1.4 \times 10^{-6}\text{ m})$. The angle will be 59°. Now, use the same geometry from part (b) to find the distance along the screen: $\tan 59° = x/(0.86\text{ m})$, so $x = 143$ cm. (Your answer will be counted correct if you rounded differently and just came close to 143 cm.)

12. (A) Converging–Convex lens. This type of lens will form a real image that can be projected on a screen (retina). Note that the image will be upside down. Your brain flips the image over!

(B) The light rays are converging too soon. We need the rays to converge farther from the eye lens. This requires a pair of diverging–concave lenses that will spread the light out a bit before it enters the eye. The eye will then focus the light on the retina.

13. (A and D) The converging optical instruments—convex lens and concave mirror—only produce virtual images if the object is inside the focal point. But when that happens, the virtual image is *larger* than the object, as when you look at yourself in a shaving mirror. But diverging optical instruments—a convex mirror and a concave lens—always produce a smaller, upright image, as when you look at yourself reflected in a Christmas tree ornament.

14. (A) Use Snell's law: $n_1 \sin \theta_1 = n_2 \sin \theta_2$. This becomes 1.0 sin 35° = 1.5 sin θ_2. Solve for θ_2 to get 22°.

(B) Use Snell's law again. This time, the angle of incidence on the water is equal to the angle of refraction in the glass, or 22°. The angle of refraction in water is 25°. This makes sense because light should bend away from normal when entering the water, because water has smaller index of refraction than glass.

(C) Important points:

- Light both refracts *and reflects* at both surfaces. You must show the reflection, with the angle of incidence equal to the angle of reflection.
- We know you don't have a protractor, so the angles don't have to be perfect. But the light must bend toward normal when entering glass, and away from normal when entering water. If you have trouble drawing this on the AP exam, just explain your drawing with a quick note to clarify it for the exam reader. This helps you and the reader!

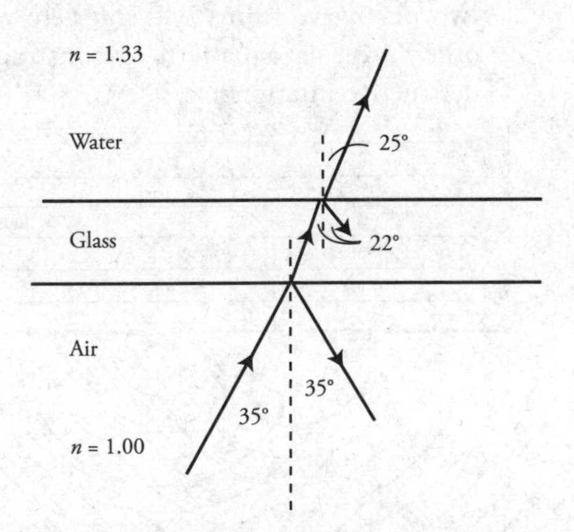

(D) The angle of incidence on the side must be measured from the normal. The angle of incidence is not 25°, then, but 90 – 25 = 65°. Using Snell's law, 1.33 sin 65° = 1.50 sin θ_2. The angle of refraction is 53°.

(E) The critical angle for glass to air is given by sin θ_c = 1.0/1.5. So θ_c = 42°. Because the angle of incidence from the glass is 53° [calculated in (D)], total internal reflection occurs.

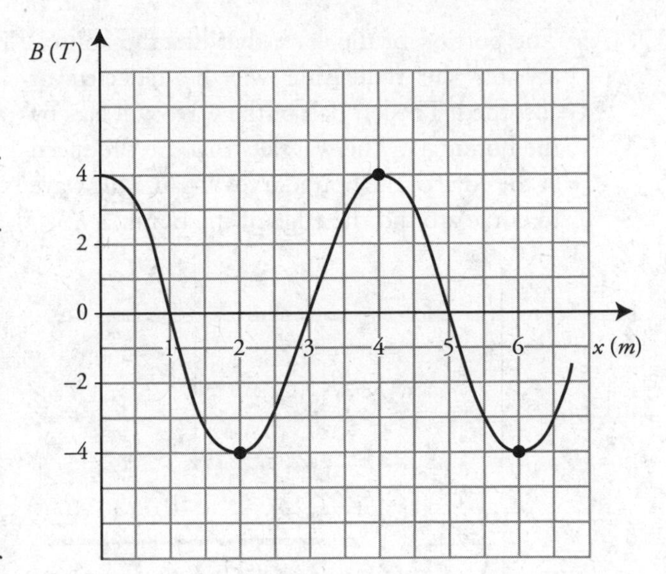

15. (A) $E = A\sin\left(\frac{2\pi t}{T}\right) = 0.3\sin\left(\frac{2\pi t}{2.4}\right)$

$= 0.3\sin(2.6t)$

(B) $B = 4.0\cos\left(\frac{\pi x}{2}\right)$. This is a cosine wave with an amplitude of 4.0 T and wavelength of 4 m.

16. Here are six of the interference patterns seen as the two waves first touch, partially overlap, completely overlap, and then pass by each other.

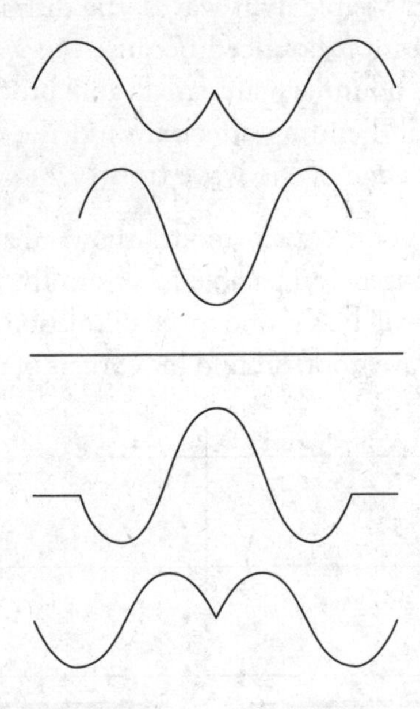

17. (A) The portion of the wave that hits the boundary on the right side will be reflected or absorbed. The left half of the wave will pass by the boundary. The wavelets that are produced at the edge of the boundary will cause the wave to curve around the edge of the boundary.

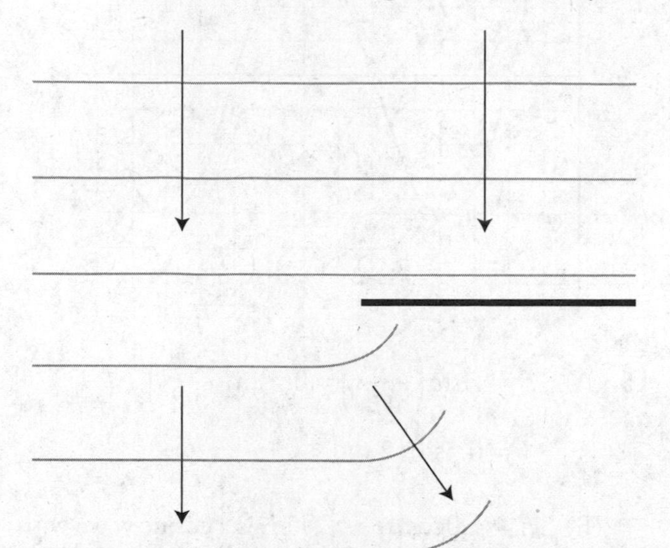

(B) Sound waves have a long wavelength that produces large wavelets near the boundary causing a pronounced diffraction effect of "bending around the corner." As the wavelength gets smaller than the boundary itself, as in visible light waves, the diffraction effect is less pronounced because the wavelets near the boundary are small and produce only a small bending effect around the corner near the edge of the wave front.

(C) The point source model shows that when the opening is comparable in size to the wavelength, there will be a pronounced diffraction (bending of the wave front) around the corners of the opening.

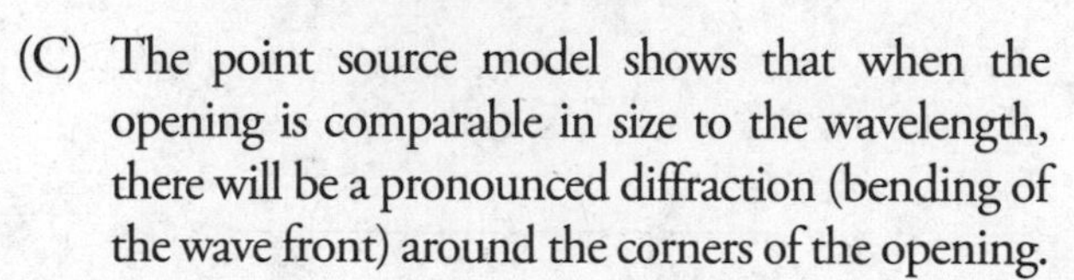

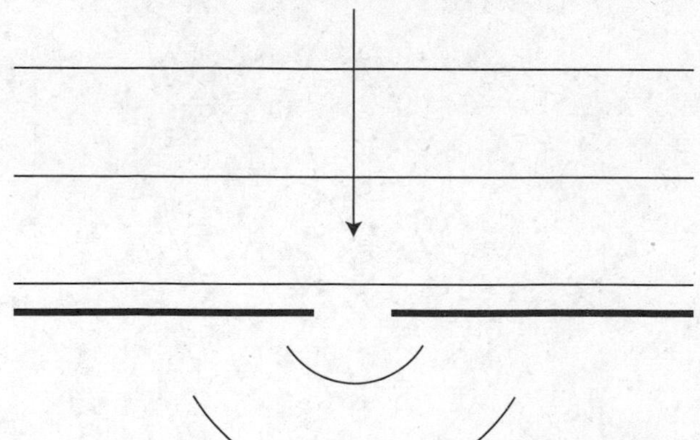

(D) When the opening is much larger than the wavelength, the point source model shows that only the edges of the wave show any bending or diffraction. The wavelets in the center overlap and continue the propagation of the plane wave unchanged.

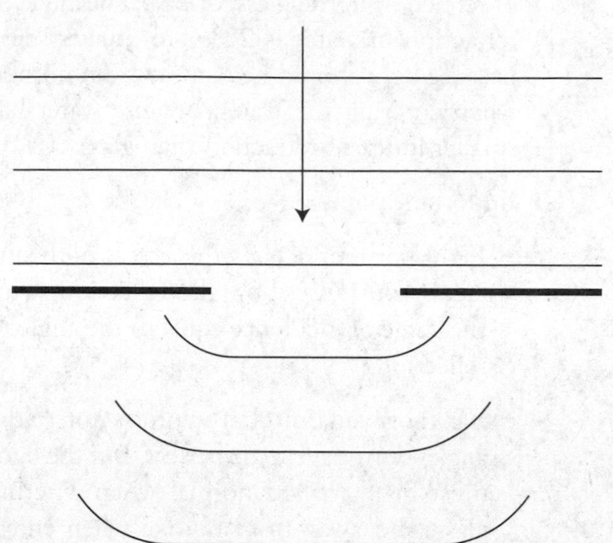

(E) The point source model shows that the two slits will produce two separate curved wave fronts passing through the openings. These two new wave fronts will interfere with each other, creating a pattern of constructive and destructive interference.

(F) The speed of light in the new medium is slower. Drawing the Huygens' wavelets, we can see that all parts of the wave will slow down at the same time. This means the wave front will not change direction but will simply travel straight through the glass with a shorter wavelength.

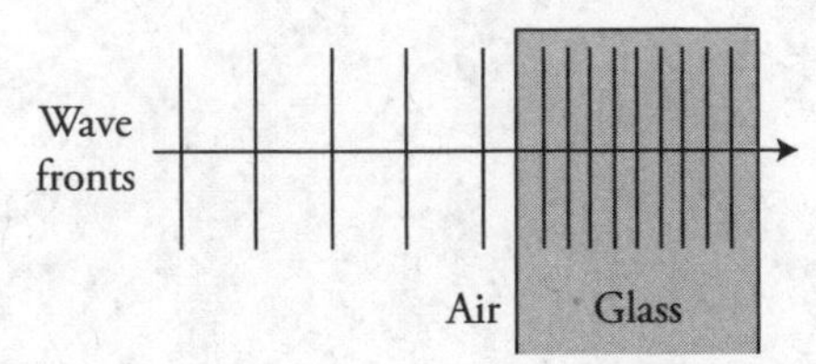

(G) Entering the new medium, the wave slows down. Drawing the wavelets in the new medium with a shorter wavelength, we see that the wave front changes direction, and the wave ray bends toward the normal line to the surface.

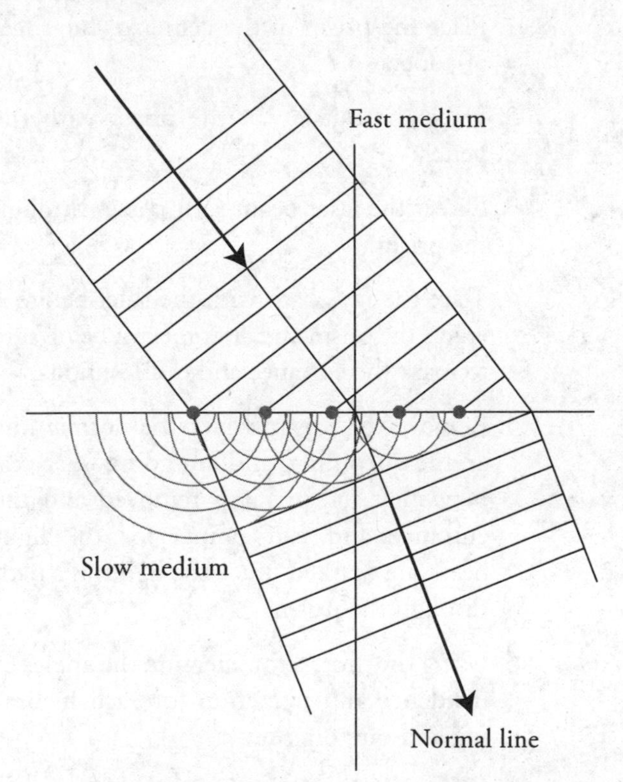

(H) Drawing wavelets for the wave that reflects off the surface, we see that the wave maintains the same speed and wavelength. However, the reflected wave front turns and travel off in a new direction. The angle of the incoming wave front measured to the normal is equal to the angle of the outbound reflected wave front measured to the normal line.

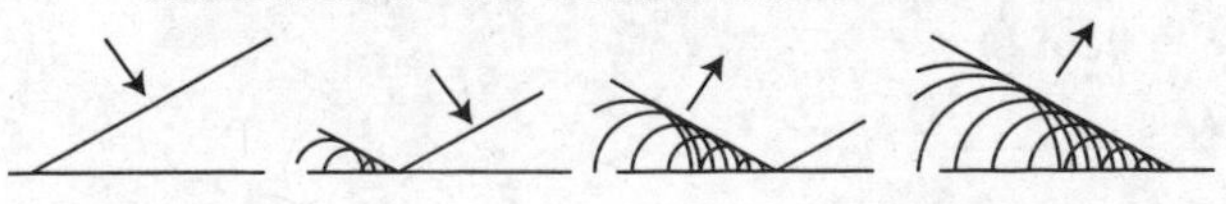

18. The light and dark pattern of double-slit diffraction is described by the equation $d \sin \theta = m\lambda$, where d is the distance between the two slits; θ is the wavelength of the light; λ is the angle from the slit to the maxima (bright spot) on the screen; and m is the maxima number. Note that $m = 0$ is the central maxima right down the center along the line of symmetry, and $m = 1$ would be the first maxima to either side of the central maxima. Solving the equation for our situation, we get $\sin \theta = \frac{m\lambda}{Z}$. Therefore, as λ goes up or Z goes down, θ increases.

(A) As W decreases, there will be no change to the pattern. Nothing in the equation changes. Note: Remember that the slit opening width and the wavelength of the green light need to be approximately the same size (order of magnitude) to produce diffraction effects.

(B) As X decreases, there will be no change to the pattern because nothing in the equation changes.

(C) As Y decreases, nothing in the equation changes. However, since the screen is closer to the barrier, there is less distance/time for the pattern to spread out before it hits the screen. Therefore, the pattern will become closer together and more tightly packed.

(D) As Z decreases, θ increases. Therefore, the pattern becomes wider and more spread out.

(E) Red light has a longer wavelength, which will increase θ, making the pattern spread out.

(F) Violet light has a shorter wavelength, which will decrease θ, making the pattern narrower.

(G) The maximas will be in the same locations but will be thinner, more point like.

19. (A)

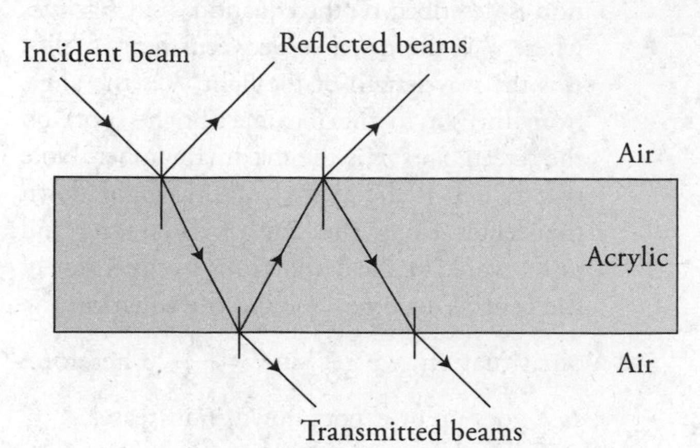

(B) The dots on the left are brighter both above and below the acrylic block. Conservation of energy tells us that less and less of the light is making it through to the right dots. At each medium interface, some of the light is reflected leaving less light energy to continue on into the next medium.

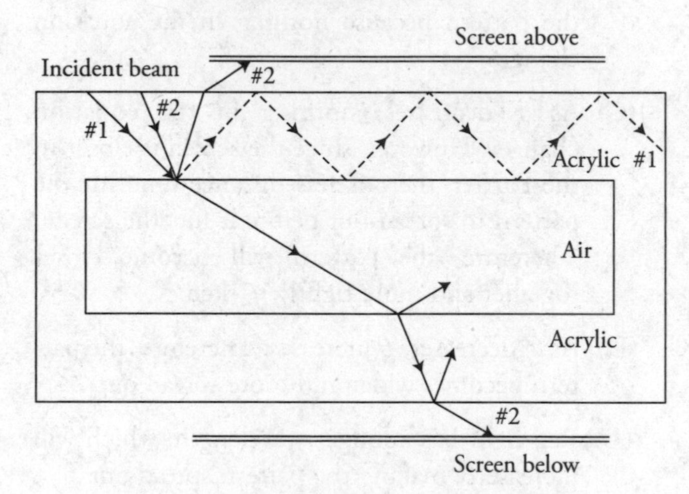

(C) Ray #1 must be striking the acrylic/air interface at an angle of incidence larger than the critical angle. This produces total internal reflection along the parallel top and bottom surfaces, as shown by the dashed line in the figure.

(D) Ray #2 must be striking the acrylic/air interface at an angle less than the critical angle. There is a partial reflection and refraction at each surface. This will produce a dot on each screen, as shown by a solid line in the figure.

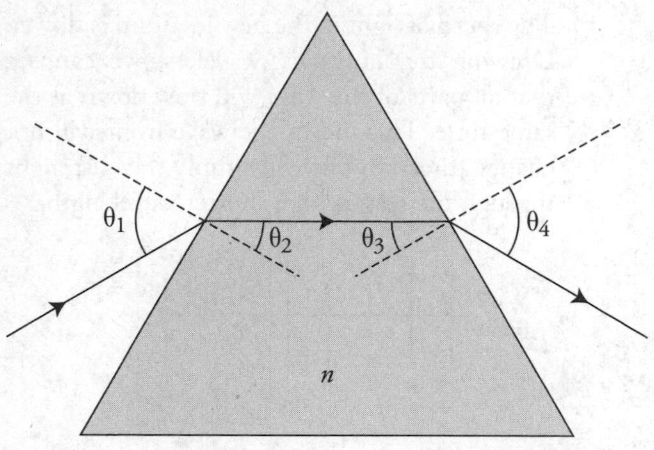

20. (A) Equipment: protractor, pencil, ruler, sheet of white paper

(B) See figure.

(C) Procedure:

1. Place the prism in the center of the sheet of paper.
2. Trace the outside of the prism with the pencil.
3. Direct the laser beam so it passes through the prism.
4. Trace the ray's path with the ruler before it enters the prism and after it exits, being sure to mark the entrance and exit locations.
5. Remove the prism. Mark the normal line at the incoming and outgoing surfaces. Now that the prism is removed and the entrance and exit points for the light beam are marked, trace the light ray's path through the prism.
6. Using the protractor, measure the angles of incidence and refraction for each incoming and outgoing ray.
7. Use Snell's law to calculate the index of refraction of the glass prism.
8. Repeat for a wide variety of angles, and average the result.

21. (A) Use Snell's law and the first set of data:

$$n_a \sin \theta_a = n_s \sin \theta_s$$

$$\sin(10°) = n_s \sin(6°)$$

$$n_s = 1.66$$

Repeating this for the rest of the data in the table, we get an average of 1.76.

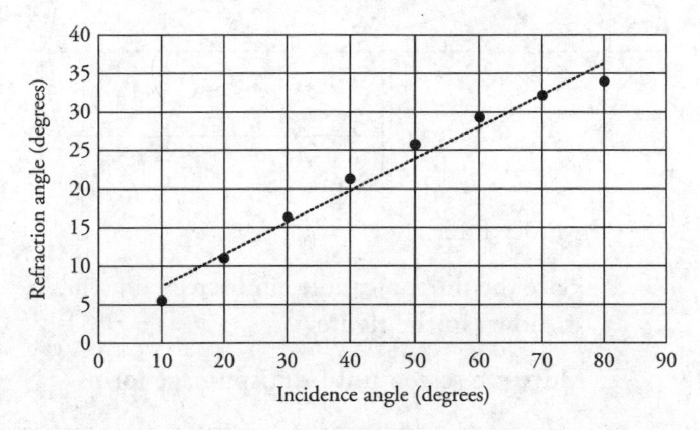

(B) We can see in the figure that the trend line does not match the data and clearly shows the data are not straight and cannot be a direct relationship.

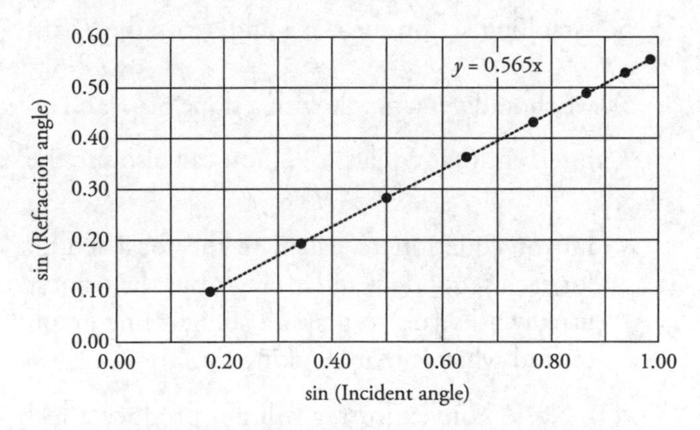

(C) Plotting the sine of the refracted angle as a function of the sine of the incident angle produces a straight line. This makes sense:

$$n_s \sin \theta_s = n_a \sin \theta_a = \sin \theta_a$$

$$n_s \sin \theta_s = \sin \theta_a$$

$$\sin \theta_s = \frac{1}{n_s} \sin \theta_a$$

$$y = mx + b$$

Matching Snell's law up to the equation of a line we can see that the slope of the line is equal to the reciprocal of the index of refraction of the sapphire. The best fit line gives a slope of 0.565. Thus, n_s = 1.77.

22. (A) Remember that at the critical angle, the refracted angle is 90°. Calculating the minimum index of refraction so light does not escape at a critical angle of 45 degrees is shown by the following calculation:

$$n_a \sin(90°) = n_{glass} \sin \theta_{critical}$$

$$\sin(90°) = n_{glass} \sin(45°)$$

$$n_{glass} = 1.41$$

(B) Originally, the light is traveling in glass and is trying to exit out into air and the light strikes the bottom of the prism beyond the critical angle. When the prism is lowered into water, the difference in the index of refraction and speed of light between the two mediums is less than it was while in air. Therefore, the light does not speed up as much going into the water as it did going into air. This reduces the bending of the light due to refraction. The light is no longer hitting the bottom of the prism beyond the critical angle. Therefore, the light is able to escape out the bottom of the prism into the water but not the air.

23. (A) See figure.
(B) See figure.
(C) See figure.

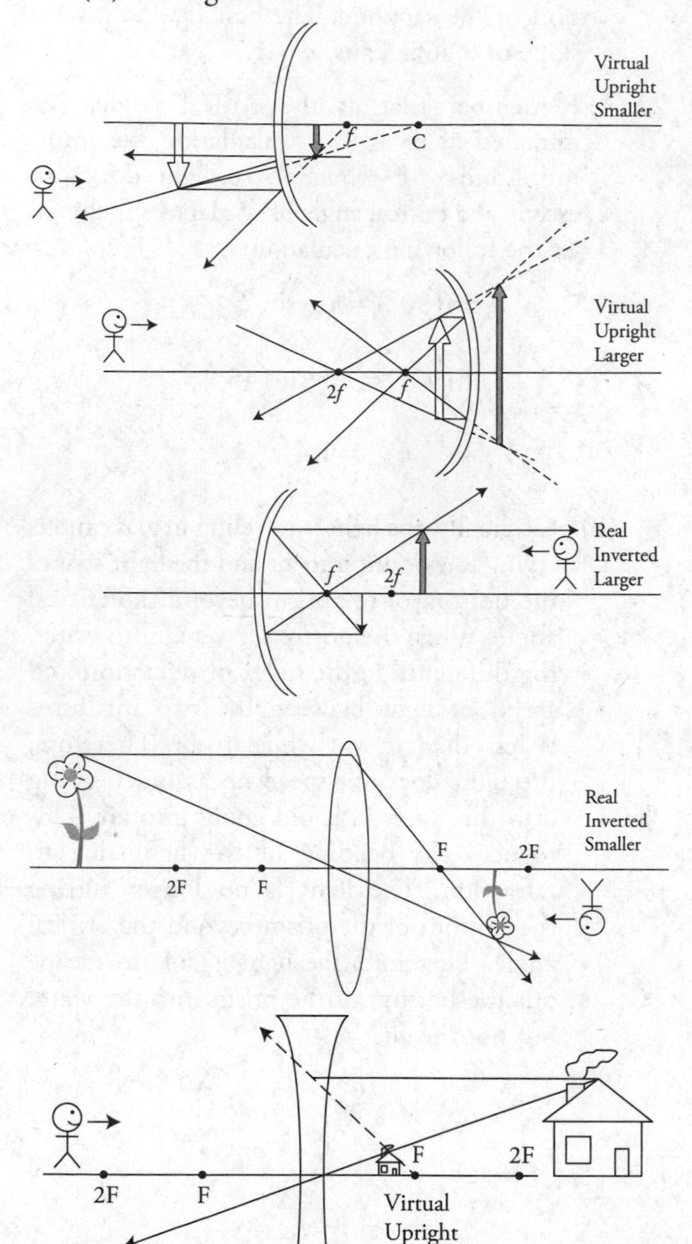

(D) Light rays converge to form real images. Light rays diverge from the location of a virtual image as if they came from that spot, even though they did not actually pass through that location.

(E) Since half of the light rays will still be passing through the lens unimpeded, the image will still form in the same spot with the same properties as before. However, the image will not be as bright because half of the rays from the flower are now blocked.

24. There are several ways to perform this lab. Here is one method.

(A) Equipment: mirror, meter stick, candle, screen

(B) See figure.

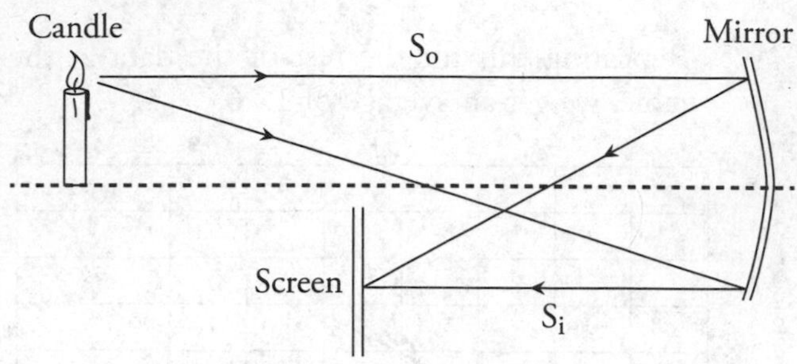

(C) Procedure:

1. Place the mirror, candle, and screen on a table, as shown in the figure.
2. Move the screen until a crisp image forms.
3. Measure the object distance from the mirror to the candle and the image distance from the mirror to the screen.
4. Repeat this process for several data points.
5. Graphing $\frac{1}{s_o}$ on the x-axis and $\frac{1}{s_i}$ on the *y*-axis, we should get a graph with a slope of -1 and an intercept that equals $\frac{1}{f}$. (You can also use the mirror equation to calculate the focal length for each set of data and average, but the AP test usually asks you to graph a straight line graph to find what you are looking for.)

(D) No! A convex mirror will not produce a real image on the screen.

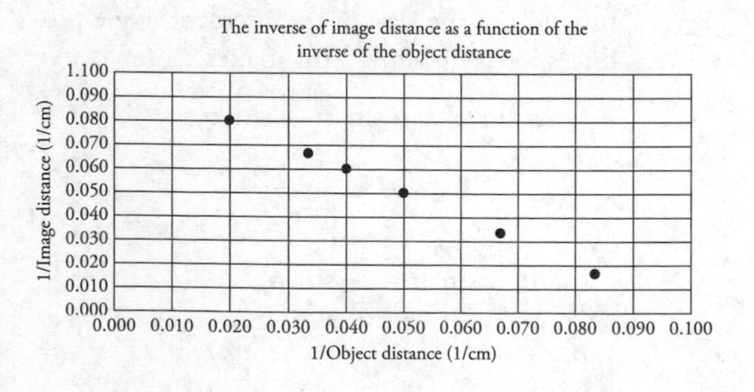

25. (A) Graph $\frac{1}{s_o}$ on the x-axis and $\frac{1}{s_i}$ on the y-axis,

we should get a graph with a straight line with a slope of −1.

(B) Rearranging the lens equation and comparing it to the equation of a line, we can see that the intercept of the line equals $\frac{1}{f}$.

$$y = mx + b$$

$$\frac{1}{s_i} = (-1)\frac{1}{s_o} + \frac{1}{f}$$

The intercept equals 0.1 (1/cm). Therefore, the focal length is 10 cm.

› Rapid Review

- Waves can be either transverse or longitudinal. Transverse waves are up–down waves, like a sine curve. Longitudinal waves are push–pull waves, like a sound wave traveling through the air.
- When two waves cross paths, they can interfere either constructively or destructively. Constructive interference means that the peaks, or troughs, of the waves line up, so when the waves come together, they combine to make a wave with bigger amplitude than either individual wave. Destructive interference means that the peak of one wave lines up with the trough of the other, so when the waves come together, they cancel each other out.
- All electromagnetic waves travel at a speed of 3×10^8 m/s in a vacuum.
- The double-slit experiment demonstrates that light behaves like a wave.
- When light travels through anything other than a vacuum, it slows down, and its wavelength decreases. The amount by which light slows down as it passes through a medium (such as air or water) is related to that medium's index of refraction.
- Thin films can cause constructive or destructive interference, depending on the thickness of the film and the wavelength of the light. When solving problems with thin films, remember to watch out for phase changes.
- Light: radio waves, microwaves, infrared, visible, ultraviolet, X-rays, and gamma rays, are all electromagnetic waves. EM waves are transverse and can be polarized.
- Diffraction (the bending of waves around obstacles), interference (constructive/destructive addition of intensity), and refraction (the bending of light when it changes speed) are wave properties of light. These properties can be explained using the point source model of waves.
- When light encounters a new medium, it can reflect, transmit (refract), or be absorbed. All three can happen at the same time.
- Snell's law describes how the direction of a light beam changes as it goes from a material with one index of refraction to a material with a different index of refraction.

- When light is directed at or beyond the critical angle, it cannot pass from a material with a high index of refraction to one with a low index of refraction; instead, it undergoes total internal reflection.
- When solving a problem involving a plane mirror, remember (1) the image is upright (it is a virtual image); (2) the magnification equals 1; and (3) the image distance equals the object distance.
- When solving a problem involving a spherical mirror, remember (1) incident rays parallel to the principal axis reflect through the focal point; (2) incident rays going through the focal point reflect parallel to the principal axis; (3) points on the same side of the mirror as the object are a positive distance from the mirror, and points on the other side are a negative distance from the mirror; and (4) the lensmaker's equation (also called the mirror equation in this case) holds.
- When solving problems involving a convex lens, remember (1) incident rays parallel to the principal axis refract through the far focal point; (2) incident rays going through the near focal point refract parallel to the principal axis; and (3) the lensmaker's equation holds, and f is positive.
- When solving problems involving a concave lens, remember (1) incident rays parallel to the principal axis refract as if they came from the near focal point; (2) incident rays going toward the far focal point refract parallel to the principal axis; and (3) the lensmaker's equation holds with a negative f.

Summary of Signs for Mirrors and Lenses

Concave Mirror	Convex Mirror	Concave Lens	Convex Lens
Like a shaving mirror or makeup compact	Like a Christmas tree ornament	Like a pair of near-sighted glasses	Like a magnifying glass
s_o is always +	s_o is always +	s_o is always +	s_o is always +
s_i is + for real images and – for virtual images	s_i is – for virtual images	s_i is – for virtual images	s_i is + for real images and – for virtual images
f is always +	f is always –	f is always –	f is always +

Quantum, Atomic, and Nuclear Physics

IN THIS CHAPTER

Summary: Our modern understanding of physics shows us that time, space, and matter are not exactly as they appear. The nano-world of atoms and subatomic particles can seem strange and does not always follow the same rules as the world we live in. But the weird and wonderful world of the ultra-small are the building blocks that make up everything we know and experience in the big world we live in.

Key Ideas

- Time and length are not constant. Both depend on the speed the object is traveling. Only the speed of light in a vacuum is constant.
- Mass can be converted into energy and vice versa: $E = mc^2$.
- Massless particles can have mass-like properties such as momentum.
- The photoelectric-effect experiment proves that light has particle properties. The particle of light is called a photon. Its mass is zero and its energy is $E = hf$.
- Subatomic particles can exhibit wave-like properties such as interference.
- Electrons in atoms can exist only in specific energy levels. To move up in energy, they must absorb a photon of light. To move down in energy, they must emit a photon.
- Conservation of mass/energy and conservation of momentum hold true in reactions involving subatomic particles, including photons.

- The strong force holds the nucleus together. It is equivalent to the nuclear mass defect.
- Some nuclei are radioactively unstable and decay into new elements.
- The time it takes for one-half of a radioactive material to decay is called half-life.
- There are many types of decay: alpha, beta, gamma, positron, and neutron are a few.
- In all nuclear reactions, four conservation laws are obeyed: conservation of mass/energy, conservation of charge, conservation of momentum, and conservation of nucleon number.

Relevant Equations

Energy of a photon: $E = hf = \frac{hc}{\lambda}$

Momentum of a photon: $p = \frac{h}{\lambda} = \frac{E}{c}$

De Broglie wavelength: $\lambda = \frac{h}{p} = \frac{h}{mv}$

Matter–energy relationship: $E = mc^2$

Photoelectric effect: $K_{max} = hf - \phi$

What Is Modern Physics?

Modern physics takes us into the bizarre world of the uber fast and the nano small. This is the realm where the distinction between waves and particles gets fuzzy, where things without mass can have momentum, where elements spontaneously break apart, and where time and space are no longer constant.

Modern physics refers to all of the physics from about 1900 on, and that is when things really got weird. In fact, things got so bizarre that Albert Einstein refused to believe some of it, even though in many ways he was the guy who got the whole ball rolling when it came to modern physics. But quantum theory, a big part of modern physics, has proven over and over again to be correct, so much so, that practically all of our modern electronic toys are based upon it. So let's dive into modern physics.

Space, Time, the Speed of Light, and $E = mc^2$

Crazy-hair Einstein changed everything when he made these two simple postulates[1] of special relativity:

1. All the laws of physics are the same in every uniformly moving frame of reference.
2. The speed of light in a vacuum is always measured to be $c = 3 \times 10^8$ m/s no matter the motion of the source of the light, or the motion of the receiver of the light.

---

[1] *Postulate* is just a fancy word for a statement that is assumed to be true based on reasoning.

The first one makes sense. In a car, you don't have to do anything special when you are drinking from a cup. But, if the driver steps on the gas or slams on the brakes, you might get wet. As long as you aren't accelerating, everything is normal.

The second postulate has really weird consequences. In order for the speed of light to always equal *c*, time and space cannot be constant! The faster you go, the slower time moves and the shorter distances become in the direction you are going. The really odd thing is that you will not notice any of these changes, because of postulate 1. All the laws of physics will be the same to you no matter what speed you travel. All the effects of relativity seem to happen to the "other guy." Confused yet? Here is an example:

> I travel 10 years near the speed of light to an exoplanet. As you watch me through a telescope you see that my ship shrinks in the direction it is traveling, I also seem to be moving really slowly and my heart rate has nearly stopped. It's as if time has stopped for me and you predict that I won't even age on the 10-year trip. I, on the other hand, don't notice anything strange happening to me. But, when I look outside, everything seems to be shrinking in the direction I am traveling. Looking back at you, I see that you are buzzing around really fast and aging like crazy. Since the distance around me has shrunk, I get to the exoplanet in no time at all. I have hardly aged a day! Cool! You, on the other hand, are 10 years older. Bummer.

That is really mind-bending because we never see this kind of stuff in everyday life. The effects of relativity don't become easily visible until we start traveling a good fraction of the speed of light.

What do you have to know for the AP exam? Bottom line: you need to know that time and distances are not constant. If two people are moving at different speeds, they will not agree on the size of something or even when things have occurred. Only the velocity of light is constant. Moving faster shrinks distances and slows time, but the effects of relativity always are observed to happen outside of your own frame of reference. Everything in your frame of reference always seems "normal."

Another gift from Einstein is the most famous equation of all time: $E = mc^2$. It is a matter-shattering idea. Literally. It means that energy and mass are just two aspects of the same thing. Mass is like a solid form of energy and it can be converted into energy. (Lots of energy—c^2 is a huge number!) Energy can be converted into mass. You heard that right. You can create matter out of energy. These things happen all the time in the nano-world. So let's take a look at the super small.

Subatomic Particles

Let's start with something familiar, the basic subatomic particles. One way to approach this is to look at things historically. Thousands of years, long before the idea of science, the accepted belief was that the universe was made of four basic things, earth, wind, fire, and water. Once humans got a little more scientific about looking at the world around themselves and started closely examining how matter behaves by breaking it down in to smaller and smaller parts, it became accepted that matter was made of supposedly indivisible particles called atoms. Further experiments revealed the structure of an atom consists of protons and neutrons in the nucleus, with tiny electrons moving around the nucleus. Also discovered were little bundles of electromagnetic energy called photons.

Name (Symbol)	Mass	Charge
Proton (p)	1.67×10^{-27} kg = 1 amu	Positive
Neutron (n)	1.67×10^{-27} kg = 1 amu	Zero
Electron (e)	9.11×10^{-31} kg	Negative
Photon (γ)	0	Zero

There are even more subatomic particles. In the 1950s neutrinos were discovered. In the 1960s quarks, the building blocks of protons and neutrons, were discovered. Just recently the Higgs particle was found. Just make sure you know what the Big 4 are from the table above and know how to find them on your equation sheet.

The Electron-Volt

The *electron-volt*, abbreviated eV, is a unit of energy that's particularly useful for problems involving subatomic particles. One eV is equal to the amount of energy needed to change the potential of an electron by 1 volt. For example, imagine an electron nearby a positively charged particle, such that the electron's potential is 4 V. If you were to push the electron away from the positively charged particle until its potential was 5 V, you would need to use 1 eV of energy.

$$1 \text{ eV} = 1.6 \times 10^{-19} \text{ J}$$

The conversion to joules shows that an eV is an itty-bitty unit of energy. However, such things as electrons and protons *are* itty-bitty particles. So the electron-volt is actually a perfectly sized unit to use when talking about the energy of a subatomic particle.

Photons

Photons are light, or more precisely, light is made up of photons. A nice way to think of a photon is a bundle of electromagnetic waves or energy. The amount of energy in that bundle is directly related to the electromagnetic wave's frequency:

$$E = hf = \frac{hc}{\lambda}$$

where:

- E is the energy of the photon.
- h is Planck's constant. Your AP Physics 2 reference table gives this to you in two forms: $h = 6.63 \times 10^{-34}$ J·s $= 4.14 \times 10^{-15}$ eV·s
- f is the frequency in hertz.
- c is the speed if light, $c = 3.0 \times 10^8$ m/s.
- λ is the wavelength in meters.

Even though you have only $E = hf$ on your reference table, it is very useful to use $c = f\lambda$ to have the same equation in terms of wavelength, $E = \frac{hc}{\lambda}$. It is especially useful since the AP people were kind enough to give you these constants to use:

$$hc = 1.99 \times 10^{-25} \text{ J·m} = 1.24 \times 10^3 \text{ eV·nm}$$

Be careful and notice that the second term is in nanometers, so your wavelength must be in nanometers.

The figure below shows the entire electromagnetic spectrum of light from radio waves to gamma rays. Looking at the photon energy equation we can see that gamma rays have the most energy because they have the highest frequency and shortest wavelength. Radio waves have the least energy and the longest wavelengths and the lowest frequencies.

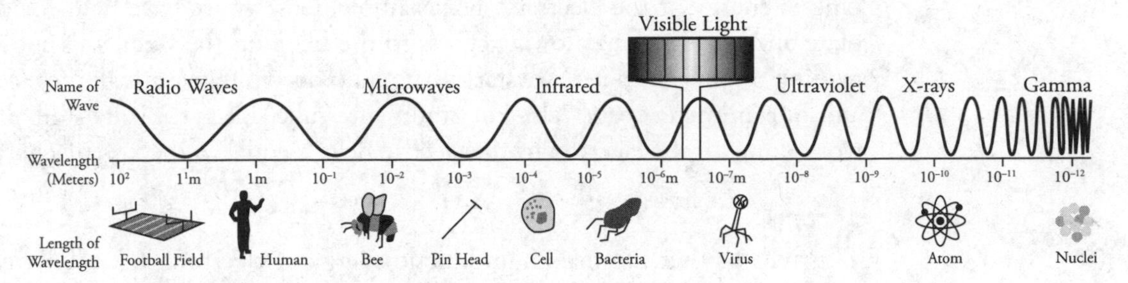

Photoelectric Effect

Before we begin, please make sure you understand this experiment. It is *really important.* This is the experiment that proves that light is a particle, even though we already know that it is a wave because light has wave properties like interference. This is the experiment that opened up the rabbit hole of wave-particle duality.

When light shines on a metal, you can get electrons to fly off the surface. Scientists call this the photoelectric effect. You might say to yourself, "What's so important about that?" There are so many important processes in nature that are governed by photoelectric effect you might be shocked. Photosynthesis, tanning, photographic film, solar panels, and cancer, for instance, are all related to the photoelectric effect.

To really understand the photoelectric effect, you have to understand the experiment that was used to study it. The next diagram shows the experimental setup. Don't worry if it looks a little scary, it'll make sense in a few minutes. In the experiment, two metal plates are sealed in a vacuum container. One of the plates is either made of, or coated with, the metal being studied. A variable potential source is connected across the two metal plates. (Think of the variable potential source as a dimmer switch in your home that varies the voltage across your lights, making them brighter or dimmer.) Incident light is directed at the plate with the metal being tested, and under the right conditions, electrons are emitted from that plate and travel to the other plate, and a current can be seen on the ammeter.

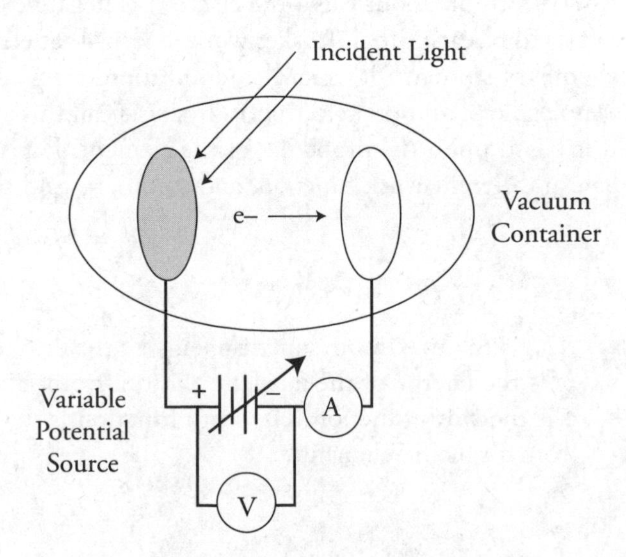

Notice that the variable potential source is oriented in such a way so that the electrons are traveling toward the negative plate and away from the positive plate. This means that the voltage across the plates is acting to stop the electrons from traveling from one side to the other. That voltage is called the stopping potential; it lets us calculate the maximum kinetic energy of the electrons. For example, let's say we have light incident upon the left plate and electrons are flowing across to the plate on the right, this shows up as a current reading on our ammeter. We start turning up our voltage until all the electrons are stopped and our ammeter reads zero. For argument's sake, I'll say 2 volts stop all the current. The work done on the electron by the electric field is equal to the kinetic energy of the electron:

$$K_{max} = \Delta U = q\Delta V = (1\ \text{e})(2\ \text{V}) = 2\ \text{eV}$$

This tells us that the maximum kinetic energy of electrons is 2 eV. Remember, when you are getting your energy in electron-volts, your charge must be in elementary charges, and an electron is exactly one elementary charge.

So, the ammeter tells us the rate the electrons are being emitted and the voltmeter tells us the maximum kinetic energy of the electrons.

The easiest way to look at this is with an example.

Blue light with a wavelength of 443 nm is incident on a potassium surface, and electrons are emitted from the surface. A current is measured until a 0.5-V stopping potential is applied. What does this tell you about potassium?

Let's first find the photon energy of the light incident on the surface:

$$E = hf = \frac{hc}{\lambda} = \frac{1.24 \times 10^3\ \text{eV·nm}}{443\ \text{nm}} = 2.8\ \text{eV}$$

Notice that since the wavelength was in nanometers, we could use for the constant *hc*. Next, let's find the kinetic energy of the emitted electrons. It takes 0.5 volt to stop the electrons so:

$$K_{max} = \Delta U = q\Delta V = (1\ \text{e})(0.5\ \text{V}) = 0.5\ \text{eV}$$

Something does not make sense. Why, when a 2.8-eV photon hits potassium, does an electron get emitted with only 0.5 eV of kinetic energy? You know enough about conservation of energy to ask: "Where did the energy go?"

Let's think about this—the electron is negative and the nucleus is positive, so they are attracted to each other. It takes work to "rip" that electron away from the nucleus. (For you chemistry students: it's similar to ionization energy, the amount of energy required to strip a molecule of an outer electron.) Physicists call this amount of energy the work function, ϕ. It is based upon the properties of the element. Potassium has one work function, gold will have a different work function, and so on. To find the work function:

$$K_{max} = hf - \phi$$

where:

- K_{max} is the maximum kinetic energy of the emitted electron.
- hf is the energy of the incident photon, from $E = hf$.
- ϕ is the work function. The work function is the energy required to remove an electron from a specific element.

Plugging in our values:

$$0.5\text{ eV} = 2.8\text{ eV} - \phi$$

$$\phi = 2.3\text{ eV}$$

OK, so now you know it takes 2.3 eV to remove an electron from potassium.

So let's change things. Suppose red light with photon energy of 1.9 eV is incident on potassium. What will happen?

It takes 2.3 eV to remove an electron, so nothing happens, no electron emission.

Let's make that same red light brighter, what will happen then?

Making the red light brighter simply means there are more red 1.9 eV photons hitting the potassium. One important thing to keep in mind is that the photons cannot gang up; two can't act together and add their energies. So no matter how bright the red light is made, there is no way a 1.9-eV photon will make potassium, with a 2.3-eV work function, emit an electron. So, blinding, skin-burning, bright-red light will never eject an electron from potassium. Never.

What if we make the original 443-nm blue light brighter, how will this change things?

The photon energy of the blue light is 2.8 eV, and the work function is 2.3 eV, so there will be electron emission, and the stopping potential will stay 0.5 eV. Making the light brighter means there are more 2.8-eV photons incident on the metal, so there will be more 0.5-eV electrons emitted, which would be seen as higher current measured by the ammeter.

Finally, let's use violet light with photon energy of 3.1 eV. How does this change things?

That's simple enough. The photons have 3.1 eV of energy and it takes 2.3 eV to remove the electron (the work function), so 3.1 eV – 2.3 eV = 0.8 eV, which is the maximum kinetic energy of the emitted electron.

A couple other quick points. You might be asked to evaluate a graph of kinetic energy of the emitted electron. There are a few things you need to remember about this graph, since it shows up on AP exams from time to time.

The next diagram shows an example of the plots you would get for two metals, potassium and iron. The first thing you might notice is that the lines are straight and parallel to each other. Let's practice matching up our photoelectric equation with the equation of a line:

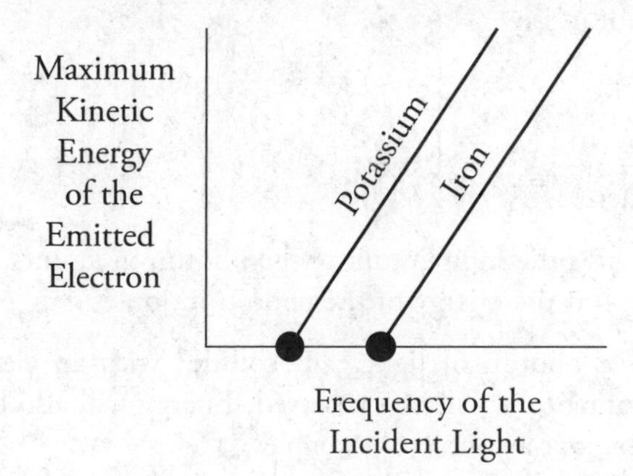

$$K_{max} = h \; f \; -\phi$$
$$y = m \; x \; +b$$

So, the slope of these lines always equals Planck's constant and the intercept will equal the work function. Notice the large points where the lines intersect the *x*-axis. This point is called the threshold frequency. It is the minimum frequency required for electron emission. Potassium has a lower threshold frequency than iron because it takes less energy to remove an electron from potassium. Another way to find the threshold frequency is to use $E = hf$ and enter the work function for your energy. Since the work function is the minimum energy needed to cause electron emission, the resulting frequency will be the lowest frequency to cause photo-electron emission.

Remember the importance of this experiment: The photoelectric effect shows us the particle behavior of light. We already discussed the double-slit experiment in Chapter 14, which demonstrates interference and the wave nature of light. So, we say that light exhibits the properties of both waves and particles. It has wave-particle duality. When we are working with energies down near the photon $E = hf$ range, we see the particle nature of light. When light interacts with objects around the size of its wavelength and smaller, we will see its wave behavior.

So when we get down to the nano-world, some of our old physics models don't work anymore. We have to invent a new model for how things behave. Electron behavior is another example of this. We will discus electrons in a little bit.

Photon Momentum

We know that the energy of a photon is $E = hf$ and that energy is related to mass: $E = mc^2$. And now we know that photons have particle properties. Is it possible for photons to have particle properties like momentum? Yes! Here is a quick derivation:

$$E = hf = mc^2$$

$$\frac{hf}{c} = \frac{mc^2}{c} = mc = p_{photon}$$

When a photon hits an atom, and that atom emits an electron as in the photoelectric effect, both energy and momentum are conserved in the process. This means that even though the photon does not have rest mass, it still has momentum. The momentum of a photon is:

$$p_{photon} = \frac{h}{\lambda} = \frac{E}{c}$$

where:

- p is the momentum of the photon in kg·m/s.
- E is the energy of the photon in joules.

So a photon of light can "collide" with an electron just like two cars collide, and the momentum will be conserved. Energy will also be conserved in the interaction. Let's take a look at a couple of examples.

Examples

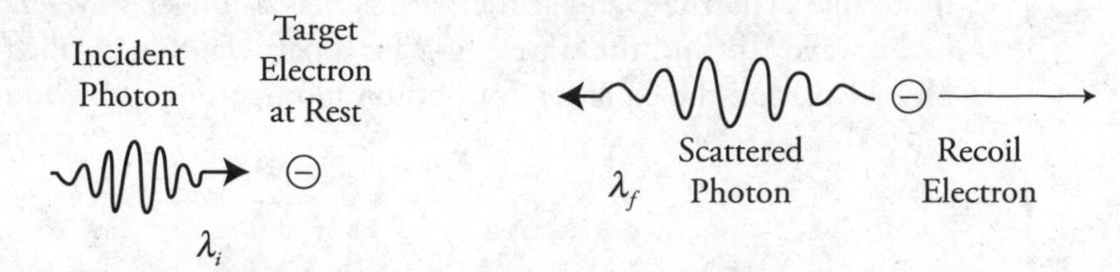

Example 1: A photon collides head-on with an electron. This is a one-dimensional interaction. The photon hits a stationary electron; afterward, the electron takes off with some of the photon's original energy and momentum. Thus the photon lost both momentum and energy. Look back at our equation. To lose momentum and energy, the wavelength of the photon must have increased and the frequency decreased. The photon has changed "color."

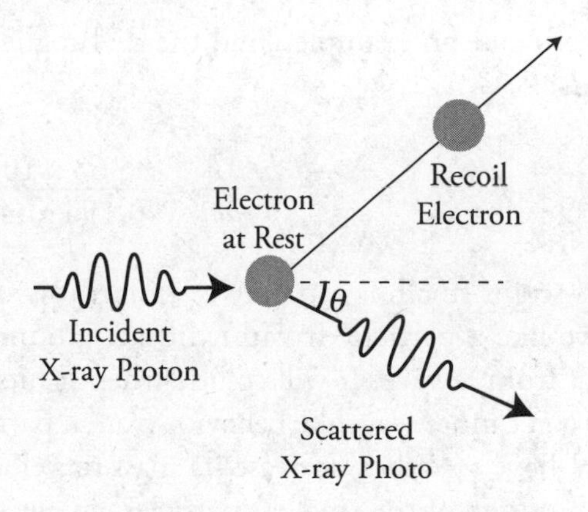

Example 2: The photon strikes an electron and the electron moves off at an angle. This is a two-dimensional momentum problem. Since the electron moves upward, the photon must move downward. The initial *y*-direction momentum of the system is zero; therefore, the final *y*-direction momentum must also equal zero. The initial and final *x*-momentums also remain constant.

De Broglie Wavelength and Wave Functions

You already know that a wave can act like a particle. Photon momentum and the photoelectric effect are examples of light acting like a particle or a photon. If that wasn't weird enough, physicists (de Broglie in particular) started to contemplate: "If waves can act like particles, then maybe particles can act like waves." This sounded ridiculous until it turned out to be true. To understand this, you need a good grasp of how the electromagnetic spectrum behaves.

- Long wavelengths of light, such as radio waves, demonstrate lots of wave behaviors like interference and diffraction. However, they have so little energy, they don't exhibit much in the way of particle properties like photon momentum or the photoelectric effect.
- Short wavelengths of light, like gamma rays, demonstrate a huge amount of photon momentum and the photoelectric effect. However, their wavelengths are so short that it's hard to measure wave properties like diffraction and interference for gamma ray photons.

Particles behave the same way. If the particle has a short wavelength, it should behave more like a particle, and if the particle has a longer wavelength, it should behave more like a wave. To find the wavelength for a particle, this is called the de Broglie wavelength, physicists took the equation for photon momentum and worked it in reverse:

$$\lambda = \frac{h}{p_{particle}} = \frac{h}{mv}$$

where:

- λ is the de Broglie wavelength in meters.
- h is Planck's constant. (This time you have no choice, you must use $h = 6.63 \times 10^{-34}$ J·s.)
- m is the mass the particle in kilograms.
- v is the velocity of the particle in meters per second.

Let's take an example. Find the de Broglie wavelength of a 0.15-kg baseball moving at 40 m/s:

$$\lambda = \frac{h}{p} = \frac{h}{mv} = \frac{6.63 \times 10^{-34}\ \text{J·s}}{(0.15\ \text{kg})(40\ \text{m/s})} = 1.1 \times 10^{-34}\ \text{m}$$

If this seems small, it certainly is. In fact it is smaller than gamma rays, which means it will behave like a particle. It won't diffract around objects or show interference as you might expect from a wave. It will collide when it hits another particle.

To get more wavelike behavior from a particle, I need to increase its wavelength. To do that, I have to either decrease its mass or velocity, or both. Let's find the de Broglie wavelength of one of the smallest particles, an electron moving at 3×10^6 m/s:

$$\lambda = \frac{h}{p} = \frac{h}{mv} = \frac{6.63 \times 10^{-34}\ \text{J·s}}{(9.11 \times 10^{-31}\ \text{kg})(3 \times 10^6\ \text{m/s})} = 2.4 \times 10^{-10}\ \text{m}$$

This might seem very small, but it is right in the wavelength range of an X-ray.

How can electrons, which are particles and have mass, show wave behavior? They can show interference through slits, just like interference patterns of light, in what is called electron crystal diffraction. The atoms in a crystal can line up to form slits. When X-rays are fired on a crystal, they undergo diffraction and interference, and an interference pattern can be seen. When electrons with the same wavelength are fired at the same crystal, a very similar interference pattern occurs, which means the electrons are undergoing diffraction and interference. Electron microscopes take advantage of the wave properties of electrons to form amazingly clear images of super-small objects.

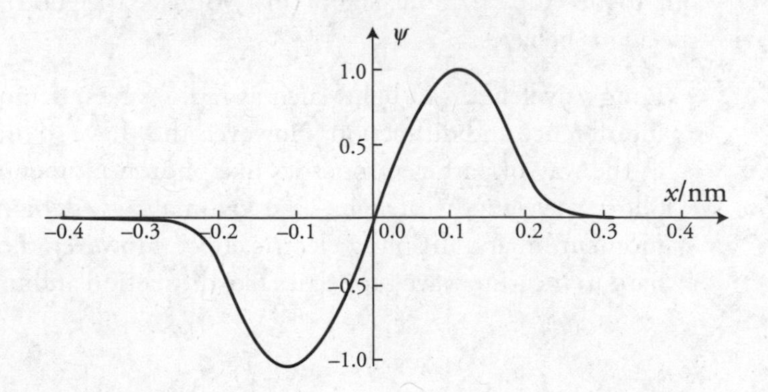

If particles have wave behaviors, what does their wave look like, and what does it mean? Particles will have a wave function, ψ, as a function of location. This can look like the example in the preceding figure. They represent the probability of finding the particle at a particular location. Here is how you read a wave function graph:

1. When the graph reads $\psi = 0$, there is no probability of ever finding the particle at that location. So in our graph, the particle will never be at location $x = 0$ and will never be found beyond ±0.3 nm.
2. The larger the amplitude (positive or negative), the more likely you will find the particle. In our graph we are most likely to find the particle at locations ±0.1 nm.
3. Is it possible to find the particle at −0.2 nm? Yes, but not as likely as finding it at ±0.1 nm.

The figure below is another example. You are most likely to find the particle at the locations marked with a circle. You won't find the particle at the locations marked with an X.

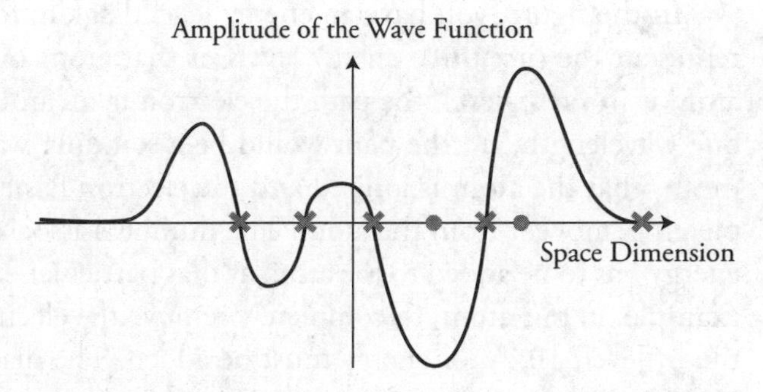

Another example of the wavy behavior of electrons is in the electron energy shells in an atom. Electrons don't orbit the nucleus in a circular "planetary" pattern. Because of their wave nature, they set up standing waves of constructive interference in whole wavelengths. See the following figure. They can exist only in orbital patterns where they have constructive interference with themselves. At different orbital radii, they will exhibit destructive interference with themselves and cancel themselves out. They simply cannot exist at destructive interference locations.

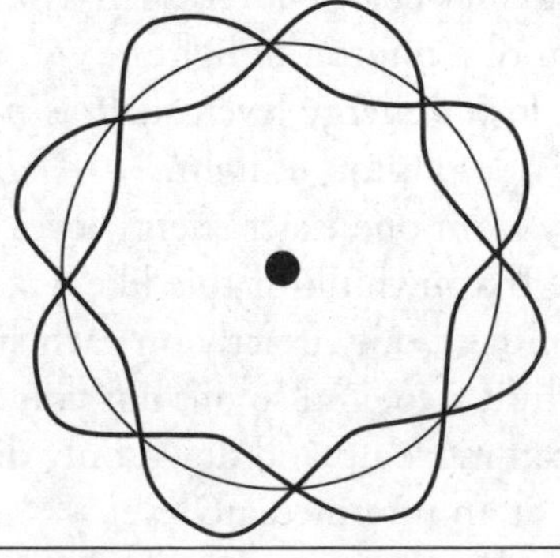

Electron forming a constructive standing wave. The electron can exist in this orbit.

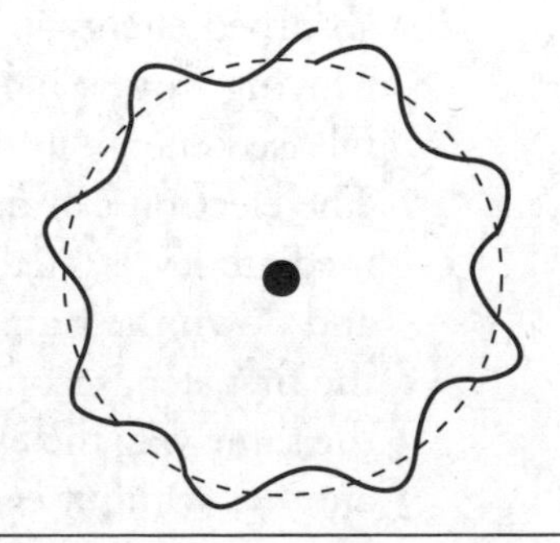

Electron in an orbit that causes destructive interference. The electron cannot exist in this orbit.

Let's examine this orbital behavior closer.

Bohr Model of the Atom and Energy Levels in an Atom

Niels Bohr discovered that electrons existed only in specific orbital energy locations due to the constructive and destructive wave nature of electrons. For an electron to move from one of these energy levels to another, the atom had to either absorb or release a little bit of energy called a photon. To make it easier to understand, physicists came up with neat charts that help explain this.

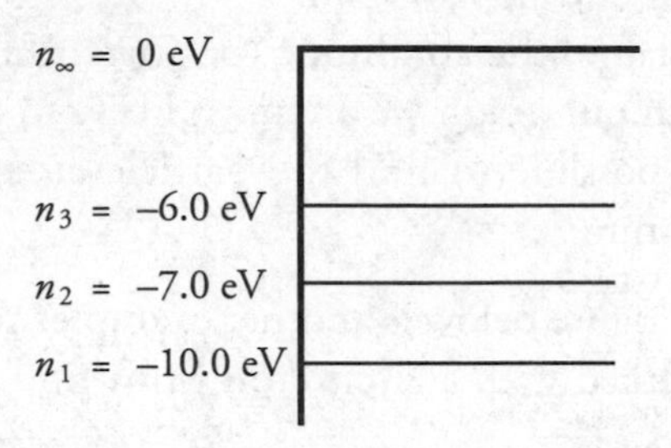

In the figure, you have an energy level diagram for a hypothetical atom—n_1, n_2, and n_3 represent the first three energy levels of that atom. Another way to look at it is that, if the atom is in the n_1 level, the path the electron takes around the atom is like a standing wave of one wavelength; n_2, the path would be a standing wave of two wavelengths; and so on. n_∞ means that the atom is ionized and the electron has had enough energy added to it to completely remove it from the atom. The numbers associated with each level tell you how much energy has to be added to the atom at that particular energy level to remove that electron. For example, in our atom, to completely remove the electron from the atom with the electron in the n_1 level, 10 eV of energy must be added. The diagram shows $n_1 = -10$ eV; the negative sign is telling you the atom needs 10 eV of energy before it can be ionized.

A nice way to think of this is that the nucleus is positive, the electron is negative, and it takes work to "pull" the electron away from the nucleus. If the electron is in a higher energy level, such as n_2, it takes less energy, only 7 eV, to remove the electron. Bohr called this little bundle of energy a quanta; Einstein later called it a photon. You may be asked to find the energy needed to move the electron from a lower to a higher energy level, or how much energy is released if the electron moves from a higher to a lower energy level. Some hints to help you out:

- n_1 is called the ground state. It is the lowest possible energy level for the electron.[2]
- Moving from a lower to a higher energy level, such as n_1 to n_2, tells you that the atom absorbed energy in the form of a photon of light.
- Moving from a higher to a lower energy level, such as n_2 to n_1, tells you that the atom released energy in the form of a photon of light.
- The electron can move only from one exact energy level to another. There are no intermediate levels, such as $n_{1.7}$. If you think of this like a set of stairs, when you move up and down the stairs, you can land only exactly on each step. So you can be standing on the first step, second step, third step, and so on, but you cannot land on the 1.7 step. In the same way the electron can move up and down only directly from one energy level to another without ever being at an intermediate level.
- Note that I listed the energy levels as n_1, n_2, and n_3. The AP exam has been known to list them as E_1, E_2, and E_3. Or simply as 1, 2, and 3. Don't let that trouble you! It all means the same thing.

[2]Yes, we did say in Chapter 14 that accelerating charges give off electromagnetic radiation. This is one way quantum physics differs from classical physics. In quantum physics, it is possible for the electron to accelerate in an orbit and not emit energy.

- To find the energy of the absorbed or released photon, all you have to do is subtract the energies of the two energy levels:

$$E_{photon} = E_{final} - E_{initial}$$

Before moving on let's take a look at some examples.

Examples

Example 1: What energy photon is absorbed or released in our hypothetical atom for the electron to move from $n = 2$ to $n = 1$?

No problem now that you understand what the energy level diagram means. Simply subtract the two energy levels:

$$E_{photon} = -7 \text{ eV} - (-10 \text{ eV}) = 3 \text{ eV}$$

Since the electron went to a lower energy level, you know that a 3-eV photon was released. Since it went from a higher to a lower energy level, that photon of light was released or emitted from the atom. Congratulations! You just solved a quantum physics problem!

Example 2: Our hypothetical atom in the ground state has a 3.5-eV photon incident upon it. Describe what happens.

We know that the electron can move only from one exact energy level to another. So for the electron to jump from $n = 1$ to $n = 2$, the atom must absorb exactly 3 eV, and to jump from $n = 1$ to $n = 3$, the atom must absorb 4 eV. Since there is no exact jump equal to 3.5 eV, the photon passes through the atom and does nothing.

One more thing to remember:

- When the atom releases energy going from a higher to a lower energy level, every possible combination of downward energy level jumps may happen.

For example:

If an atom is in the $n = 4$ level and is returning to the $n = 1$ level, how many different photons may be released?

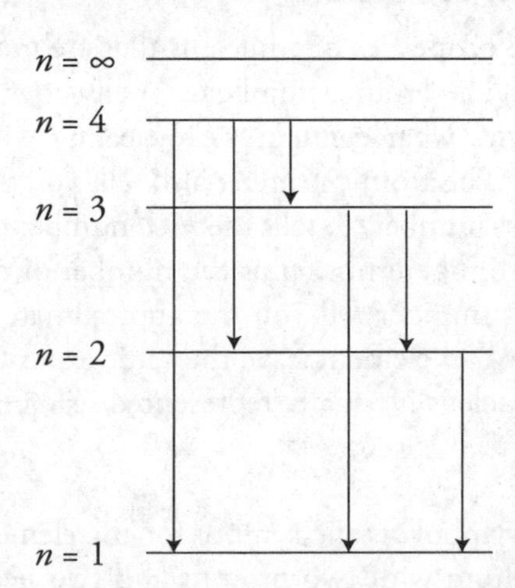

Electrons can jump directly from the $n = 4$ level to the $n = 1$ level, or jump from $n = 4$ to 3 to 2 to 1, releasing three photons. Jumping down from $n = 4$ to 2 to 1 is another possibility. So every possible jump down from every level can happen, as shown in the diagram above. There are six possible photons that could be emitted in this process.

You know that the electron can jump only from one energy level to another. But what happens if there's more than enough energy to ionize the atom?

- If a photon is incident upon the atom with an energy equal to or greater than the energy needed to ionize the atom, the electron will be ejected from the atoms. The excess photon energy will be converted into kinetic energy of the released electron.

Let's look at another example.

Example 3: Our hypothetical atom is again in the ground state and a 12-eV photon is incident on the atom. What happens?

It takes 10 eV to ionize the atom, so there's more than enough energy to remove the electron. Assuming the photon gets completely absorbed, the electron will be released with a kinetic energy of 12 eV – 10 eV = 2 eV.

Three Types of Nuclear Decay Processes

When physicists first investigated nuclear decay in the early twentieth century, they didn't quite know what they were seeing. The subatomic particles (protons, neutrons, electrons, and so forth) had not been definitively discovered and named. But, physicists did notice that certain kinds of particles emerged repeatedly from their experiments. They called these particles *alpha*, *beta*, and *gamma* particles.

Years later, physicists found out what these particles actually are:

- Alpha particle, α: two protons and two neutrons stuck together. This is the same thing as a helium nucleus.
- Beta particle, β: an electron or a positron[3]
- Gamma particle, γ: a gamma ray photon

Be sure you can recall the properties of these particles. A couple of observations: The alpha particle is by far the most massive; the gamma particle is both massless and chargeless.

Nuclear Notation

The two properties of a nucleus that are most important are its *atomic number* and its *mass number*. The atomic number, *Z*, tells how many protons are in the nucleus; this number determines what element we're dealing with because each element has a unique atomic number. The atomic number also tells you the net positive elementary charge of the nucleus. The mass number, *A*, tells the *total* number of nuclear particles a nucleus contains. It is equal to the atomic number plus the number of neutrons in the nucleus. Like the name implies, the mass number tell you the approximate mass of the nucleus in atomic mass units "u". *Isotopes* of an element have the same atomic number, but different mass numbers.

A nucleus is usually represented using the following notation:

$$^{A}_{Z}\textit{Symbol}$$

where "symbol" is the symbol for the element we're dealing with. For example, $^{4}_{2}\text{He}$ represents helium with two protons and two neutrons. A different isotope of helium might be $^{5}_{2}\text{He}$, which contains three neutrons.

[3]The *positron* is the antimatter equivalent of an electron. It has the same mass as an electron, and the same amount of charge, but the charge is positive. Since there are two different types of beta particles, we often write them as β^+ for the positron and β^- for the electron.

This is not a chemistry exam, you don't have to memorize the periodic table, nor do you have to remember what element is number 74.[4] But you *do* have to recognize the significance of the atomic number and the mass number. Both numbers are conserved in nuclear reactions. This is explained in detail in the following sections.

Alpha Decay

Alpha decay happens when a nucleus emits an alpha particle. Since an alpha particle has two neutrons and two protons, then the *daughter nucleus* (the nucleus left over after the decay) must have two fewer protons and two fewer neutrons than it had initially.

Example: $^{238}_{92}U$ undergoes alpha decay. Give the atomic number and mass number of the nucleus produced by the alpha decay.

The answer is found by simple arithmetic. The atomic number decreases by two, to **90**. The mass number decreases by four, to **234**. (The element formed is thorium, but you don't have to know that.) This alpha decay can be represented by the equation below:

$$^{238}_{92}U \rightarrow \, ^{234}_{90}Th + \, ^{4}_{2}\alpha$$

Beta Decay

In beta decay, a nucleus emits either a positron (β^+ decay) or an electron (β^- decay). Because beta particles have very little mass and do not reside in the nucleus, the total mass number of the nucleus will stay the same after beta decay. The total charge must also stay the same—charge is a conserved quantity. So, consider an example of neon (Ne) undergoing β^+ decay:

$$^{19}_{10}Ne \rightarrow \, ^{19}_{9}F + e^+$$

Here e^+ indicates the positron. The mass number stayed the same. But look at the total charge present. Before the decay, the neon nucleus has a charge of +10. After the decay, the total net charge must still be +10, and it is: +9 for the protons in the fluorine (F), and +1 for the positron. Effectively, then, in β^+ decay, a proton turns into a neutron and emits a positron.

For β^- decay, a neutron turns into a proton, as in the decay process important for carbon dating:

$$^{14}_{6}C \rightarrow \, ^{14}_{7}N + e^-$$

The mass number of the daughter nucleus didn't change. But the total charge of the carbon nucleus was initially +6. Thus, a total charge of +6 has to exist after the decay, as well. This is accounted for by the electron (charge –1) and the nitrogen (charge +7).

In beta decay, a *neutrino* or an *antineutrino* also must be emitted to carry off some extra energy. But that doesn't affect the products of the decay, just the kinetic energy of the products.

Gamma Decay

A gamma particle is a photon, a particle of light. It has no mass and no charge. So a nucleus undergoing gamma decay does not change its outward appearance:

$$^{238}_{92}U \rightarrow \, ^{238}_{92}U + \gamma$$

---

[4]Tungsten: symbol "W." Used in lightbulb filaments.

However, the photon carries away some energy and momentum, so the nucleus recoils.

Here is a decay we have not discussed—neutron decay.

Fill in the blank in the following diagram with what is missing.

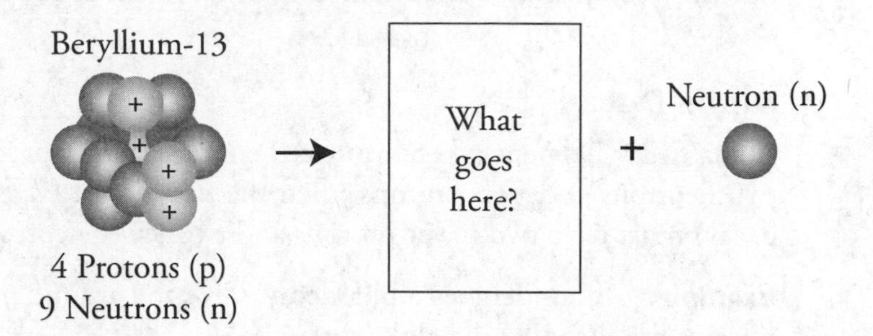

No worries. A neutron, which has a mass number of 1 and an atomic number of zero because it has no charge, has been ejected. Therefore, what remains behind must have 4 protons and 8 neutrons. So, it has to be a different isotope of beryllium: $^{12}_{4}Be$.

$E = mc^2$ and Conservation During Nuclear Reactions

We can imagine that the nuclei in the last few examples are at rest before emitting alpha, beta, or gamma particles. But when the nuclei decay, the alpha, beta, or gamma particles come whizzing out at an appreciable fraction of light speed.[5] So, the daughter nucleus must recoil after decay in order to conserve momentum.

But now consider energy conservation. Before decay, no kinetic energy existed. After the decay, both the daughter nucleus and the emitted particle have gobs of kinetic energy.[6] Where did this energy come from? Amazingly, it comes from mass.

The total mass present before the decay is very slightly greater than the total mass present in both the nucleus and the decay particle after the decay. That slight difference in mass is called the *mass defect*, often labeled Δm. This mass is destroyed and converted into kinetic energy.

How much kinetic energy? Use Einstein's famous equation to find out. Multiply the mass defect by the speed of light squared:

$$E = (\Delta m)c^2$$

And that is how much energy is produced.

Let's review before moving on. Things that are conserved during nuclear reactions:

1. Conservation of mass/energy—In every naturally occurring nuclear reaction, mass of the original nucleus is lost to other forms of energy such as photons or kinetic energy, and the final mass is less than the original mass. Less mass means more nuclear stability.
2. Conservation of momentum—When a nucleus undergoes decay and shoots off a particle, there is a recoil just like when a bullet is fired from a gun.
3. Conservation of charge—The sum of the atomic numbers before and the sum of the atomic numbers after the nuclear reaction are equal.
4. Conservation of nucleons—The sum of the neutrons and protons (atomic mass number) before and after must be the same.

---

[5] 100% of the speed of light in the case of a gamma particle!

[6] No, the energy doesn't cancel out, because the nucleus and the particle move in opposite directions; energy is a scalar, so direction is irrelevant.

Half-Life

Radioactive isotopes decay at different rates. The decay rate is called half-life, and it is the time it takes for half of the radioactive atoms to decay (transmute) into a new nucleus. Uranium-238 has a half-life of 4.5 billion years. Some elements have a half-life of less than a second! Don't blink . . . or it will already be transmuted into something new.

Which material is more dangerous? One with a long half-life like uranium or one with a short half-life?

A short half-life = fast decay rate = lots of radiation in the short term and then it is gone. So, short half-life materials are likely to kill you quick and then disappear /transmute.

A long half-life = slow decay rate = small radiation emissions = little danger to you. But the material will be around for a long time.

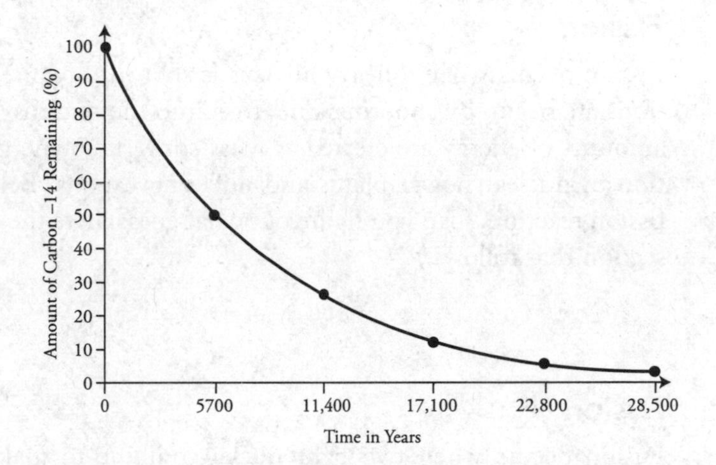

Carbon-14 has a half-life of 5700 years. Look at the graph above. Starting at time = 0, we have 100% of the original carbon-14. After 5700 years, only 50% is left. After two half-lives, only 25% is left (that's ½ times ½). After 28,500 years, or five half-lives, there is only about 3% of the original material left. Five half-lives is $\frac{1}{2} \times \frac{1}{2} \times \frac{1}{2} \times \frac{1}{2} \times \frac{1}{2} = \frac{1}{32} = 3.125\%$.

Decay rates appear to be independent of external conditions like temperature, electric or magnetic fields, and even chemical reactions. How do scientists measure half-life? Well, they can't actually count the atoms because they are tiny, and we can't see atoms. So, scientists measure the radioactive activity with a Geiger counter. A Geiger counter literally counts the number of decay products like alpha and beta particles that are emitted by the radioactive material.

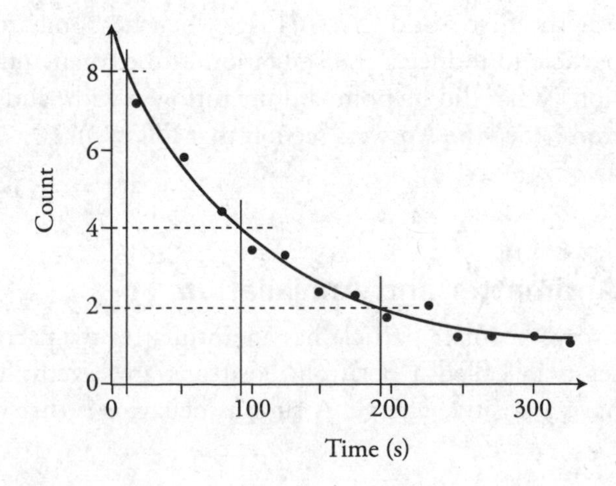

Look at the preceding activity count graph. What is the half-life of this material? Look to see how long it takes the count to be cut in half. The time to go from 8 to 4 counts is about 80 seconds. This is experimental data, which always has some errors or uncertainty, so let's check with another data point. The time it takes to go from 4 to 2 counts is about 100 s. So, the half-life is approximately 90 seconds.

Other Nuclear Reactions

Decay is not the only type of nuclear reaction. There are lots of different kinds. The following are a few that you may run across on the exam.

Fission

Fission occurs when a heavy nucleus is split into two relatively large chunks. Fission reactions are begun by shooting a neutron into the nucleus, which initiates the reaction. Large amounts of energy are created as mass converts to energy in the reaction. This is the reaction in nuclear power plants and nuclear weapons. Below is an example of a plutonium fission reaction. Can you figure out what goes where the question mark is? (See the Answers section that follows.)

$$^{239}_{94}\text{Pu} + ^{1}_{0}\text{n} \rightarrow ^{100}_{40}\text{Zr} + ^{137}_{54}\text{Xe} + ? + \textit{Energy}$$

Fusion

Fusion occurs when two light nuclei combine to make a new heavier and more stable nucleus. Large amounts of energy are released in the reaction. This is what happens in the sun. Here is an example of fusion. Notice how one of the products of the reaction is larger than any of the originals. Where did all the energy come from, and what does all that energy do? (See the Answers section that follows.)

$$^{2}_{1}\text{H} + ^{3}_{2}\text{He} \rightarrow ^{4}_{2}\text{He} + ^{1}_{1}\text{H} + 18.3\text{ MeV}$$

Induced Reaction

Sometimes scientists bombard a nucleus with high-speed particles to see what will happen. This is called an *induced reaction* because you are causing a reaction to happen. Rutherford was the first to do this in 1919 when he bombarded a stable isotope of nitrogen with a particle to induce a transmutation into oxygen and a proton. Here is the Rutherford reaction. What did he bombard the nitrogen with, and where do you think he got this particle from? (See the Answers section that follows.)

$$? + ^{14}_{7}\text{N} \rightarrow ^{17}_{8}\text{O} + ^{1}_{1}\text{p}$$

Antimatter and Annihilation

Every "normal" particle has an antimatter twin. For instance, an electron has an antimatter twin called a positron. Positrons are exactly like electrons in every way except they have a positive charge. A curious behavior is that when matter and antimatter meet, they annihilate each other. All the mass is converted into energy in the form of photons. For example: an electron and positron greet each other and shake hands. Oh no, annihilation! The electron and positron disappear and turn into photon energy. How much energy will be converted into photon energy? Both turn into energy. Both have the same mass:

$$E_{electron} + E_{positron} = (2m_{electron})c^2 = hf$$

Question: Why does the electron–positron annihilation not just produce one photon? Think about all the conservation laws we have to abide by. (See the Answers section that follows.)

Answers

Fission: 3 neutrons

Fusion: The energy produced in the reaction comes from a loss of mass $E = mc^2$. As for what all that energy does . . . it creates photons of light energy that are emitted. This is why the sun glows and keeps us toasty warm.

Induced Reaction: He bombarded it with an alpha particle or helium nucleus. Rutherford got his alpha particles from the alpha decay of other radioactive isotopes, though he could have stripped helium gas of its electrons as well.

Antimatter and Annihilation: Due to conservation of momentum, two photons of the same energy traveling in the opposite direction have to be created so that the final momentum is zero, just like the initial momentum.

Mass Defect, Binding Energy, and the Strong Nuclear Force

Carbon-12 is made up of 6 protons and 6 neutrons. We know the mass of a proton. We know the mass of a neutron. So, you would think that if we added the mass of 6 protons to the mass of 6 neutrons, we would get the mass of a carbon-12 nucleus. Remarkably, the mass of the carbon-12 is a little less than we expect! Scientists call this a mass defect, Δm.

What has happened to the lost mass? The small amount of missing mass has been converted into energy ($E = \Delta mc^2$) that holds the nucleus together. The mass defect has become the nuclear binding energy and is equivalent to the strong nuclear force holding the nucleus together. If you wanted to break the carbon-12 apart into separate protons and neutrons, you would need to supply enough energy to the nucleus to re-create the mass that is missing $\Delta m = \frac{E}{c^2}$.

Because physicists so often need to convert matter into energy and back again, they took Einstein's equation and a bunch of conversions to give you the following relationship:

$$1 \text{ u} = 1.66 \times 10^{-27} \text{ kg} = 931 \text{ MeV}/c^2$$

Where *u* is called the unified atomic mass unit. Remember, that MeV is a mega-electron-volt. Don't worry so much about the c^2, so long as you know that 1 u is 931 MeV when converted to energy.

Let's take a look at an example.

Example

Given the following masses, what is the mass defect and binding energy for the helium atom?

Particle (Symbol)	Mass
Proton (p)	1.0073 u
Neutron (n)	1.0087 u
Helium atom (He, α)	4.0016 u

Let's first find the mass defect. It's really simple to do, just subtract the mass of the atom from the mass of its separate particles.

The helium atom is made of 2 protons plus 2 neutrons, so:

$$2(1.0073 \text{ u}) + 2(1.0087 \text{ u}) = 4.0320 \text{ u}$$

Now subtract the mass of the atom from its separate parts:

$$4.0320 \text{ u} - 4.0016 \text{ u} = 0.0304 \text{ u}$$

This is the mass defect. It is the amount of matter that's converted to binding energy and released from the atom when its component nucleons[7] are put together.

Now we want to convert that amount of matter into energy. You could convert the mass into kilograms, use $E = mc^2$ to change the mass into energy, and then convert the energy from joules into mega-electron-volts, but it is so much easier to use the conversion listed above and on the reference table:

$$0.0304 \text{ u}\left(\frac{931 \text{ MeV}}{1 \text{ u}}\right) = 28.3 \text{ MeV}$$

So there is 28.3 MeV of energy that would have to be added back to the helium nucleus to tear it apart into four separate nucleons.

The mass of a nucleus is less than the mass of its constituent parts. The bigger the mass defect, the harder it will be to break the nucleus apart, and the more stable the nucleus will be. In essence, the mass defect is a measure of how large a nuclear force holding the nucleus together is. Remember that the nucleus is populated with protons that repel each other. The force holding it all together must be very strong. Thus the name given to this binding force is: the strong nuclear force. (May the force be with you!)

[7]Remember that a nucleon is a proton or neutron.

› Practice Problems

Multiple Choice

1. Which of the following lists types of electromagnetic radiation in order from least to greatest energy per photon?

(A) ultraviolet, infrared, red, green, violet
(B) red, green, violet, infrared, ultraviolet
(C) infrared, red, green, violet, ultraviolet
(D) ultraviolet, violet, green, red, infrared

Questions 2 and 3

In a nuclear reactor, uranium fissions into krypton and barium via the reaction

$$n + {}^{235}_{92}U \rightarrow {}^{141}_{56}Ba + Kr + 3n$$

2. What are the mass number A and atomic number Z of the resulting krypton nucleus?

	A	Z
(A)	92	36
(B)	90	36
(C)	94	36
(D)	92	33

3. How much mass is converted into the kinetic energy of the resulting nuclei?

(A) 1 amu
(B) 2 amu
(C) zero
(D) much less than 1 amu

4. ${}^{15}_{8}O$ decays via β^+ emission. Which of the following is the resulting nucleus?

(A) ${}^{15}_{9}F$
(B) ${}^{16}_{8}O$
(C) ${}^{15}_{7}N$
(D) ${}^{16}_{7}N$

5. A physicist has two samples of radioactive material with these properties:

Sample 1: mass = 52 g half-life = 3 days

Sample 2: mass = 200 g half-life = unknown

After 12 days, both samples have approximately the same mass of radioactive material remaining. The half-life of sample 2 is most nearly which of the following?

(A) 6 days
(B) 4 days
(C) 3 days
(D) 2 days

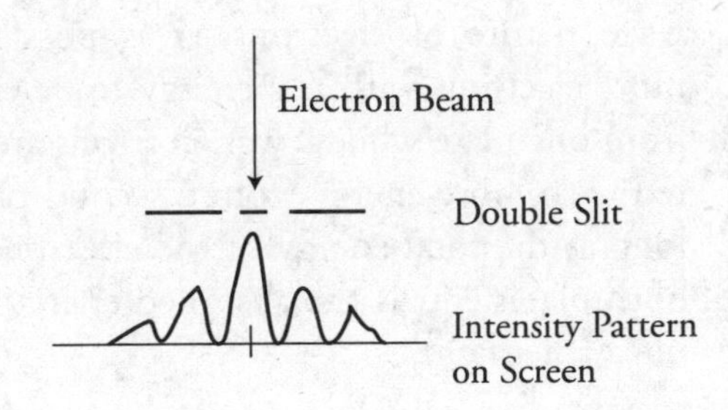

6. An electron beam is shot through a double slit, which produces the intensity pattern shown in the figure. Which of the following diagrams best represents the intensity pattern resulting when the velocity of the electron beam is reduced?

(A)

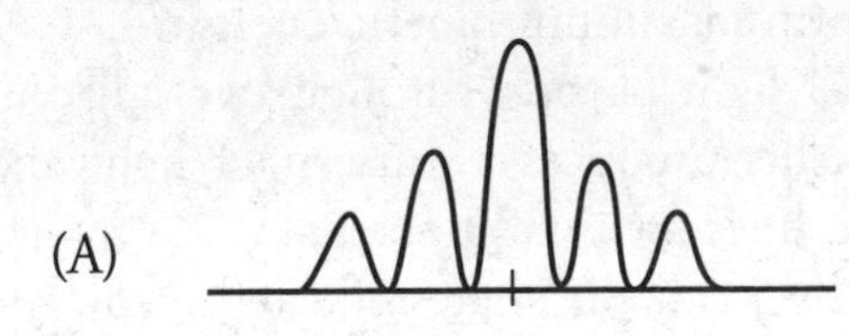

(B)

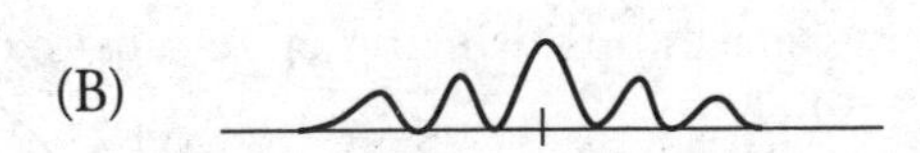

(C)

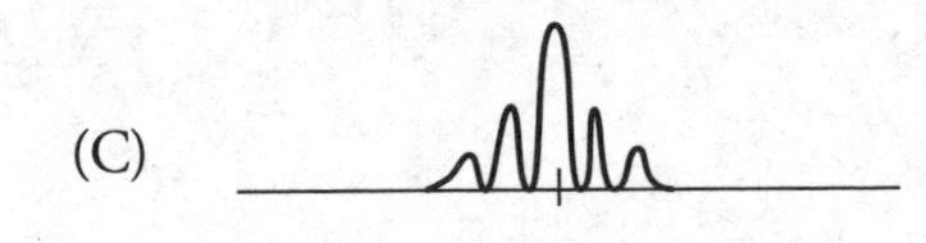

(D)

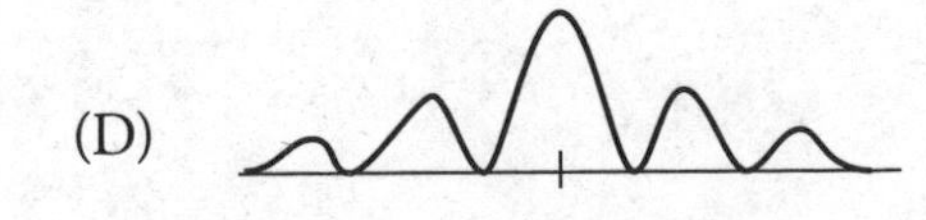

7. A uniform ultraviolet light source shines on two metal plates causing electrons to be emitted from each plate. Plate A emits twice as many electrons as plate B. However, the electrons emitted from plate B have a higher maximum velocity. Which of the following describes a plausible explanation for the differences in electron emission? **(Select two answers.)**

(A) Plate A must have a larger work function than plate B.
(B) The higher velocity electrons in case B would be produced by placing plate B closer to the light source, where plate B would receive more intense ultraviolet light from the source.
(C) More electrons would be produced from plate A if it were larger in area than plate B.
(D) Due to the unpredictability inherent in the wave nature of electrons, it is possible for more electrons with less energy to be emitted from one plate while fewer electrons are emitted with more energy from a second plate, as long as the total energy of the electrons from both plates equals the absorbed energy of the ultraviolet light.

8. In which of the cases listed below would it be important to consider the particle nature of electromagnetic radiation? **(Select two answers.)**

(A) Ultraviolet light of various intensities shining on a sodium sample produce ejected electrons of uniform maximum kinetic energy.
(B) Red laser light passing through two adjacent narrow slits produces a pattern of light and dark red lines on a white screen.
(C) A car driver can eliminate reflected road glare from the sun by wearing polarized glasses.
(D) NASA engineers have developed a method for propelling spaceships utilizing large sails that reflect sunlight.

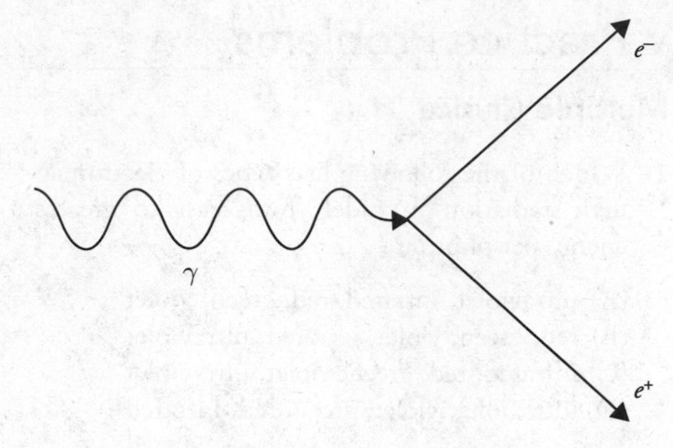

9. Under the right conditions, a photon of light can be converted into a positron and electron, as shown in the figure and represented in the following equation: $\gamma \rightarrow e^+ + e^-$. Which of the following correctly explain why this interaction is possible? **(Select two answers.)**

(A) The positron is the antimatter particle of the electron, so the mass of the two particles cancel.
(B) The signs of the positron and the electron cancel.
(C) The momentum of the two particles sum to that of the photon.
(D) The kinetic energy of the two particles sum to that of the photon.

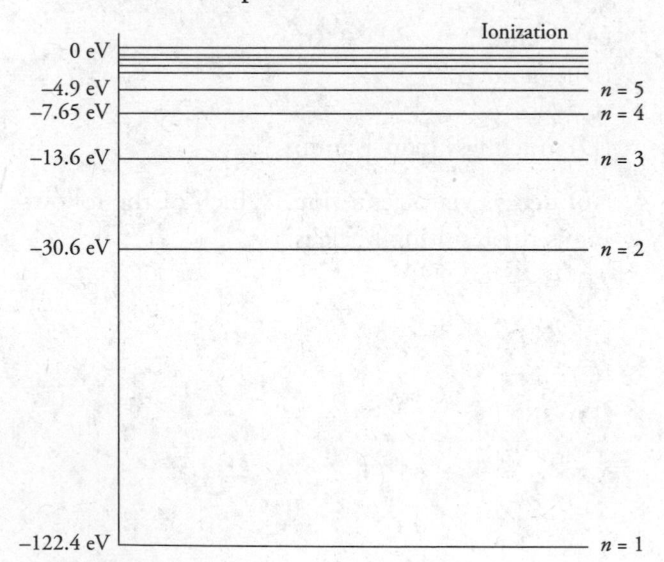

10. The figure shows energy levels for a gas. Electrons in the ground state are excited by a continuous range of electrical energy from 100 eV to 115 eV. After the electrical input is turned off, the gas emits light in discrete frequencies. Which of the following correctly indicates the electron transition from

the gas that would produce the highest frequency light?

(A) $n = 5$ to $n = 4$
(B) $n = 4$ to $n = 3$
(C) $n = 3$ to $n = 2$
(D) $n = 5$ to $n = 2$

11. ${}^{12}_{6}C + {}^{1}_{1}H \rightarrow {}^{13}_{7}N + \gamma$ Which of the following correctly expresses the energy of the released gamma ray in the reaction represented by the equation shown?

(A) $m_C + m_H - m_N$
(B) $(m_C + m_H - m_N)c^2$
(C) $(m_C + m_H - m_N)c^2 - hf_\gamma$
(D) $(m_C - m_H - m_N)c^2$

Free Response

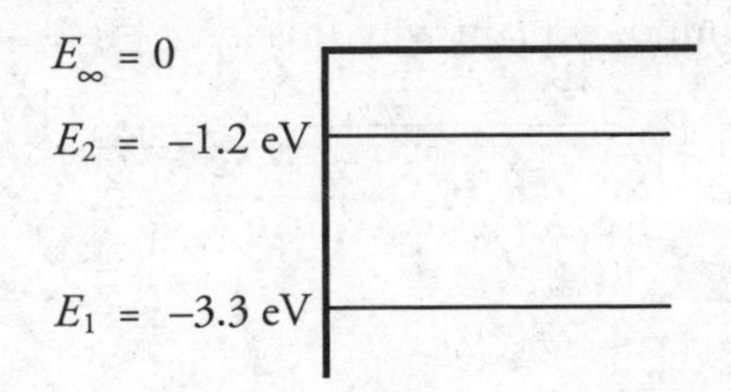

12. A hypothetical atom has two energy levels, as shown above.

(A) What wavelengths of electromagnetic radiation can be absorbed by this atom? Indicate which of these wavelengths, if any, represents visible light.
(B) Now, monochromatic 180-nm ultraviolet radiation is incident on the atom, ejecting an electron from the ground state. What will be
(i) the ejected electron's kinetic energy
(ii) the ejected electron's speed
(iii) the incident photon's speed

For parts (C) and (D), imagine that the 180-nm radiation ejected an electron that, instead of being in the ground state, was initially in the –1.2 eV state.

(C) Would the speed of the ejected electron increase, decrease, or stay the same? Justify your answer briefly.
(D) Would the speed of the incident photon increase, decrease, or stay the same? Justify your answer briefly.

13. Explain why ultraviolet (UV) light can discharge a negatively charged electroscope but not a positively charged electroscope.

14. Sodium has a work function of 2.4 eV. Iron has a work function of 4.7 eV. Cesium has a threshold frequency of 4.7×10^{14} Hz.

(A) Calculate the threshold frequency of sodium and iron. Calculate the work function of cesium.
(B) Plot photoelectron energy as a function of photon frequency for the three elements. Label the slope.
(C) Ultraviolet light (200 nm) shines on cesium. What is the maximum energy of ejected electrons?

15. Which of the following behaviors of light support the wave model of light, and which support the particle model of light? Justify your response in each case.

(A) Light that passes through a double slit produces an alternating pattern of intensity.
(B) X-rays can be used to ionize gas.
(C) X-rays can be directed at crystals to produce interference patterns.
(D) Photons of light can impart momentum to electrons in a collision.
(E) Long wavelength radio waves can bend around hills and buildings.
(F) Light from distant stars is redshifted.
(G) Infrared (IR) light does not generate electricity in a solar cell no matter how intense the light is.

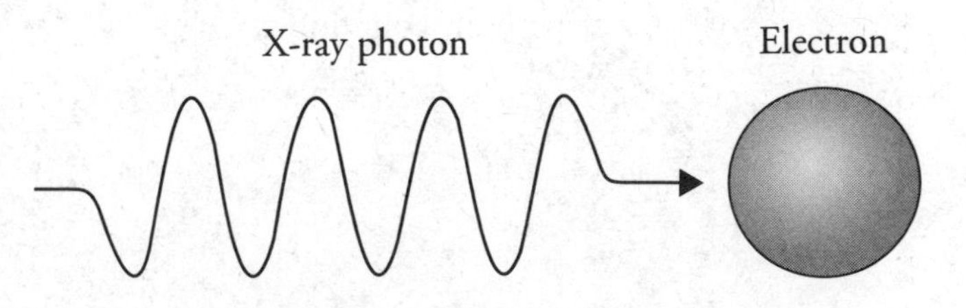

16. A photon moving to the right collides with a stationary electron, as shown in the figure. The photon recoils to the left along the same path.

(A) Describe any changes to the photon. Justify your claims.
(B) Describe any changes to the electron. Justify your claims.

17. Visible light shines on an unknown gas. The gas is found to absorb light of a 400-nm wavelength. When the light is turned off, the gas is seen to emit both 400-nm and 600-nm wavelengths of light.

(A) Draw an energy level diagram for this gas. Label the energy levels in electron volts.

(B) Is there another wavelength emission that is not in the visible spectrum? Justify your answer with your diagram.

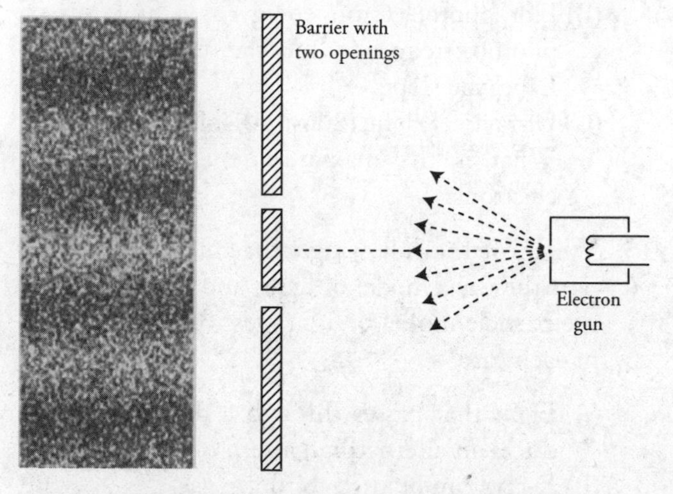

18. A gun fires electrons at a barrier with two small openings. A screen registers electron impact locations as shown in the figure.

(A) What does this prove about the nature of electrons?

(B) What happens to the pattern when the voltage used to accelerate the electrons is increased? Justify your answer with an equation.

(C) What happens to the pattern if the distance between the openings is decreased? Justify your answer with an equation.

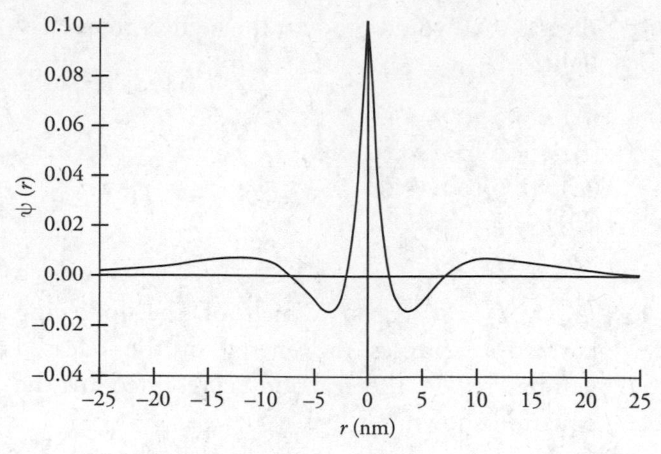

19. The figure represents the wave function of an unknown particle.

(A) On the figure, circle the locations where the particle is most likely to be found. Rank the locations from most likely to least likely.

(B) Place an X on the horizontal axis where the particle will never be found. If there are no locations where the particle will never be found, explain why this is.

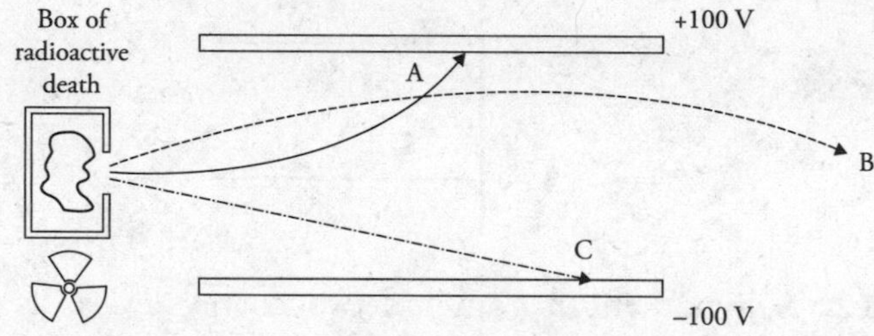

20. Scientists in the early 1900s discovered that some materials (radioactive substances) emitted strange, unknown rays, originally called Becquerel rays. In an effort to understand the nature of these rays, scientists sent them through electric fields and observed the results. The figure shows the path of three rays that exit a box of radioactive material.

(A) Determine the direction of the electric field through which the rays are being sent. Justify your claim.

(B) What can be learned about rays A, B, and C from this experiment? Explain.

(C) What could each of the rays be? Justify your answer.

21. Derive a mathematical equation for the mass defect (Δm) of a helium nucleus. Justify your equation.

22. Decide if the following nuclear reactions are possible. If not, explain why not.

(A) ${}^{6}_{3}\text{Li} + {}^{4}_{2}\text{He} \rightarrow {}^{12}_{7}\text{N} + 2{}^{0}_{-1}\beta + \text{energy}$

(B) ${}^{3}_{1}\text{H} + {}^{2}_{1}\text{H} \rightarrow {}^{4}_{2}\text{He} + {}^{1}_{1}\text{H} +$

23. Complete the following reactions and classify as either decay (specify the type), fission, or fusion.

(A) $2{}^{1}_{1}\text{H} + 2{}^{1}_{0}\text{n} \rightarrow \text{X} + \text{energy}$

(B) ${}^{137}_{55}\text{Cs} \rightarrow \text{X} + {}^{0}_{-1}\text{e}$

(C) ${}^{238}_{92}\text{U} \rightarrow {}^{234}_{90}\text{Th} + \text{X} + \text{energy}$

(D) ${}^{1}_{0}\text{n} + {}^{235}_{92}\text{U} \rightarrow {}^{144}_{54}\text{Xe} + {}^{90}_{38}\text{Sr} + \text{X}$

(E) ${}^{12}_{6}\text{C} \rightarrow {}^{12}_{6}\text{C} + \text{X}$

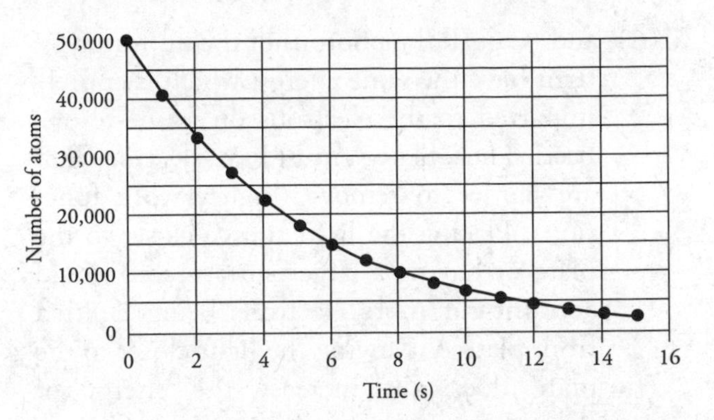

24. (A) Use the figure to estimate the half-life of the radioactive sample.

(B) Explain what the term *half-life* means. Be sure to include reactants and products in your discussion.

❯ Solutions to Practice Problems

1. (D) The radiation with the highest frequency (or shortest wavelength) has the highest energy per photon by $E = hf$. In the visible spectrum, red has the longest wavelength and violet has the shortest. Outside the visible spectrum, infrared radiation has a longer wavelength than anything visible, and ultraviolet has a shorter wavelength than anything visible. So, infrared has the smallest energy per photon, and so on up the spectrum to ultraviolet with the most energy per photon.

2. (A) The total number of protons + neutrons is conserved. Before the reaction, we have one free neutron plus 235 protons and neutrons in the uranium, for a total of 236 amu. After the reaction, we have 141 amu in the barium plus 3 free neutrons for a total of 144 amu . . . leaving 92 amu for the krypton.

Charge is also conserved. Before the reaction, we have a total charge of +92 from the protons in the uranium. After the reaction, we have 56 protons in the barium. Since a neutron carries no charge, the krypton must account for the remaining 36 protons.

3. (D) Einstein's famous equation is written $\Delta E = \Delta mc^2$, because it is only the lost mass that is converted into energy. Since we still have a total of 236 amu after the reaction, an entire amu of mass was *not* converted to energy. Still, the daughter particles have kinetic energy because they move. That energy came from a very small mass difference, probably about a million times less than one amu.

4. (C) In β^+ emission, a positron is ejected from the nucleus. This does not change the total number of protons + neutrons, so the atomic mass A is still 15. However, charge must be conserved. The original O nucleus had a charge of +8, so the products of the decay must combine to a charge of +8. These products are a positron, charge +1, and the daughter nucleus, which must have charge +7.

5. (D) Twelve days is four half-lives for sample 1. In four half-lives, only $\frac{1}{16}$ (3.25 g) of the original 52-g sample will remain. For sample 2 to reach 3.25 grams in the same time, it must have a shorter half-life. Answer choice D is the only choice shorter than 3 days. However, to confirm the answer, 3.25 grams is approximately $\frac{1}{64}$ of the original 200 grams. That is six half-lives of decay, or a half-life of two days.

6. (D) Reducing the velocity also reduces the momentum of the electron. This will increase the wavelength of the electron $(\lambda = \frac{h}{p})$. Increasing the wavelength also increases the pattern spacing ($d \sin\theta = m\lambda$).

7. (A) and (C) All the photons of the uniform UV light have the same energy, which is entirely imparted to the electrons on a one-to-one basis. Thus, plate A must have electrons that are harder to remove (higher work function). Placing the light source closer to the plate or having a larger surface area would account for more electrons being emitted from plate A. Moving the light closer to the plate does not increase the energy of the individual photons and will not increase the energy of the ejected electrons.

8. (A) and (D) The behavior of ejected electrons from a surface due to electromagnetic radiation (the photoelectric effect) was the experiment that first proved light waves have particle properties. Light waves reflecting off "solar sails" to propel spacecraft can only be understood by modeling light as a photon particle that collides with the sail, imparting momentum to the sail during the collision.

9. (B) and (C) Conservation of charge is satisfied because the net charge before and after is zero. Conservation of momentum tells us that the initial momentum of the gamma ray must be equal to the net momentum of the two particles. The gamma ray has x-direction momentum and no y-direction momentum. This appears to be stratified in the figure. The answer choice D is incorrect; we have to use conservation of mass/energy because most of the gamma ray energy has been converted into mass, not just kinetic energy of the two particles.

10. (C) Adding the excitation energy of 100 eV −115 eV to the ground state energy, we get the electron energy range of (−22.4 eV) − (−7.4 eV). This will take the electrons of the gas from the ground state only to the $n = 3$ and 4 energy levels. The highest frequency photon occurs when the electron jumps down in energy by the largest amount: $E = hf$. Therefore, the transition from $n = 3$ to $n = 2$ is the answer. Answer choices A and D are not possible because the electrons were not excited to the $n = 5$ level.

11. (B) The energy of the gamma ray comes from the mass that is lost in the reaction. Adding the mass of the reactants, subtracting the mass of the products, and converting into energy via $E = mc^2$ gives us the mass defect that was converted into the energy of the gamma ray. This is an application of conservation of mass/energy.

12. (A) $\Delta E = hc/\lambda$, so $hc/\Delta E = \lambda$. $hc = 1240$ eV·nm, as found on the equation sheet. ΔE represents the difference in energy levels. An electron in the ground state can make either of two jumps: it could absorb a 2.1-eV photon to get to the middle energy level, or it could absorb a 3.3 eV photon to escape the atom. An electron in the middle state could absorb a 1.2-eV photon to escape the atom. That makes three different energies. Convert these energies to wavelengths using $\Delta E = hc/\lambda$, so $hc/\Delta E = 1$. $hc = 1240$ eV·nm, as found on the equation sheet; ΔE represents the energy of the absorbed photon, listed above. These photons thus have wavelengths of 590 nm for the E_1 to E_2 transition; 380 nm or less for the E_1 to E_∞ transition; and 1030 nm or less for the E_2 to E_∞ transition. Only the 590 nm wavelength is visible because the visible spectrum is from about 400–700 nm.

(B) (i) Find the energy of the incident photon using $\Delta E = hc/\lambda$, getting 6.9 eV. This is the total energy absorbed by the electron, but 3.3 eV of this is used to escape the atom. The remaining 3.6 eV is kinetic energy.

(ii) To find speed, set kinetic energy equal to $\frac{1}{2}mv^2$. However, to use this formula, the kinetic energy must be in standard units of joules. Convert 3.6 eV to joules by multiplying by 1.6×10^{-19} J/eV (this conversion is on the constant sheet), getting a kinetic energy of 5.8×10^{-19} J. Now solve the kinetic energy equation for velocity, getting a speed of 1.1 × 106 m/s. This is reasonable—fast, but not faster than the speed of light.

(iii) A photon is a particle of light. Unless it is in an optically dense material (which the photons here are not), the speed of a photon is always 3.0 × 108 m/s.

(C) The electron absorbs the same amount of energy from the incident photons. However, now it takes only 1.2 eV to get out of the atom, leaving 5.7 eV for kinetic energy. With a larger kinetic energy, the ejected electron's speed is greater than before.

(D) The speed of the photon is still the speed of light, 3.0×10^8 m/s.

13. Negativly charged objecs have more electrons than protons. Ultraviolet photons have enough energy to dislodge electrons from a object, which can eliminate the excess of electrons from the negatively charged object. This will neutralize the object's charge. On the other hand, positively charged objects have more protons than electrons. The only way to discharge a positively charged object is to add electrons to the object or remove protons from the object. A UV photon cannot do either of these.

14. (A) At the threshold frequency, the kinetic energy of the photoelectrons is zero:

$$K_{max} = 0 = hf_T - \Phi$$

$$f_{T\,sodium} = 5.8 \times 10^{14}\,\text{Hz},\ f_{T\,iron} = 1.1 \times 10^{15}\,\text{Hz},$$

$$\Phi_{cesium} = 1.9\ \text{eV}$$

(B)

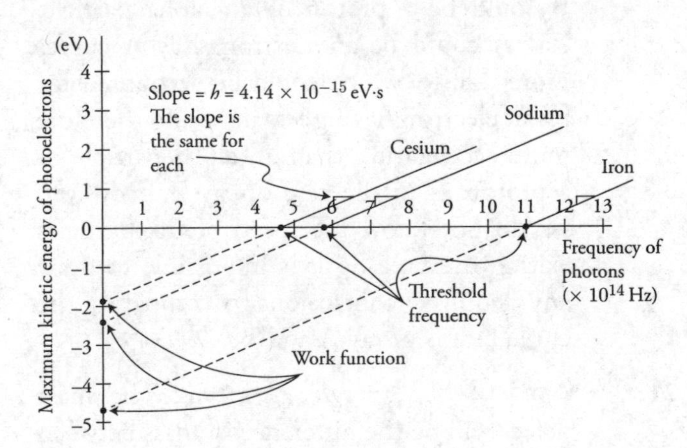

(C) $$K_{max} = hf - \Phi = \frac{hc}{\lambda} - \Phi$$

$$K_{max} = \frac{1240\ \text{eV}\cdot\text{nm}}{200\ \text{nm}} - 1.9\ \text{eV} = 4.3\ \text{eV}$$

15. (A) Wave property: Only waves exhibit interference.

(B) Particle property: Light behaves like a particle by colliding with an electron, thus imparting momentum to the electron and knocking it from the atom.

(C) Wave property: Interference is a wave property.

(D) Particle property: Collisions and momentum are particle properties.

(E) Wave property: Diffraction, the bending of waves around boundaries, is a wave property.

(F) Wave property: The Doppler effect is a property of waves.

(G) Particle property: Solar cells, or photovoltaic cells, are a practical application of the photoelectric effect, which demonstrates the particle nature of light.

16. (A) The X-ray will lose energy and momentum. Conservation of momentum must be obeyed in the collision. Since the electron gained momentum in the collision, the photon must lose momentum. Also, the electron gains kinetic energy in the collision. Therefore, the photon must lose energy. Losing momentum and energy means the frequency of the X-ray has decreased: $E = hf = pc$.

(B) The electron gains both kinetic energy and momentum to the right from the collision with the X-ray.

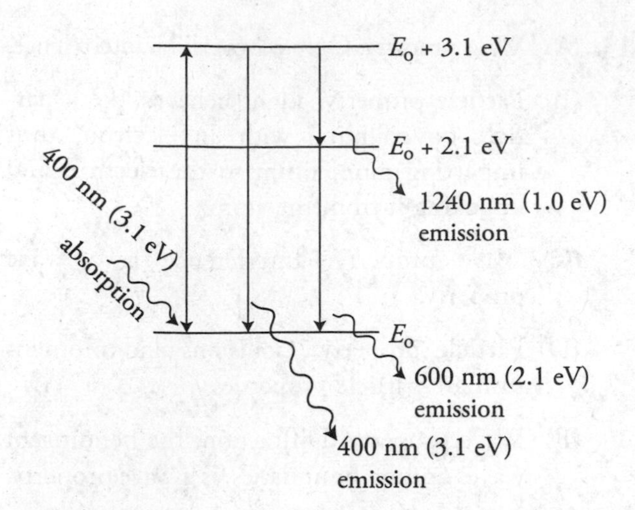

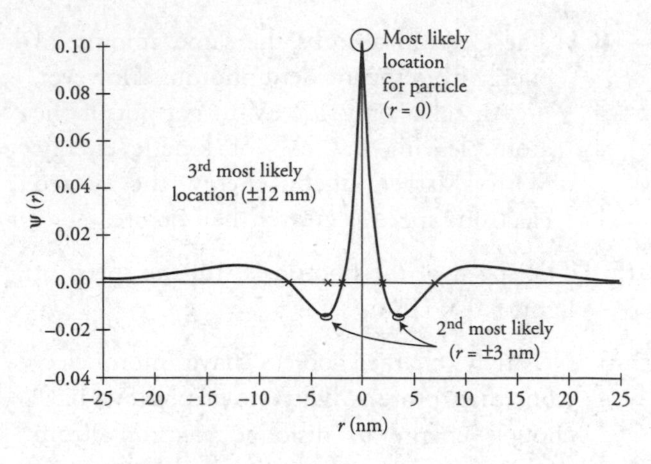

17. (A) The problem does not tell us the starting position of the electron or its initial energy. So, let's just label it at E_0. Using the equation $\Delta E = hf = \frac{hc}{\lambda}$, we can calculate the change in energy of the electrons in the atom associated with the photons:

$$\Delta E = \frac{(1240\,\text{eV}\,\Delta\text{nm})}{\lambda}.$$

The wavelengths of 400 nm and 600 nm give us energies of 3.1 eV and 2.1 eV, respectively. See figure.

(B) Yes! The figure shows that when the electron is in the E_0 + 3.1 eV energy level, it can fall to the E_0 + 2.1 eV energy level. This will produce a 1240 nm/1.0 eV photon that is not in the visible spectrum, so it will not be seen.

18. (A) This is an interference pattern. Therefore, electrons are exhibiting wave properties (wave-particle duality).

(B) When the accelerating voltage increases, the electron velocity increases: $\Delta U = q\Delta V = \frac{1}{2}mv^2$

When electron velocity increases, the de Broglie wavelength of the electron decreases:

$$\lambda = \frac{h}{p} = \frac{h}{mv}$$

When wavelength decreases, the pattern gets tighter. The angle of the maxima decreases: $d\sin\theta = m\lambda$.

(C) The interference pattern will become more spread out as the distance d decreases: $\sin\theta = \frac{m\lambda}{d}$.

19. (A) and (B) See figure.

20. (A) This is a capacitor. The top plate has a higher electric potential. The electric field is directed straight down toward the bottom plate. The electric field is uniform in strength and direction between the plates except for at the edges of the capacitor.

(B) Ray C must not have a charge because it is not affected by the electric field. Ray B curves in the direction of the electric field. Therefore, it must be receiving a force from the field and must be positively charged. Ray A must be negative as it receives a force in the opposite direction of the electric field and curves upward.

(C) Ray C could be a neutron or a gamma ray because neither has an electric charge. Ray B could be a proton or an alpha particle. Ray A could be an electron. Ray A curves more tightly, which could be accounted for by an electron having less mass and therefore more acceleration than an alpha particle or a proton as in the case of ray B. However, we do not know the speed of the particles exiting the box. So it is impossible to draw any definitive conclusions by comparing the curve radius of ray A and B.

21. $\Delta m = (2m_{\text{neutron}} + 2m_{\text{proton}}) - m_{\text{He}}$. The mass defect will be the difference in mass between the helium nucleus and the masses of all the parts that are used to construct the nucleus (two neutrons and two protons).

22. (A) Not possible! The number of nucleons is not conserved.

(B) Not possible! Conservation of charge is violated.

23. (A) $2{}_{1}^{1}\text{H} + 2{}_{0}^{1}\text{n} \rightarrow {}_{2}^{4}\text{He} + \text{energy}$

(B) ${}_{55}^{137}\text{Cs} \rightarrow {}_{56}^{137}\text{Ba} + {}_{-1}^{0}\text{e}$

(C) ${}_{92}^{238}\text{U} \rightarrow {}_{90}^{234}\text{Th} + {}_{2}^{4}\alpha + \text{energy}$

(D) ${}_{0}^{1}\text{n} + {}_{92}^{235}\text{U} \rightarrow {}_{54}^{144}\text{Xe} + {}_{38}^{90}\text{Sr} + 2{}_{0}^{1}\text{n}$

(E) ${}_{6}^{12}\text{C} \rightarrow {}_{6}^{12}\text{C} + \gamma$

24. (A) Approximately 3.5 years.

(B) Half-life is the time it takes for one half of a radioactive sample to decay into a new element. After two half-lives, a quarter of the original radioactive substance will remain. This means that three quarters of the original material has transmuted into a new element.

❯ Rapid Review

- Time and space (length) are not constants. They depend on the speed of the object or person.
- Mass and energy are two aspects of the same thing and can be converted back and forth.
- The photoelectric effect experiment proved that light has particle properties.
- Photons are particles of light. Their energy depends on their frequency. Higher frequency = higher energy.
- Photons have momentum and can interact with particles, creating collisions similar to objects colliding. When photons lose momentum and energy, their frequency decreases and their wavelength gets longer.
- Particles can display wave properties. The wave function of a particle is interpreted to be the probability of finding the particle at a location.
- Atoms contain a nucleus, made of protons and neutrons, and one or more electrons that orbit that nucleus. Protons, neutrons, and electrons all have mass. By contrast, photons are subatomic particles without mass.
- The electron-volt is a unit of energy that's convenient to use when solving atomic physics problems.
- The electrons that surround an atom can have only certain, specific amounts of energy because of the wave nature of electrons. To go from a low-energy level to a high-energy level, an electron absorbs a photon. To go from a high-energy level to a low-energy level, an electron emits a photon.
- If an electron absorbs a photon that has a higher energy than the electron's work function, the electron will be expelled from the atom.
- Moving particles have a characteristic wavelength, found by the de Broglie equation.
- Nuclei can undergo several types of decay. In alpha decay, a nucleus emits an alpha particle, which consists of two protons and two neutrons. In beta decay, a nucleus emits either a positron or an electron. In gamma decay, a nucleus emits a photon.

- When solving nuclear problems, remember to conserve nucleon number, nuclear charge, mass/energy, and momentum.
- During nuclear decay, mass is converted to energy. The relationship between the mass defect and the gained energy is found by Einstein's famous formula, $E = (\Delta m)c^2$.
- Half-life is the time it takes one-half of the radioactive material to decay.

Build Your Test-Taking Confidence

AP Physics 2: Practice Exam 1 Section 1 (Multiple Choice)
AP Physics 2: Practice Exam 1 Section 2 (Free Response)
Solutions: Section 1 (Multiple Choice)
Solutions: Section 2 (Free Response)
How to Score Practice Exam 1

AP Physics 2: Practice Exam 2 Section 1 (Multiple Choice)
AP Physics 2: Practice Exam 2 Section 2 (Free Response)
Solutions: Section 1 (Multiple Choice)
Solutions: Section 2 (Free Response)
How to Score Practice Exam 2

The AP Physics 2 Practice Exams

Taking the Practice Exams

Taking a full-length practice test is one of the most important things you can do to prepare yourself for the AP Physics 2 exam. Besides helping build your test-taking confidence, you'll be able to practice pacing yourself to make the most efficient use of your limited time. Of course, it will also help to familiarize yourself with the test so that there are no surprises on test day. Finally, taking the practice exam provides a good review of the physics content on which you will be tested; in fact, you may find a topic or two about which you'll want to go back and reread a section of this book. In the pages that follow, you'll find two practice exams, each followed by complete explanations for all questions. Be sure to take both tests.

Structure of the Exam

Remember the structure of the test described in Chapter 2? To refresh your memory, you may want to take another look at this chart. The practice exams in this book follow exactly the structure and timing of the real test. Remember that you may use a calculator, the equation sheet, and the table of information. The equations and information table can be found in the appendix. However, I recommend that you download and use the most current one from the College Board at: https://secure-media.collegeboard.org/digitalServices/pdf/ap/ap-physics-2-equations-table.pdf.

Section	Number of Questions	Time Limit
1. Multiple-Choice Questions	45 with a single correct answer	90 minutes
	5 with two correct answers	
Total	50 Multiple-Choice Questions	
2. Free-Response Questions	1 Experimental Design Question	90 minutes
	1 Qualitative/Quantitative Translation Question	
	2 Short-Answer Questions (including a paragraph-length response)	
Total	4 Free-Response Questions	

Tips for Taking the Practice Exams

When taking a practice test, try to replicate, as much as possible, the actual test-taking experience. Here are some tips to make the most of the practice exams:

- Set aside a block of time to take each part of the test. There should be no distractions or interruptions. Make sure your cell phone won't disturb your concentration.
- Carefully time yourself—90 minutes for the multiple-choice section and 90 minutes for the free-response section. Check the time occasionally to practice pacing yourself, but don't become overly obsessive watching the clock!
- Tear out the answer sheet provided and mark your answers in pencil on the grid, just like on the actual test.
- Check your answers against the solutions provided to see how well you did. You can do this after you finish each 90-minute section. Don't just read the explanations for the questions you missed, be sure you also look at the explanations for the ones you got right, especially those you weren't sure of. Even when you got a question right, you need to understand why the answer you selected is correct and why the other answer choices are wrong.

Test-Taking Strategies

Remember the strategies you learned in Chapters 6–8? These will help you use the limited testing time more efficiently, allowing you to get the best score you can. Here are brief summaries of the most important strategies for taking the AP Physics 2 exam.

Strategies for the Multiple-Choice Section

- Work all the easy questions first and leave the harder ones for later.
- Answer every question. You won't lose points for guessing.
- Read the stem of the question and look at the answer choices before you jump into trying to answer the question.
- Cross out answer choices that you know can't be correct. Choose the best answer from what is left.
- Don't be afraid of the multiple-correct questions, which are the last five questions.
- Keep calm and determined. Nobody gets all the multiple-choice questions correct. Not even AP Physics 2 teachers! You are shooting for 40%–45% for a 3 and 75% for a 5.

Strategies for the Free-Response Section

- Skim all the questions and start with the one that looks the easiest and work your way to the hardest one.
- Don't leave any part on any question blank. You are trying to earn partial credit. You can't earn points when you leave it blank.
- Show all your work. Show the steps you are taking to come to a solution. Go for partial credit.
- Don't try to snow the exam reader with a bunch of malarkey. It will only lower your score. Show your knowledge of physics.
- The AP Physics readers consistently tell us that students need to be more precise in their explanations. Remember, you are not writing poetry that is up to the interpretation of the reader. You need to use "physics language" and avoid the use of pronouns. Instead of saying "It gets pushed away." Be more precise! Say something more like this: "The electron receives a force to the right due to the electric field." Or, "The electric force pushes the electron to the right." Speak in "Physics" and your grade will improve.
- Keep your written answers short and to the point. Remember to be CLEVeR! Make your **CL**aim. Give your **EV**idence. Explain your **R**easoning and move on.
- For heaven's sake, write clear, orderly, and legibly! What can't be read gets a zero.
- Keep calm and determined. Nobody earns all of the points. Not even AP Physics 2 teachers! You are shooting for 40%–45% for a 3 and 75% for a 5.

AP Physics 2: Practice Exam 1

Multiple-Choice Questions

ANSWER SHEET

1 (A) (B) (C) (D)	18 (A) (B) (C) (D)	35 (A) (B) (C) (D)
2 (A) (B) (C) (D)	19 (A) (B) (C) (D)	36 (A) (B) (C) (D)
3 (A) (B) (C) (D)	20 (A) (B) (C) (D)	37 (A) (B) (C) (D)
4 (A) (B) (C) (D)	21 (A) (B) (C) (D)	38 (A) (B) (C) (D)
5 (A) (B) (C) (D)	22 (A) (B) (C) (D)	39 (A) (B) (C) (D)
6 (A) (B) (C) (D)	23 (A) (B) (C) (D)	40 (A) (B) (C) (D)
7 (A) (B) (C) (D)	24 (A) (B) (C) (D)	41 (A) (B) (C) (D)
8 (A) (B) (C) (D)	25 (A) (B) (C) (D)	42 (A) (B) (C) (D)
9 (A) (B) (C) (D)	26 (A) (B) (C) (D)	43 (A) (B) (C) (D)
10 (A) (B) (C) (D)	27 (A) (B) (C) (D)	44 (A) (B) (C) (D)
11 (A) (B) (C) (D)	28 (A) (B) (C) (D)	45 (A) (B) (C) (D)
12 (A) (B) (C) (D)	29 (A) (B) (C) (D)	46 (A) (B) (C) (D)
13 (A) (B) (C) (D)	30 (A) (B) (C) (D)	47 (A) (B) (C) (D)
14 (A) (B) (C) (D)	31 (A) (B) (C) (D)	48 (A) (B) (C) (D)
15 (A) (B) (C) (D)	32 (A) (B) (C) (D)	49 (A) (B) (C) (D)
16 (A) (B) (C) (D)	33 (A) (B) (C) (D)	50 (A) (B) (C) (D)
17 (A) (B) (C) (D)	34 (A) (B) (C) (D)	

AP Physics 2: Practice Exam 1

Section 1 (Multiple Choice)

Directions: The multiple-choice section consists of 50 questions to be answered in 90 minutes. You may write scratch work in the test booklet itself, but only the answers on the answer sheet will be scored. You may use a calculator, the equation sheet, and the table of information. These can be found in the appendix or you can download the official ones from the College Board at: https://secure-media.collegeboard.org/digitalServices/pdf/ap/ap-physics-2-equations-table.pdf.

Questions 1–45: Single-Choice Items

Choose the single best answer from the choices provided, and mark the answer with a pencil on the answer sheet.

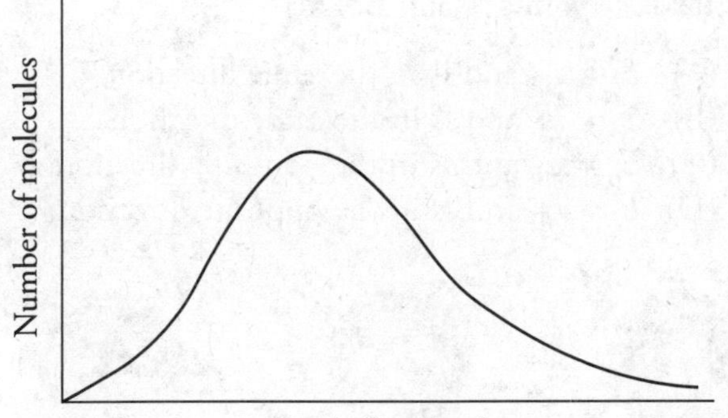

1. The graph shows the distribution of speeds for one mole of hydrogen at temperature T, pressure P, and volume V. How would the graph change if the sample was changed from one mole hydrogen to one mole of argon at the same temperature, pressure, and volume?

(A) The peak will shift to the left.
(B) The peak will shift upward and to the left.
(C) The peak will shift to the right.
(D) The peak will shift downward and to the right.

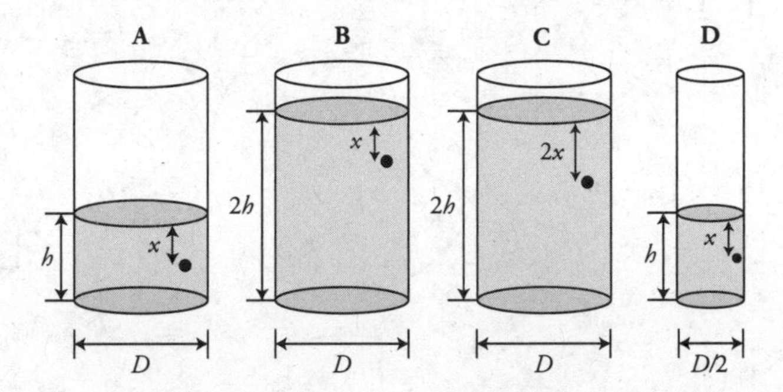

2. The figure shows four cylinders of various diameters filled to different heights with water. A hole in the side of each cylinder is plugged by a cork. All cylinders are open at the top. The corks are removed. Which of the following is the correct ranking of the velocity of the water (v) as it exits each cylinder?

(A) $v_A > v_D > v_C > v_B$
(B) $v_A = v_D > v_C > v_B$
(C) $v_B > v_C > v_A = v_D$
(D) $v_C > v_A = v_B = v_D$

3. An observer can hear sound from around a corner but cannot see light from around the same corner. Which of the following helps to explain this phenomenon?

(A) Sound is a longitudinal wave, and light is an electromagnetic wave.
(B) Sound is a mechanical wave, and light is a transverse wave.
(C) Light travels at a speed much faster than that of sound.
(D) Light has a much smaller wavelength than sound.

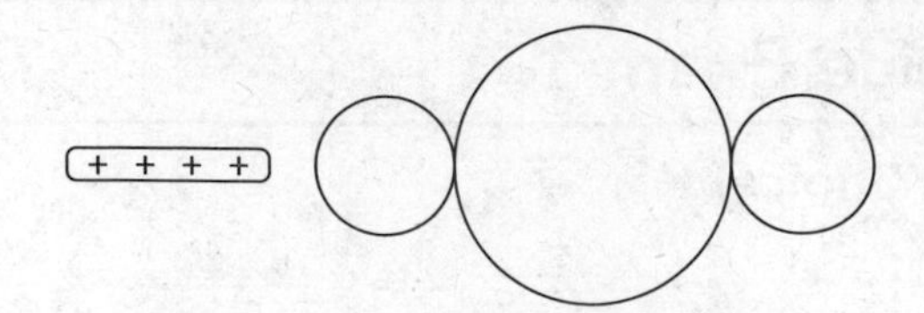

4. A positively charged rod is brought near to but not touching three metal spheres that are in contact with each other, as shown in the figure. Which is the best representation of the charge arrangement inside the three spheres?

(A)

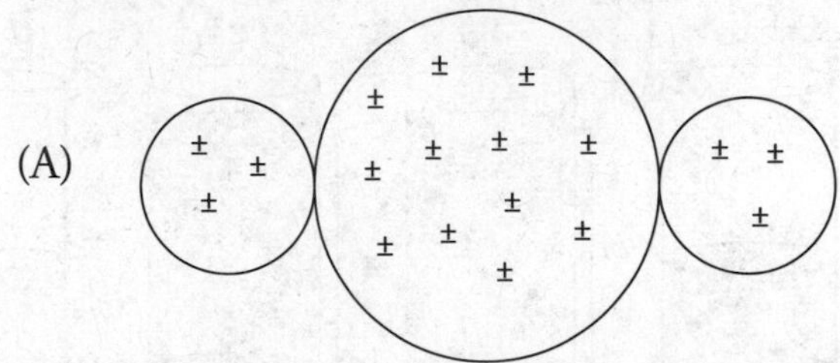

(B)

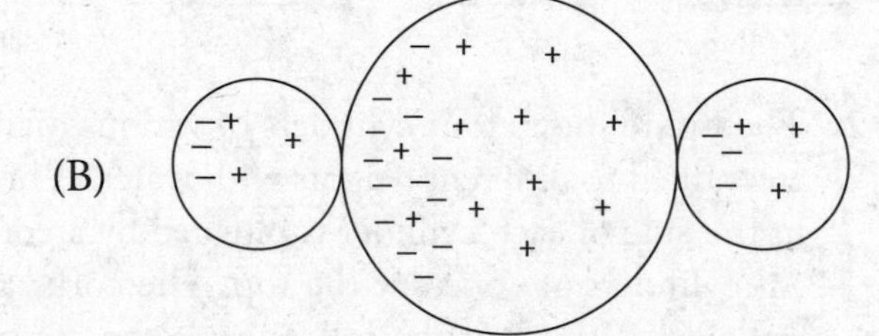

(C)

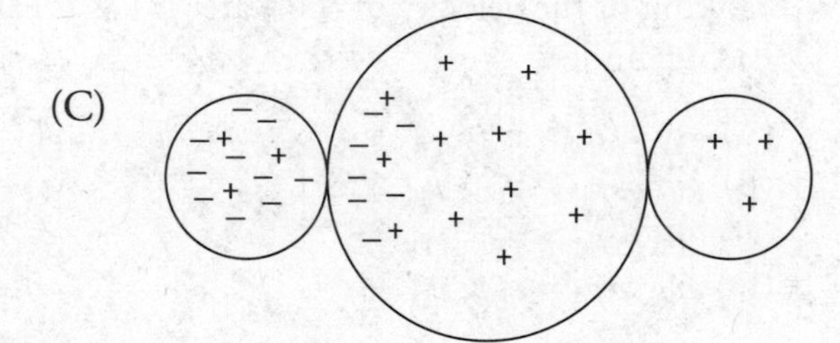

(D)

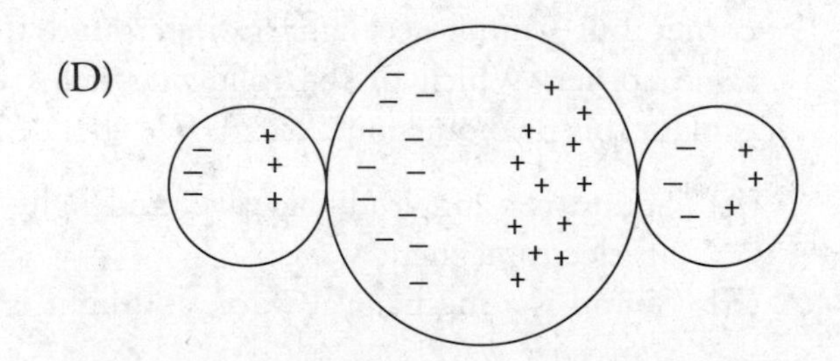

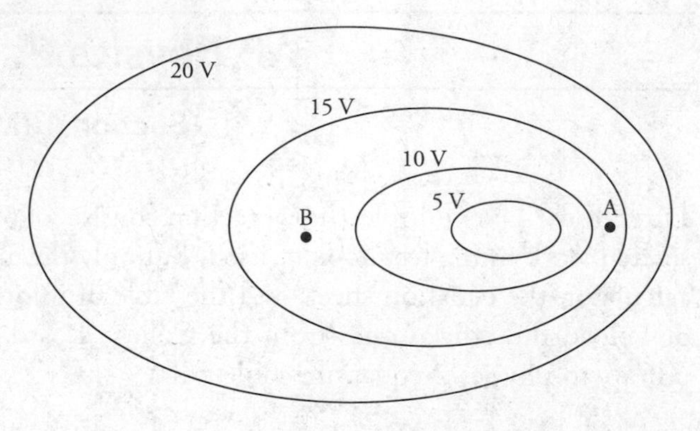

5. Isolines of equal electric potential in a region of space are shown in the figure. Points A and B are in the plane of the isolines. Which of the following correctly describes the relationship between the magnitudes and directions of the electric fields at points A and B?

(A) $E_A = E_B$ and is in the same direction.
(B) $E_A \neq E_B$ and is in the same direction.
(C) $E_A = E_B$ and is in the opposite direction.
(D) $E_A \neq E_B$ and is in the opposite direction.

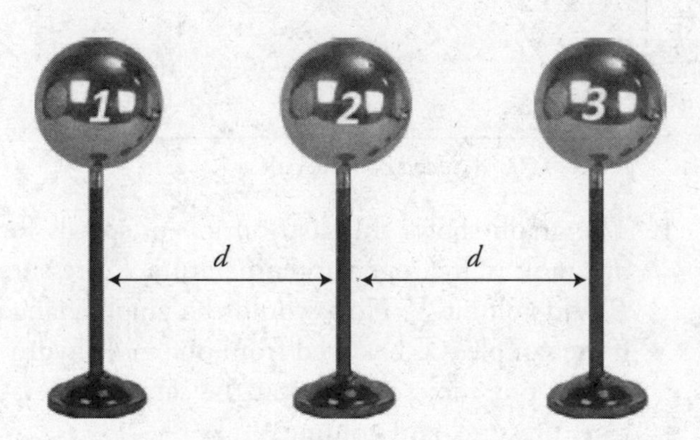

6. The metal spheres on insulating stands 1, 2, and 3 are all identical and situated as shown in the figure. Spheres 1 and 2 have a charge of $-Q$, and sphere 3 has a charge of $+2Q$. The force of sphere 1 on sphere 2 is $+F$. What is the magnitude of the net force on sphere 3 in terms of F?

(A) 3/2 F
(B) 2 F
(C) 5/2 F
(D) 3 F

GO ON TO THE NEXT PAGE

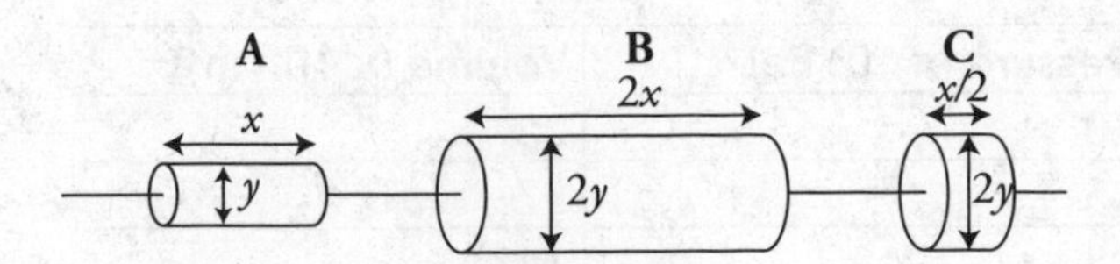

7. Three cylindrical resistors made of the same material but different dimensions are connected, as shown in the figure. A battery is connected to produce current through the resistors. Which is the correct ranking of the potential differences across the individual resistors?

(A) $V_A = V_B = V_C$
(B) $V_A > V_B > V_C$
(C) $V_A = V_B > V_C$
(D) $V_C > V_B > V_A$

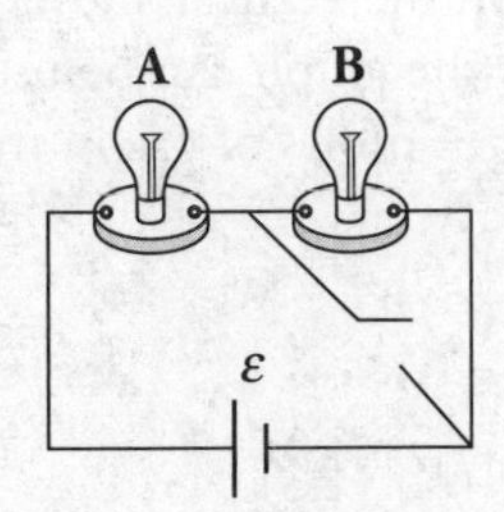

8. The figure shows two bulbs connected to a battery in a circuit with a switch that is originally in the closed position. What happens to the brightness of the bulbs when the switch is opened?

	Bulb A	Bulb B
(A)	Four times brighter	Goes out
(B)	Same brightness	Glows as brightly as bulb A
(C)	Half as bright	Glows as brightly as bulb A
(D)	Quarter as bright	Glows as brightly as bulb A

9. An astronaut in a rocket is passing by a space station at a velocity of 0.33 *c*. Looking out the window, the astronaut sees a scientist on the space station fire a laser at a target. The laser is pointed in the same direction that the astronaut is traveling. On which of the following observations will the astronaut and scientist agree?

(A) The length of the rocket
(B) The time it takes the laser to hit the target
(C) The speed of the laser beam
(D) The astronaut and scientist will not agree on any of these measurements.

10. A proton is moving toward the top of the page when it encounters a magnetic field that changes its direction of motion. After encountering the magnetic field, the proton's velocity vector is pointing out of the page. What is the direction of the magnetic field? Assume gravitational effects are negligible.

(A) Toward the bottom of the page
(B) To the right
(C) To the left
(D) Into the page

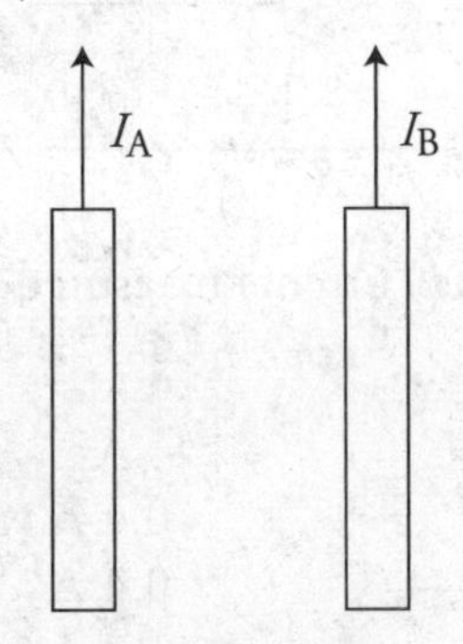

11. Two long parallel wires carry currents (I_A and I_B), as shown in the figure. Current I_A in the left wire is twice that of current I_B in the right wire. The magnetic force on the right wire is *F*. What is the magnetic force on the left wire in terms of *F*?

(A) *F* in the same direction
(B) *F* in the opposite direction
(C) *F*/2 in the same direction
(D) *F*/2 in the opposite direction

GO ON TO THE NEXT PAGE

Questions 12 and 13

Four identical resistors of resistance R are connected to a battery, as shown in the figure. Ammeters A_1 and A_2 measure currents of 1.2 A and 0.4 A, respectively.

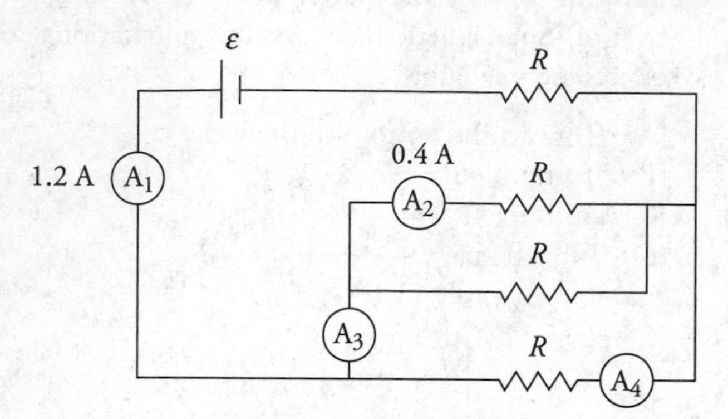

12. What are the currents measured by ammeters A_3 and A_4?

	A_3	A_4
(A)	0.4 A	0.4 A
(B)	0.8 A	0.4 A
(C)	0.4 A	1.2 A
(D)	0.8 A	1.2 A

13. What is the equivalent resistance of the circuit?

(A) ¼ R
(B) 4/3 R
(C) 5/2 R
(D) 4 R

Pressure ($\times 10^5$ Pa)	Volume ($\times 10^{-3}$ m^3)
1.0	25
1.5	17
1.8	14
2.2	11
2.6	9.6
3.3	7.6

14. In an experiment, a gas is confined in a cylinder with a movable piston. Force is applied to the piston to increase the pressure and change the volume of the gas. Each time the gas is compressed, it is allowed to return to a room temperature of 20°C. The data gathered from the experiment is shown in the table. What should be plotted on the vertical and horizontal axes so the slope of the graph can be used to determine the number of moles of gas in the cylinder?

(A) P and V^2
(B) P and V
(C) P and $(V)^{½}$
(D) P and $1/V$

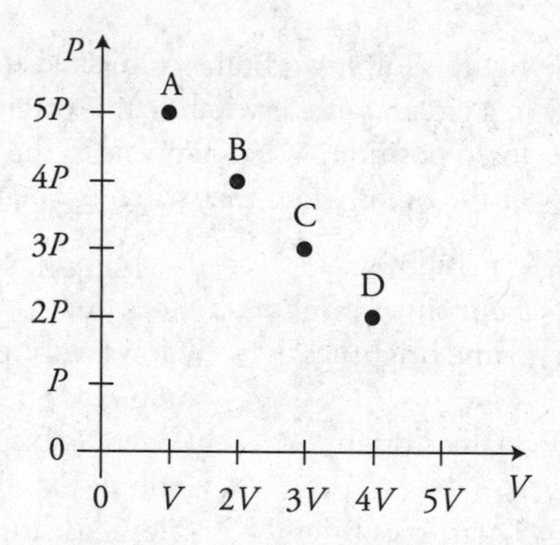

15. The figure shows the pressure and volume of a gas at four different states. Which of the following correctly ranks the temperature of the gas at the different states?

(A) $T_A > T_B > T_C > T_D$
(B) $T_B = T_C > T_A = T_D$
(C) $T_C > T_B = T_D > T_A$
(D) $T_D > T_C > T_B > T_A$

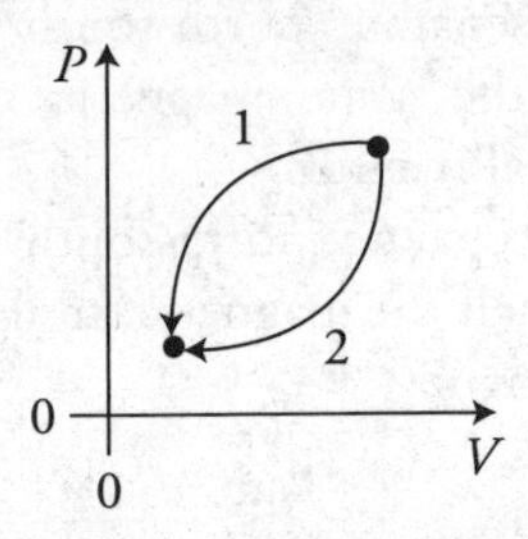

16. The figure shows the pressure and volume of three moles of gas being taken through two different processes. Which of the following is correct concerning the two processes shown in the figure?

(A) $\Delta U_1 = \Delta U_2$ and $W_1 = W_2$
(B) $\Delta U_1 = \Delta U_2$ and $W_1 > W_2$
(C) $\Delta U_1 > \Delta U_2$ and $W_1 = W_2$
(D) $\Delta U_1 > \Delta U_2$ and $W_1 > W_2$

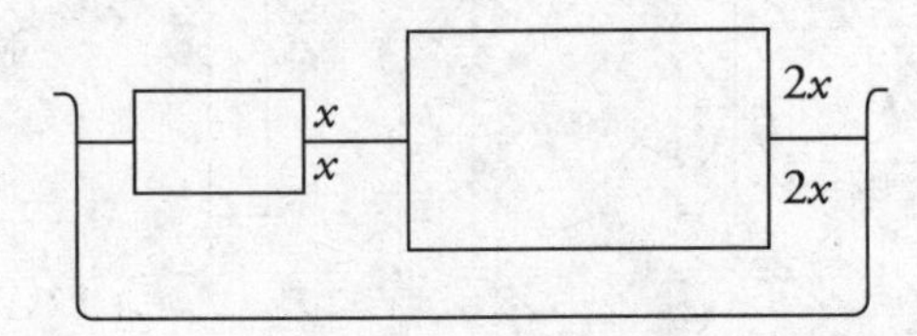

17. Two blocks of different sizes and masses float in a tray of water. Each block is half submerged, as shown in the figure. Water has a density of 1,000 kg/m^3. What can be concluded about the densities of the two blocks?

(A) The two blocks have different densities, both of which are less than 1,000 kg/m^3.
(B) The two blocks have the same density of 500 kg/m^3.
(C) The two blocks have the same density, but the density cannot be determined with the information given.
(D) The larger block has a greater density than the smaller block, but the densities of the blocks cannot be determined with the information given.

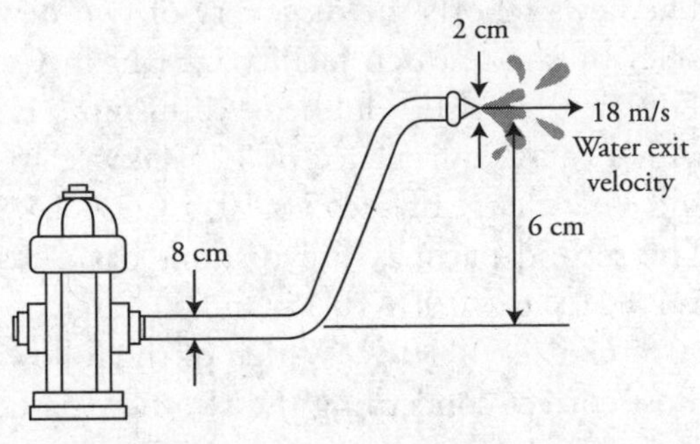

18. Firefighters use a hose with a 2 cm diameter exit nozzle connected to a hydrant with an 8 cm diameter opening to attack a fire on the second floor of a building 6 m above the hydrant, as shown in the figure. What pressure must be supplied at the hydrant to produce an exit velocity of 18 m/s? (Assume the density of water is 1,000 kg/m^3, and the exit pressure is 1×10^5 Pa.)

(A) 1.7×10^5 Pa
(B) 2.0×10^5 Pa
(C) 2.6×10^5 Pa
(D) 3.2×10^5 Pa

19. Two electrons exert an electrostatic repulsive force on each other. Is it possible to arrange the two electrons so the gravitational attraction between them is large enough to cancel out the electric repulsive force?

(A) No, the charge of the electrons squared is much larger than the mass of the electrons squared.
(B) No, there is no gravitational force between subatomic particles.
(C) Yes, reducing the radius between the electrons will increase the gravitational force as it is proportional to the inverse of the radius squared.
(D) Yes, increasing the distance between the electrons will reduce the electrostatic repulsion until it is equal to the gravitational force.

20. The news reports the discovery of two new particles by the research facility CERN in Geneva. The first particle, dubbed Alithísium, is large with a mass equivalence of 125 GeV ± 15 GeV and a net charge of -1.55×10^{-18} C ± 0.1×10^{-18}. The second particle, Psevdísium, has a mass of 5.4×10^{-4} u ± 0.1×10^{-4} u and a charge of 1.6×10^{-20} C ± 0.5×10^{-20}. Which of the following is most correct concerning the two new particles?

(A) Both particles appear reasonable.
(B) Alithísium appears reasonable, but Psevdísium does not.
(C) Psevdísium appears reasonable, but Alithísium does not.
(D) Neither particle appears reasonable.

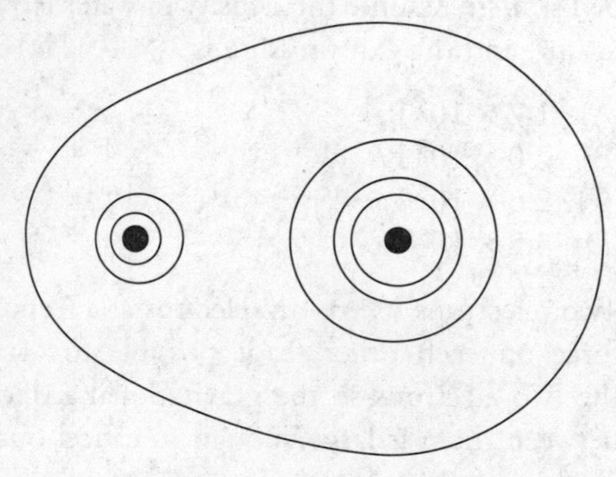

21. The figure shows isolines of constant electric potential surrounding two charges. Which of the following correctly describes the two charges?

(A) The charges are the same magnitude and the same sign.
(B) The charges are the same magnitude but different signs.
(C) The charges are different magnitudes but the same sign.
(D) The charges are different magnitudes and different signs.

22. An iron magnet is broken in half at the midpoint between its north and south ends. What is the result?

(A) A separate north pole and south pole, each with the same magnetic strength as the original magnet
(B) A separate north pole and south pole, each with half the magnetic strength of the original magnet
(C) Two separate north-south magnets, each with the same magnetic strength as the original magnet
(D) Two separate north-south magnets, each with half the magnetic strength of the original magnet

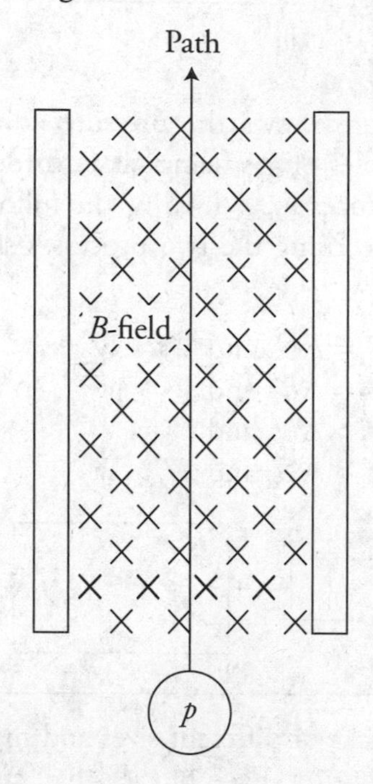

23. A magnetic field, directed into the page, is placed between two charged capacitor plates, as shown in the figure. The magnetic and electric fields are adjusted so a proton moving at a velocity of v will pass straight through the fields. The speed of the proton is doubled to $2v$. Which of the following force diagrams most accurately depicts all the forces acting on the proton when traveling at $2v$?

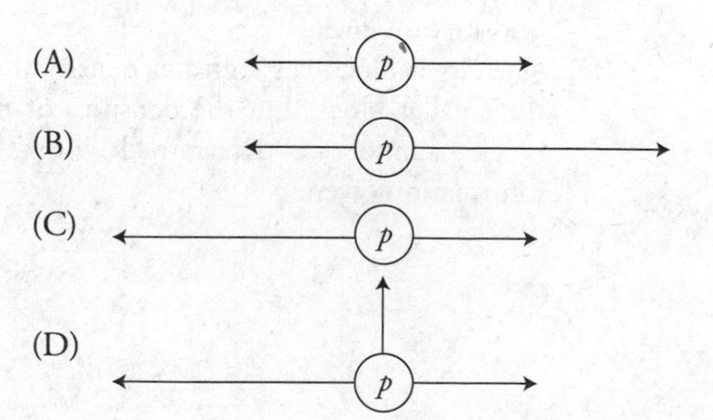

GO ON TO THE NEXT PAGE

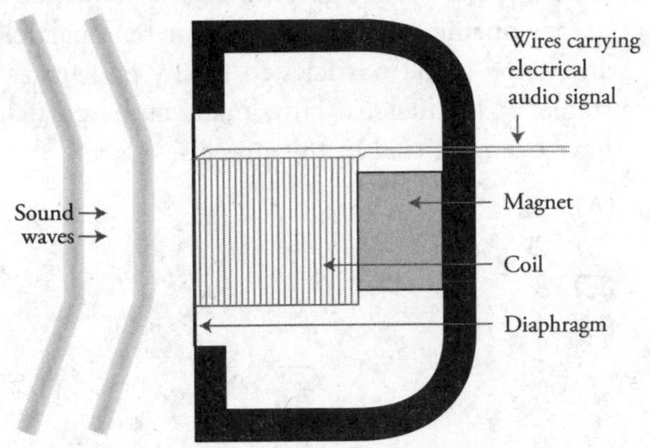

24. A dynamic microphone contains a magnet and a coil of wire connected to a movable diaphragm, as shown in the figure. Sound waves directed at the diaphragm generate a current in the wires leading from the coil. Which of the following helps to explain why this occurs?

(A) The area of the coil changes.
(B) The magnitude of the magnetic field produced by the magnet changes.
(C) The angle between the plane of the coil and the magnetic field produced by the magnet change.
(D) The strength of the magnetic field in the plane of the coil changes.

25. A lens and a mirror both have a focal length of f in air. Both are submerged in water, and the focal length f_{water} is measured for both. How does the focal length under water compare to the focal length in air?

	Lens	Mirror
(A)	$f = f_{water}$	$f = f_{water}$
(B)	$f = f_{water}$	$f < f_{water}$
(C)	$f < f_{water}$	$f = f_{water}$
(D)	$f < f_{water}$	$f < f_{water}$

26. Which of the following correctly describes the motion of the electric and magnetic fields of a microwave transmitted by a cell phone?

(A) Both the electric and magnetic fields oscillate in the same plane and perpendicular to the direction of wave propagation.
(B) Both the electric and magnetic fields oscillate perpendicular to each other and to the direction of wave propagation.
(C) The electric field oscillates perpendicular to the direction of wave propagation. The magnetic field oscillates parallel to the direction of wave propagation.
(D) Both the electric and magnetic fields oscillate parallel to the direction of wave propagation.

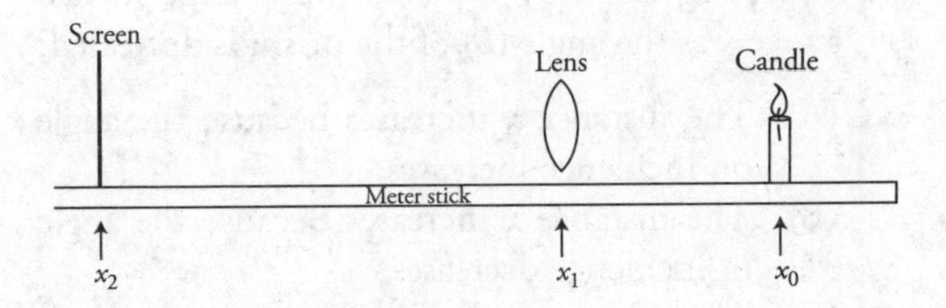

27. An optics bench is set up on a meter stick, as shown in the figure. The light source is a candle placed at x_0. The lens is located at x_1. The screen is moved until a sharp image appears at location x_2. The data is recorded in a table, the lens is moved to a new location (x_1), and the screen is adjusted until the image is sharp again. Which of the following procedures will allow a student to determine the focal length of the lens?

(A) Plot x_2 as a function of x_0. The focal length will be the vertical axis intercept.
(B) Plot $(x_2 - x_1)$ as a function of $(x_0 - x_1)$. The focal length will be the vertical axis intercept.
(C) Plot $1/x_2$ as a function of $1/x_0$. The focal length will be the inverse of the vertical axis intercept.
(D) Plot $1/(x_2 - x_1)$ as a function of $1/(x_0 - x_1)$. The focal length will be the inverse of the vertical axis intercept.

GO ON TO THE NEXT PAGE

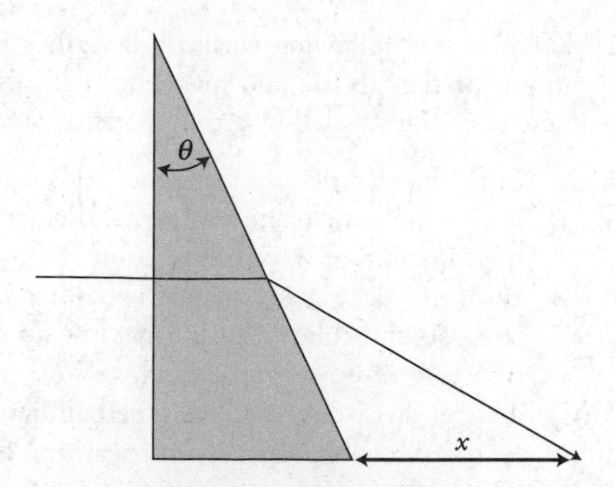

28. A laser beam passes through a prism and produces a bright dot of light a distance of x from the prism, as shown in the figure. Which of the following correctly explains the change in distance x as the angle (θ) of the prism is decreased?

(A) The distance x increases because the angle on incidence increases.
(B) The distance x increases because the angle of incidence decreases.
(C) The distance x decreases because the angle on incidence increases.
(D) The distance x decreases because the angle of incidence decreases.

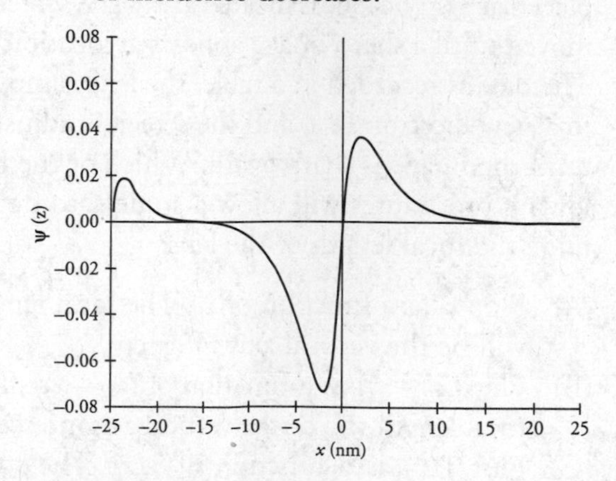

29. The graph shows the wave function of a particle as a function of x in the region between −25 nm < x < +25 nm. At which of the following positions is the probability of finding the particle greatest?

(A) −24 nm
(B) −15 nm and 0.0 nm
(C) −2 nm
(D) 2 nm

30. A nucleus of $^{237}_{93}$Np goes through a sequence of decays during which it emits four beta particles and some alpha particles to finally end up as a stable $^{250}_{81}$Tl nucleus. How many alpha particles have been emitted in this process?

(A) 32
(B) 26
(C) 8
(D) 4

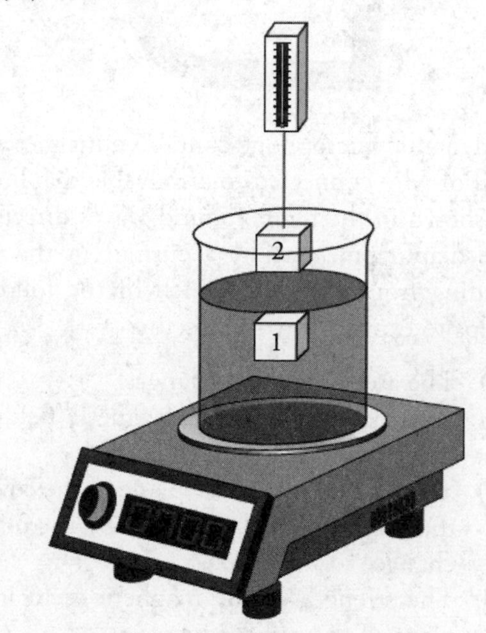

31. A beaker of water sits on a balance. A metal block with a mass of 190 g is held suspended in the water by a spring scale in position 1, as shown in the figure. In this position, the reading on the balance is 1,260 g, and the spring scale reads 120 g. When the block is lifted from the water to position 2, what are the readings on the balance and spring scale?

	Balance reading	Spring scale reading
(A)	1,190 g	120 g
(B)	1,190 g	190 g
(C)	1,260 g	190 g
(D)	1,330 g	120 g

GO ON TO THE NEXT PAGE

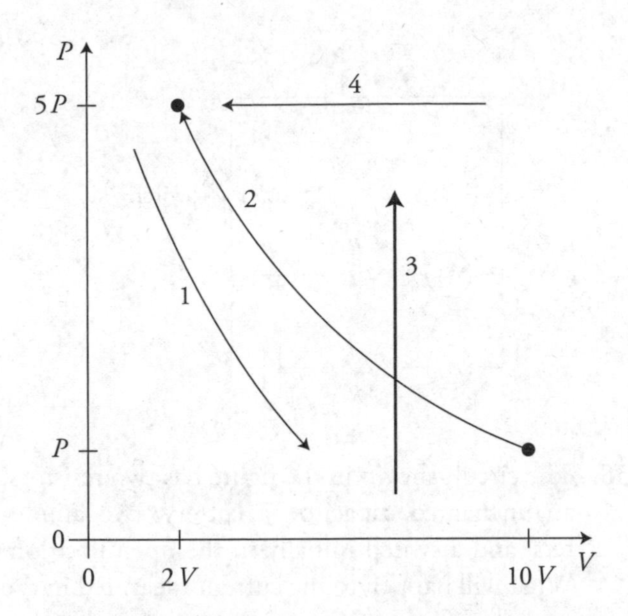

32. The figure shows four samples of gas being taken through four different processes. Process 1 is adiabatic. In which process is heat being transferred to the gas sample from the environment?

(A) 1
(B) 2
(C) 3
(D) 4

33. Two sealed cylinders holding different gases are placed one on top of the other so heat can flow between them. Cylinder A is filled with hydrogen. Cylinder B is filled with helium moving with an average speed that is half that of the hydrogen atoms. Helium atoms have four times the mass of hydrogen atoms. Which of the following best describes the transfer of heat between the two containers by conduction?

(A) Net heat flows from cylinder A to cylinder B, because heat flows from higher kinetic energy atoms to lower kinetic energy atoms.
(B) Net heat flows from cylinder B to cylinder A, because heat flows from higher kinetic energy atoms to lower kinetic energy atoms.
(C) There is no net heat transfer between the two cylinders, because both gases have the same average atomic kinetic energy.
(D) There is no net heat transfer between the two cylinders, because heat conduction requires the movement of atoms between the cylinder, and the cylinders are sealed.

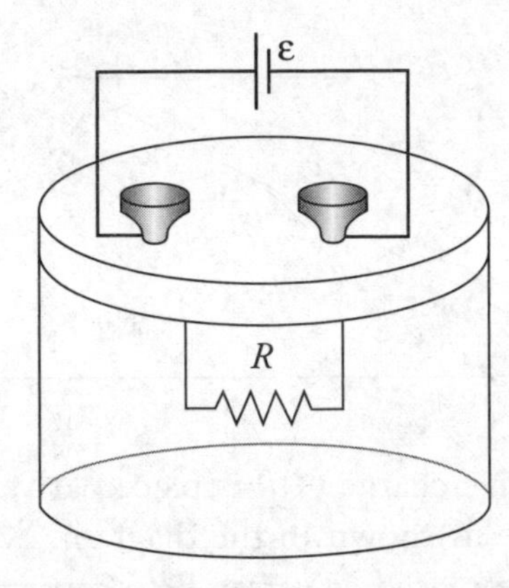

34. A resistor of resistance (R) is sealed in a closed container with n moles of gas inside. A battery of emf (ε) is connected to the resistor. Which of the following graphs shows the correct relationship between the gas atoms' average velocity (v_{avg}) and electrical energy (E) supplied to the resistor?

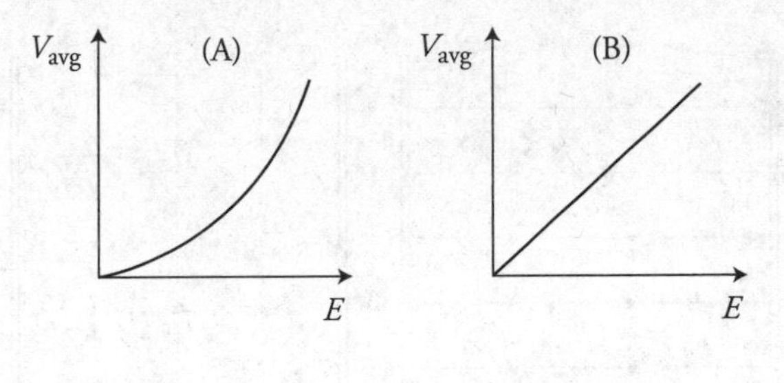

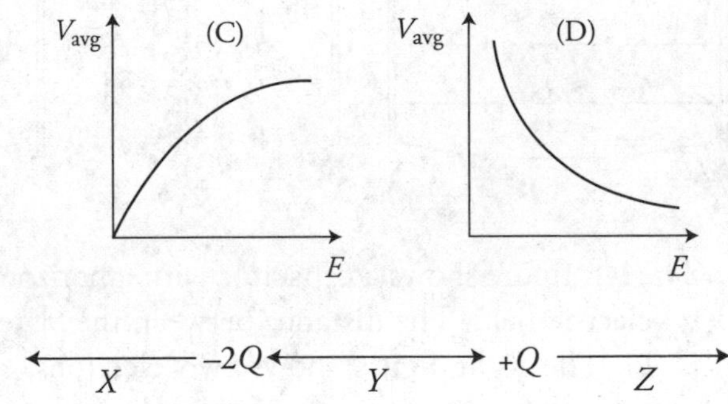

35. Two charges ($-2Q$ and $+Q$) are located as shown in the figure. Three regions are designated in the figure: X is to the left of $-2Q$; Y is between the two charges; and Z is to the right of $+Q$. Which of the following correctly ranks the magnitude of electric field in the three regions?

(A) $E_X > E_Y > E_Z$
(B) $E_Y > E_X > E_Z$
(C) $E_Y > E_X = E_Z$
(D) It is not possible to rank the magnitudes of the electric fields without more information.

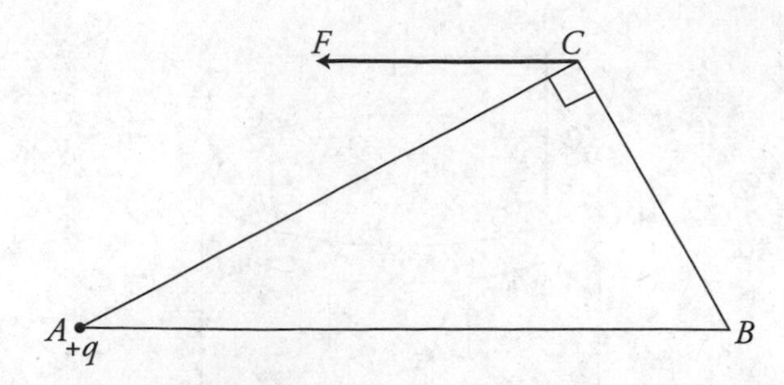

36. A positive charge ($+q$) is placed at vertex A of a triangle, as shown in the diagram. What charge must be placed at vertex B to cause an electron placed at vertex C to receive a force as shown?

(A) Positive and smaller than $|+q|$
(B) Positive and larger than $|+q|$
(C) Negative and smaller than $|+q|$
(D) Negative and larger than $|+q|$

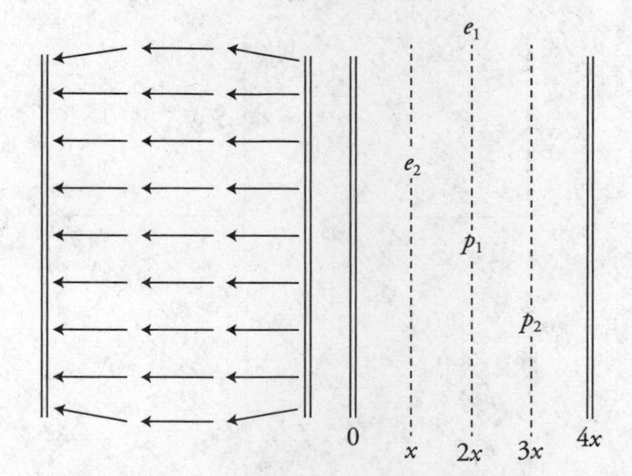

37. The left figure shows a capacitor with a horizontal electric field. The distance between the plates is $4x$. The right figure shows two electrons, e_1 and e_2, and two protons, p_1 and p_2, which are placed between the plates at the locations shown. Which of the following is a correct statement about the forces on the charges?

(A) The forces on e_1 and e_2 are not the same in magnitude but are the same in direction.
(B) All four particles receive the same magnitude of force but not all in the same direction.
(C) The force on p_1 is the largest in magnitude because it is in the middle of the capacitor where the electric field is strongest.
(D) The forces on e_2 and p_2 are the largest in magnitude because they are closer to the charged plates.

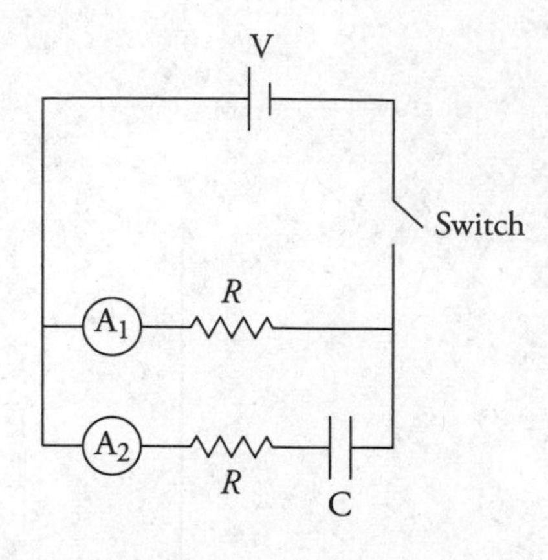

38. The circuit shown in the figure has two resistors, an uncharged capacitor, a battery, two ammeters, and a switch initially in the open position. What will happen to the current measured in the ammeters from the instant the switch is closed to a long time after the switch is closed?

	Ammeter 1	Ammeter 2
(A)	Reading remains constant	Reading remains constant
(B)	Reading remains constant	Reading will change
(C)	Reading will change	Reading remains constant
(D)	Reading will change	Reading will change

GO ON TO THE NEXT PAGE

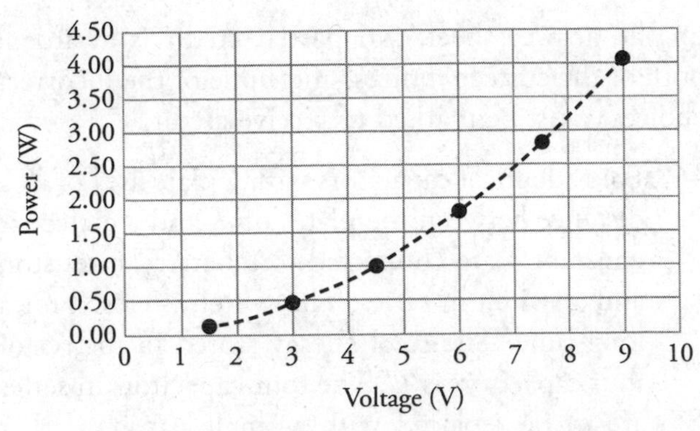

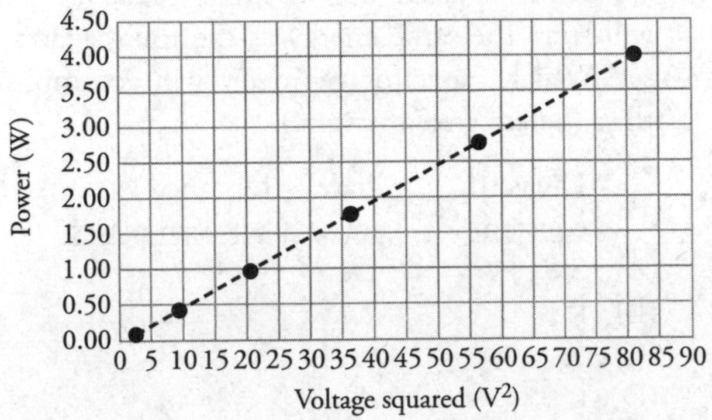

39. A single resistor is connected to a voltage source that consists of batteries with the same voltage connected in series. The power dissipated by the resistor for various voltages is shown in the two graphs. What is the resistance of the resistor?

(A) 0.05 Ω
(B) 0.22 Ω
(C) 4.5 Ω
(D) 20 Ω

40. Compasses are arranged in a tight circle around a long wire that is perpendicular to the plane of the compasses. The wire is represented in the figures by a dot. The wire carries a large current directly into the page. Which of the following best depicts the orientation of the compass needles?

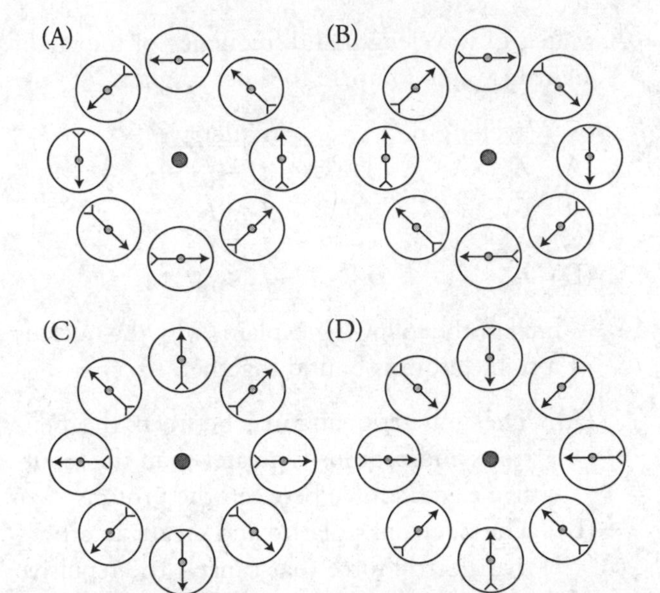

41. A mirror produces an upright image one-half the height of the object when the object is 12 cm from the mirror's surface. What is the focal length of the mirror?

(A) −12 cm
(B) −4 cm
(C) 4 cm
(D) 6 cm

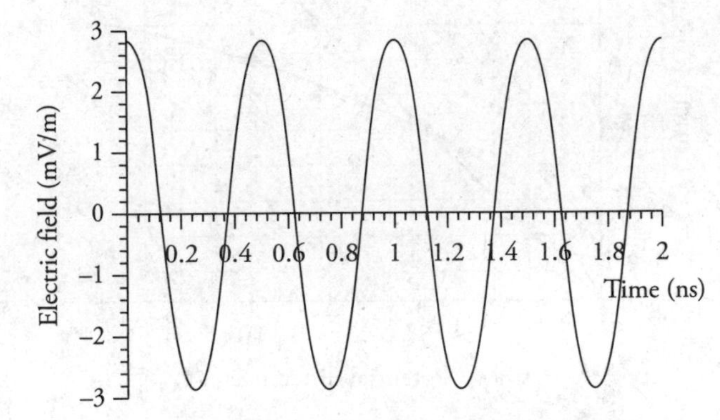

42. Which of the following best represents the electric field (E) measured in mV/m as a function of time measured in nanoseconds (ns)?

(A) $E = 6\cos(25.1t)$
(B) $E = 6\cos(12.6t)$
(C) $E = 3\cos(12.6t)$
(D) $E = 3\cos(25.1t)$

43. A light ray with a wavelength of λ_w and a frequency of f_w in water ($n = 1.33$) is incident on glass ($n = 1.61$). In the glass, the wavelength and frequency of the light is λ_g and f_g. How do the values of wavelength and frequency of the ray of light in water compare to those in glass?

	Wavelength	Frequency
(A)	$\lambda_w > \lambda_g$	$f_w = f_g$
(B)	$\lambda_w > \lambda_g$	$f_w > f_g$
(C)	$\lambda_w < \lambda_g$	$f_w = f_g$
(D)	$\lambda_w < \lambda_g$	$f_w < f_g$

44. Which of the following explains why the nucleus of a stable atom is bound together?

(A) The gravitational force between the neutrons and protons is greater than the repulsive electric force between the protons.
(B) The neutrons polarize and create an attractive electric force that cancels the repulsive electrostatic force of the protons.
(C) The orbit of electrons creates a magnetic force on the protons that is greater than the repulsive electric force.
(D) The strong force between nucleons is greater than the repulsive electric force of the protons.

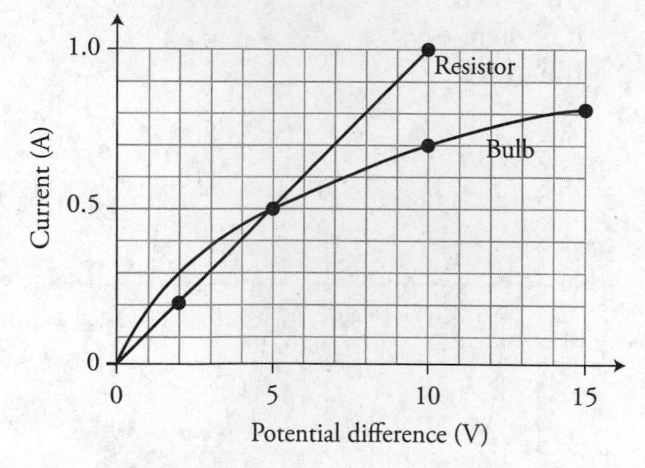

45. The figure shows current as a function of electric potential difference for a resistor and bulb. Are the devices ohmic?

	Resistor	Bulb
(A)	Ohmic	Ohmic
(B)	Ohmic	Non-ohmic
(C)	Non-ohmic	Ohmic
(D)	Non-ohmic	Non-ohmic

Questions 46–50: Multiple-Correct Items

Directions: Identify exactly two of the four answer choices as correct, and mark the answers with a pencil on the answer sheet. No partial credit is awarded; both of the correct choices, and none of the incorrect choices, must be marked to receive credit.

46. Four identical capacitors with a plate area of A, a distance between the plates of d, and a dielectric constant k are connected to a battery, a resistor, and a switch in series. The switch is closed for a long time. The total energy stored in the set of four capacitors is U. The four capacitors in series are to be replaced with a single capacitor that will store the same energy as the four-capacitor set. Which capacitor geometry will accomplish this? **(Select two answers.)**

	Dielectric constant	Plate area	Distance between plates
(A)	2κ	$2A$	d
(B)	κ	$2A$	$2d$
(C)	½ κ	½ A	d
(D)	κ	A	$4d$

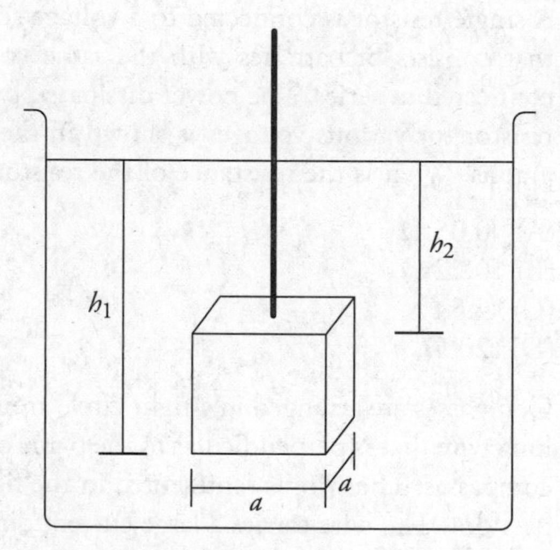

47. A mass (m) is suspended in a fluid of density (ρ) by a thin string, as shown in the figure. The tension in the string is T. Which of the following is an appropriate equation for the buoyancy force? **(Select two answers.)**

(A) $F_b = mg$
(B) $F_b = mg - T$
(C) $F_b = a^2 \rho g h_1$
(D) $F_b = a^2 \rho g (h_1 - h_2)$

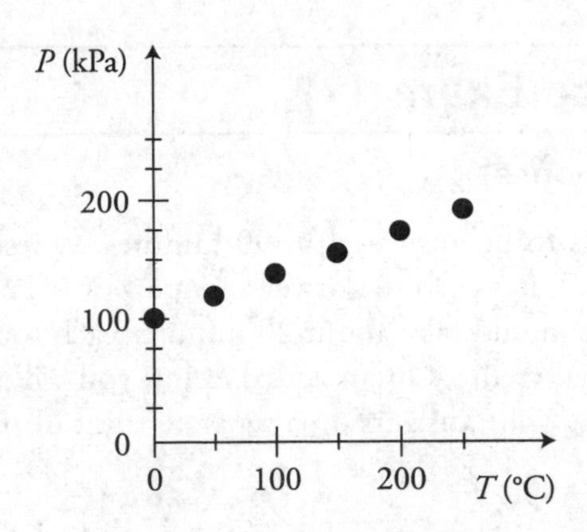

48. In an experiment, a sealed container with a volume of 100 ml is filled with hydrogen gas. The container is heated to a variety of temperatures, and the pressure is measured. The data from the experiment is plotted in the figure. Which of the following methods can be used to determine additional information regarding the gas? **(Select two answers.)**

(A) The slope can be used to calculate the number of atoms in the gas.
(B) The area under the graph can be used to calculate the work done by the gas.
(C) The vertical axis can be used to calculate the force the gas exerts on the container.
(D) The x-intercept can be used to estimate the value of absolute zero.

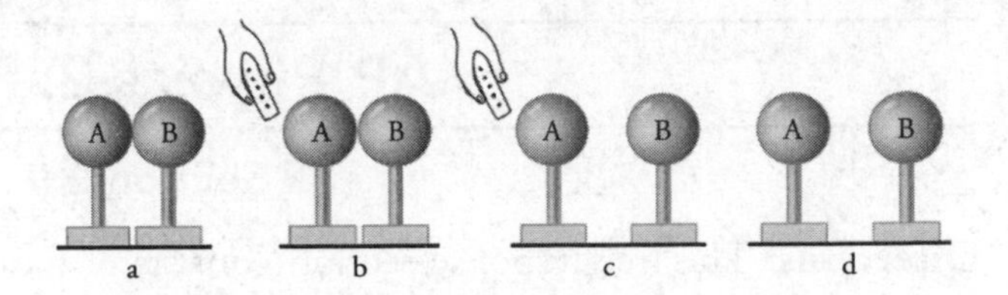

49. Two neutral metal spheres on insulating stands are placed so they touch, as shown in figure a. A positive rod is brought close to sphere A, as shown in figure b. Sphere B is moved to the right, as shown in figure c. The positive rod is then removed, as shown in figure d. Which of the following correctly describes the situation after the rod is removed? **(Select two answers.)**

(A) The net charge of the system that includes both spheres remains neutral.
(B) The net charge of sphere B is negative.
(C) Spheres A and B attract each other.
(D) The electric field between the spheres points to the right.

50. In each of the answer choices below, either a proton or an electron is moving toward the top of the page through either an electric or a magnetic field. In which case does the charged particle experience a force to the right? **(Select two answers.)**

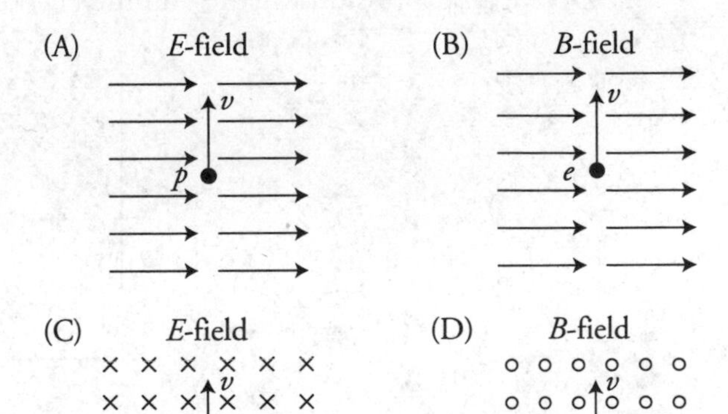

STOP: End of AP Physics 2 Practice Exam, Section 1 (Multiple-Choice)

AP Physics 2: Practice Exam 1

Section 2 (Free Response)

Directions: The free-response section consists of four questions to be answered in 90 minutes. Questions 1 and 4 are longer free-response questions that require about 25 minutes each to answer and are worth 12 points each. Questions 2 and 3 are shorter free-response questions that should take about 20 minutes each to answer and are worth 10 points each. Show all your work to earn partial credit. On an actual exam, you will answer the questions in the space provided. For this practice exam, write your answers on a separate sheet of paper.

1. (12 points—suggested time 25 minutes)

An air bubble is formed at the bottom of a swimming pool and then released. The air bubble ascends toward the surface of the pool.

(A) In a clear, coherent, paragraph-length response, describe any changes in the bubble size and describe the motion of the bubble as it ascends to the surface. Explain the factors that affect the size of the bubble and the bubble's motion. Include a description of any forces acting on the bubble from the time it is at the bottom of the pool until the bubble is just below the surface of the pool.

(B) On the figure, draw a vector for each force acting on the bubble. Make sure all vectors are drawn in correct proportion to each other.

(C) The bubble does not collapse under the pressure of the water. Explain how the behavior of the gas atoms keeps the bubble from collapsing.

(D) The bubble begins at a depth of D below the surface of the water where the bubble has an initial volume of V_D. The atmospheric pressure at the surface of the pool is P_S. The density of the water in the pool is ρ. Assume that the air temperature in the bubble remains constant as it rises to the surface. Derive an expression for the volume (V_S) of the bubble when it reaches the surface of the pool.

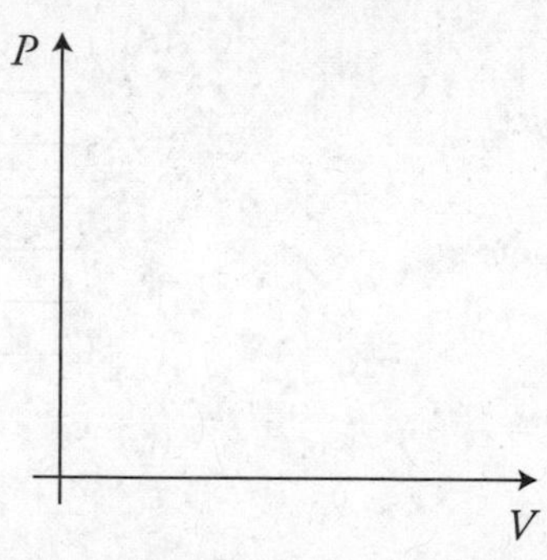

(E) In part (D) it was assumed that the temperature of the bubble remains constant. Now assume that the air temperature in the bubble can change but that the bubble rises so quickly to the surface that there is negligible thermal energy transfer between the bubble and the swimming pool water. Base your answers on this assumption.

i. Sketch the process on the PV diagram. Indicate on the axis the initial and final pressures and volumes.

ii. How does the value $P_S V_S$ compare to the value $P_D V_D$?

__Greater than $P_D V_D$ __Equal to $P_D V_D$ __Less than $P_D V_D$

Justify your answer.

2. (10 points—suggested time 20 minutes)

Some students are investigating how the geometry of the cylindrical shaft of graphite in a wooden pencil influences the resistance of the graphite. The students use a 9 V battery as an emf source. In the first part of the investigation, the students choose to investigate the influence of length on the resistance of the graphite conductive pathway.

(A) i. Besides the graphite and battery, what additional equipment would you need to gather the data needed to determine the influence of length on the resistance of the graphite conductive pathway?

ii. Using standard symbols for circuit elements, draw a schematic diagram of the circuit the students could use to determine the influence of length on the resistance of the conductive pathway. Include the appropriate locations and electrical connection of all equipment including any measuring devices. Clearly label your diagram.

iii. Describe the procedure you would use with your circuit to gather enough data to determine the influence of length on the resistance of the conductive pathway. Make sure your procedure is detailed enough that another student could perform the experiment.

iv. The 9 V battery used in the experiment has a sizable internal resistance. Would you need to change your procedure in part iii? Justify your answer.

Next the students investigate how the geometry of Play-Doh influences the resistance of cylindrical lengths of Play-Doh used as a conductive pathway. The investigation results in the data in the table.

Trial	Diameter (m)	Length (m)	Current (A)	Voltage across Play-Doh (V)
1	0.002	0.1	0.003	9.0
2	0.002	0.2	0.001	9.0
3	0.002	0.3	0.001	9.0
4	0.002	0.4	0.001	9.0
5	0.002	0.5	0.001	9.0
6	0.003	0.1	0.006	9.0
7	0.003	0.2	0.003	9.0
8	0.003	0.3	0.002	9.0
9	0.004	0.1	0.011	9.0
10	0.004	0.2	0.006	9.0
11	0.006	0.1	0.025	8.9
12	0.008	0.1	0.045	8.9
13	0.010	0.1	0.069	8.8

GO ON TO THE NEXT PAGE

(B) i. Which subset of data would be most useful in creating a graph to determine the relationship between the resistance and diameter of the Play-Doh? If the data chosen are incomplete, fill in the needed data in the extra columns provided in the table.

ii. Plot the subset of data you chose on the axis, being sure to label the axis. Draw a line or curve that best represents the relationship between the variables.

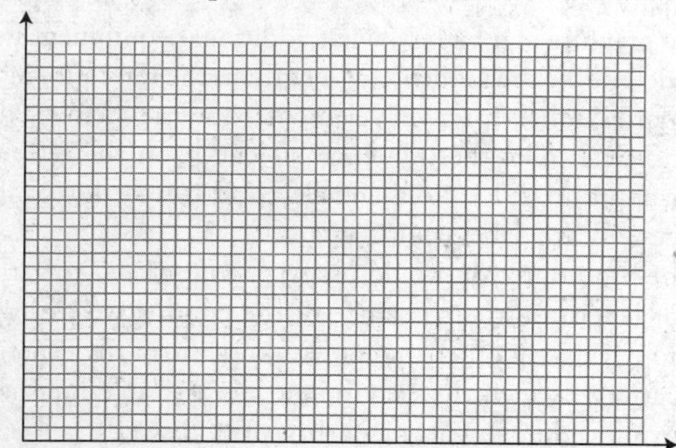

iii. What can you conclude from your line or curve about the relationship between the resistance and diameter of the conductive pathway?

iv. How can you prove that the relationship you suspect between resistance and diameter is correct?

v. The students who produced this data set said, "It took a long time to measure these data, and the Play-Doh was noticeably drier by the end of the lab." Will this influence the validity of the relationship you concluded in part iii?

Justify your answer.

3. (10 points—suggested time 20 minutes)

A gas with a ground state of –8.0 eV is illuminated by a broad ultraviolet spectrum of light and is found to absorb 248 nm of light. When the ultraviolet light is turned off, the gas sample emits three different wavelengths of light: 248 nm, 400 nm, and 650 nm.

(A) On the axis provided, construct and label an energy level diagram that displays the process of both the absorption and emissions by the gas. Show your supporting calculations below.

(B) The light emitted by the gas is directed toward a sample of tin. It is found that the 248-nm emission from the gas causes electrons to be ejected from the tin, but that the 400-nm emission does not. Will the 650-nm emission eject electrons from the tin? If so, explain how it could be accomplished. If not, explain why it is not possible.

(C) The light from the gas is now directed at a sample of potassium that subsequently ejects electrons with a maximum energy of 2.71 eV.

i. Calculate the de Broglie wavelength of the maximum energy electrons.

ii. These electrons are directed at two small openings spaced 2 nm apart. Will this result in the formation of an interference pattern? Justify your answer using appropriate physics principles.

4. (12 points—suggested time 25 minutes)

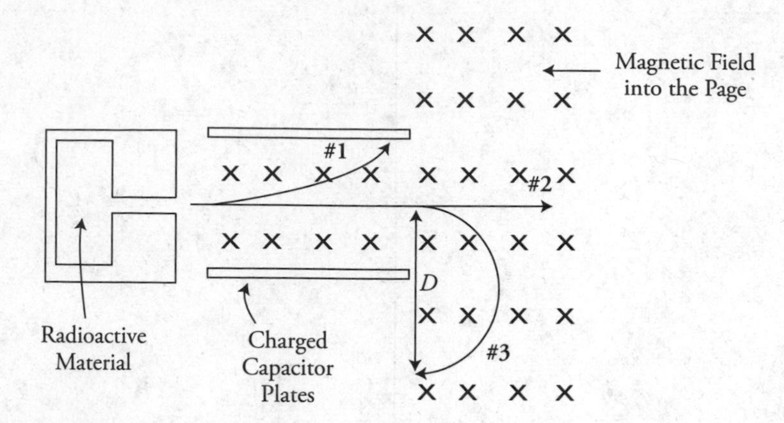

A sample of radioactive sources is enclosed within a lead-shielded container with a narrow exit aperture that ensures that any ejected particles will exit the container directly to the right. The ejected particles pass between charged parallel conductive plates and a region of magnetic field that is directed into the page. Three particles exit the container and follow paths as shown in the figure. Students observing the particles make these statements:

Student A: Both particles 2 and 3 pass through the capacitor region undeflected so there is no net force on them. They both must be neutral particles.

Student B: The path of particle 3 implies it has a negative charge. Therefore, the bottom plate of the capacitor must be negatively charged.

Student C: Particle 1 is positive because it curves upward.

(A) List all parts of the students' statements that are correct. Explain your reasoning for each.

(B) List all parts of the students' statements that are incorrect. Explain your reasoning for each.

(C) On the figure, sketch the electric field vectors between the capacitor plates that are consistent with the motion of the particles.

(D) Particle 3 is detected at distance D from its exit point from the capacitor plates. Using this information, derive an expression for the charge to mass ratio (q/m) of particle 3 in terms of D, E (the electric field between the capacitor plates), and B (the magnetic field).

(E) Using the equation derived in (D), a charge to mass ratio (q/m) of 1.76×10^{11} C/kg was calculated. Explain how this number could be used to determine the set of possible masses of particle 3.

(F) One of the radioactive sources enclosed in the shielded container is fermium-257, which has 100 protons and a half-life of 100.5 days. Fermium transmutes into Californium (Cf) by emitting an alpha particle with a velocity of 2.0×10^7 m/s.

(i) Write the complete nuclear equation of this decay reaction.

(ii) Write, but do not solve, a symbolic expression that could be used to calculate the mass released to energy in one single fermium–257 decay.

(iii) Assuming the fermium–257 is isolated and stationary, calculate the velocity of the Californium nucleus after the alpha particle is ejected.

STOP: End of AP Physics 2 Practice Exam, Section 2 (Free Response)

Solutions: Section 1 (Multiple Choice)

Questions 1–45: Single-Choice Items

1. **B**—The mass of argon is greater. Thus, at the same temperature, argon has the same kinetic energy but a slower average speed. This shifts the curve to the left. Since the number of moles is the same, the peak of the argon graph must be higher to accommodate the same number of atoms.

2. **D**—The exit velocity is proportional to the water pressure at the hole, which is proportional to the depth of the water: $P = P_0 + \rho gh$.

3. **D**—The point source model shows us that the larger the wavelength, the greater the bending of the wave around the corner. We can also see this in the equation for diffraction: $\sin\theta = \frac{m\lambda}{d}$. As the wavelength gets smaller, the diffraction bending angle also gets smaller. Visible light has a very small wavelength and only bends a tiny amount around corners.

4. **C**—Only electrons can move in the metal spheres. Protons (positive charges) are stuck in the nuclei of atoms and cannot move. In the presence of the positively charged rod, the three-sphere system will polarize with excess electrons moving toward the positively charged rod. Answer choice A shows no charge polarization. Choice B shows each individual sphere polarized, which cannot happen since the conductors are in contact and act as one system. Choice D shows protons moving, which also cannot happen.

5. **D**—Electric field lines are perpendicular to the isolines and point from higher potential to lower potential. The average electric field strength is $E = \frac{\Delta V}{\Delta r}$. Therefore, the electric field at point A will be stronger than at point B.

6. **C**—The force from sphere 1 on sphere 3 is ½ F to the left. The force from sphere 2 on sphere 3 is 2 F to the left. The sum is (5/2) F.

7. **B**—The current is the same; therefore, the higher the resistance, the greater the electric potential. Use the resistance equation to find the relative magnitude of the three resistors: $R = \frac{\rho L}{A}$.

8. **D**—Originally only bulb A is lit, and it experiences all the emf of the battery. When the switch is opened, the emf of the battery is split evenly between the bulbs. $P = \frac{V^2}{R}$; therefore, the power dissipation by bulb A is one-quarter of the original value.

9. **C**—Special relativity tells us that when two observers are moving relative to each other, they will not necessarily agree on length and time. This becomes evident when we get up near the speed of light. We start to easily notice the effect around 0.1 c and faster. The only constant that all observers will agree on is the speed of light.

10. **C**—The original velocity is toward the top of the page. The force on the proton is out of the page. Therefore, by the right-hand rule, the magnetic field is directed to the left.

11. **B**—Newton's third law.

12. **B**—The three resistors in the bottom right corner are in parallel. Since they all have the same resistance and the same electric potential across them, they must also have the same current. Ammeter #3 is the sum of the currents in the top two resistors and will be twice as large as ammeter #4.

13. **B**—The three resistors in parallel add up to a resistance of ⅓ R. Adding these in series with the resistor in the main line, we get 4/3 R.

14. **D**—The data in the table suggest that gas pressure and volume are inversely related: $P \propto \frac{1}{V}$. This is also seen in the Ideal Gas model: $P = \frac{nRT}{V}$. Therefore, plotting P on the vertical axis and $1/V$ on the horizontal axis will produce a straight line from which the number of moles could be calculated knowing that the slope will equal nRT.

15. C—The ranking will be based on the $P \times V$ value: $T_C > T_B = T_D > T_A$

16. B—Both paths start and end at the same point. Therefore, the initial and final temperatures are the same, as are the initial and final thermal energies. Process 1 has a higher average pressure for the same volume change. Another way to think about it is to compare the area under the curves. Graph 1 has more area underneath and, therefore, a larger magnitude of work.

17. B—The weight of the blocks is balanced by the buoyancy force. Since the blocks are half submerged, their densities are half that of water:

$$F_g = F_b$$
$$(mg)_{\text{block}} = (\rho V g)_{\text{water}}$$
$$(\rho V g)_{\text{block}} = (\rho V g)_{\text{water}}$$
$$(\rho V)_{\text{block}} = \left(\rho\left(\frac{1}{2}V_{\text{block}}\right)\right)_{\text{water}}$$
$$\rho_{\text{block}} = \frac{1}{2}\rho_{\text{water}}$$

18. D—Using the conservation of mass/continuity equation, we see that the water must be slower at the hydrant than exiting the nozzle:

$$A_1 v_1 = A_2 v_2$$

The area of the hose is proportional to the diameter squared:

$$\pi r_1^2 v_1 = \pi r_2^2 v_2$$
$$\pi\left(\frac{8\text{cm}}{2}\right)^2 v_1 = \pi\left(\frac{2\text{cm}}{2}\right)^2 18\frac{\text{m}}{\text{s}}$$

This gives us a velocity in the hose of 1.125 m/s.

Using conservation of energy/Bernoulli's equation and assuming the exit pressure is atmospheric,

$$\left(P + \rho g y + \frac{1}{2}\rho v^2\right)_1 = \left(P + \rho g y + \frac{1}{2}\rho v^2\right)_2$$
$$\left(P + 0 + \frac{1}{2}(1{,}000\ \text{kg/m}^3)(1.125\ \text{m/s})^2\right)_1$$
$$= \left(100{,}000\ \text{Pa} + (1{,}000\ \text{kg/m}^3)(10\ \text{m/s}^2)(6\ \text{m}) + \frac{1}{2}(1{,}000\ \text{kg/m}^3)(18\ \text{m/s})^2\right)_2$$

19. A—Gravity is proportional to the product of the two masses, and the electric force is proportional to the product of the two charges. Since the mass of an electron is on the order of 10^{-31} kg, and the charge is on the order of 10^{-19} C, the electric force will be much larger than the gravitational force. In addition, the universal gravitational constant is much smaller than the Coulomb's law constant. This makes the gravitational force between the electrons negligible compared to the electric force.

20. B—The charge of Psevdísium is smaller than the electron charge, which calls this particle's existence into doubt. Within the level of uncertainly listed, the charge of Alithísium is ten times the charge of the electron. We would expect the charge to be a whole integer multiple of the electron charge. The mass of Alithísium is listed as an energy equivalent. This is perfectly acceptable: $E = mc^2$.

21. C—Electric field vectors are perpendicular to the equipotential lines. The pattern of the electric field vectors indicates that both charges are the same sign. Additionally, there is no zero potential line separating the two charges, indicating that they have the same sign.

22. D—It is not possible to separate a north pole from a south pole. All magnets are dipoles. When you break a magnet in half, you get two weaker magnets. If they stayed the same magnitude as the original, we would be violating conservation of energy.

23. C—By the right-hand rule, the magnetic force on the proton is to the left. Originally, the electric and magnetic forces were equal. Since the velocity has increased, the magnetic force is now larger than the electric force that is to the right.

24. D—The diaphragm vibrates back and forth along the axis of the magnet, changing the magnetic field strength through the coil area.

25. C—The law of reflection is not influenced by the water. Snell's law of refraction depends on the indices of refraction of the two materials. The speed of light changes less going from water to lens then going from air to lens. This means there will be less refraction in water, making the focal length larger.

26. B—Electromagnetic waves are transverse, with both the E- and B-fields oscillating perpendicular to the direction of motion and each other.

27. D—The lens equation can be rearranged to produce a straight line:

$$\frac{1}{s_i} = (-1)\frac{1}{s_o} + \frac{1}{f}$$

$$y = mx + b$$

Thus, if we plot $1/s_o$ on the x-axis and $1/s_i$ on the y-axis, we should get a graph with a slope of −1 and an intercept of $1/f$. The image distance is $x_2 - x_1$, and the object distance is $x_0 - x_1$.

28. B—When the angle of the prism decreases, the right side of the prism becomes more vertical, and the angle of incidence with the normal becomes smaller. This creates less refraction and the distance (x) increases. Look at the extreme case when θ becomes zero. Then the angle of incidence is zero, and there is no refraction at all. The beam will pass straight through the "prism" because it has become flat like a window, and the distance (x) becomes infinite.

29. C—The particle is most likely to be found at the highest positive/negative amplitude location of Ψ as a function of x. The particle will not be found at locations −15 nm and 0.0 nm.

30. C

$$^{237}_{93}\text{Np} \rightarrow\ ^{205}_{81}\text{Tl} + 4\,^{0}_{-1}\beta +\ ^{A}_{Z}\alpha.$$

Solve for A and Z. $A = 32$ and $Z = 16$. Knowing that alpha particles have two protons and four nucleons, we can divide 32 by 4 or divide 16 by 2 to find out that we need 8 alpha particles to balance the equation.

31. B—When the metal block is lifted out of the water, the buoyancy force between the block and the water disappears. This makes the spring scale reading go up by 70 g and the balance reading go down by 70 g. Note that mass of the block is 190 g. Therefore, the spring scale reading must be 190 g when the block is lifted out of the water!

32. C—In process 3, the work done is zero: $W = 0$. The temperature and the internal kinetic energy of the gas are increasing. Therefore, thermal energy must be entering the gas: $\Delta U = Q$.

33. C

$$K_{\text{average of the gas}} = \frac{3}{2}k_B T \text{ and } K = \frac{1}{2}mv^2$$

The helium is moving at half the speed and has four times the mass of hydrogen. This means they both have the same average kinetic energies and the same temperatures:

$$K_{\text{helium}} = \frac{1}{2}(4m_{\text{hydrogen}})\left(\frac{1}{2}v_{\text{hydrogen}}\right)^2$$

$$= \frac{1}{2}m_{\text{hydrogen}}v^2_{\text{hydrogen}} = K_{\text{hydrogen}}$$

Objects with the same temperature are in thermal equilibrium and do not transfer any net thermal energy between them.

34. C—Adding heat to the gas from the resistor will increase the kinetic energy and temperature of the gas. The kinetic energy of the gas is proportional to the velocity squared:

$$E_{\text{added to gas from resistor}} = \Delta K_{\text{average of the gas}} = \frac{1}{2}mv^2$$

Thus, the average velocity of the gas is proportional to the square root of the energy added to the gas (E).

35. D—Without knowing the exact locations, it is impossible to know the exact electric field strength. Since electric field is proportional to $\frac{1}{r^2}$, the electric field strength varies in strength in the three regions.

36. C—For the electron to receive a force to the left, the electric field must be pointing to the right. The electric field from the charge at vertex B must be smaller in magnitude than the field produced by the charge at vertex A. To accomplish this, the charge at point B must be negative and smaller than $|+q|$.

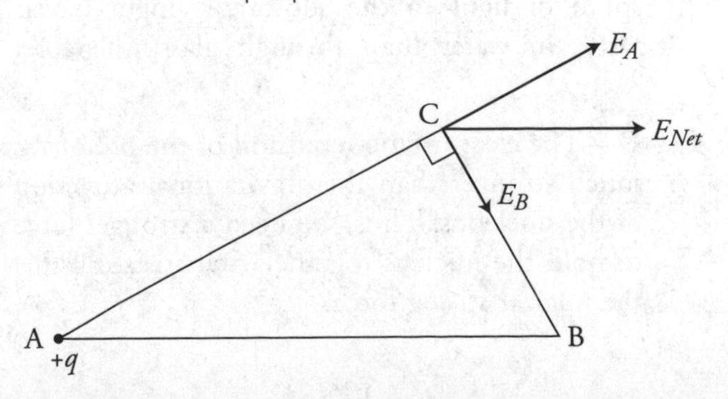

37. A—Near the edges of the capacitor, the electric field is still pointing in the same direction but is weaker than between the two plates.

38. B—The reading in ammeter 1 never changes because the potential difference and resistance in that line does not change. The capacitor behaves like a wire when uncharged and like an open switch when fully charged. This behavior changes the current through ammeter 2 as time passes.

39. D—$P = \frac{V^2}{R}$. The slope of the right graph will equal the reciprocal of the resistance.

40. B—Use the right-hand rule for magnetic fields around current-carrying wires.

41. A—The image is upright and smaller. This means the image is virtual, the image distance is negative, and the mirror must be diverging/convex.

$$M = \frac{1}{2} = \frac{-s_i}{s_o} = \frac{-s_i}{12\text{ cm}}$$

$$s_i = -6\text{ cm}$$

$$\frac{1}{f} = \frac{1}{s_i} + \frac{1}{s_o}$$

$$\frac{1}{f} = \frac{1}{-6\text{ cm}} + \frac{1}{12\text{ cm}}$$

$$f = -12\text{ cm}$$

42. C—The amplitude is 3 mV/m, and the time period is 0.5 ns. Using the wave equation, we get

$$E = A\cos\left(\frac{2\pi t}{T}\right) = 3\cos\left(\frac{2\pi t}{0.5}\right) = 3\cos(4\pi t)$$

$$= 3\cos(12.6t)$$

43. A—The frequency remains the same. The wavelength of light is directly proportional to the speed of light in the substance. Light travels faster in water than through glass; therefore, $\lambda_w > \lambda_g$.

44. D—The electrostatic repulsion of the protons is much stronger than the gravitational attraction of the nucleons. Thus, we need a stronger force to hold the nucleus together. Scientists call this the nuclear strong force.

45. B—Ohmic materials will have a constant ratio of current to potential difference. This shows up on a graph as a straight line.

Questions 46–50: Multiple-Correct Items
(You must indicate both correct answers; no partial credit is awarded.)

46. C and D—Four capacitors connected in series result in a capacitance one-quarter the size of the individual capacitors. $C = \frac{\kappa\varepsilon_0 A}{d}$. To get a single capacitor with one-quarter the capacitance, we need to get a factor of four in the denominator.

47. B and D—The sum of the forces must equal zero. Therefore, the buoyancy force upward must equal gravity downward minus the tension upward. Buoyancy force is $\rho V g$, and the volume of the block is $a^2(h_1 - h_2)$.

48. A and D—The Ideal Gas model ($PV = nRT$) shows us that gas pressure is directly related to gas temperature ($P \propto T$). This is seen in the straight line data represented in the graph. The slope of this graph will be equal to $\frac{nR}{V}$. Since the volume of the gas is given, the number of moles—and thus, the number of atoms—in the gas can be calculated using the slope. The *x*-intercept represents the temperature of the gas when the volume reaches zero. This point is absolute zero. The area of a *PV* diagram would represent work. We cannot calculate the force from the pressure because we do not know the surface area of the container. We are only given the volume.

49. A and C—Sphere A is negative, and sphere B is positive. The rod polarizes the system that consists of the two spheres, pulling excess electrons to the left and leaving the right side with an excess positive charge of equal magnitude.

50. A and D—Be careful! Make sure you are paying attention to which of these is a magnetic field and which is an electric field. Electric forces are along the axis of the field. E-fields push positive charges in the direction of the field. Negative charges are pushed in the opposite direction of the E-field. Magnetic forces abide by the right-hand rule. Only moving charges experience forces from B-fields.

Solutions: Section 2 (Free Response)

Your answers will not be word-for-word identical to what is written in this key. Award points for your answer as long as it contains the correct physics explanation *and* as long as it does not contain incorrect physics or contradict the correct answer.

Question 1

Part (A) 5 points total for a well-constructed and easy to read explanation that contains the following:

1 point—The amount of air inside the bubble remains the same during ascent.

1 point—The external pressure from the water on the bubble decreases as the bubble rises.

1 point—The bubble increases in volume as it moves upward.

1 point—As the bubble volume increases, the buoyancy force increases, but the gravity force on the bubble remains constant.

1 point—The bubble accelerates upward at an increasing rate as it ascends.

Part (B)

1 point—For buoyancy force upward and gravity force downward. Buoyancy must be drawn larger than gravity. If extra vectors are drawn, this point is not awarded.

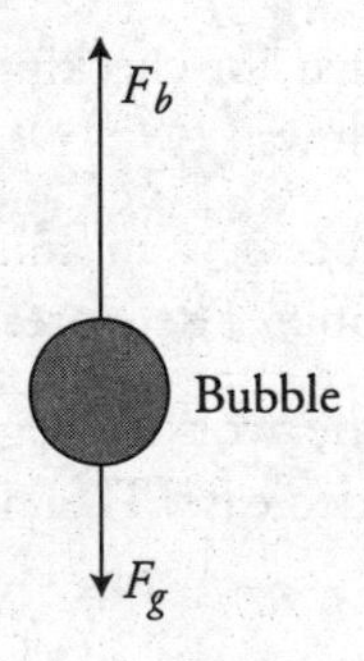

Part (C)

1 point—The atoms inside the gas collide with and bounce off the water molecules. This causes a change in momentum of the gas atoms. This generates an equal and opposite force between the water and gas that keeps the bubble from collapsing.

Part (D)

1 point—For the correct relationship between the pressure at the surface of the water and at a depth D: $P_D = P_S + \rho g D$.

1 point—For the correct application of the ideal gas law and the correct final expression:

$$P_S V_S = P_D V_D$$

$$P_S V_S = (P_S + \rho g D) V_D$$

$$V_S = \frac{(P_S + \rho g D) V_D}{P_S}$$

Part (E)

(i)

1 point—The sketch should have the shape of an adiabatic process, with an upward concave curve similar to that for an inverse relationship. The initial and ending points should be marked as seen in the figure. Pressure decreases and volume increases.

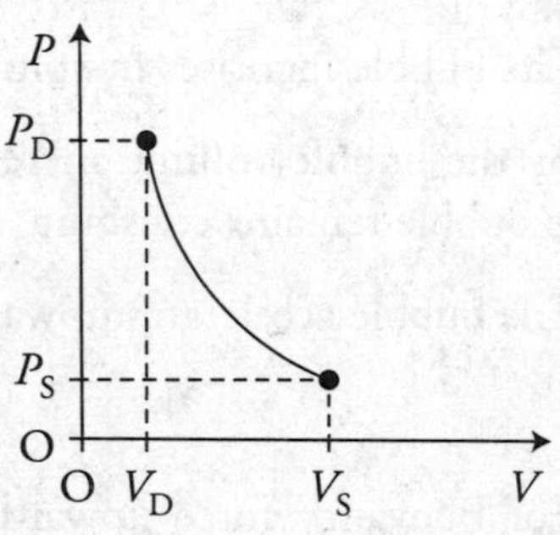

(ii)

No point is awarded for checking the blank "Less than $P_D V_D$." Points are awarded for the justification as follows:

1 point—The process is an adiabatic expansion. There is no heat transfer between the water and the bubble. Therefore: $Q = 0$, and $\Delta U = W$.

1 point—Since the work is negative as the bubble expands, the internal energy of the bubble must also decrease. Therefore, the temperature and the PV value must also decrease.

Question 2

Part (A)

(i)

1 point—Minimum equipment: wires, voltmeter, and ammeter. Note that the equipment that is listed must be used in the lab procedure.

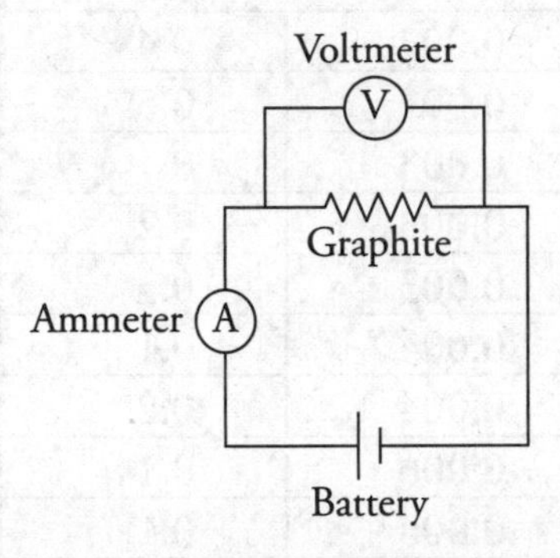

(ii)

1 point—There are several ways to draw this. The key is that the ammeter is in series with the battery and the graphite. The voltmeter must be in parallel with the graphite to measure the correct potential difference. The graphite should be drawn as a resistor and the entire schematic needs to be labeled.

(iii)

1 point—Procedure:

1. Connect the graphite, ammeter, and battery in series.
2. Connect the voltmeter in parallel with the graphite.
3. Measure the current and voltage.
4. Repeat for several lengths of graphite.

(iv)

1 point—Since the current and voltage are measured for the graphite specifically, the internal resistance of the emf source is irrelevant.

Part (B)

(i)

1 point—Trials: 1, 6, 9, 11, 12, and 13 are the best because they give the widest quantity and spread of data for a constant length segment of Play-Doh with differing diameters.

1 point—The resistance is missing and needs to be calculated. (See the table.)

Trial	Diameter (m)	Length (m)	Current (A)	Voltage across Play-Doh (V)	Resistance (Ω)
1	0.002	0.1	0.003	9.0	**3,200**
2	0.002	0.2	0.001	9.0	
3	0.002	0.3	0.001	9.0	
4	0.002	0.4	0.001	9.0	
5	0.002	0.5	0.001	9.0	
6	0.003	0.1	0.006	9.0	**1,400**
7	0.003	0.2	0.003	9.0	
8	0.003	0.3	0.002	9.0	
9	0.004	0.1	0.011	9.0	**800**
10	0.004	0.2	0.006	9.0	
11	0.006	0.1	0.025	8.9	**350**
12	0.008	0.1	0.045	8.9	**200**
13	0.010	0.1	0.069	8.8	**130**

(ii)

1 point—Data point must be correctly plotted. The axes must be correctly labeled. The graph should be a curve.

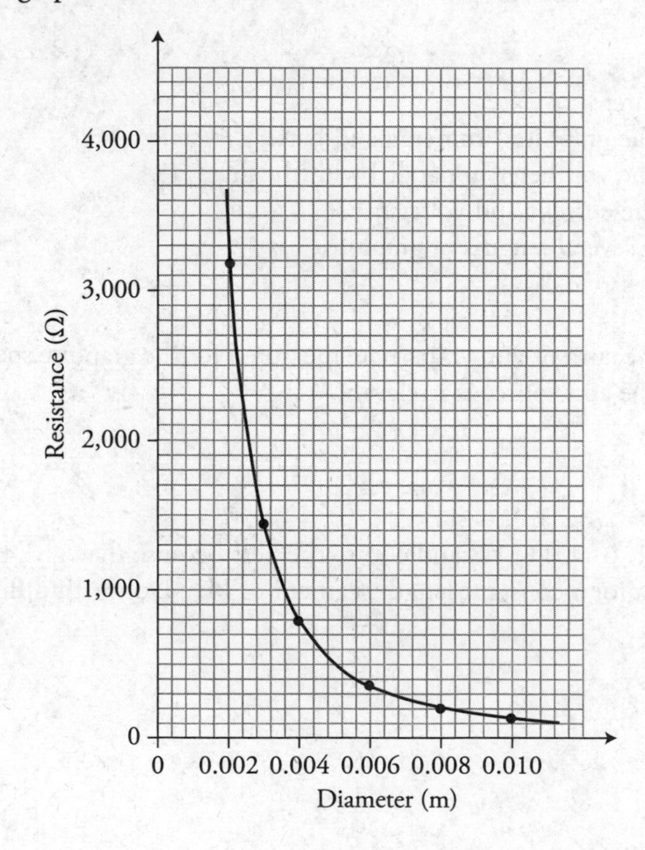

(iii)

1 point—For a statement that the resistance appears to be inversely related to the diameter (or the diameter squared).

(iv)

1 point—To determine if the relationship between resistance and diameter is inverse (or inverse squared), we need to graph $R - vs - 1/d$ (or $R - vs - 1/d^2$) and see if this graph produces a straight line.

(v)

1 point—The changing water content of Play-Doh could change the resistivity of the material and the resistance-diameter relationship, which damages the validity of the lab. It is a variable that is not held constant and calls the validity of the experiment into question.

Question 3

Part (A)

There are two ways to construct this energy level diagram. Both are shown below.

1 point—For converting the wavelengths of light in energy in electron volts.

$$E = hf = \frac{hc}{\lambda}$$

248 nm = 5 eV, 400 nm = 3.1 eV, 650 nm = 1.9 eV

1 point—For drawing energy levels at both −8 eV and −3 eV. (See figure below.)

1 point—For drawing an additional energy level at either −4.9 eV or −6.1 eV. (See figure below.)

1 point—For drawing an arrow upward from the −8 eV energy level to the −3 eV energy level to indicate the absorption of the 248 nm photon. (See figure below.)

1 point—For drawing three downward arrows as shown in the diagram below and indicating the emission of the 248 nm, 400 nm, and 650 nm photons.

Note: Only one of the diagrams below needs to be drawn by the student.

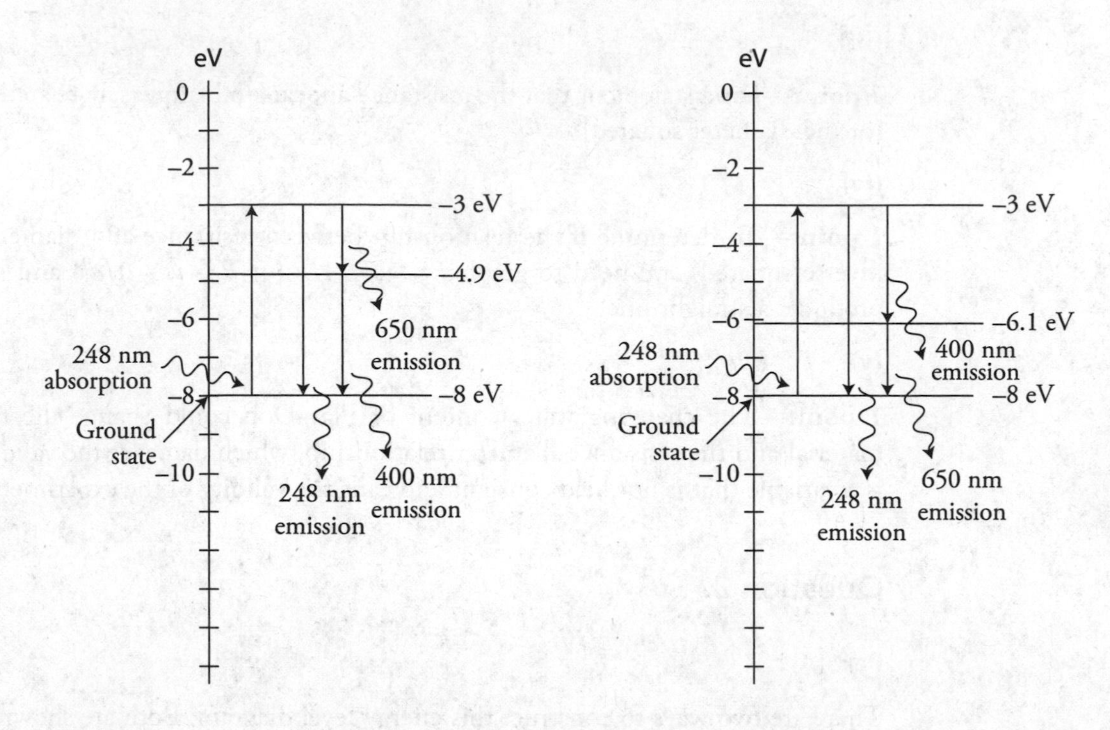

Part (B)

1 point—For indicating that the 400 nm photon is below the threshold energy (work function) needed to eject an electron.

1 point—For indicating that the 650 nm photon has even less energy than the 400 nm photon.

Part (C)

(i)

1 point—For converting the electron energy from eV into joules of kinetic energy and then into an electron velocity.

Electron kinetic energy = (2.71 eV)(1.6×10^{-19} J/eV) = 4.34×10^{-19} J

$$K = \frac{1}{2} m_e v^2$$

$$v = \sqrt{\frac{2K}{m_e}} = 9.76 \times 10^5 \text{ m/s}$$

1 point—For calculating the de Broglie wavelength.

$$\lambda = \frac{h}{p} = \frac{h}{m_e v} = 7.46 \times 10^{-10} \text{ m}$$

(ii)

1 point—For indicating that an interference pattern will form with appropriate justification. Here is an example explanation:

The electrons will form an interference pattern, as the waves are on the same order of magnitude and smaller than the opening spacing. The equation $\sin\theta = \frac{m\lambda}{d}$ shows us that the ratio $\frac{\lambda}{d}$ needs to be less than 1 for an interference pattern to form. Any larger ratio will cause the angle θ to be larger than 90 degrees, and no pattern will form. This means the wavelength must be smaller than the spacing, but not too small or the angle will be so small that the pattern will be too small to see.

Question 4

Part (A)

1 point—Student A is correct that particles 2 and 3 have no net force acting on them while between the charged plates. This is evident in the fact that they travel in a straight line through the charged plate.

AND

Student A is correct in stating that particle 2 is neutral as it travels in a straight line through the magnetic field. Knowing that a charged particle will experience a force while passing through a magnetic field, particle 2 must be uncharged.

1 point—Student B is correct in stating that particle 3 must have a charge, since a moving charged particle will experience a force when passing through a magnetic field. Student B is also correct in stating that the path implies a negative charge. According to the right-hand rule for a charged particle in a magnetic field, the direction of force on a moving charged particle will be perpendicular to both the magnetic field and the direction of positive charge motion. Using this right-hand rule, we find a circular path directed upward for a positive charge, thus the downward curve of particle 3 suggests a negative charge.

AND

Student B is correct in stating that the bottom plate must be negatively charged. Since particle 3 is negative as previously stated, the negative charge will experience a downward force from the magnetic field. In order to maintain a straight line and thus no acceleration, particle 3 must experience an upward force to balance the downward force. This can be achieved only if the bottom plate repels the negative charge upward, indicating that the bottom plate is negatively charged.

Part (B)

1 point—Student A is incorrect in stating that particle 3 is neutral as evidenced by the fact that it experiences a force exerted by the magnetic field and arcs downward as soon as it leaves the electric field between the charged plates.

1 point—Student C is incorrect in stating that particle 1 is positively charged. Although the direction of the magnetic force on a positive charge would be upward, the bottom plate is negative and the top plate is positive. Thus, the electric force on particle 1 is downward. It is not possible to tell from the given information what the charge of particle 1 is. We can only say for sure that particle 1 has a net charge.

Part (C)

1 point—For indicating that the electric field is pointing directly downward between the plates and uniform in strength.

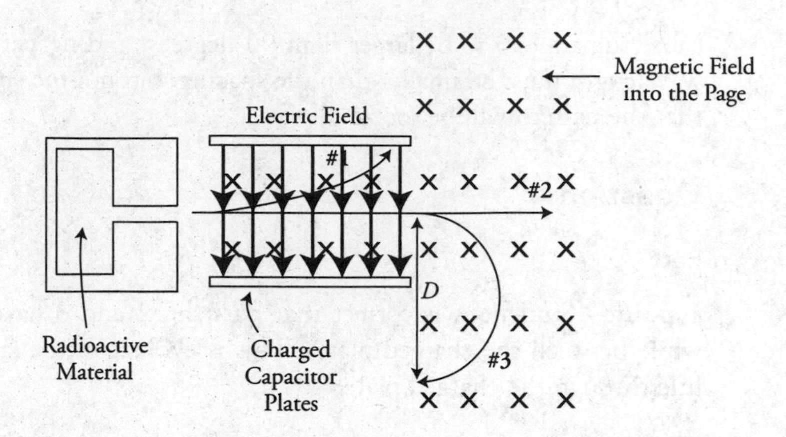

Part (D)

1 point—For finding the velocity of particle 3 using the magnetic and electric forces between the plates:

$$F_M = F_E$$

$$qvB = Eq$$

$$v = \frac{E}{B}$$

1 point—For setting the magnetic force equal to the centripetal force in the magnetic field:

$$F_B = m\frac{v^2}{r}$$

1 point—For the correct expression for charge to mass ratio:

$$F_B = m\frac{v^2}{r}$$

$$qvB = m\frac{v^2}{\left(\frac{D}{2}\right)}$$

$$qB = 2m\frac{v}{D}$$

$$\frac{q}{m} = \frac{2v}{BD}$$

Substituting in the equation for velocity from above:

$$\frac{q}{m} = \frac{2E}{B^2D}$$

Part (E)

1 point—For clearly explaining that only multiples of the electron charge can be used to find the possible masses of particle 3.

Example: The mass can be found by dividing the charge by the charge to mass ratio (q/m). However, we know that particles come only in multiples of the electron charge. So, dividing the multiple of the electron charge by the charge to mass ratio will give us a set of possible masses for particle 3.

Part (F)

1 point—For the correct nuclear equation showing Californium as the end product with the correct atomic number and atomic mass number:

$$^{257}_{100}\text{Fm} \rightarrow {}^{253}_{98}\text{Cf} + {}^{4}_{2}\alpha + \text{Energy}$$

1 point—For a correct expression of masses and utilizing $E = mc^2$:

$$m_{\text{Fm}} - m_{\text{Cf}} - m_{\alpha} = \Delta mc^2$$

$$\Delta m = \frac{m_{\text{Fm}} - m_{\text{Cf}} - m_{\alpha}}{c^2}$$

1 point—For a correct velocity of the Californium nucleus:

$$P_1 = P_2$$

$$0 = P_2$$

$$m_{\text{Cf}} v_{\text{Cf}} = m_{\alpha} v_{\alpha}$$

$$(253\text{u}) v_{\text{Cf}} = (4\text{u})(2 \times 10^7 \text{ m/s})$$

$$v_{\text{Cf}} = 316{,}000 \text{ m/s}$$

How to Score Practice Exam 1

The practice exam cut points are based on historical exam data and will give you a ballpark idea of where you stand. The bottom line is this: If you can achieve a 3, 4, or 5 on the practice exam, you are doing great and will be well prepared for the real exam in May. This is the curve I use with my own students, and it is has been a good predictor of their actual exam scores.

Multiple-Choice Score: Number Correct ____________ (50 points maximum)

Free-Response Score: Problem 1 ____________ (12 points maximum)

Problem 2 ____________ (10 points maximum)

Problem 3 ____________ (10 points maximum)

Problem 4 ____________ (12 points maximum)

Free-Response Total: ____________ (44 points maximum)

Calculating Your Final Score

Final Score = (1.136 × Free-Response Total) + (Multiple-Choice Score)

Final Score: ____________ (100 points maximum)

Round your final score to the nearest point.

Raw Score to AP Grade Conversion Chart

Final Score on Practice Exam	AP Grade	
70–100	**5**	Outstanding! Keep up the good work.
55–69	**4**	
40–54	**3**	Great job! Practice and review to secure a score of 3 or better.
25–39	**2**	Almost there . . . concentrate on your weak areas to improve your score.
0–24	**1**	Don't give up! Study hard and move up to a better grade.

AP Physics 2: Practice Exam 2

Multiple-Choice Questions

ANSWER SHEET

1 (A) (B) (C) (D)
2 (A) (B) (C) (D)
3 (A) (B) (C) (D)
4 (A) (B) (C) (D)
5 (A) (B) (C) (D)
6 (A) (B) (C) (D)
7 (A) (B) (C) (D)
8 (A) (B) (C) (D)
9 (A) (B) (C) (D)
10 (A) (B) (C) (D)
11 (A) (B) (C) (D)
12 (A) (B) (C) (D)
13 (A) (B) (C) (D)
14 (A) (B) (C) (D)
15 (A) (B) (C) (D)
16 (A) (B) (C) (D)
17 (A) (B) (C) (D)
18 (A) (B) (C) (D)
19 (A) (B) (C) (D)
20 (A) (B) (C) (D)
21 (A) (B) (C) (D)
22 (A) (B) (C) (D)
23 (A) (B) (C) (D)
24 (A) (B) (C) (D)
25 (A) (B) (C) (D)
26 (A) (B) (C) (D)
27 (A) (B) (C) (D)
28 (A) (B) (C) (D)
29 (A) (B) (C) (D)
30 (A) (B) (C) (D)
31 (A) (B) (C) (D)
32 (A) (B) (C) (D)
33 (A) (B) (C) (D)
34 (A) (B) (C) (D)
35 (A) (B) (C) (D)
36 (A) (B) (C) (D)
37 (A) (B) (C) (D)
38 (A) (B) (C) (D)
39 (A) (B) (C) (D)
40 (A) (B) (C) (D)
41 (A) (B) (C) (D)
42 (A) (B) (C) (D)
43 (A) (B) (C) (D)
44 (A) (B) (C) (D)
45 (A) (B) (C) (D)
46 (A) (B) (C) (D)
47 (A) (B) (C) (D)
48 (A) (B) (C) (D)
49 (A) (B) (C) (D)
50 (A) (B) (C) (D)

AP Physics 2: Practice Exam 2

Section 1 (Multiple Choice)

Directions: The multiple-choice section consists of 50 questions to be answered in 90 minutes. You may write scratch work in the test booklet itself, but only the answers on the answer sheet will be scored. You may use a calculator, the equation sheet, and the table of information. These can be found in the appendix or you can download the official ones from the College Board at: https://secure-media.collegeboard.org/digitalServices/pdf/ap/ap-physics-2-equations-table.pdf.

Questions 1–45: Single-Choice Items

Choose the single best answer from the choices provided, and mark the answer with a pencil on the answer sheet.

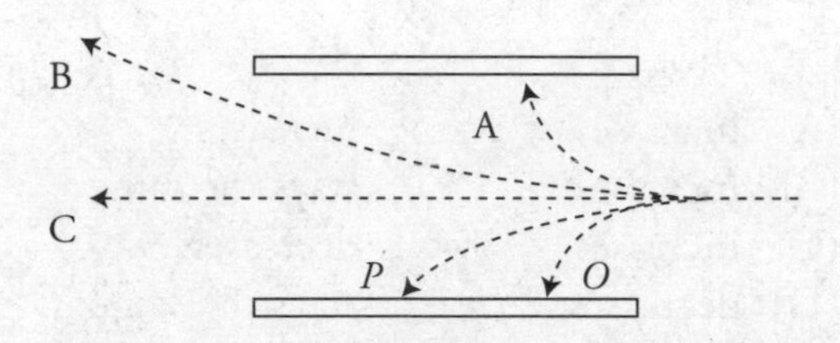

1. A beam of various particles is launched into the space between two oppositely charged parallel plates as shown in the figure. It is known that the particles are all traveling at the same initial velocity as they enter the region, and that particle P is a proton. What else can be inferred?

(A) Particle A could be a β-particle.
(B) Particle B could be an α-particle.
(C) Particle C has too much mass to be affected by the external force field.
(D) Particle D could be an electron.

Questions 2 and 3 refer to the following material.

The figure shows an electric potential field created by charges that are not shown.

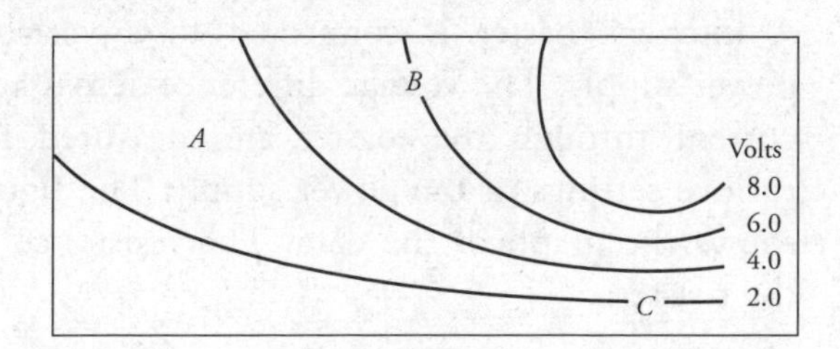

2. A small negative charge is placed at the location labeled A. Which one of the following vectors most closely depicts the direction of the electric force on the charge?

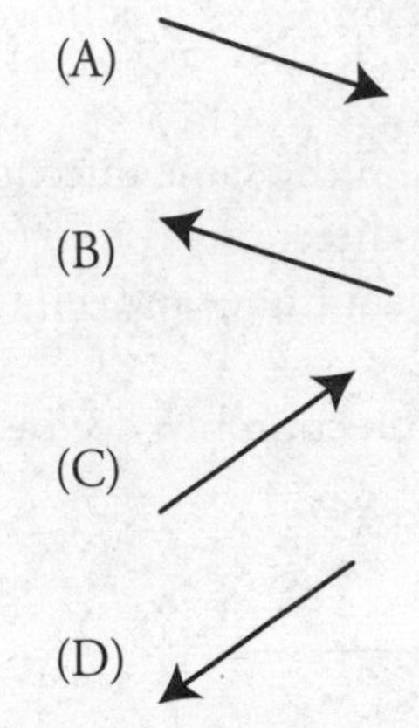

(A)

(B)

(C)

(D)

3. At which location is the electric field strongest, and which direction is the electric field?

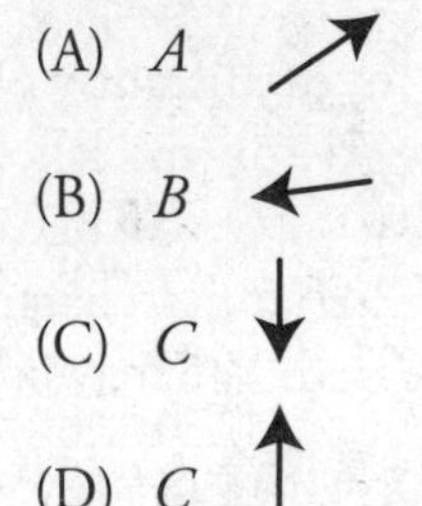

(A) A

(B) B

(C) C

(D) C

GO ON TO THE NEXT PAGE

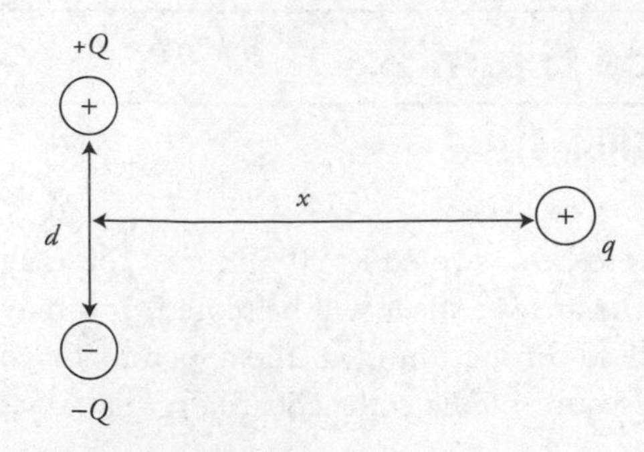

4. Two spheres with charges $+Q$ and $-Q$ of equal magnitude are placed a vertical distance d apart on the y-axis as shown in the figure. A third charge $+q$ is brought from a distance x, where $x >> d$, horizontally toward the midpoint between $+Q$ and $-Q$. The net force on $+q$ as it is moved to the left along the x-axis:

(A) increases and remains in the same direction
(B) increases and changes direction
(C) remains the same magnitude and in the same direction
(D) remains in the same direction but decreases to zero

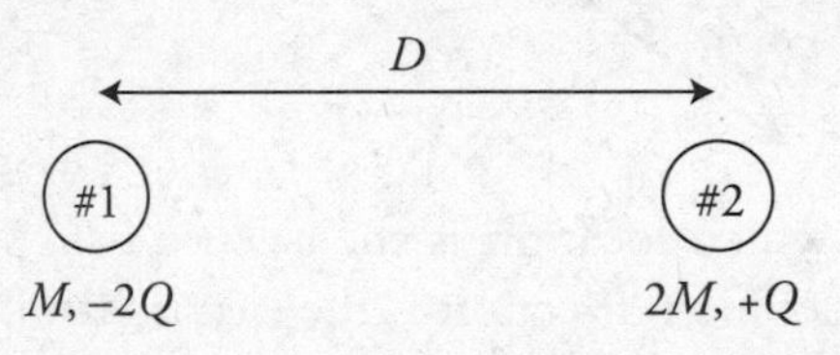

5. Two charges are separated by a distance D as shown in the figure. Charge #1, on the left, has a mass of M and a charge of $-2Q$. Charge #2, on the right, has a mass of $2M$ and a charge of $+Q$. The two charges are released. Where will the charges collide and why?

(A) The charges will collide closer to charge #1 because charge #1 has a larger magnitude charge and will exert a larger force on charge #2 giving charge #2 a larger acceleration.
(B) The charges will collide in the middle because the larger charge of #1 cancels the larger mass of #2.
(C) The charges will collide in the middle because the force between the charges will be equal and opposite in direction.
(D) The charges will collide closer to charge #2 because the center of mass is closer to charge #2 and there are no other forces on the system.

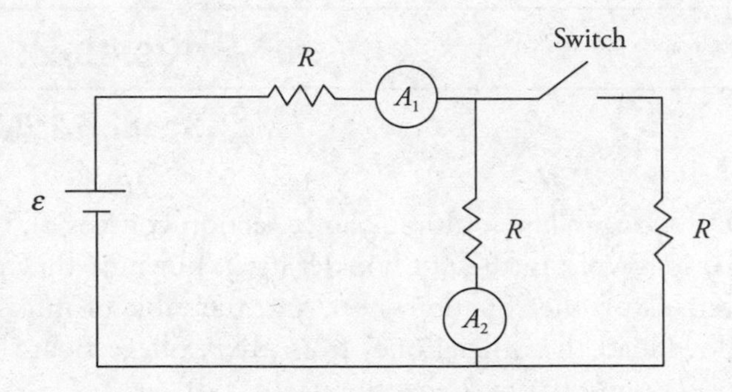

6. The circuit shown has a battery of emf $\mathcal{E}$; three identical resistors, R; two ammeters, A_1 and A_2; and a switch that is initially in the open position as shown in the figure. When the switch is closed, what happens to the current reading in the two ammeters?

	A_1	A_2
(A)	increases	increases
(B)	increases	stays the same
(C)	increases	decreases
(D)	decreases	stays the same

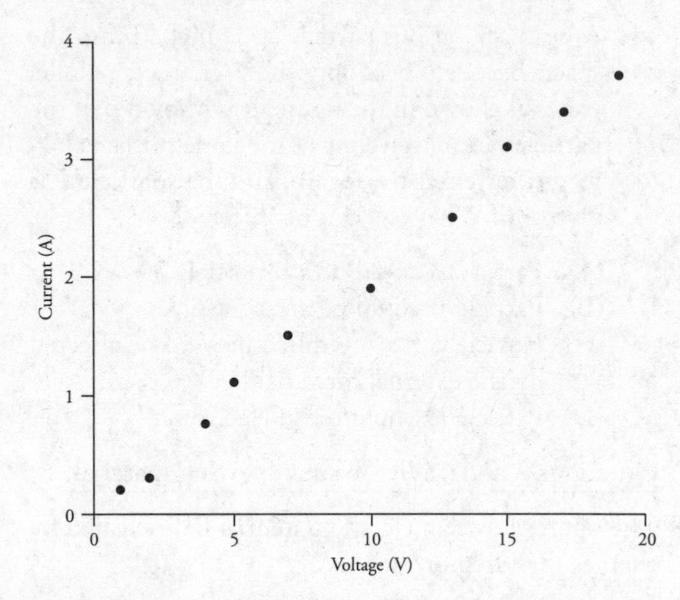

7. A kitchen toaster is connected to a variable power supply. The voltage difference across and current through the toaster are measured for various settings of the power supply. The figure shows the graph of the data. The resistance of the toaster:

(A) varies up and down with voltage
(B) increases linearly with input voltage
(C) is constant at 0.2 Ω
(D) is constant at 5.0 Ω

GO ON TO THE NEXT PAGE

Questions 8 and 9 refer to the following material.

The figure shows three resistors, X, Y, and Z, which have different shapes but are made of the same material. Resistors X and Y have a length L. Resistor Z has a length of $3L/2$. The resistors as shown are connected to a voltage source.

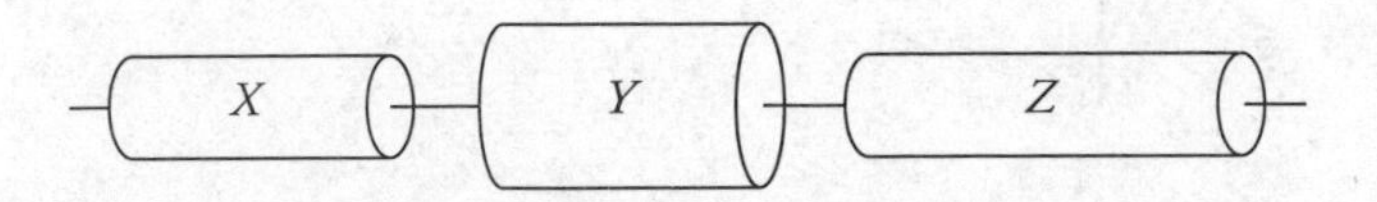

8. Which of the following correctly ranks the current in the resistors?

(A) $I_X = I_Y = I_Z$
(B) $I_X > I_Y > I_Z$
(C) $I_Y > I_X > I_Z$
(D) $I_Z > I_X > I_Y$

9. Which of the following correctly ranks the potential difference across the resistors?

(A) $V_X = V_Y = V_Z$
(B) $V_X > V_Y > V_Z$
(C) $V_Y > V_X > V_Z$
(D) $V_Z > V_X > V_Y$

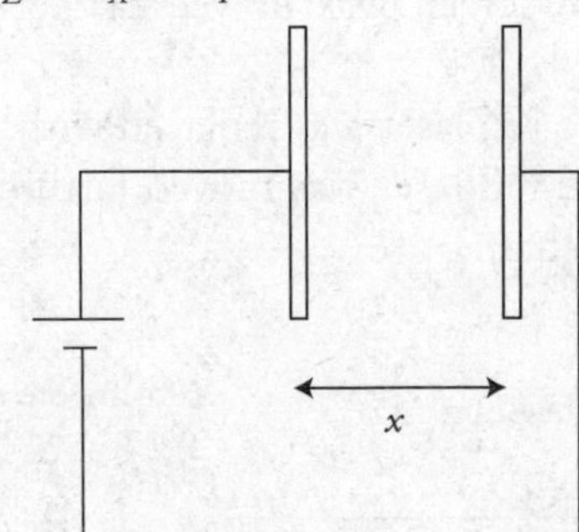

10. A capacitor with movable parallel plates is connected to a battery. The plates of the capacitor are originally a distance x apart as shown in the figure. With the battery still connected, what happens to the energy stored in the capacitor and the electric field between the capacitor plates, if the plates are moved to a new distance $x/2$?

	Energy	Electric Field
(A)	increases	increases
(B)	increases	decreases
(C)	decreases	increases
(D)	decreases	decreases

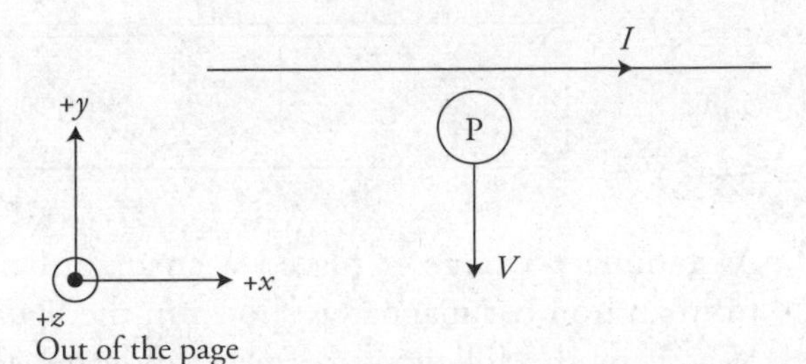

11. A proton with a velocity v is moving directly away from a wire carrying a current I directed to the right in the $+x$ direction as shown in the figure. The proton will experience a force in which direction?

(A) $-x$
(B) $-z$
(C) $+x$
(D) $+z$

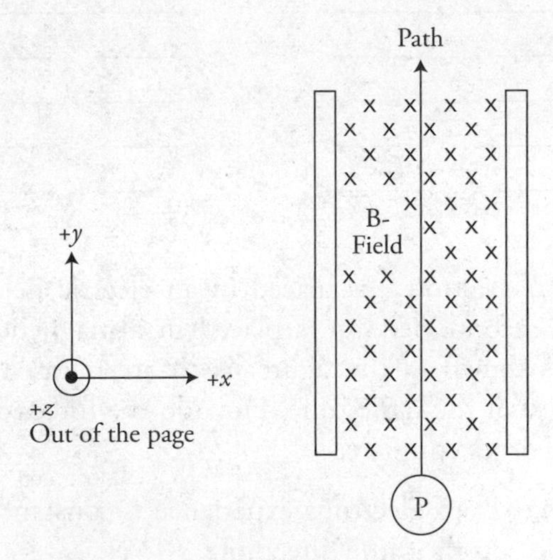

12. A magnetic field directed into the page in the $-z$ direction is placed between the plates of a charged capacitor as shown in the figure. The magnetic and electric fields are adjusted so that a particle of charge +1 e moving at a velocity of v will pass straight through the fields in the $+y$ direction. Which of the following changes will cause the particle to deflect to the left as it passes through the fields?

(A) Increasing the electric field strength
(B) Changing the sign of the charge to -1 e
(C) Increasing the velocity v of the particle
(D) Decreasing the magnetic field strength

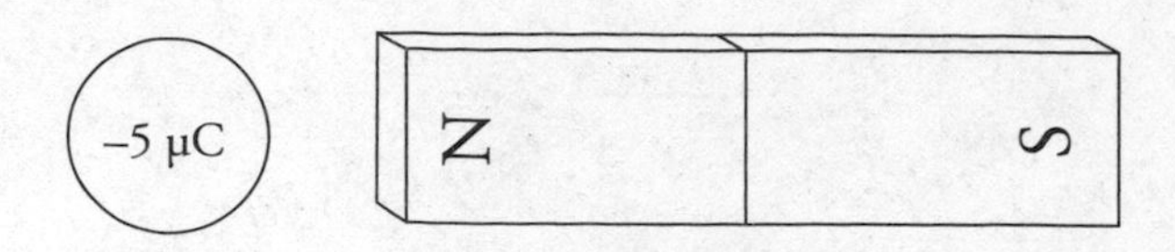

13. A stationary charge is placed a small distance from an iron bar magnet as shown in the figure. Which of the following correctly indicates the cause and direction of the force on the charge?

(A) The charge experiences an electric force to the right.
(B) The charge experiences a magnetic force to the right.
(C) The charge experiences a magnetic force to the left.
(D) The charge experiences no force.

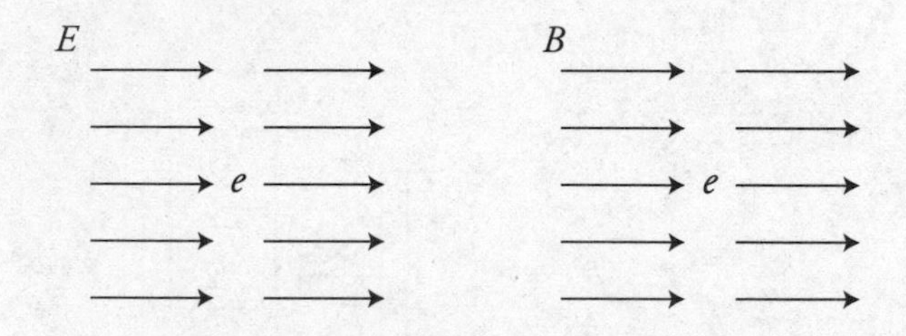

14. An electron *e* is placed in an electric field and a second electron is placed in a magnetic field as shown in the figure. Both are released from rest at the same time. How do the forces on the charges compare?

(A) Both electrons experience a constant force in the same direction.
(B) Both electrons experience a constant force but in different directions.
(C) One electron experiences a constant force in a constant direction, while the other experiences a constant force that changes direction.
(D) One electron experiences a constant force, while the other experiences no force.

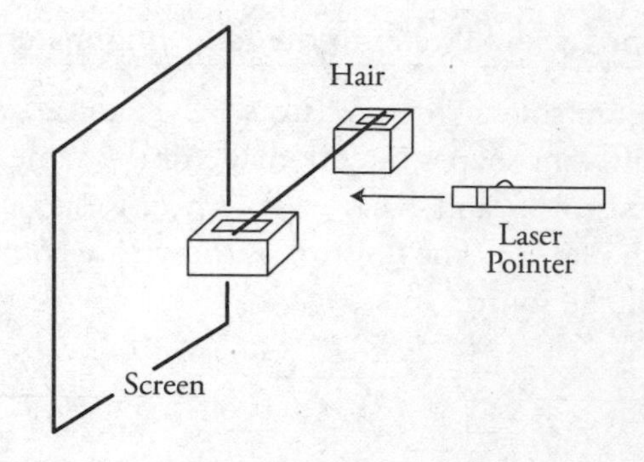

15. A technician sets up a red laser pointer and directs light at a strand of hair suspended between two blocks in such a way that the light falls on a screen behind the hair as shown in the figure. This produces a pattern of alternating light and dark fringes on the screen. The technician makes one change to the setup and repeats the procedure. This change produces a similar pattern, but the light and dark fringes are spaced farther apart. Which of the following could account for this different pattern?

(A) The hair was replaced with one of larger diameter.
(B) The screen was moved farther away from the hair.
(C) The red laser was replaced with a blue laser.
(D) The red laser was moved farther away from the hair.

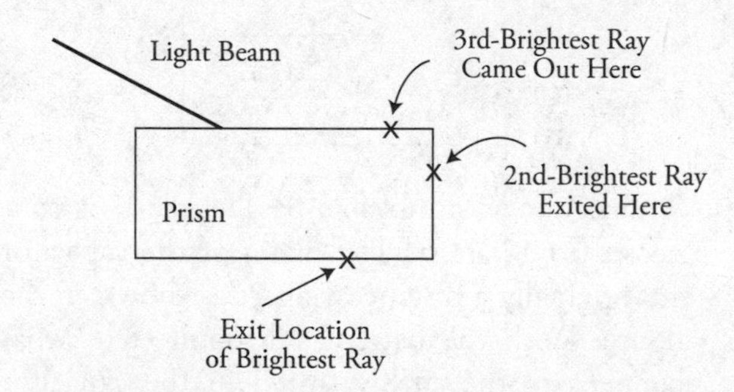

16. Your lab group has been asked to determine the index of refraction of a plastic rectangular prism. Your partner places the prism on a sheet of paper, traces its shape, shines a beam of light through the prism, and traces the incoming ray. The light exits the prism at multiple locations. Your partner marks the exit locations of the three brightest rays on the side of the prism as

GO ON TO THE NEXT PAGE

shown in the figure. Is this enough information to determine the index of refraction of the prism, and why?

(A) Yes, by measuring ray angles with a protractor and using the concepts of refraction
(B) Yes, by measuring the internal reflected angles with a protractor and using the concepts of total internal reflection
(C) No, the exit rays from the prism should have been traced. It is impossible to utilize the concepts of refraction without having the exit ray directions.
(D) No, without knowing the velocity of the beam inside the prism, you cannot use the index of refraction equation.

17. A very slow-moving positron interacts with a stationary electron. Which of the following statements correctly describes a possible outcome of this reaction and why it would occur?

(A) Conservation of mass indicates that if a single new particle were created in the reaction, it must have a total mass equal to the combined masses of the electron and positron.
(B) Conservation of charge indicates that all new particles created in the reaction would have no electric charge.
(C) Conservation of momentum indicates that two identical gamma rays moving off in opposite directions could be created.
(D) Conservation of energy indicates that the antimatter positron could annihilate into energy, leaving the stationary electron behind.

18. ${}^{14}_{6}C$ decays into ${}^{14}_{7}N$ and an electron. The half-life of carbon-14 is 5730 years. Which of the following statements is correct?

(A) Carbon-14 has the same number of protons as carbon-11.
(B) Carbon-14 is used to determine the age of 120-million-year-old dinosaur bones.
(C) The combined mass of nitrogen-14 and an electron equals the mass of carbon-14.
(D) Carbon-14 and carbon-11 will have different chemical properties.

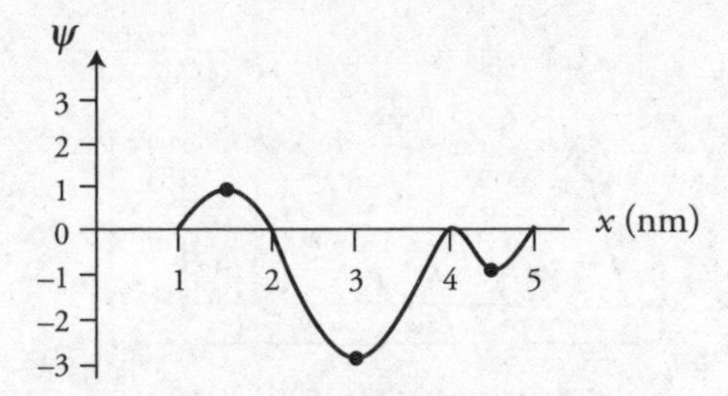

19. The figure shows the wave functions $\Psi(x)$ of a particle moving along the x-axis. Which of the following statements correctly interprets this graph?

(A) The particle is oscillating in charge from positive to negative.
(B) The lowest probability of finding the particle is at 3.0 nm.
(C) There is an equal probability of finding the particle at 1.5 nm as at 4.5 nm.
(D) The length of the particle is 4 nm.

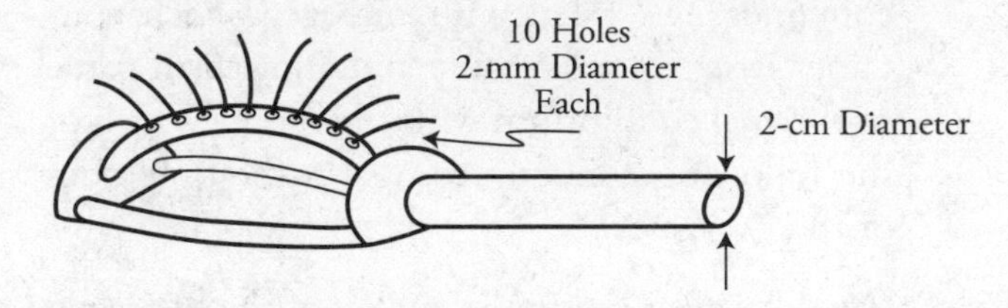

20. A 2-cm diameter hose leads to a lawn sprinkler with ten 2-mm exit holes as shown in the figure. The velocity of the water in the hose is v_1. What is the velocity of the water exiting the holes?

(A) $0.1\ v_1$
(B) v_1
(C) $2.5\ v_1$
(D) $10\ v_1$

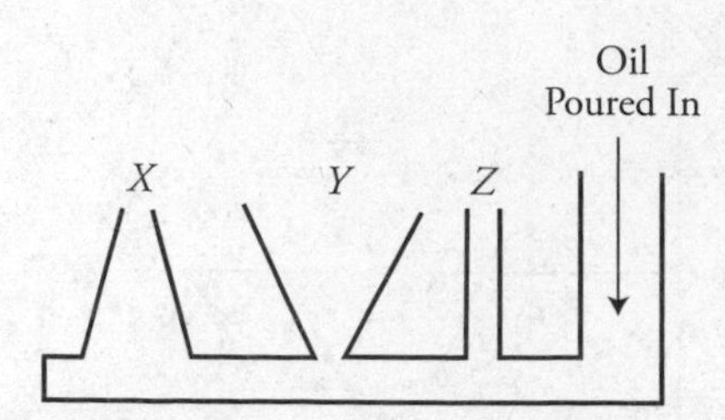

21. Oil is very slowly poured into the container shown in the figure until the fluid enters the three open tubes X, Y, and Z. The oil does not overflow any of the tubes. Which of the following correctly ranks the height of the oil in each of the three tubes?

(A) $Z > Y > X$
(B) $Z > X > Y$
(C) $X > Z > Y$
(D) $X = Y = Z$

22. Two identical containers are filled with different gases. Container 1 is filled with hydrogen and container 2 is filled with nitrogen. Each container is set on a lab table and allowed to come to thermal equilibrium with the room. Which of the following correctly compares the properties of the two gases?

(A) The average kinetic energy of the hydrogen gas is greater than the nitrogen gas.
(B) The average force exerted on the container by the hydrogen gas is greater than the nitrogen gas.
(C) The density of the hydrogen gas is less than the nitrogen gas.
(D) The pressures of the gases cannot be compared without knowing the number of molecules in each container.

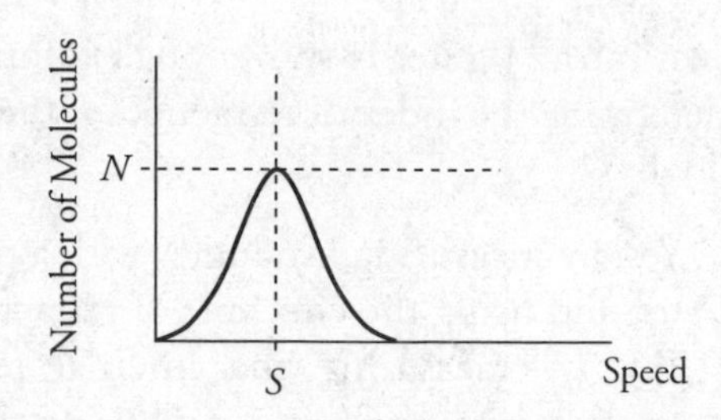

23. The graph shows the distribution of speeds for a gas sample at temperature T. The gas is heated to a higher temperature. Which of the following depicts a possible distribution after the gas has been heated? (Horizontal line N and vertical line S are shown for reference.)

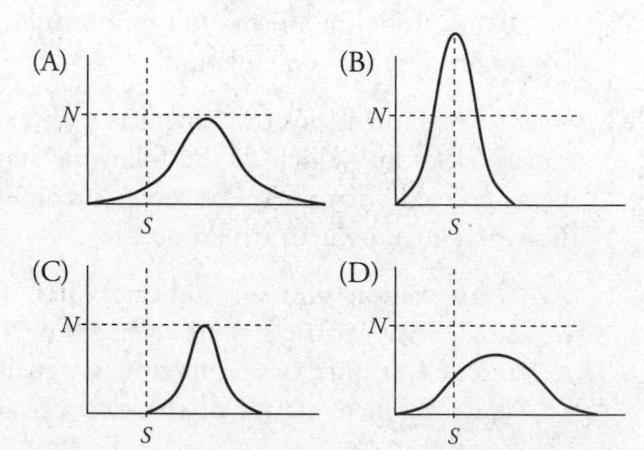

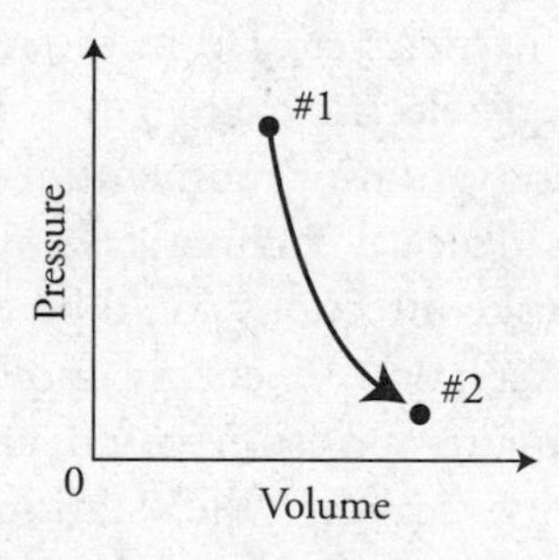

24. The graph shows the pressure and volume of a gas being taken from state #1 to state #2, in a very quick process, where there is not enough time for heat to either leave or enter the gas. Which of the following correctly indicates the sign of the work done on the gas, and the change in temperature of the gas?

	Work Done	Δ Temperature
(A)	+	+
(B)	+	–
(C)	–	+
(D)	–	–

GO ON TO THE NEXT PAGE

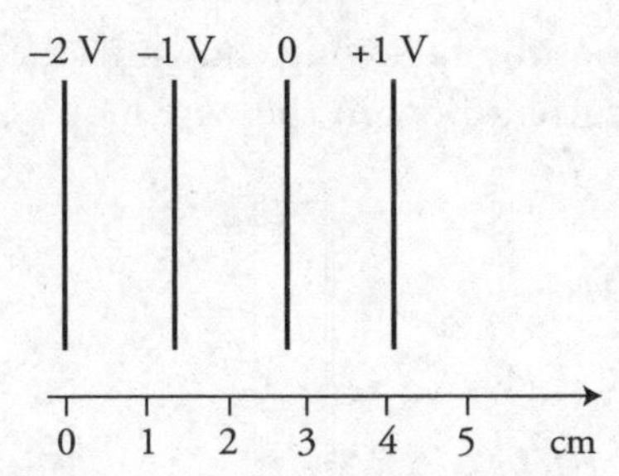

25. In a region of space, there is an electric potential field with isolines as shown in the figure. Which of the following is the most accurate description of the electric field in the region?

(A) An electric field is directed to the left with a uniform strength of 3 V/m.
(B) An electric field is directed to the left with a uniform strength of 75 V/m.
(C) An electric field is directed to the left that increases in strength from left to right with an average value of 75 V/m.
(D) An electric field is directed to the right that increases in strength from right to left with an average value of −75 V/m.

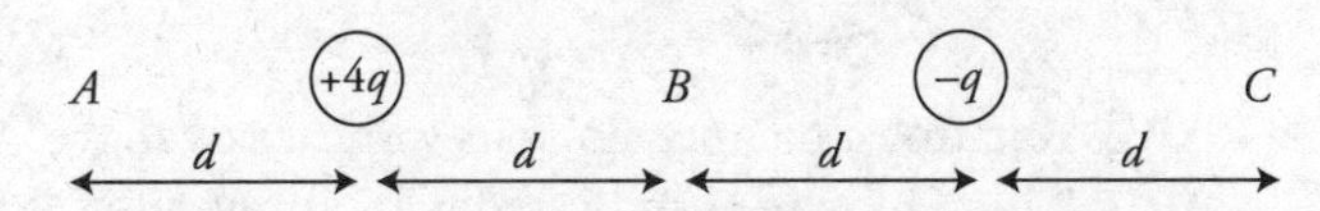

26. Which of the following correctly represents the relationship between the electric field strengths at locations A, B, and C?

(A) $E_A > E_B > E_C$
(B) $E_A = E_B > E_C$
(C) $E_B > E_A > E_C$
(D) $E_B > E_A = E_C$

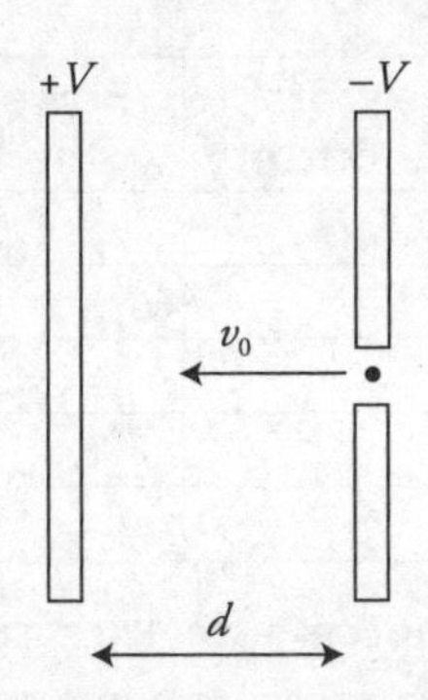

27. An electron of charge $-e$ and mass m is launched with a velocity of v_0 through a small hole in the right plate of a parallel plate capacitor toward the opposite plate a distance d away. The electric potential of both plates are equal in magnitude, but opposite in sign $\pm V$ as shown in the figure. What is the kinetic energy of the electron as it reaches the left plate?

(A) $\frac{1}{2}mv_0 - 2\ Ve$
(B) $\frac{1}{2}mv_0$
(C) $\frac{1}{2}mv_0 + Ve$
(D) $\frac{1}{2}mv_0 + 2\ Ve$

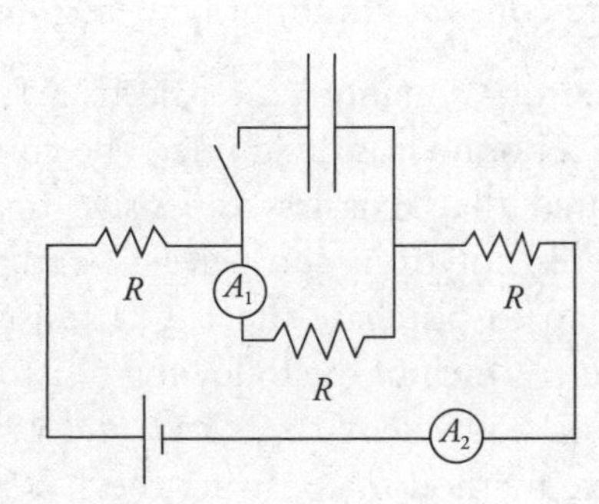

28. The circuit shown in the figure consists of three identical resistors, two ammeters, a battery, a capacitor, and a switch. The capacitor is initially uncharged and the switch is open.

What happens to the readings of the ammeters immediately after the switch is closed?

	A_1	A_2
(A)	decrease	decrease
(B)	decrease	stay the same
(C)	decrease	increase
(D)	stay the same	stay the same

Number of Resistors Connected to the Battery	Power Delivered to the Resistors (W)
1	5.00
2	2.50
3	1.67
4	1.25
5	1.00

29. A student connects a single resistor to a 10-V battery and takes measurements to find the power delivered by the battery to the resistor. The student then adds another resistor to the circuit and measures the power delivered to both resistors. This process is repeated, adding one resistor at a time, until the battery is connected to five resistors. The data from the experiment is given. Which of the following can be concluded from the data?

(A) The resistors are all identical and are connected in parallel.
(B) The resistors are all identical and are connected in series.
(C) The resistors are not identical, but they must be connected in series.
(D) The resistors are nonohmic and how they are connected cannot be determined.

30. An electronics manufacturer needs a 1.2-Ω resistor for a phone it is designing. The company has calculated that it is less expensive to build the 1.2-Ω resistor from cheaper 1-Ω and 2-Ω resistors than to purchase the 1.2-Ω resistor from a supplier. Which of the following resistor arrangements is equivalent to 1.2 Ω and is the most effective method to construct the 1.2-Ω resistor?

(A) 1 Ω in series with 2 Ω, 2 Ω, 2 Ω, 2 Ω, 2 Ω in parallel

(B) 2 Ω, 1 Ω, 2 Ω in parallel

(C) 1 Ω, 2 Ω in series

(D) 1 Ω and 2 Ω in series, in parallel with 2 Ω

31. Which of the following shown in the figure will cause current flow in the wire loop?

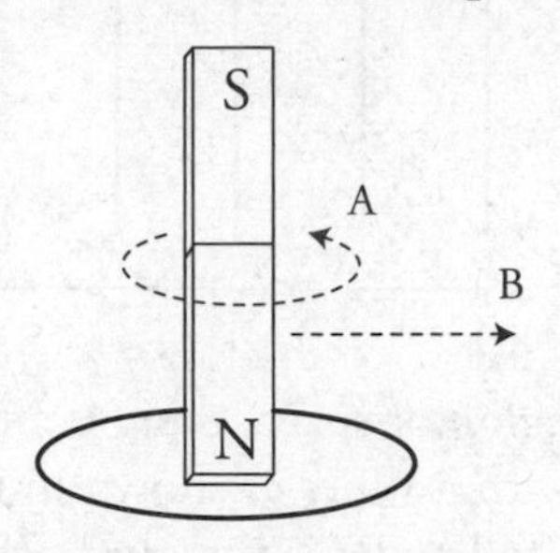

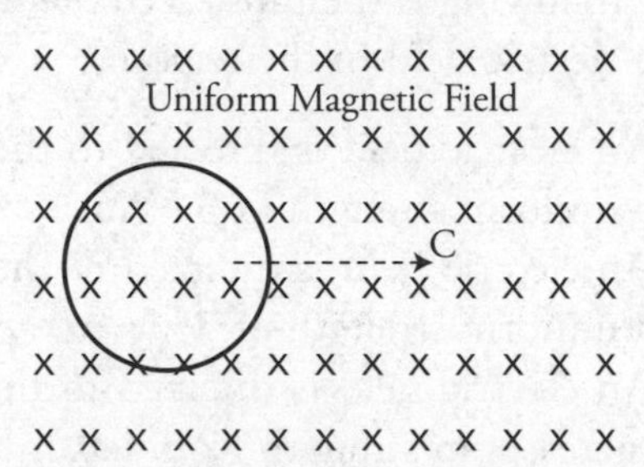

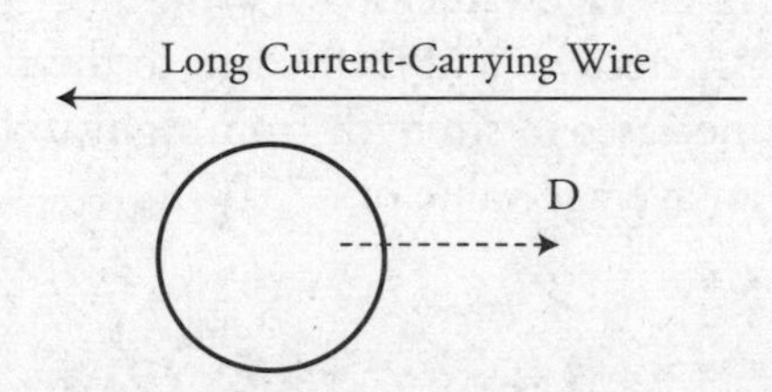

(A) Rotating a magnet along its vertical axis above a loop of wire as indicated with arrow A
(B) Moving a magnet from above a loop of wire to the right as indicated with arrow B
(C) Moving a loop of wire to the right in a uniform magnetic field as indicated by arrow C
(D) Moving a loop of wire to the right along a long current-carrying wire as indicated by arrow D

32. Under which of the following conditions would it be appropriate to neglect the gravitational force on a charge in a magnetic field?

(A) $G << \mu_0$
(B) $g << B$
(C) $mg << qvB$
(D) $v = 0$

GO ON TO THE NEXT PAGE

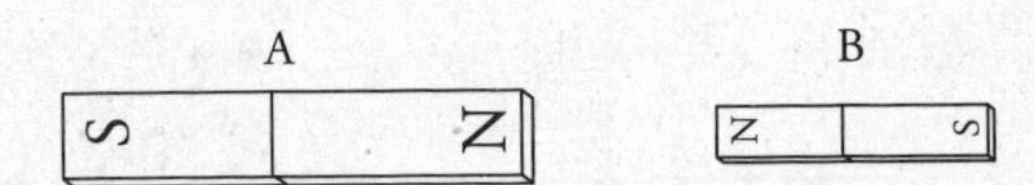

33. A child playing with two magnets places them close together and holds them in place to keep them from moving, as shown in the figure. Magnet A has a larger magnetic field and a larger mass than magnet B. What will happen when the magnets are released and free to move?

(A) Magnet A will not move, while magnet B will accelerate away because magnet B is smaller.
(B) Magnet A will accelerate away more slowly than magnet B because magnet B exerts a smaller force on magnet A.
(C) Magnet A will accelerate away more slowly than magnet B because magnet A has a larger mass.
(D) Both magnets will accelerate away at the same rate because the force between them will be the same magnitude.

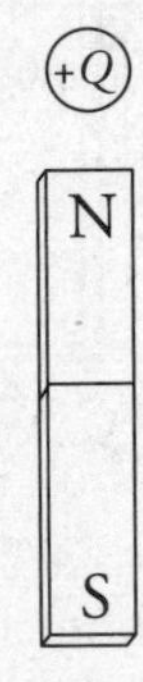

34. A charge $+Q$ is positioned close to a bar magnet as shown in the figure. If the charge is experiencing a force into the page, which way must the charge be moving?

(A) To the right
(B) To the left
(C) Toward the top of the page
(D) Out of the page

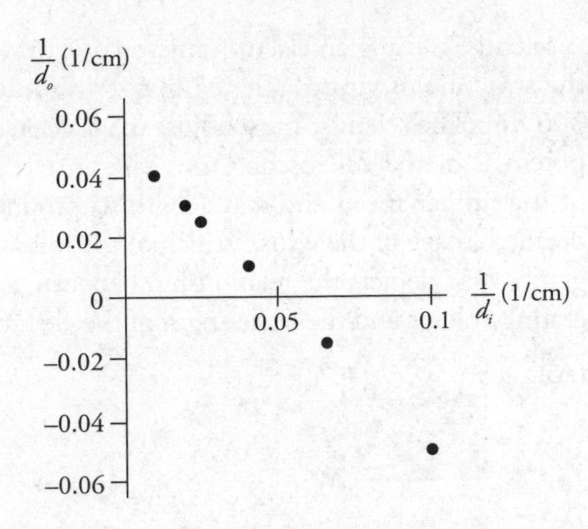

35. You were absent when your AP Physics 2 class performed a lens lab. When you return to school, the teacher says to get the data from a friend and complete the assignment of calculating the focal length of the lens used in the lab. Your friend hands you the graph shown in the figure where d_o and d_i are the object and image distances. Which of the following is equal to the focal length of the lens?

(A) The y-intercept of the best-fit line
(B) The inverse of the y-intercept of the best-fit line
(C) The slope of the best-fit line
(D) The inverse of the slope of the best-fit line

36. A scientist is using an electron microscope to study the structure of viruses ranging in size from 20 to 300 nm. The scientist must adjust the accelerating potential of the microscope to create an electron of the proper speed and wavelength to produce a detailed image of the virus. Which of the following graphs best depicts the relationship between accelerating voltage and the wavelength of the electrons?

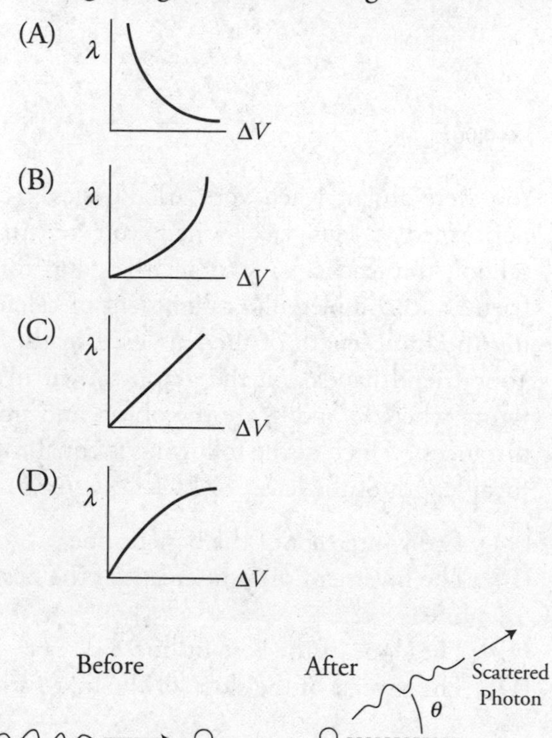

37. A photon of wavelength λ collides with a stationary electron glancing off at an angle of θ, while the electron moves off with an angle of ϕ measured from the original path of the photon as shown in the figure. Which of the following statements correctly states the effect on the wavelength of the scattered photon and why?

(A) The wavelength of the scattered photon decreases because the photon transfers energy to the electron in the interaction.
(B) The wavelength of the scattered photon decreases because the photon transfers momentum to the electron in the interaction.
(C) The wavelength of the scattered photon remains the same because the photon is acting as a particle during the collision.
(D) The wavelength of the scattered photon increases because the photon transfers energy to the electron in the interaction.

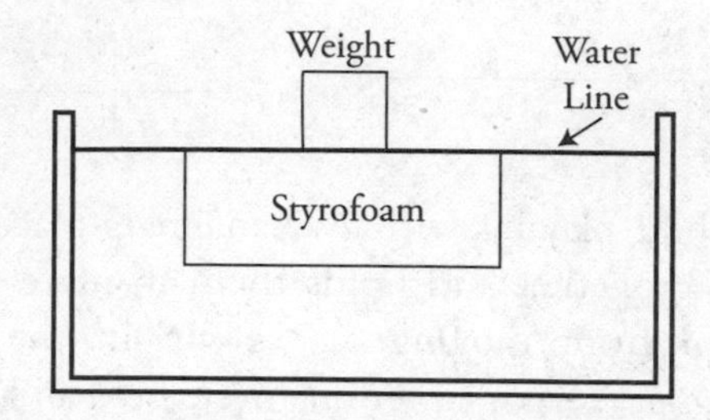

38. A mass is fixed to the top of a Styrofoam block that floats in a container of water. The mass is large enough to make the water line flush with the top of the Styrofoam as shown in the figure. What will happen if the Styrofoam is inverted so that the mass is now suspended under the block?

(A) The whole contraption sinks.
(B) The contraption floats with the waterline still flush with the top of the Styrofoam.
(C) The contraption floats with the waterline below the top of the Styrofoam.
(D) It is impossible to determine without knowing the density of the mass and Styrofoam.

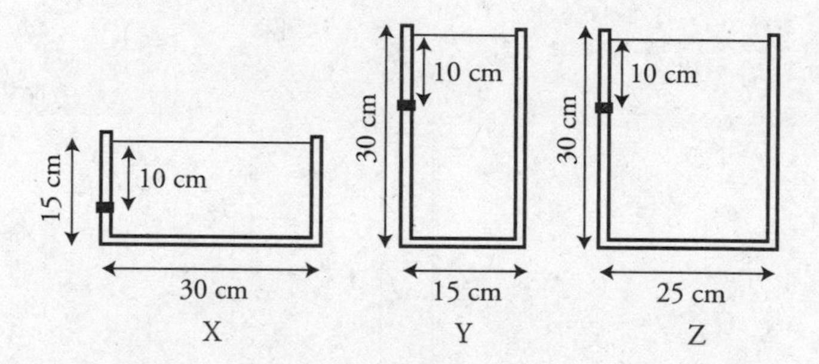

39. Three beakers of different sizes are filled with water as shown. Each beaker has a rubber stopper of the identical size and shape fitted to a drain hole in the side. Which correctly ranks the force applied to the stoppers by the water?

(A) X = Y = Z
(B) X > Y = Z
(C) X > Z > Y
(D) Z > Y = X

GO ON TO THE NEXT PAGE

40. A person can stand outside on a cold day for hours without ill effect, but falling into a cold lake can kill a person in a matter of minutes. Which of the following is the primary reason for this phenomenon?

(A) The molecules of the person are, on average, moving faster than those of the surroundings.
(B) Thermal energy moves from high concentration areas (hot) to low concentration areas (cold).
(C) As heat flows out of the person and warms the fluid surrounding the person, the warmer fluid rises, allowing fresh cool fluid to come in contact with the person and increasing the rate of heat transfer.
(D) Water has more molecules per volume than air, increasing molecular contact with the person.

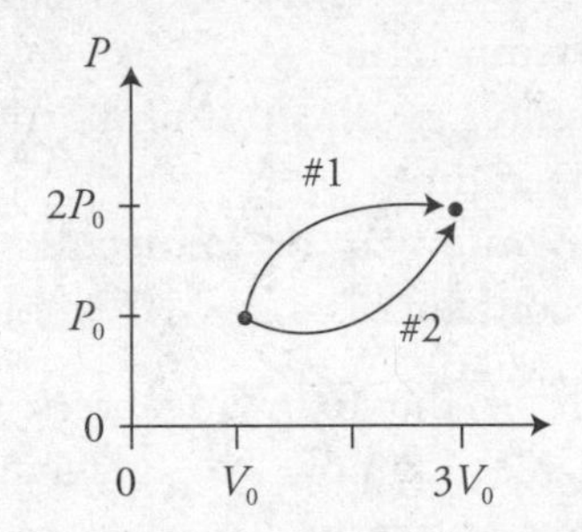

41. A sample of gas can be taken from an initial pressure P_0 and volume V_0 to a final pressure and volume along either path 1 or 2 as shown in the figure. Your lab partner states: "Moving along either path won't make any difference because both paths start and end at the same places. So, everything about the gases during both processes 1 and 2 will be the same." Which of the following is a proper analysis of your lab partner's statement?

(A) The change in temperature is the same, but the change in internal energies will be different.
(B) The change in internal energies will be the same, but the thermal energy transferred to the gas will be different.
(C) The thermal energy transferred to the gas will be the same, but the work done by the gas will be different.
(D) The work done by the gas will be the same, but the change in temperature will be different.

42. A glass lens of focal length $f = 80$ cm in air is submerged in oil. How will submerging the lens in oil affect the focal length of the lens and why?

(A) $f > 80$ cm because there is now a smaller change in velocity as light passes from oil into the glass.
(B) $f = 80$ cm because the index of refraction of glass has remained unchanged by placing the lens into the oil.
(C) $f = 80$ cm because the shape of the lens has remained unchanged by placing the lens into the oil.
(D) $f < 80$ cm because light travels slower in oil.

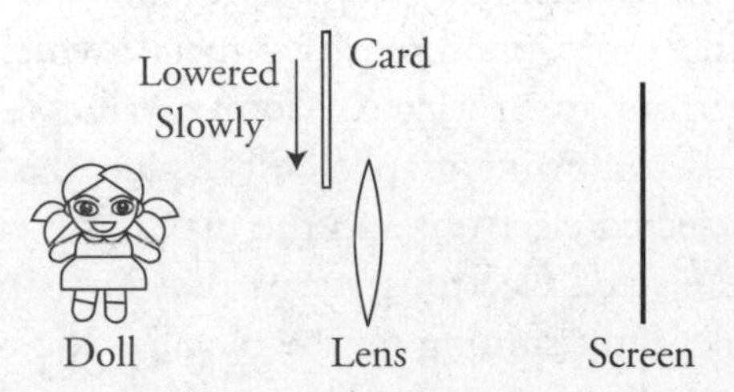

43. A lens is placed between a doll and a white sheet of paper in such a way as to produce an image on the paper. On the side of the lens facing the doll, a dark card is slowly lowered to cover the lens as shown in the figure. Which of the following correctly explains what will happen to the image of the doll?

(A) As the descending card blocks more and more of the lens, the focal point of the lens shifts. This causes the image to become increasingly blurry until the image disappears from the screen.
(B) The image remains clear, but the head of the doll in the image disappears first, followed by the feet, as the light from the top of the object is blocked before the light from the feet.
(C) The image remains clear, but since the doll projects an inverted real image on the screen, the doll's feet will disappear from view first, followed by the head last, as light from the object is blocked.
(D) As the card blocks light from the object, the image remains clear but becomes increasingly dim until the image disappears.

GO ON TO THE NEXT PAGE

$$^{12}_{6}\text{C} + ^{1}_{1}\text{H} \rightarrow ^{13}_{7}\text{N} + \gamma$$

44. Which of the following expressions correctly relates the masses of the constituent particles involved in the nuclear reaction shown?

(A) $m_C + m_H - m_N = 0$

(B) $m_C + m_H - m_N - \frac{hf_\gamma}{c^2} < 0$

(C) $m_C + m_H - m_N - \frac{hf_\gamma}{c^2} = 0$

(D) $m_C + m_H - m_N - \frac{hf_\gamma}{c^2} > 0$

45. In an experiment, monochromatic violet light shines on a photosensitive metal, which causes electrons to be ejected from the metal. Which of the following graphs best depicts the number of ejected electrons and the maximum energy of the ejected electrons versus the intensity of the violet light shining on the metal?

(A)

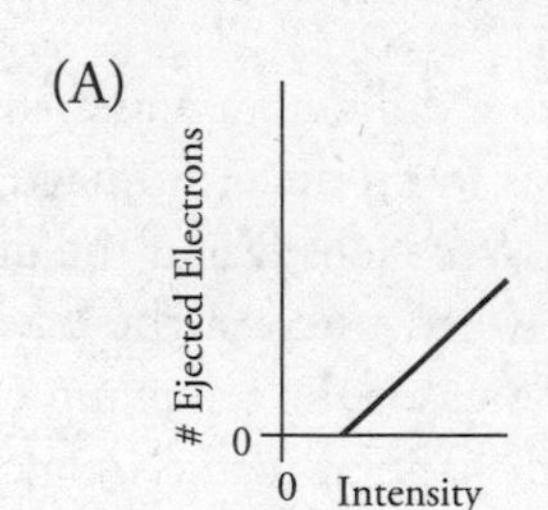

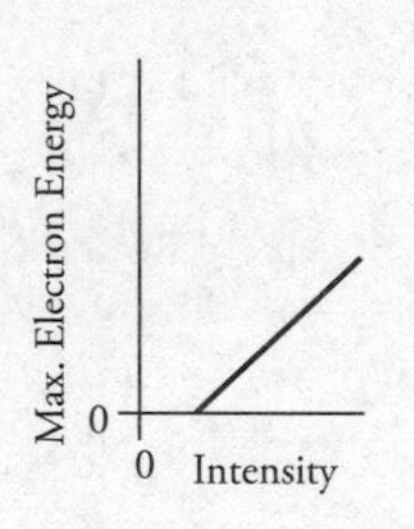

(B)

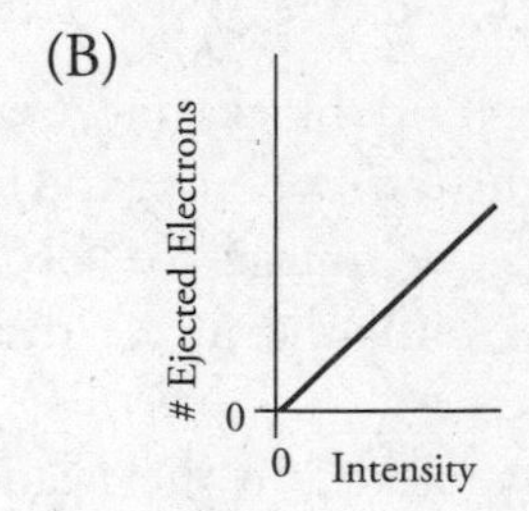

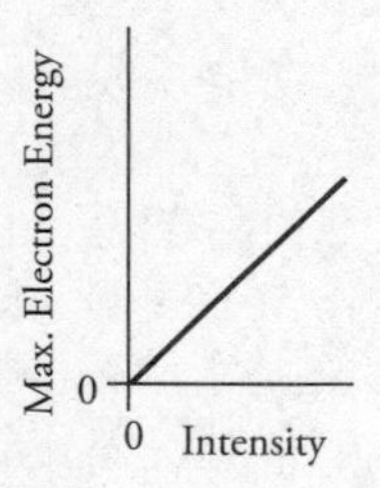

(C)

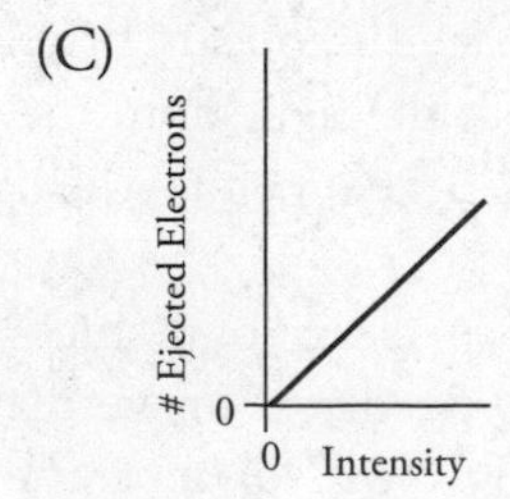

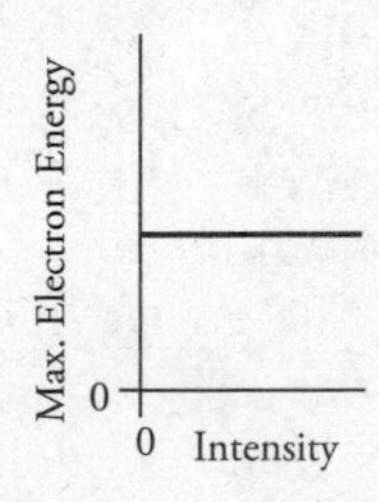

(D)

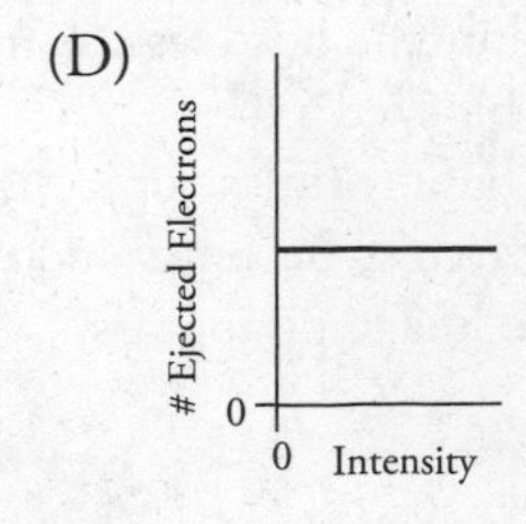

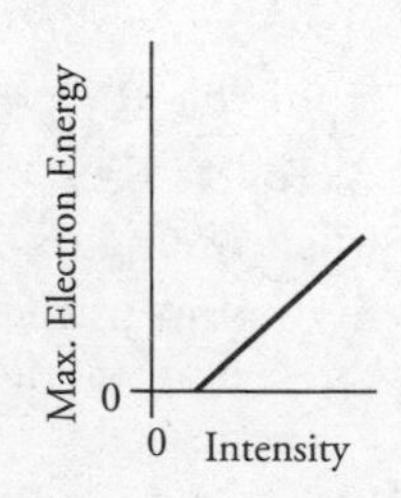

Questions 46–50: Multiple-Correct Items

Directions: Identify exactly two of the four answer choices as correct, and mark the answers with a pencil on the answer sheet. No partial credit is awarded; both of the correct choices, and none of the incorrect choices, must be marked to receive credit.

Volume (cm^3)	5.0	5.0	5.0	5.0
Pressure (kPa)	200	210	220	230
Temperature (°C)	0	20	40	60

46. A technician is experimenting with a sample of gas in a closed container and produces this set of data. What can be concluded from this data? (**Select two answers.**)

(A) Pressure is inversely proportional to the volume.

(B) Volume is directly proportional to the temperature.

(C) Pressure is directly proportional to the temperature.

(D) The number of molecules of gas in the container is 4.4×10^{-4} moles.

47. In which of the following cases would it be appropriate to ignore the gravitational force? (**Select two answers.**)

(A) A teacher charges a balloon with her hair and demonstrates how the balloon can be stuck to the ceiling by the electrostatic force.

(B) A scientist suspends a charged droplet of oil between two charged horizontal capacitor plates.

(C) Electrons move through a wire in an electrical circuit.

(D) A physics student is asked to calculate the force between the proton and electron in a hydrogen atom.

GO ON TO THE NEXT PAGE

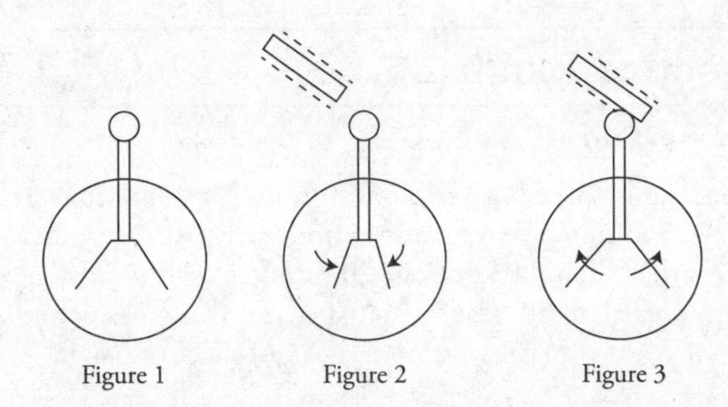

48. An electroscope is shown with its movable metal leaves in a sequence of events. Originally the electroscope is in a position with the leaves at an outward angle as shown in Figure 1. A negatively charged rod is brought close to the electroscope and the leaves swing downward as shown in Figure 2. Finally, the rod touches the electroscope and the leaves spring outward as shown in Figure 3. Which of the following statements correctly describe this behavior? (**Select two answers.**)

(A) In Figure 1, the electroscope has a net positive charge.
(B) In Figure 2, the leaves move closer together because the rod discharges the electroscope.
(C) In Figure 2, negative charges are pushed toward the leaves at the bottom of the electroscope.
(D) In Figure 3, positive charges from the electroscope migrate onto the rod, leaving the electroscope negatively charged.

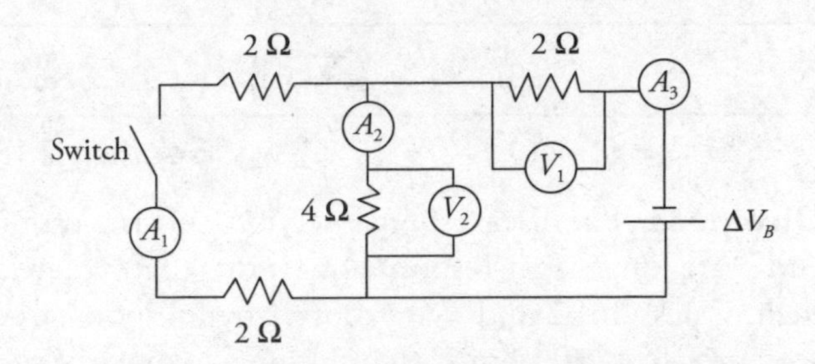

49. The circuit shown has a battery of negligible internal resistance, resistors, and a switch. There are voltmeters, which measure the potential differences V_1 and V_2, and ammeters A_1, A_2, A_3, which measure the currents I_1, I_2, and I_3. The switch is initially in the closed position. With the switch still closed, which of the following relationships are true? (**Select two answers.**)

(A) $I_1 + I_2 - I_3 = 0$
(B) $\Delta V_B - V_1 - V_2 = 0$
(C) $V_1 > V_2$
(D) $I_2 = I_3$

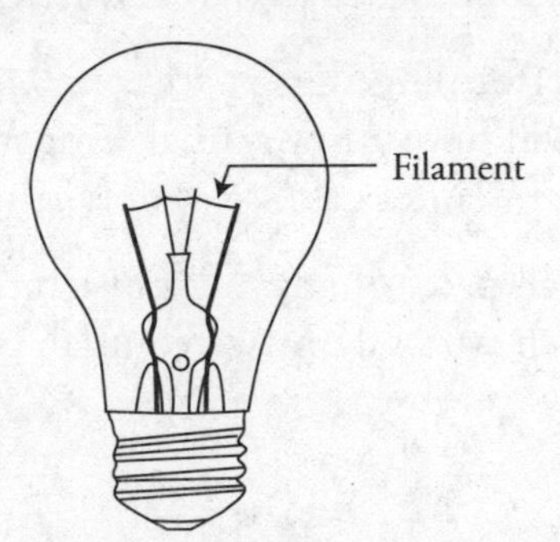

50. An incandescent bulb is shown in the figure. When connected to the same voltage source, which of the following would make this bulb brighter? (**Select two answers.**)

(A) Increase the length of the filament
(B) Increase the thickness of the filament
(C) Connect two of the existing filaments in series
(D) Connect two of the existing filaments in parallel

STOP: End of AP Physics 2 Practice Exam, Section 1 (Multiple-Choice)

AP Physics 2: Practice Exam 2

Section 2 (Free Response)

Directions: The free-response section consists of four questions to be answered in 90 minutes. Questions 1 and 3 are longer free-response questions that require about 25 minutes each to answer and are worth 12 points each. Questions 2 and 4 are shorter free-response questions that should take about 20 minutes each to answer and are worth 10 points each. Show all your work to earn partial credit. On an actual exam, you will answer the questions in the space provided. For this practice exam, write your answers on a separate sheet of paper.

1. (12 points—suggested time 25 minutes)

A cylinder, with a sealed movable piston, is filled with a gas and sits on a lab station table in thermal equilibrium with the lab room. The original pressure, volume, and temperature of the gas are P_0, V_0, and T_0, respectively. You have been instructed by the teacher to experimentally determine the work required to compress the gas to half its original volume.

(A) Describe an experimental procedure you could use to determine the work. Include the following:

- all the equipment needed
- a labeled diagram of the setup
- a clear indication of which variables will be manipulated and which will be measured
- a clear explanation of how the work will be calculated from your data
- enough detail so that another student could carry out the procedure

While performing the experiment, your lab partner makes this claim: "If we compress the gas very slowly, the gas will have a lower final temperature than if we compress the gas very quickly. But, the work will be the same in either case, so it doesn't matter which way we do it."

(B) Which parts, if any, of your lab partner's claim are correct? Justify your response.

(C) Which parts, if any, of your lab partner's claim are incorrect? Justify your response.

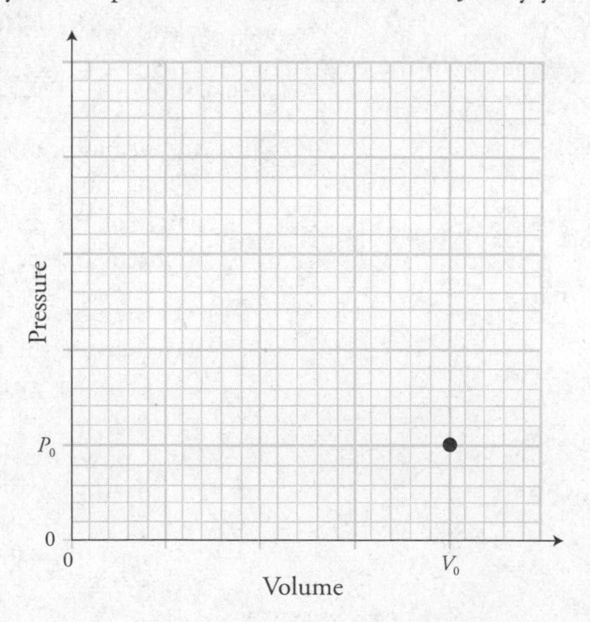

(D) On the graph, sketch the pressure as a function of volume as the gas is compressed from its original pressure P_0 and volume V_0 (shown by a dot on the graph) to half its original volume. Assume that the gas remains in thermal equilibrium with the lab room throughout the compression.

(E) Utilizing your answer from (D), explain how the graph could be utilized to find the work required to compress the gas.

GO ON TO THE NEXT PAGE

2. (10 points—suggested time 20 minutes)

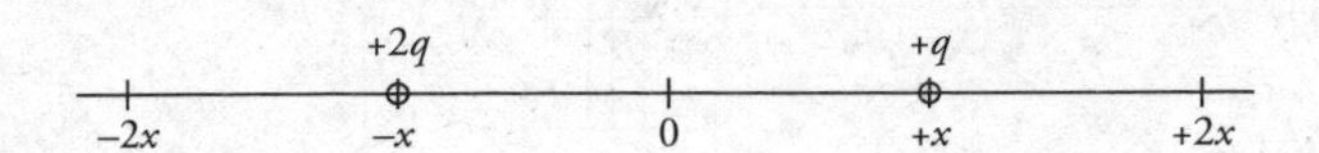

Two charges ($-2q$ and $+q$) are situated along the x-axis as shown in the figure.

(A) What is the direction of the electric field at point $+2x$? Justify your answer.
(B) Write an expression for the electric field at $-2x$. Show all your work.
(C) On the axis, sketch a graph of the electric field E along the x-axis as a function of position x. Electric fields to the right are defined as positive.

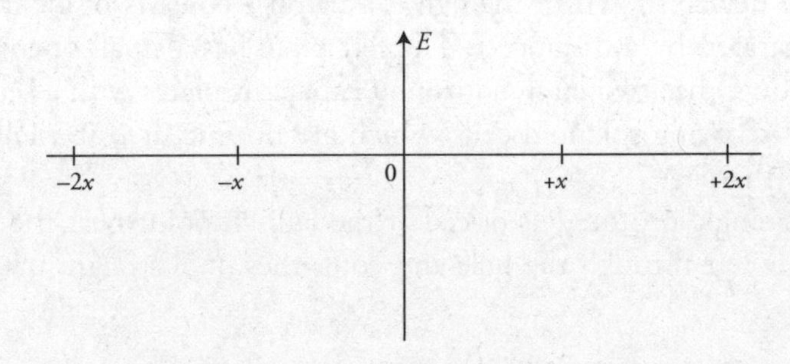

(D) Is there a spot on the x-axis where the electric field will have a magnitude of zero? If so, give the general location of where it will be and explain why it will occur in that location. If not, explain why not.
(E) A charge of $-5q$ is placed at $-2x$. Calculate the magnitude and direction of the force on $5q$.

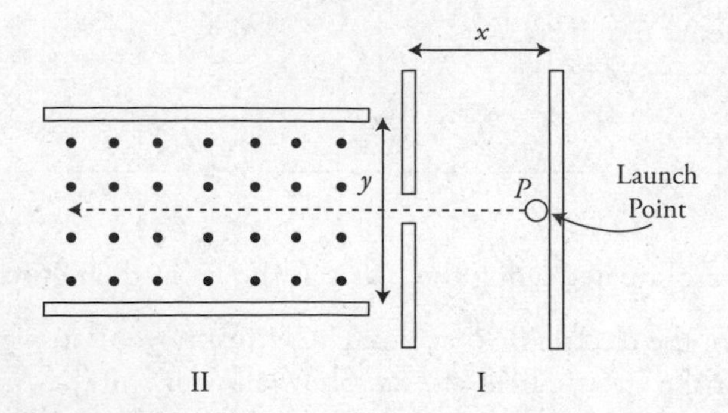

3. (12 points—suggested time 25 minutes)

A scientist constructs a device shown in the figure. Region I consists of a charged parallel plate capacitor with vertical plates separated by a distance x. The left plate has a small opening that accelerated particles can pass through. Region II has two large horizontal capacitor plates with a separation of y and a magnetic field created by current-carrying solenoid coils, which are not pictured in the figure. The magnetic field is directed upward out of the page.

In an experiment a single proton P is placed at the launch point near the right plate in Region I. The proton accelerates to the left through the hole and continues on a straight path through Region II as seen in the figure.

(A) What is the direction of the electric field in Region I? Justify your answer.
(B) Which capacitor plate has the higher potential in Region II? Justify your answer.
(C) The experiment is repeated, replacing the proton with an alpha particle that has a mass approximately four times larger than the proton and a charge two times larger than the proton. The alpha particle is released from the launch point and passes through the hole in the left plate. Compare the motion of the alpha particle to the motion of the proton through Regions I and II. Justify your reasoning.

The Region I capacitor plates have a potential difference of 5400 V and a plate separation x of 0.14 m. The Region II capacitor plates have a separation y of 0.060 m and a magnetic field of 0.50 T.

(D) A proton is again placed at the launch point near the right plate in Region I. Derive an algebraic expression for the velocity of the proton as it passes through the hole in the left plate. Use your expression to calculate the numerical value of the velocity.
(E) Using your work from (D), derive an algebraic expression for the potential difference that must be applied to the capacitor in Region II so that the proton moves in a straight line through the region. Use your expression to calculate the numerical value of the potential difference.
(F) Keeping the potential difference the same as the calculated value from (E), the scientist places an unknown particle at the launch point and observes that it travels straight through both Regions I and II just as the proton did. Discuss what the scientist can deduce about the unknown particle. Justify your answer, making appropriate reference to the algebraic expressions derived in (D) and (E).

GO ON TO THE NEXT PAGE

4. (10 points—suggested time 20 minutes)

The figures show two different representations of the same plane wave traveling through medium #1 and approaching a boundary between two transparent media. The left figure shows a light ray representation. The right figure shows a wave front representation. The index of refraction of medium #1 is greater than that of medium #2.

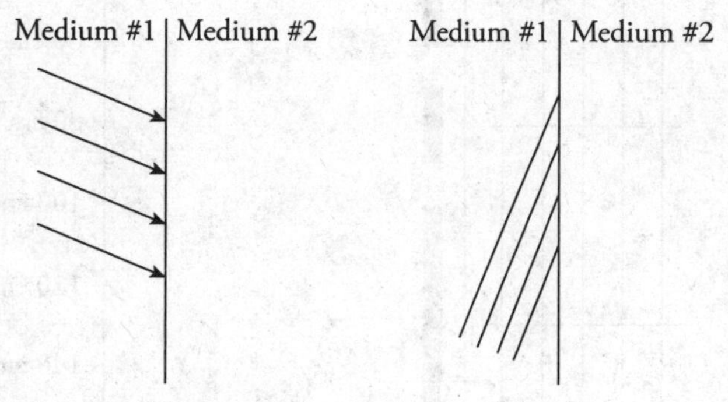

(A) i. On the left figure, complete the diagram by sketching the path of all four rays in medium #2.
ii. On the right figure, complete the diagram by sketching all four wave fronts in medium #2.

(B) The figure represents another series of plane waves traveling toward and incident on a barrier in its path. One wave model treats every point on a wave front as a point source.

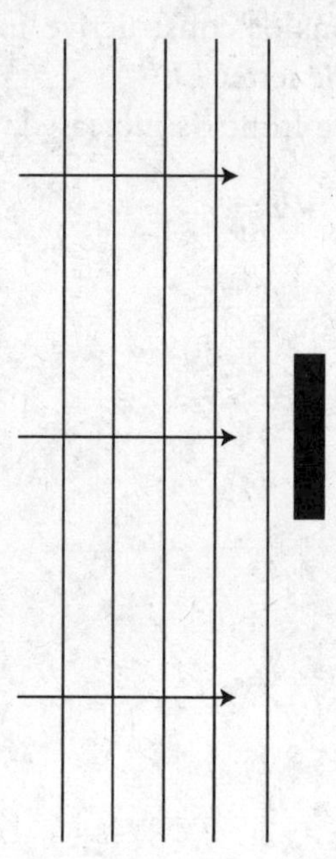

i. Use the point source model to help you sketch the wave as it passes the barrier.
ii. In a clear, coherent paragraph-length response, describe how this point source model explains the shape of the wave as it passes the barrier and why an interference pattern is produced beyond the barrier on the right.

(C) The figure represents another series of plane waves traveling toward and incident on a barrier with two identical openings.

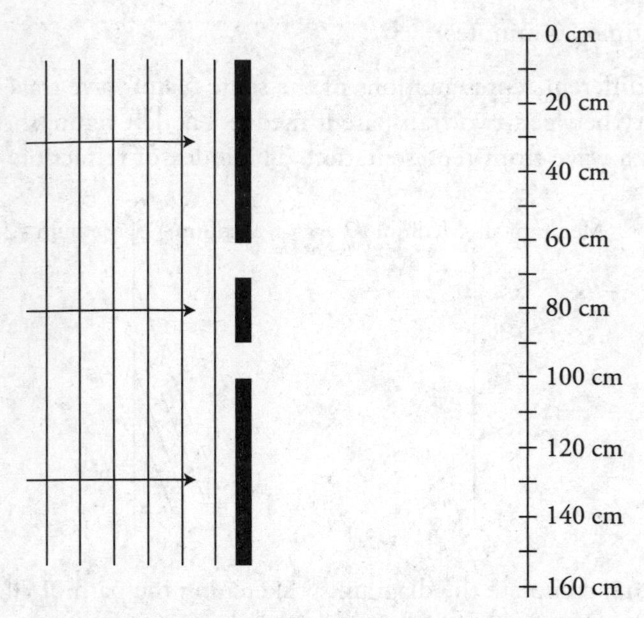

i. Sketch a representation of the interference pattern produced on the wall beyond the barrier. The wall is marked with centimeters for your reference. The 80-cm mark is aligned with the center of the barrier. Indicate locations of constructive interference with the letter C and locations of destructive interference with the letter D.

ii. The distance between the wave fronts is increased. How does this affect the interference pattern? Explain your answer.

STOP: End of AP Physics 2 Practice Exam, Section 2 (Free Response)

Solutions: Section 1 (Multiple Choice)

Questions 1–45: Single-Choice Items

1. **A**—Particle A bends in the opposite direction with a tighter radius, indicating that it is negative with a smaller mass. A β-particle would satisfy these requirements. Particle B must be negative and have a larger mass. All charges would be affected in some way by the electric field; thus particle C must have no charge. Particle D must be positive and have a smaller mass than a proton. Perhaps it is a positron?

2. **C**—The electric field is always perpendicular to the potential isolines, pointing away from higher potential and toward lower potential, which would be downward and to the left. However, an electron would receive a force in the opposite direction to the field.

3. **C**—Electric fields always point toward decreasing electric potential, are perpendicular to the potential isolines, and are strongest where the change in potential is greatest: $E = \frac{\Delta V}{\Delta r}$ (i.e., isolines of potential are closer together).

4. **A**—In its original location x, the two electrostatic forces on $+q$ add as shown in the figure.

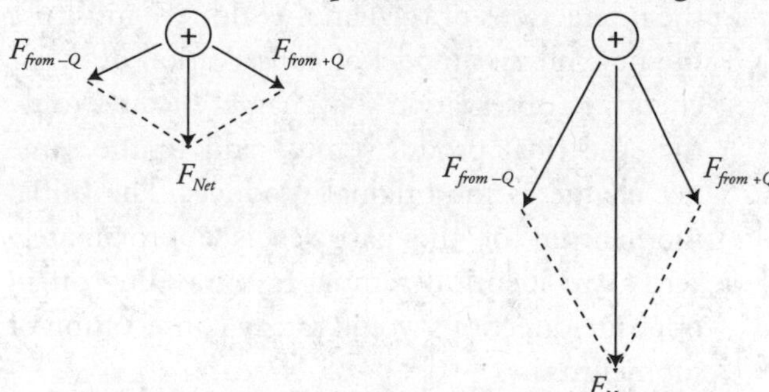

As the charge $+q$ is moved toward $+Q$ and $-Q$, the electrostatic forces increase in strength, and their direction also shifts as shown in the figure. Thus, the net force increases in strength, but continues to point directly downward due to the symmetry of the charge arrangement.

5. **D**—The force between the charges is identical but in opposite directions (Newton's third law). The more massive object will have a smaller acceleration, so they collide closer to it. Or, using the ideas of conservation of momentum, the center of mass of the system is closer to the right mass, and the original velocity of the center of mass is zero. Since there are no external forces on the system, the two charges will collide at the center of mass.

6. **C**—With the switch open, the circuit is a simple series circuit with an equivalent resistance of $2R$. Both ammeters will receive the same current:

$$I = \frac{V}{R} = \frac{\varepsilon}{2R}$$

When the switch is closed, the two resistors in parallel on the right add to give $\frac{1}{2}R$, which, when added in series to the resistor in the main line, gives a new equivalent resistance for the circuit of $\frac{3}{2}R$. This will give a new, larger total current passing through the battery and ammeter A_1:

$$I_T = I_{A_1} = \frac{V}{R} = \frac{\varepsilon}{\frac{3}{2}R} = \frac{2\varepsilon}{3R}$$

Ammeter A_2, however, receives only half of this total current as the total current splits evenly to pass through each of the parallel section on the right of the circuit:

$$I_P = I_{A_2} = \frac{I_{A_1}}{2} = \frac{2\varepsilon}{6R} = \frac{1\varepsilon}{3R}$$

Thus, the current in A_2 decreases when the switch is closed.

7. **D**—There is some data scatter, but the data appears to be linear with a slope of about 1/5. Current, as a function of voltage, is given by: $I = \frac{\Delta V}{R} = \frac{1}{R}\Delta V$. Thus, the slope of the line should equal $1/R$. Therefore, the resistance is approximately 5 Ω.

8. **A**—The currents must be the same because there are no branching paths for the current to take.

9. **D**—$\Delta V = IR$. The currents are all the same. Thus, the potential difference depends on the resistance of the three shapes. Higher resistance = Higher potential difference ($R = \frac{\rho l}{A}$).

Therefore, the longest/skinniest shape has the largest potential difference, and the shortest/fattest shape has the smallest potential difference.

10. **A**—Capacitance for a parallel plate capacitor is: $C = \kappa\varepsilon_0 \frac{A}{d}$. Since distance decreases, capacitance must have gone up. Energy stored in a capacitor is $U_C = \frac{1}{2}C(\Delta V)^2$. Capacitance has increased while the voltage has stayed the same, because the capacitor is still connected to the battery. Therefore, the energy has increased. The electric field of a capacitor is $E = \frac{Q}{\varepsilon_0 A}$ and the charge on a capacitor can be found from $\Delta V = \frac{Q}{C}$. We already know that capacitance has increased but the voltage has stayed the same. This means that the charge on the plates Q has increased. Therefore, the electric field must have also increased.

11. **C**—Using the right-hand rule (RHR) for a current-carrying wire, we can determine that the magnetic field around the wire points in the $-z$ direction in the vicinity of the proton. Using the RHR for forces on moving charges, we can determine that the proton will experience a magnetic force in the $+x$ direction.

12. **C**—For a charged particle to cross undeflected straight through crossed, perpendicular, magnetic, and electric fields, the electric and magnetic forces on the charge must cancel:

$$F_{Electric} = F_{Magnetic}$$

$$E\overline{q} = \overline{q}v\ B \sin \theta$$

Note that since the velocity is perpendicular to the magnetic field: $\sin \theta = \sin (90°) = 1$. Therefore, $E = vB$ when the charge travels in a straight line through the fields, the electric force on the proton is directed to the right, while the magnetic force is directed to the left. Therefore, to make the charge deflect to the left, one of the following must happen: the electric field must decrease, the magnetic field must increase, or the velocity of the charge must increase. Note that changing the charge will have no effect as the charge q cancels out of the equation above.

13. **A**—Nonmoving charges do not experience a magnetic force. However, the charge will polarize the magnet. This charge separation will cause an electric attraction force between the two.

14. **D**—Stationary charges do not experience a magnetic force. Electric fields always exert a force on charges.

15. **B**—From the equation that models this interference pattern ($d \sin \theta = m\lambda$), we see that in order to make the fringe spacing farther apart, we need the angle θ to get larger. This could be accomplished by decreasing d (the width of the hair) or increasing λ (the wavelength of the laser). Neither of these is listed. The spacing could also be spread farther apart by simply moving the screen farther away from the hair.

16. **A**—This is plenty of information! Just trace a line from the entrance location of the light beam to the exit location of the brightest ray. Measure the angles of incidence and refraction and use Snell's law to calculate the index of refraction of the plastic rectangle.

17. **C**—A and D are not correct. The matter and antimatter particles would annihilate into energy (EM waves). In addition, neither conservation of mass nor conservation of energy properly model the interactions of the nano-world. We must use the new physics model of conservation of mass/energy. Conservation of charge dictates only that the final products must add to the same net charge as the original products. The initial momentum of the particles is approximately zero. Two identical gamma rays traveling off in opposite directions would satisfy conservation of momentum.

18. **A**—Different isotopes of atoms have the same number of protons and electrons. Having the same electron configuration means the two isotopes have the same chemical properties. The half-life of carbon-14 is too short to date the age of dinosaur bones. There is little to no measurable carbon-14 left in bones that are millions of years old. During a spontaneous decay of a nucleus, energy is released and the mass of the daughter nuclei is less than the original nucleus.

19. C—The amplitude of the wave function (positive or negative) indicates the probability of finding the particle at that location. The probability of finding the particle is the same at locations 1.5 nm and 4.5 nm.

20. D—Using the conservation of mass/continuity equation: $A_1 v_1 = A_2 v_2$.

The area is proportional to the diameter squared $A = \pi\left(\frac{d}{2}\right)^2$, there are 10 openings on the sprinkler, and $d_1 = 10d_2$. This gives us the equation: $\pi\left(\frac{10\,d_2}{2}\right)^2 v_1 = 10\left[\pi\left(\frac{d_2}{2}\right)^2\right] v_2$. Canceling terms and solving for v_2, the velocity of the water exiting the sprinkler holes is 10 v_1.

21. D—Since the fluid is poured in slowly, we can assume the fluid to be static. In a static fluid, the pressure must be the same along any horizontal line. If this were not true, the differential pressure would cause the fluid to move. Since all the tubes are open to the same atmosphere at the top, and the pressure at the bottom of each tube must be the same, each tube must have the same height of fluid: $p = p_0 + \rho g h$.

22. D—At the same temperature, the average kinetic energies are the same for each gas even though the hydrogen will have a higher average velocity due to its lighter mass ($K_{avg} = \frac{3}{2}k_b T$). Without knowing the amount of gas in each container, we cannot determine the pressure ($PV = nRT$) or the density of the gases $\left(\rho = \frac{m}{V} = \frac{\text{(number of moles)(molar mass)}}{V}\right)$. Just because the containers are the same size does not mean they contain the same number of molecules!

23. D—The average speed of the molecules will increase with temperature, with some still moving slow but others moving faster. This moves the peak of the graph to the right and spreads the graph out more than before. Since the number of molecules has not changed, the peak of the graph is lower than before but the total area under the graph should be the same.

24. D—The gas is expanding indicating that it is doing work on the environment, thus losing energy to the environment ($-W$). Heat transfer is zero. Therefore, by the first law of thermodynamics ($\Delta U = Q + W$), ΔU is also negative. This indicates that the gas has lost energy and has decreased in temperature.

25. B—Electric fields are always perpendicular to the electric potential isolines and directed from more positive potentials toward more negative potentials. The electric field strength is calculated with the equation $E = \frac{\Delta V}{\Delta r} = \frac{3\text{ V}}{0.04\text{ m}} = 75\frac{\text{V}}{\text{m}}$.

26. C—The electric fields from the two charges add up at location B, making it the strongest. The electric field from the two charges are in opposite directions at A and C. However, at location A, the electric field must be larger than that at location C, due to its proximity to the largest charge.

27. D—Using conservation of energy:

$$E_i = E_f$$

$$(K + U_E)_i = (K + U_E)_f$$

$$K_f = K_i + \Delta U_E$$

$$K_f = \tfrac{1}{2}mv_0^2 + \Delta Vq + \tfrac{1}{2}mv_0^2 + 2Ve$$

28. C—The capacitor is originally uncharged. When the switch is closed, the capacitor acts as a short circuit wire and effectively takes the middle resistor out of the circuit. All current through A_1 will stop. The total resistance of the circuit temporarily decreases and the current through A_2 increases.

29. B—$P = I\Delta V = I^2 R = \frac{\Delta V^2}{2}$. Since voltage is being held constant and resistance is being varied, it seems most beneficial to consider the equation: $P = \frac{\Delta V^2}{R}$. Each time a resistor is added, the power decreases. Therefore, the resistance must be increasing, which implies that resistors are being added in series. To confirm this suspicion, use the power equation and the voltage of the battery to calculate that one resistor is 20 Ω, Two resistors is 40 Ω, three resistors is 60 Ω, and so on.

30. D—Adding the 1-Ω and 2-Ω resistors in series produces an equivalent resistance of 3 Ω. Next, add the resulting 3-Ω resistor to the 2-Ω resistor in parallel to produce the required 1.2-Ω resistor:

$$\frac{1}{R_p} = \frac{1}{2\,\Omega} + \frac{1}{3\,\Omega} = \frac{5}{6\,\Omega}$$

$$R_p = \frac{6\,\Omega}{5} = 1.2\,\Omega$$

Note that both answer choices A and D produce an equivalent resistance of 1.2 Ω! However, answer choice D only needs two 2-Ω resistors while answer choice A needs five 2-Ω resistors, making answer D a more effective and cheaper solution.

31. **B**—To produce current in the wire, we need to create a change in magnetic flux through the wire loop. This is accomplished by changing one of the following: magnetic field strength, orientation of the magnetic field through the loop, or the magnetic flux area. By moving the magnet to the right, away from the loop, we change the flux by changing the magnetic field strength through the wire loop.

32. **C**—The gravitational force can only be neglected when it is very small in comparison to the magnetic force.

33. **C**—The force between the magnets is equal in size and opposite in direction, but magnet A is more massive and will, therefore, have a smaller acceleration.

34. **B**—According to the right-hand rule for magnetic forces on moving charges and because the magnetic field is upward at the location of the charge, the charge must be moving to the left.

35. **B**—First off, choosing a point on the graph and plugging it into the lens equation will get you a number, but you will be wasting all the other data from the lab. This is not the best use of the data. This line was obtained by rearranging the lens equation to produce a straight line (linearization). This is a great way to eliminate error and get a more accurate answer. Rearrange the lens equation to linearize the data: $\frac{1}{d_0} = -\frac{1}{d_i} + \frac{1}{f}$. From this we can see that, if we plot $1/d_0$ on the y-axis and $1/d_0$ on the x-axis, we should get a graph with a –1 slope and a y-intercept of $1/f$. So, draw a best-fit line, find the y-intercept, and take the reciprocal, and you will have the focal length of the lens.

36. **A**—The wavelength of a proton is given by: $\lambda = \frac{h}{p} = \frac{h}{mv}$. Thus, the faster the electron is moving, the smaller its wavelength; the greater the accelerating potential, the higher the velocity will be. Therefore, as the accelerating potential increases, the wavelength decreases (an inverse relationship).

37. **D**—During the collision, the photon transferred energy and momentum to the electron. This means the wavelength must have increased: $E = hf = \frac{hc}{\lambda}$.

38. **C**—The mass and Styrofoam block are originally floating as a unit, and must be displacing enough volume of water to produce a buoyancy force equal to their combined weight. Because both shapes are solid and cannot be filled with water, turning the contraption over will not change the fact that they can still displace enough water to float. However, the submerged mass displaces water, requiring less Styrofoam water displacement to produce the required buoyancy force. Thus, the contraption floats with some of the Styrofoam sticking out of the water.

39. **A**—All stoppers have the same height of fluid above them and thus all have the same static fluid pressure.

40. **D**—All the answer choices are true, but the primary reason cold water can kill a person so quickly is that water is denser than air. Water has more molecules per unit volume, facilitating more collisions between slower-moving molecules of water and faster-moving molecules in the person. This conducts heat away faster.

41. **B**—The beginning and ending points are the same. Therefore, change in temperature and change in internal energies will be the same. The areas under the graphs are different. Therefore, the work will not be the same for the two paths. Since the ΔU are the same but the W is not the same for the two paths, the heat Q must also be different for the two paths ($\Delta U = Q + W$).

42. A—When submerged in water, the light will not refract as much passing through the lens, because the speed of the light will not change as much traveling from oil to glass as it did traveling from air to glass. This will move the focal point farther from the lens. In fact, if the oil has the same index of refraction as the glass, light will not change direction as it passed through the lens, because it did not change speed passing through the lens.

43. D—Blocking part of the lens does not keep other parts of the lens from producing an image. It just makes the image dimmer because less of the lens is being used to produce the image. This is what happens when you squint to block a bright light. You can still see everything clearly. You just make the image dimmer.

44. C—Mass/energy is conserved in nuclear reactions. The mass of the carbon + mass of the hydrogen = mass of the nitrogen + the mass equivalent of the gamma ray $m_\gamma = \frac{E_\gamma}{c^2} = \frac{hf}{c^2}$.

45. C—Higher intensity (brighter light) simply means more photons. Therefore, higher-intensity light will eject more electrons. Lower-intensity light ejects fewer electrons. When the intensity is zero, no electrons will be emitted. So we have answer choices B and C to choose from. The light is monochromatic (same wavelength and frequency); therefore, the maximum energy of each ejected photon is the same no matter how bright the light intensity is. $K_{max} = hf - \phi$.

Questions 46–50: Multiple-Correct Items

(You must indicate both correct answers; no partial credit is awarded.)

46. C and D—The volume is being held constant. Therefore, no relationships can be found from this set of data concerning the volume. From the data, we can see that each time pressure increases 10 kPa the temperature increases 20°C. This indicates a direct relationship. The number of moles can be calculated using the ideal gas law. Just remember to convert the units of volume to m^3 and temperature to K.

47. C and D—The gravitational force can only be neglected when it has little influence on the behavior of the object, or when it is very small in comparison to other forces. Even though the electric force is larger than the gravitational force acting on the balloon, gravity is still a sizable component of the overall forces on the balloon. The gravity force and electric force must be equal for the droplet of oil to levitate. However, the gravitational force on subatomic particles is negligible compared to the electric force experienced in electrical circuits and between charged particles in an atom.

48. A and C—The electroscope is initially charged in Figure 1 because the leaves repel each other. When the negatively charged rod is brought close, but not in contact with the electroscope, it repels electrons toward the bottom of the electroscope. Since this brings the leaves closer together, the original charge must have been positive. Note that the rod cannot "discharge" the electroscope without touching it. Positive charge carriers do not move through solid objects, because they are buried in the nucleus.

49. A and B—Applying Kirchhoff's junction rule to the node above the A_2 ammeter we get:

$$\Sigma I_{junction} = I_1 + I_2 - I_3 = 0$$

Applying Kirchhoff's current rule to the loop on the right of the circuit containing the battery we get:

$$\Sigma \Delta V_{loop} = \Delta V_B - \Delta V_1 - \Delta V_2 = 0$$

$I_2 < I_3$: Ammeter 3 is in the main circuit line and carries the total current. This current splits with half going through ammeter 1 and 2.

$V_1 = V_2$: If you work out the math you will see that these voltages are the same.

50. B and D—$P = IV$ to increase power, the current must increase and/or the voltage must increase. Increasing the current can be accomplished by decreasing the resistance of the bulb's filament by making it thicker: $R = \frac{\rho l}{A}$ another method to decrease the resistance would be to connect two filaments in parallel. Remember that the net resistance decreases in parallel.

Solutions: Section 2 (Free Response)

Your answers will not be word-for-word identical to what is written in this key. Award points for your answer as long as it contains the correct physics explanation *and* as long as it does not contain incorrect physics or contradict the correct answer.

Question 1

Part (A)

5 points total. There is more than one way to accomplish this lab. Here are two possibilities:

Method #1

1 point—Choose a force sensor and a ruler.

1 point—Force will be manipulated by pushing down on the piston with the force sensor, and the distance the gas in the cylinder is compressed will be measured.

1 point—Labeled diagram that shows the piston, how the force will be measured, and how the distance the gas is compressed will be measured.

1 point—Work will be calculated by using $W = F\Delta x$ = area under the graph of force versus distance the gas is compressed.

1 point—Procedure: Cylinder is placed vertically on the lab table. Force sensor is used to push the piston down in 0.01-m increments until we have reached half the original volume. Force and distance compressed are recorded for each. Plot force as a function of distance compressed on a graph. The work will equal the area under the graph.

Method #2

1 point—Choose a pressure sensor and a ruler.

1 point—The pressure is manipulated and the distance the gas in the cylinder is compressed will be measured.

1 point—Labeled diagram that shows the piston, how the force will be applied and the pressure is measured, how compression distance will be measured.

1 point—Work is calculated using: $W = P\Delta V$ = area under the graph of pressure versus volume of gas inside the cylinder.

1 point—Procedure: Cylinder is placed vertically on the lab table. Push the piston down in 0.01-m increments until we have reached half the original volume. Pressure and height of the cylinder are recorded for each increment. Calculate volume using V = (area of piston) (height of gas). Plot pressure as a function of volume on a graph. The work will equal the area under the graph.

Note: As the gas is compressed, the pressure inside the gas does not remain constant. The third and fourth points are not awarded unless the student demonstrates an acceptable method for taking this into account while calculating the work. Here are two possible methods a student could use:

- Calculating work by using the area under the F versus x graph or P versus V graph
- A student could state that the force/pressure is not constant and that they will estimate an average force/pressure.

Part (B)

1 point—Correct statement: The temperatures will be different depending on the speed of compression.

1 point—Justification: When compressing the gas very slowly, heat has time to escape, lowering the final temperature. If the compression is done quickly, there is no time for heat to escape and the final temperature will be higher.

Part (C)

1 point—Incorrect statement: The work will be the same in either case.

1 point—Justification: Compressing the gas quickly will not allow time for heat to escape to the environment. This will give the gas a higher final temperature and thus a higher final pressure for the same final volume. This means more force is required to compress the gas quickly and thus more work is required as well.

Part (D)

1 point—Sketching a path that starts at P_0, V_0, ends at 2 P_0, $V_0/2$.

1 point—The sketch is concave upward: an isothermal line.

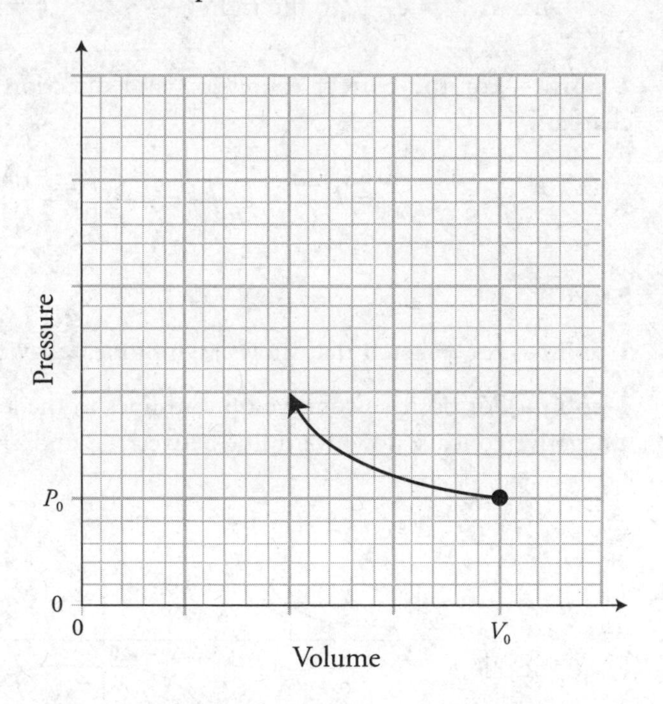

Part (E)

1 point—The work would be calculated by finding the area under the sketched line.

Question 2

Part (A)

1 point—For stating that the net electric field will be to the right.

1 point—For a correct explanation. For example: The electric field is directly proportional to the size of the charge and inversely proportional to the radius squared ($E \propto \frac{q}{r^2}$). Since the $-2q$ is twice the charge and three times the distance of the $+q$ charge, the electric field will be dominated by the $+q$ charge.

Part (B)

1 point—For any statement that shows that the net electric field is the vector addition of the two separate electric fields created by $-2q$ and $+q$. AND that the two fields are in the opposite direction.

1 point—For a correct expression for the magnitudes of the electric fields from both charges: (Note that the correct directions are not needed to earn this point.)

$$E_{-2q} = k\frac{2q}{x^2} \text{ to the right} \qquad E_{+q} = k\frac{q}{(3x)^2} = k\frac{q}{9x^2} \text{ to the left}$$

1 point—For the correct expression and direction of the net electric field from both charges:

$$E_{\text{Total}} = k\frac{2q}{x^2} - k\frac{q}{9x^2} = k\frac{q}{x^2}\left(2 - \frac{1}{9}\right) = \frac{17kq}{9x^2} \text{ to the right}$$

Part (C)

1 point—For a sketch that shows asymptotic behavior near $-x$ and $+x$. See figure.

1 point—For positive concave upward lines to the left of $-x$ and to the right of $+x$. AND a negative concave downward line between $-x$ and $+x$. See figure.

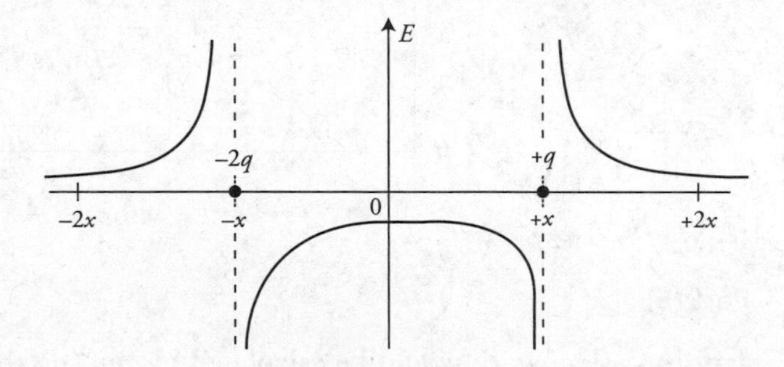

Part (D)

1 point—For indicating that there will be a zero electric field somewhere to the right of the $+q$ charge.

1 point—For a complete explanation of why there will be a zero electric field somewhere to the right of the $+q$ charge. For example: The electric fields from the two charges are in the opposite directions beyond $+x$ and $-x$. Since the left charge is larger than the right charge, there can only be a zero electric field location to the right of the $+q$ charge. This means that our graph in part (d) will eventually cross the horizontal axis and become negative somewhere to the right of point $+q$.

Part (E)

1 point—For using the equation $F_E = Eq$, substituting the answer form part (A) for the electric field, and indicating that the direction will be to the left. Work must be shown to receive this point.

Question 3

Part (A)

Electric field is to the left.

1 point—For correct justification: The electric field points in the direction of force on a positive charge. The proton accelerates to the left; therefore, the force is to the left and the electric field is also to the left. *No credit is awarded without justification.*

Part (B)

The upper plate must have a higher potential.

1 point—For utilizing the right-hand rule for forces on moving positive charges, and showing that the proton receives a magnetic force upward from the magnetic field.

1 point—For indicating that the electric force is equal and opposite in direction to the magnetic force so that the proton will travel in a straight line.

1 point—For indicating that electric fields point from higher potential to lower potential. Thus, the upper plate is a higher potential.

Part (C)

The alpha particle will curve downward in Region II.

1 point—For explaining that the alpha particle will exit Region I with a slower velocity. The electric potential energy has doubled, but the mass has quadrupled leading to a smaller exit velocity. (Or, for explaining that in Region I the electric force on the charge has doubled, but the mass has quadrupled, leading to a smaller acceleration and exit velocity.)

1 point—For explaining that the electric force will now be larger than the magnetic force causing the alpha particle to arc downward.

The electric force will double as the change doubles: $F_E = q_E$

The magnetic force will not increase as much as the electric force. Even though the charge doubles, the velocity is now slower: $F_M = qvB$.

Part (D)

Conservation of energy: $\Delta U_E = K$

1 point—For the correct expression of conservation of energy: $\Delta Vq = \frac{1}{2}mv^2$

1 point—For the correct numerical answer:

$$v = \sqrt{\frac{2\Delta Vq}{m}} = \sqrt{\frac{2(5400\text{ V})(1.6 \times 10^{-19}\text{ C})}{1.67 \times 10^{-27}\text{ kg}}} = 1.02 \times 10^6\text{ m/s}$$

Part (E)

To move in a straight line: $F_E = F_M$

1 point—For the correct expression of magnetic force equal to electric force: $E\overline{q} = \overline{q}vB$

$$E = \frac{\Delta V}{y} = vB$$

1 point—For the correct equations with substitutions:

$$\Delta V = yvB = (0.060)(1.02 \times 10^6\text{ m/s})(0.50\text{ T})$$

$$\Delta V = 30{,}600\text{ V}$$

Part (F)

1 point—From the equation derived in (E) $\Delta V = yvB$, we know that the velocity of the particle is the same because ΔV, y, and B are all the same.

1 point—From the equation in (D) $v = \sqrt{\frac{2\Delta Vq}{m}}$, we know that both v and ΔV are the same. Thus, we can deduce the charge to mass ratio q/m of the unknown particle.

Question 4

Part (A)

(i)

1 point—The left figure rays should all be parallel and pointing downward, with the angle of refraction larger than angle of incidence.

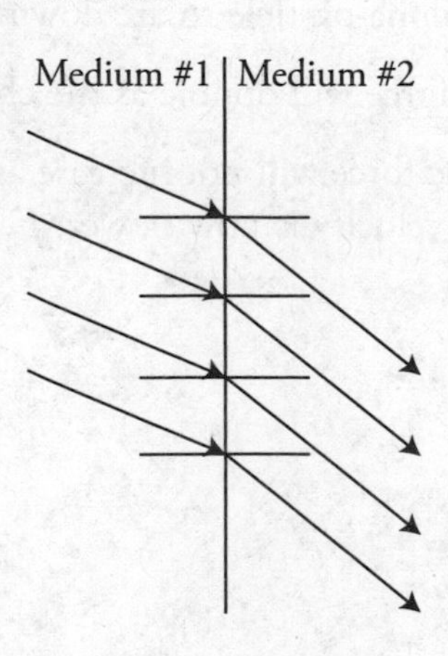

(ii)

1 point—The right figure should also show the waves traveling in a more downward direction. All wave fronts should be parallel, and the wavelength between the wave fronts should be larger in medium #2.

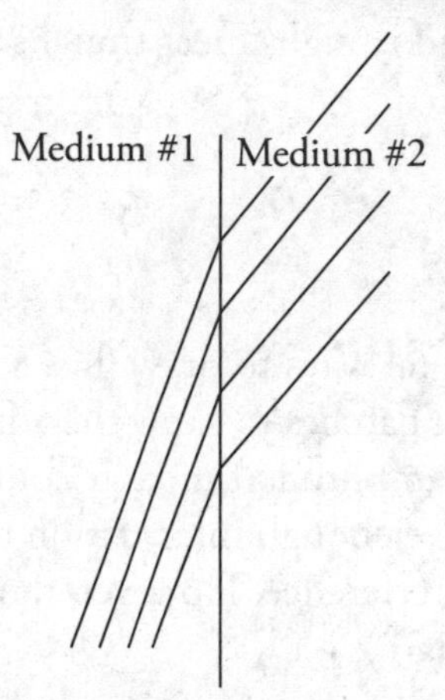

Part (B)

(i)

1 point—The sketched waves should bend around the boundary and overlap in circular paths that maintain the same wavelength.

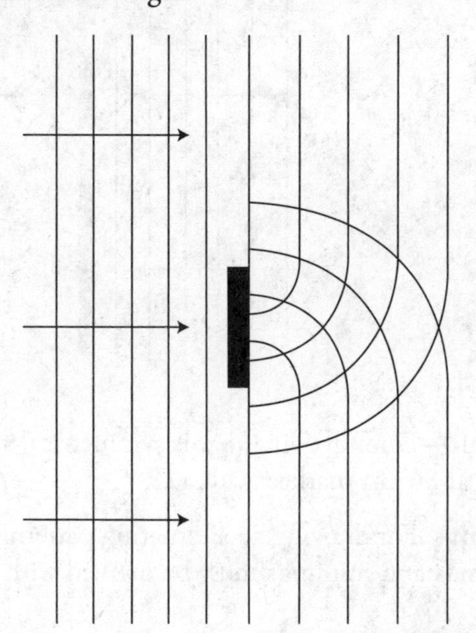

(ii)

1 point—The point source model says that every point on the wave is the source of a new wavelet that propagates outward.

1 point—For the plane waves approaching the barrier, the sum of all the wavelets produces another plane wave in front of the last one.

1 point—The barrier blocks the center wavelets. The plane waves on either side of the boundary continue forward, but the wavelets on the end of the blocked wave produce curved waves that propagate inward to fill the central area beyond the boundary.

1 point—These curved waves will overlap and form constructive interference, where crests meet crests and troughs meet troughs. They will create destructive interference where crests meet troughs.

Part (C)

(i)

There are several ways to draw this. The interference pattern could be drawn as alternating dark and light patches to show the constructive interference points as shown in this sample diagram. An amplitude function could also be used to represent the interference pattern. We do not have enough information to calculate the exact locations of the constructive and destructive interference. However, the patterns should be symmetrical about the center line (the 80-cm mark).

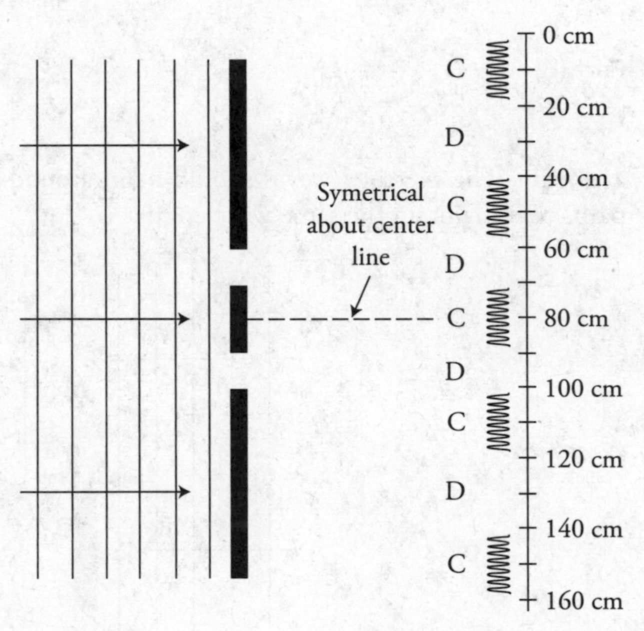

1 point—For a drawing that is symmetrical along central axis and showing a central maximum at 80 cm marked with a C.

1 point—For drawing an alternating pattern of evenly spaced maxima's and minima's. The maxima's and minima's must be marked with C and D as seen in the figure.

(ii)

1 point—For indicating that the pattern will get wider and spread out from the central maximum with a coherent explanation. For example: As the distance between the wave fronts is increased, the wavelength gets larger. This means the pattern will spread out as the angle θ gets larger: $d \sin \theta = m\lambda$.

How to Score Practice Exam 2

The practice exam cut points are based on historical exam data and will give you a ballpark idea of where you stand. The bottom line is this: If you can achieve a 3, 4, or 5 on the practice exam, you are doing great and will be well prepared for the real exam in May. This is the curve I use with my own students, and it is has been a good predictor of their actual exam scores.

Multiple-Choice Score: Number Correct ______________ (50 points maximum)

Free-Response Score: Problem 1 ______________ (12 points maximum)

Problem 2 ______________ (10 points maximum)

Problem 3 ______________ (12 points maximum)

Problem 4 ______________ (10 points maximum)

Free-Response Total: ______________ (44 points maximum)

Calculating Your Final Score

Final Score = (1.136 × Free-Response Total) + (Multiple-Choice Score)

Final Score: ____________ (100 points maximum)

Round your final score to the nearest point.

Raw Score to AP Grade Conversion Chart

Final Score on Practice Exam	AP Grade	
70–100	5	Outstanding! Keep up the good work.
55–69	4	Outstanding! Keep up the good work.
40–54	3	Great job! Practice and review to secure a score of 3 or better.
25–39	2	Almost there . . . concentrate on your weak areas to improve your score.
0–24	1	Don't give up! Study hard and move up to a better grade.

AP Physics 2: Practice Exam 3

Multiple-Choice Questions

ANSWER SHEET

1 Ⓐ Ⓑ Ⓒ Ⓓ	18 Ⓐ Ⓑ Ⓒ Ⓓ	35 Ⓐ Ⓑ Ⓒ Ⓓ
2 Ⓐ Ⓑ Ⓒ Ⓓ	19 Ⓐ Ⓑ Ⓒ Ⓓ	36 Ⓐ Ⓑ Ⓒ Ⓓ
3 Ⓐ Ⓑ Ⓒ Ⓓ	20 Ⓐ Ⓑ Ⓒ Ⓓ	37 Ⓐ Ⓑ Ⓒ Ⓓ
4 Ⓐ Ⓑ Ⓒ Ⓓ	21 Ⓐ Ⓑ Ⓒ Ⓓ	38 Ⓐ Ⓑ Ⓒ Ⓓ
5 Ⓐ Ⓑ Ⓒ Ⓓ	22 Ⓐ Ⓑ Ⓒ Ⓓ	39 Ⓐ Ⓑ Ⓒ Ⓓ
6 Ⓐ Ⓑ Ⓒ Ⓓ	23 Ⓐ Ⓑ Ⓒ Ⓓ	40 Ⓐ Ⓑ Ⓒ Ⓓ
7 Ⓐ Ⓑ Ⓒ Ⓓ	24 Ⓐ Ⓑ Ⓒ Ⓓ	41 Ⓐ Ⓑ Ⓒ Ⓓ
8 Ⓐ Ⓑ Ⓒ Ⓓ	25 Ⓐ Ⓑ Ⓒ Ⓓ	42 Ⓐ Ⓑ Ⓒ Ⓓ
9 Ⓐ Ⓑ Ⓒ Ⓓ	26 Ⓐ Ⓑ Ⓒ Ⓓ	43 Ⓐ Ⓑ Ⓒ Ⓓ
10 Ⓐ Ⓑ Ⓒ Ⓓ	27 Ⓐ Ⓑ Ⓒ Ⓓ	44 Ⓐ Ⓑ Ⓒ Ⓓ
11 Ⓐ Ⓑ Ⓒ Ⓓ	28 Ⓐ Ⓑ Ⓒ Ⓓ	45 Ⓐ Ⓑ Ⓒ Ⓓ
12 Ⓐ Ⓑ Ⓒ Ⓓ	29 Ⓐ Ⓑ Ⓒ Ⓓ	46 Ⓐ Ⓑ Ⓒ Ⓓ
13 Ⓐ Ⓑ Ⓒ Ⓓ	30 Ⓐ Ⓑ Ⓒ Ⓓ	47 Ⓐ Ⓑ Ⓒ Ⓓ
14 Ⓐ Ⓑ Ⓒ Ⓓ	31 Ⓐ Ⓑ Ⓒ Ⓓ	48 Ⓐ Ⓑ Ⓒ Ⓓ
15 Ⓐ Ⓑ Ⓒ Ⓓ	32 Ⓐ Ⓑ Ⓒ Ⓓ	49 Ⓐ Ⓑ Ⓒ Ⓓ
16 Ⓐ Ⓑ Ⓒ Ⓓ	33 Ⓐ Ⓑ Ⓒ Ⓓ	50 Ⓐ Ⓑ Ⓒ Ⓓ
17 Ⓐ Ⓑ Ⓒ Ⓓ	34 Ⓐ Ⓑ Ⓒ Ⓓ	

AP Physics 2: Practice Exam 3

Section 1 (Multiple Choice)

Directions: The multiple-choice section consists of 50 questions to be answered in 90 minutes. You may write scratch work in the test booklet itself, but only the answers on the answer sheet will be scored. You may use a calculator, the equation sheet, and the table of information. These can be found in the appendix or you can download the official ones from the College Board at: https://secure-media.collegeboard.org/digitalServices/pdf/ap/ap-physics-2-equations-table.pdf.

Questions 1–45: Single-Choice Items

Choose the single best answer from the choices provided, and mark your answer with a pencil on the answer sheet.

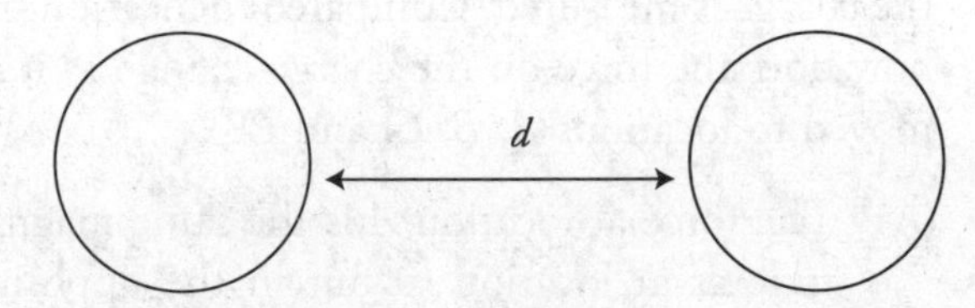

1. Two identical insulating spheres of mass m are separated by a distance d as shown in the scale drawing. Both spheres carry a uniform charge distribution of Q. The magnitude of Q is much larger than the magnitude of m. What additional information is needed to write an algebraic expression for the net force on the spheres?

(A) No other information is needed.
(B) The sign of the charge on the spheres
(C) The mass of the spheres
(D) The radius of the spheres

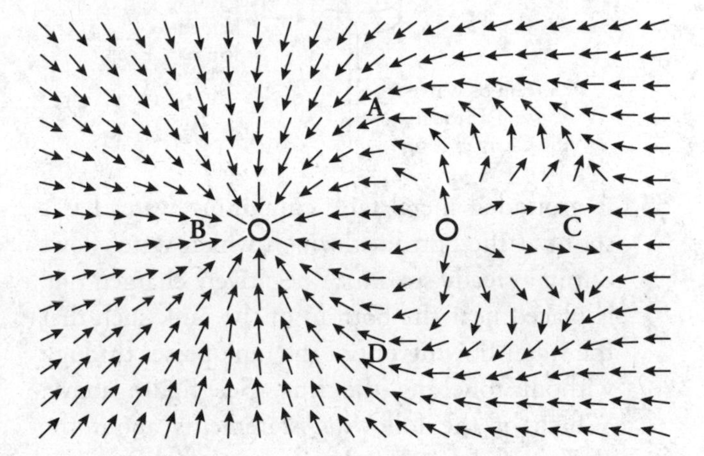

2. The figure shows the electric field in a region surrounding two charges. The vectors in the diagram are not scaled to represent the strength of the electric field but show only the direction for the field at that point. Which two points have electric fields of the same magnitude?

(A) A and B
(B) B and C
(C) C and D
(D) D and A

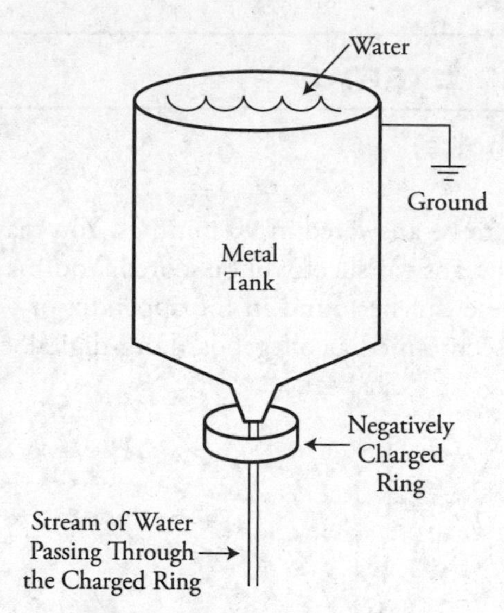

3. A grounded metal tank containing water has a spout at the bottom through which water flows out in a steady stream. A negatively charged ring is placed near the bottom of the tank such that the water flowing out of the tank passes through without touching the ring. See figure above. Which of the following statements about the charge of the water flowing out of the tank is correct?

(A) Water exiting the tank is neutral because the tank is grounded.
(B) Water exiting the tank is negative because charges jump from the ring to the water.
(C) Water exiting the tank will be positive because the ring repels negative charges in the water and tank toward the top of the tank and ground.
(D) Water exiting the tank will be positive at first but negative as the water level in the tank goes down. The ring attracts positive charges to the bottom of the tank and pushes negative charges to the top of the tank. The positive water flows out first and negative water flows out last.

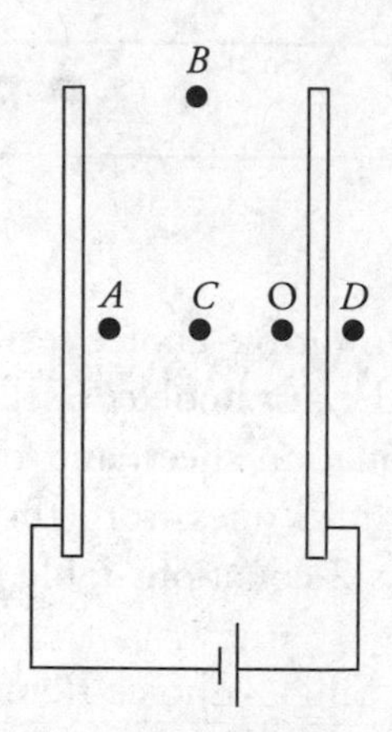

4. Two parallel metal plates are connected to a battery as shown in the figure. A small negative charge is placed at location O and the force on the charge is measured. Compared to location O, how does the force on the charge change as it is moved to locations *A*, *B*, *C*, and *D*?

(A) The force at location *A* is the same magnitude as at location O but in the opposite direction.
(B) The force at location *B* is smaller than at location O but in the same direction.
(C) The force at location *C* is smaller than at location O but in the same direction.
(D) The force at location *D* is the same magnitude as at location O but in the opposite direction.

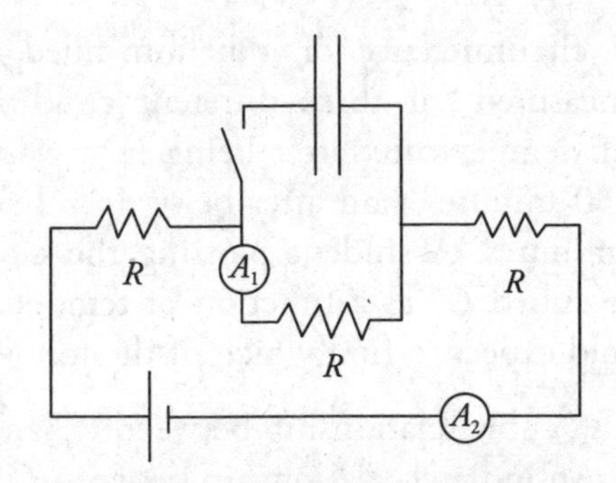

5. The circuit shown in the figure consists of three identical resistors, two ammeters, a battery, a capacitor, and a switch. The capacitor is initially uncharged and the switch is open. Which of the following correctly compares the original open switch readings of the ammeters to their readings after the switch has been closed for a very long time?

	A_1	A_2
(A)	decreases	decreases
(B)	decreases	stays the same
(C)	decreases	increases
(D)	stays the same	stays the same

6. A battery is connected to a section of wire bent into the shape of a square. The lower end of the loop is lowered over a magnet. Which of the following orientations of the wire loop and magnet will produce a force on the wire in the positive *x* direction?

(A)

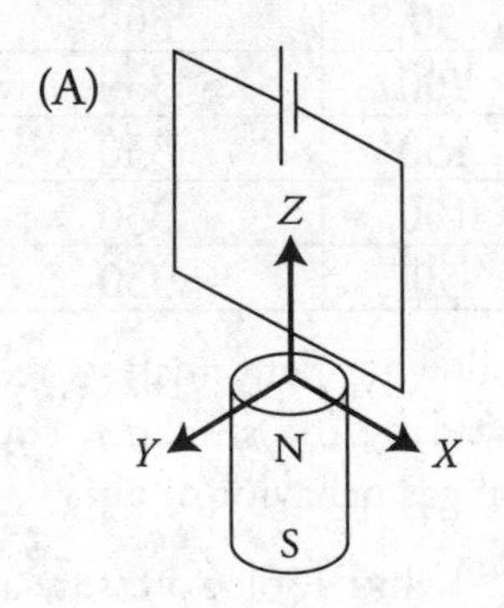

(B)

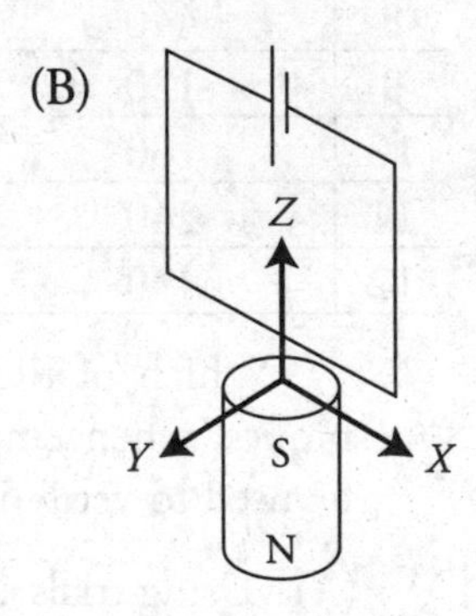

(C)

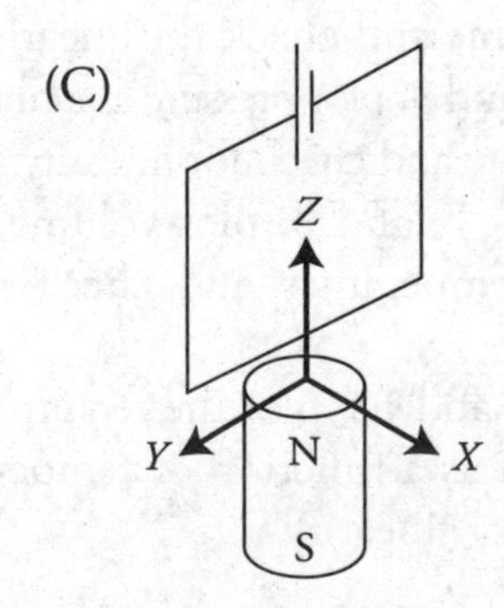

(D)

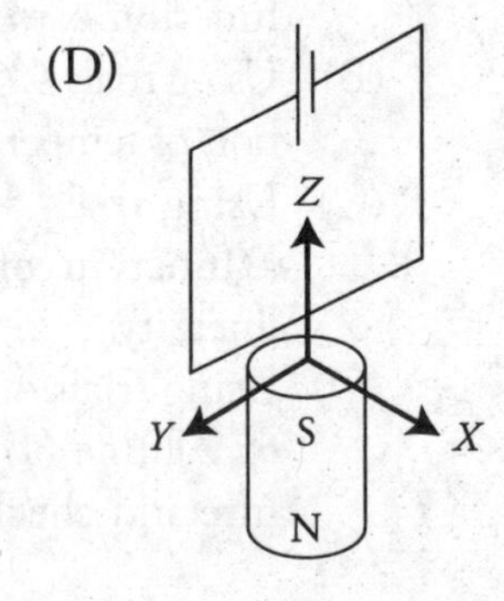

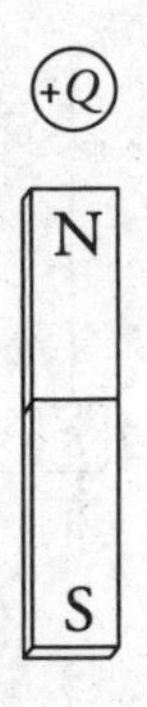

7. A charge $+Q$ is positioned close to a bar magnet as shown in the figure. Which way should the charge be moved to produce a magnetic force into the page on the bar magnet?

(A) To the right
(B) Toward the top of the page
(C) Out of the page
(D) Moving the charge will not produce a force on the magnet.

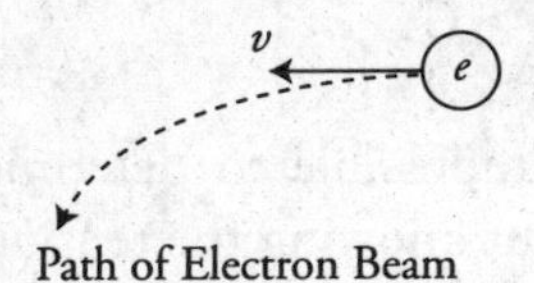

8. In an experiment, a scientist sends an electron beam through a cloud chamber and observes that the electron accelerates in a downward direction at a constant rate as shown in the figure. Which of the following could the scientist conclude?

(A) Earth's gravitational field is causing the beam to change direction.
(B) A uniform electric field is causing the beam to change direction.
(C) A uniform magnetic field is causing the beam to change direction.
(D) The electron beam collided with another particle.

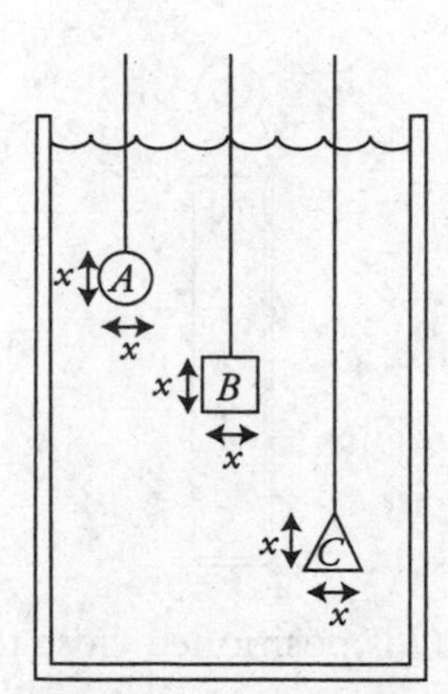

9. A sphere, cube, and cone are each suspended stationary from strings in a large container of water as shown in the figure. Each has a width and height of x. Which of the following properly ranks the buoyancy force on the objects? (Assume the vertical distance between points A and B is small.)

(A) $A > B > C$
(B) $B > A > C$
(C) $C > B > A$
(D) It is impossible to determine the ranking without knowing the tension in the strings.

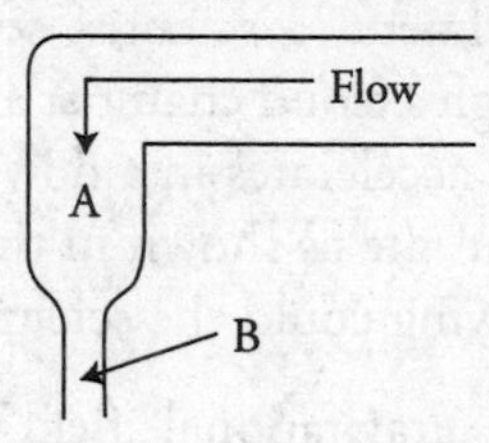

10. Bubbles in a carbonated liquid drink dispenser flow through a tube as shown in the figure. Which of the following correctly describes the behavior of the bubbles as they move from point A to point B? The vertical distance between points A and B is small.

(A) The bubbles increase in speed and expand in size.
(B) The bubbles increase in speed and decrease in size.
(C) The bubbles decrease in speed and expand in size.
(D) The bubbles decrease in speed and decrease in size.

11. The circumference of a helium-filled balloon is measured for three different conditions: at room temperature, after being in a warm oven for 30 minutes, and after being in a freezer for 30 minutes. A student plotting the circumference cubed C^3 as a function of temperature T, should expect to find which of the following?

(A) A cubic relationship between C^3 and T
(B) An indirect relationship between C^3 and T
(C) An extrapolated temperature T where C^3 reaches zero
(D) A maximum C^3 as the temperature T increases

12. A group of physics students has been asked to confirm that air exhibits properties of an ideal gas. Using a sealed cylindrical container with a movable piston, baths of cool and warm water, a thermometer, a pressure gauge, and a ruler, the students are able to produce this table of data.

Trial	Absolute Gas Pressure (kPa)	Volume (cm^3)	Temperature (K)
1	100	200	270
2	130	150	270
3	200	100	270
4	400	50	270
5	110	200	300
6	150	150	300
7	220	100	300
8	440	50	300
9	120	200	330
10	160	150	330
11	240	100	330
12	490	50	330

Which of the following data analysis techniques, when employed by the students, could be used to verify ideal gas behavior of air?

(A) Using trials 1, 2, 3, and 4, plot pressure as a function of volume and check for linearity.
(B) Using trials 1, 5, and 9, plot pressure as a function of temperature and check for linearity.
(C) Using trials 5, 6, 7, and 8, plot volume as a function of temperature and check for linearity.
(D) Using trials 4, 8, and 12, plot the reciprocal of volume ($1/V$) as a function of temperature and check for linearity.

GO ON TO THE NEXT PAGE

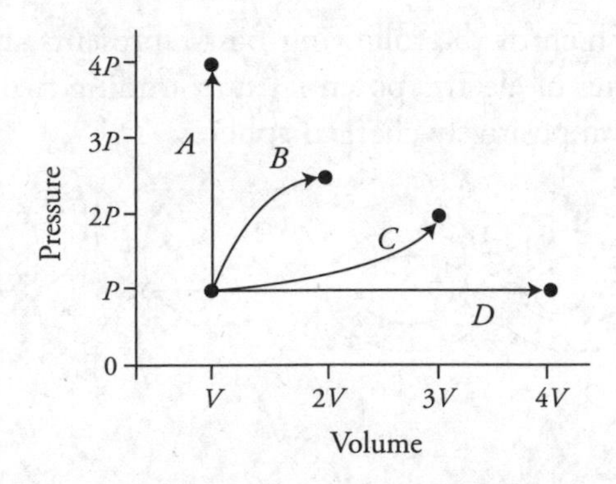

13. A gas is initially at pressure P and volume V as shown in the graph. Along which of the labeled paths could the gas be taken to achieve the greatest increase in temperature?

(A) Path A
(B) Path B
(C) Path C
(D) Path D

14. It is observed that sounds can be heard around a corner but that light cannot be seen around a corner. What is a reasonable explanation for this observation?

(A) Light travels at 3×10^8 m/s, which is too fast to change direction around a corner.
(B) Sound has a longer wavelength, which increases the diffraction.
(C) Light is an electromagnetic wave that is behaving as a particle.
(D) Sound is a mechanical wave that can change direction in its propagation media.

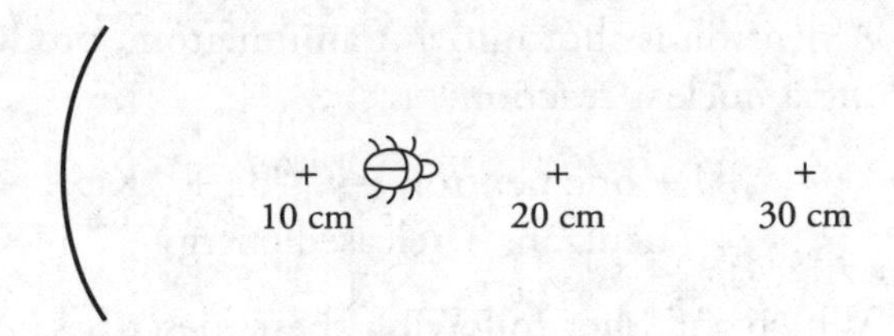

15. A bug crawls directly away from a mirror of focal length 10 cm as shown in the figure. The bug begins at 13 cm from the mirror and ends at 20 cm. What is happening to the image of the bug?

(A) The image inverts.
(B) The image gets larger in size.
(C) The image is becoming the same size as the bug.
(D) The image is moving away from the lens to the right.

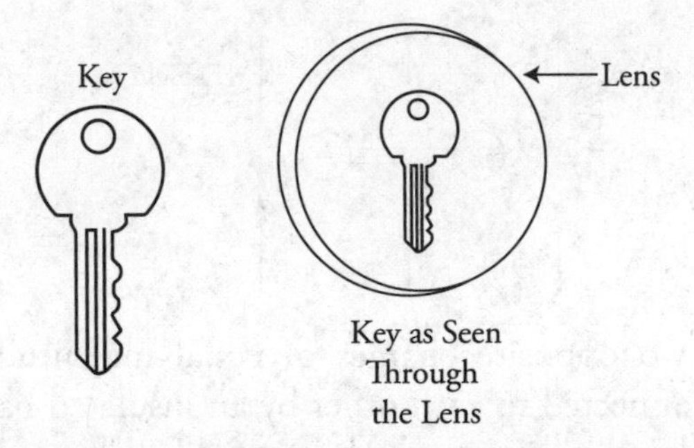

16. A student looks at a key through a lens. When the lens is 10 cm from the key, what the student sees through the lens is shown in the figure. The student estimates that the image is about half the size of the actual key. What is the approximate focal length of the lens being used by the student?

(A) –10 cm
(B) –0.1 cm
(C) 0.3 cm
(D) 3.0 cm

17. A neutron is shot into a uranium atom, producing a nuclear reaction:

$$^{235}_{92}U + \text{one neutron} \rightarrow\ ^{142}_{56}Ba +\ ^{91}_{36}Kr + \text{neutrons} + \text{released energy}$$

Which of the following best describes this reaction?

(A) The reaction products include two neutrons.
(B) Combining uranium with a neutron is characteristic of nuclear fusion.
(C) The released energy in the reaction is equal to the kinetic energy of the neutron shot into the uranium.
(D) The combined mass of uranium-235 and a neutron will be greater than the sum of the mass of the reaction products.

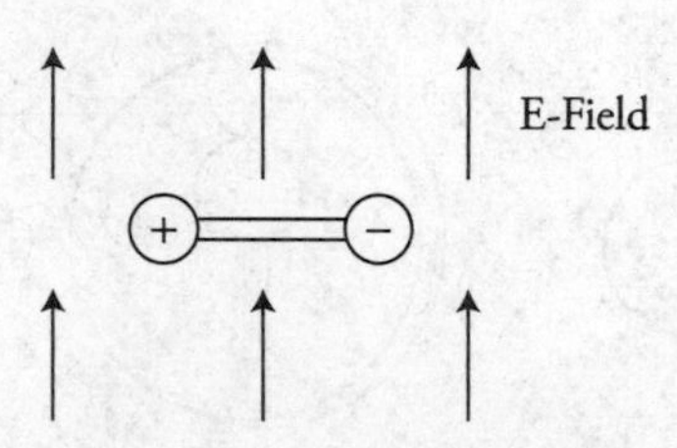

18. Two opposite charges of equal magnitude are connected to each other by an insulated bar and placed in a uniform electric field as shown in the figure. Assuming the object is free to move, how will the object move and why?

(A) It will remain stationary because the object has a net charge of zero.
(B) It will rotate clockwise at a constant rate because both charges and the electric field are constant.
(C) It will rotate at a constant rate until aligned with the electric field and then stop rotating because the net force will equal zero when aligned with the field.
(D) It will rotate back and forth clockwise and counterclockwise because the torque changes as the object rotates.

19. Which of the following best represents the isolines of electric potential surrounding two identical positively charged spheres?

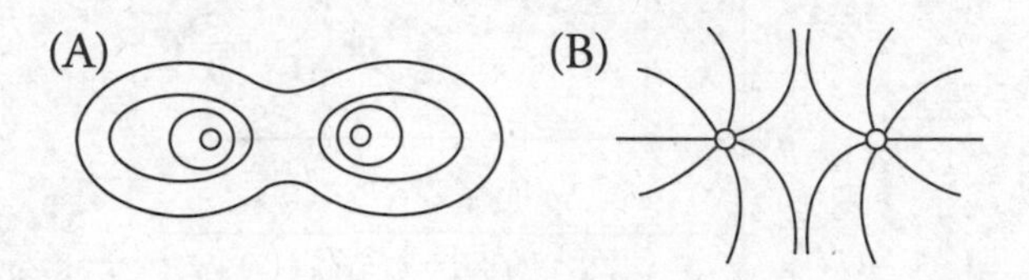

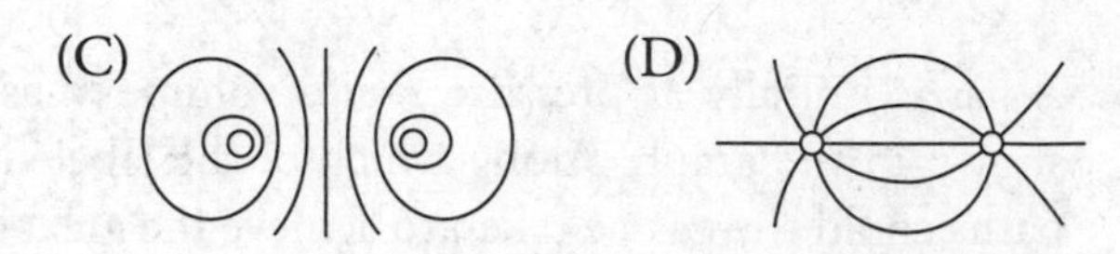

Questions 20 and 21 refer to the following material.

The diagram shows a circuit that contains a battery with a potential difference of V_B and negligible internal resistance; five resistors of identical resistance; three ammeters A_1, A_2, A_3; and a voltmeter.

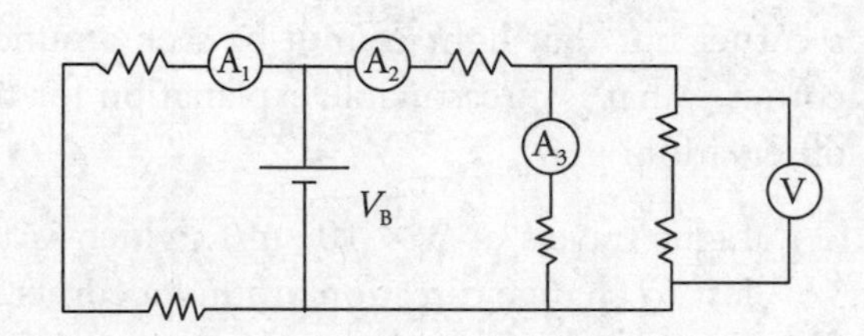

20. Which of the following correctly ranks the readings of the ammeters?

(A) $A_1 = A_2 = A_3$
(B) $A_1 = A_2 > A_3$
(C) $A_1 > A_2 > A_3$
(D) $A_2 > A_1 > A_3$

21. What will be the reading of the voltmeter?

(A) $\frac{1}{5}V_B$
(B) $\frac{1}{3}V_B$
(C) $\frac{2}{5}V_B$
(D) $\frac{3}{5}V_B$

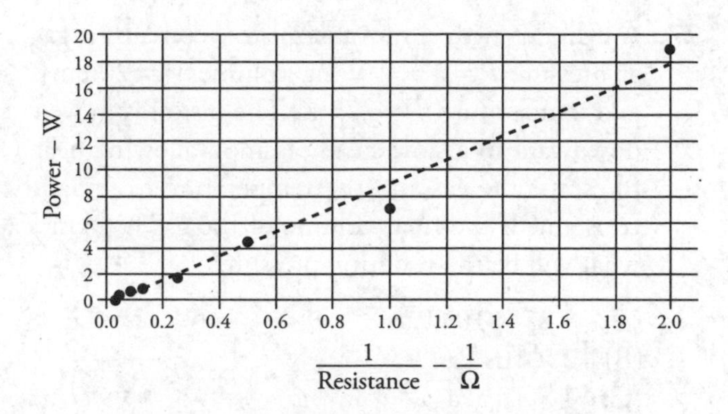

22. A battery of unknown potential difference is connected to a single resistor. The power dissipated in the resistor is calculated and recorded. The process is repeated for eight resistors. A plot of the data with a best-fit line was made and is displayed in the figure. The potential difference provided by the battery is most nearly:

(A) 3 V
(B) 6 V
(C) 18 V
(D) 36 V

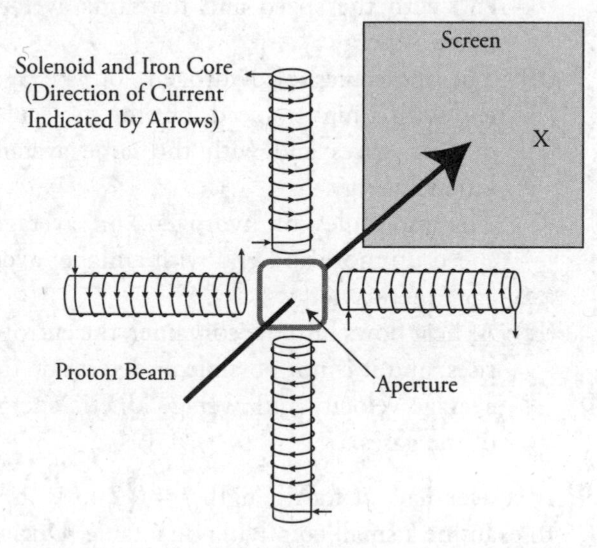

23. A high-energy proton beam is used in hospitals to treat cancer patients. The beam is shot through a small aperture surrounded by four solenoids with iron cores that are used to direct the beam at cancer cells to kill them. (The direction of the current around the solenoids is indicated by the arrows.) During routine maintenance, technicians calibrate the machine by pointing the beam at the center of a screen and then directing it toward designated target points. Which of the four solenoids will the technician use to direct the beam toward the target "X" on the right side of the screen?

(A) The top solenoid
(B) The right solenoid
(C) The bottom solenoid
(D) The left solenoid

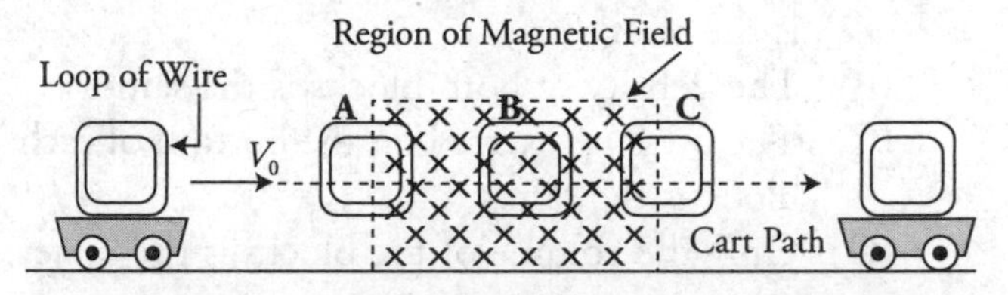

24. A lab cart with a rectangular loop of metal wire fixed to its top travels along a frictionless horizontal track as shown. While traveling to the right, the cart encounters a region of space with a strong magnetic field directed into the page. The cart travels through locations A, B, and C on its way to the right as shown in the figure. Which of the following best describes any current that is induced in the loop or wire?

(A) Current is induced in the loop at all three locations A, B, and C.
(B) Current is induced in the loop only at locations A and C.
(C) Current is induced in the loop only at location B.
(D) No current is induced in the loop, because the area of the loop, the magnetic field strength, and the orientation of the loop with respect to the magnetic field all remain constant while the cart moves to the right.

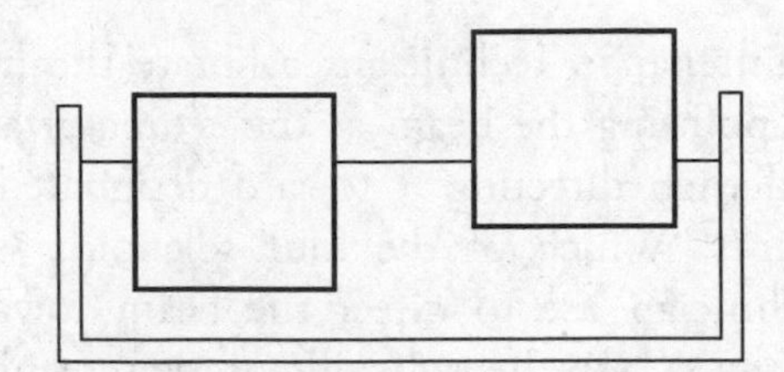

25. Two blocks of the same size are floating in a container of water as shown in the figure. Which of the following is a correct statement about the two blocks?

(A) The buoyancy force exerted on both blocks is the same.
(B) The density of both blocks is the same.
(C) The pressure exerted on the bottom of each block is the same.
(D) Only the volume of the blocks is the same.

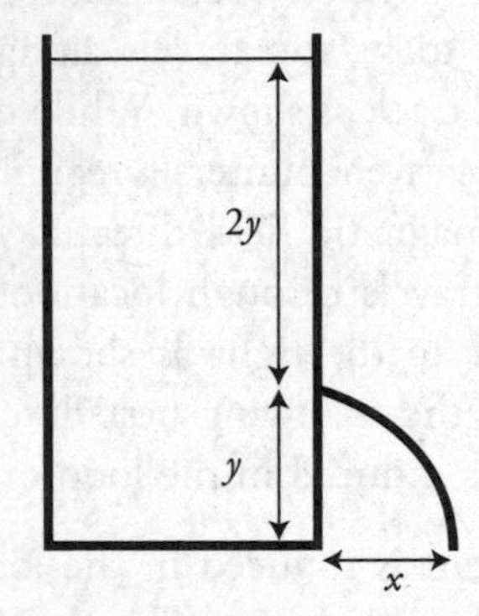

26. A large container of water sits on the floor. A hole in the side a distance y up from the floor and $2y$ below the surface of the water allows water to exit and land on the floor a distance x away as shown in the figure. If the hole in the side was moved upward to a distance $2y$ from the floor and y below the surface of the water, where would the water land?

(A) $\frac{x}{2}$
(B) $\frac{x}{\sqrt{2}}$
(C) x
(D) $2x$

27. A cylinder with a movable piston contains a gas at pressure $P = 1 \times 10^5$ Pa, volume $V = 20$ cm^3, and temperature $T = 273$ K. The piston is moved downward in a slow steady fashion allowing heat to escape the gas and the temperature to remain constant. If the final volume of the gas is 5 cm^3, what will be the resulting pressure?

(A) 0.25×10^5 Pa
(B) 2×10^5 Pa
(C) 4×10^5 Pa
(D) 8×10^5 Pa

28. An equal number of hydrogen and carbon dioxide molecules are placed in a sealed container. The gases are initially at a temperature of 300 K when the container is placed in an oven and brought to a new equilibrium temperature of 600 K. Which of the following best describes what is happening to the molecular speeds and kinetic energies of the gases' molecules as they move from 300 K to 600 K?

(A) The molecules of both gases, on average, end with the speed and the same average kinetic energy.
(B) The molecules of hydrogen, on average, end with a higher speed, but the molecules of both gases end with the same average kinetic energy.
(C) The molecules of hydrogen, on average, speed up more and end with a higher average kinetic energy.
(D) As heat flows into the container, the entropy rises, and it is not possible to determine the average velocity and average kinetic energy of the gases.

29. A convex lens of focal length $f = 0.2$ m is used to examine a small coin lying on a table. During the examination, the lens is held a distance 0.3 m above the coin and is moved slowly to a distance of 0.1 m above the coin. During this process, what happens to the image of the coin?

(A) The image continually increases in size.
(B) The image continually decreases in size.
(C) The image gets smaller at first and then bigger in size.
(D) The image flips over.

GO ON TO THE NEXT PAGE

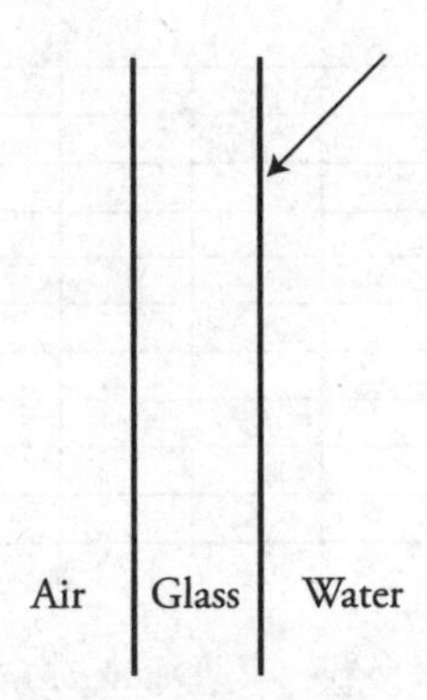

30. Light from inside an aquarium filled with water strikes the glass wall as shown in the figure. Knowing that $n_{water} = 1.33$ and $n_{glass} = 1.62$, which of the following represents a possible path that the light could take?

(A) (B)

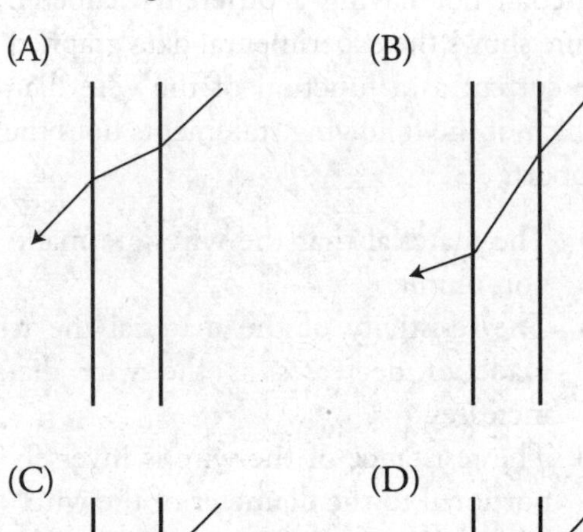

(C) (D)

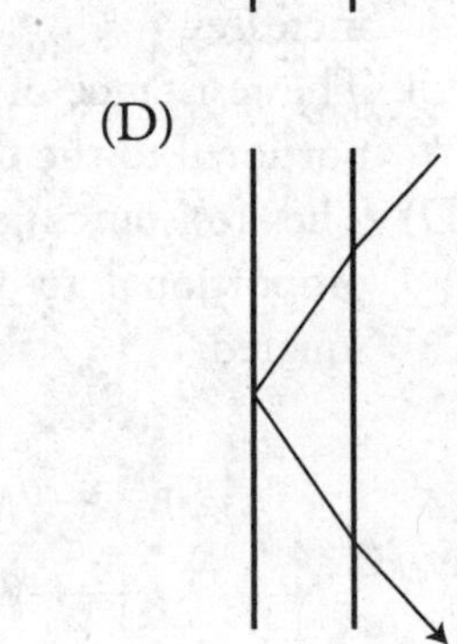

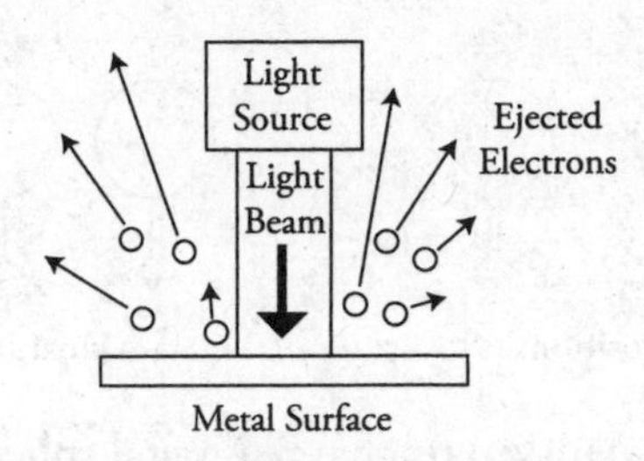

31. A beam of ultraviolet light shines on a metal plate, causing electrons to be ejected from the plate as shown in the figure. The velocity of the ejected electrons varies from nearly zero to a maximum of 1.6×10^6 m/s. If the brightness of the beam is increased to twice the original amount, what will be the effect on the number of electrons leaving the metal plate and the maximum velocity of the electrons?

	Number of Electrons Ejected	Maximum Velocity of Ejected Electrons
(A)	increases	increases
(B)	increases	remains the same
(C)	remains the same	increases
(D)	remains the same	remains the same

Energy
0
−1 eV
−2 eV
−4 eV
−6 eV
−8 eV
−12 eV
A *B* *C*

32. Scientists shine a broad spectrum of electromagnetic radiation through a container filled with gas toward a detector. The detector indicates that three specific wavelengths of the radiation were absorbed by the gas. The figure shows the energy level diagram of the electrons that absorbed the radiation. Which of the following correctly ranks the wavelengths of the absorbed electromagnetic radiation?

(A) $A = B > C$
(B) $A > B = C$
(C) $A > C > B$
(D) $B > C > A$

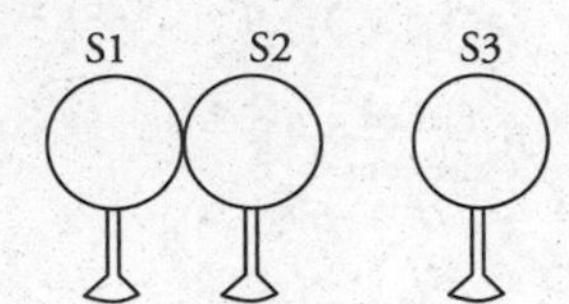

33. Three identical uncharged metal spheres are supported by insulating stands. They are placed as shown in the left figure with S1 and S2 touching. A sequence of events is then performed.

- S3 is given a negative charge.
- S1 is moved to the left away from S2.
- S3 is brought into contact with S2 and then placed back in its original position.

This leaves the spheres in the positions shown in the right figure. Which of the following most closely shows the signs of the final net charge on the spheres?

(A) 

(B) 

(C) 

(D) 

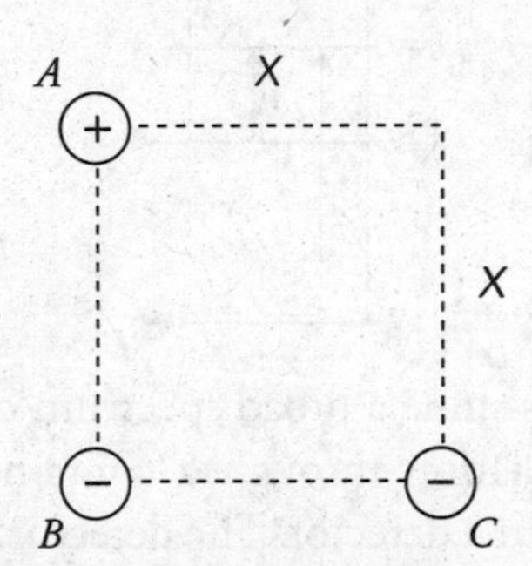

34. Three identical objects with an equal magnitude of charge are placed on the corners of a square with sides of length x as shown in the figure. Which of the following correctly expresses the magnitude of the net force F acting on each charge due to the other two charges?

(A) $F_A = F_C$
(B) $F_A > F_C$
(C) $F_B = F_C$
(D) $F_B < F_C$

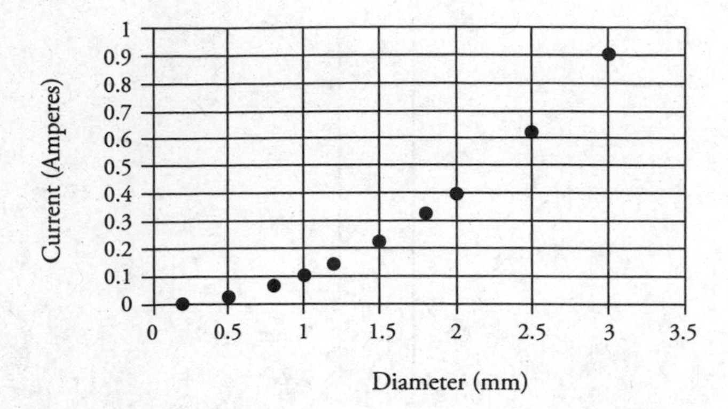

35. In an experiment, a long wire is connected to a battery and the current passing through the wire is measured. The wire is then removed and replaced with new wire of the same length and material, but having a different diameter. The figure shows the experimental data graphed with the current as a function of the wire diameter. Which of the following statements does the data support?

(A) The material that the wires are made of is non-ohmic.
(B) The resistivity of the material the wire is made of decreases as the wire diameter increases.
(C) The resistance of the wire is inversely proportional to the diameter of the wire.
(D) The resistance of the wire is inversely proportional to the diameter of the wire squared.

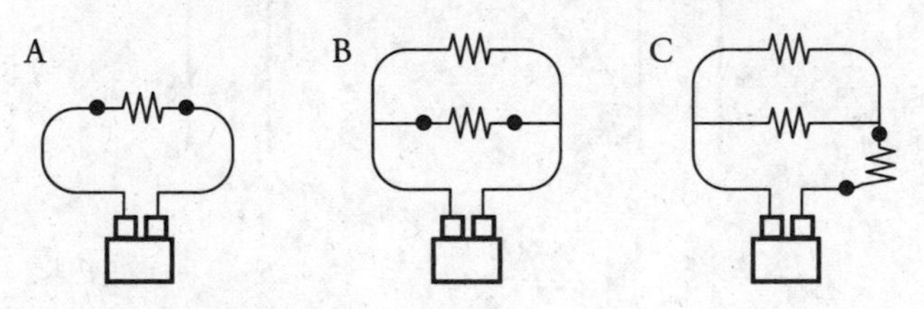

36. Four identical batteries of negligible resistance are connected to resistors as shown. A voltmeter is connected to the points indicated by the dots in each circuit. Which of the following correctly ranks the potential difference measured by the voltmeter?

(A) $\Delta V_A = \Delta V_B > \Delta V_C$
(B) $\Delta V_A > \Delta V_B > \Delta V_C$
(C) $\Delta V_B > \Delta V_A > \Delta V_C$
(D) $\Delta V_C > \Delta V_A = \Delta V_B$

GO ON TO THE NEXT PAGE

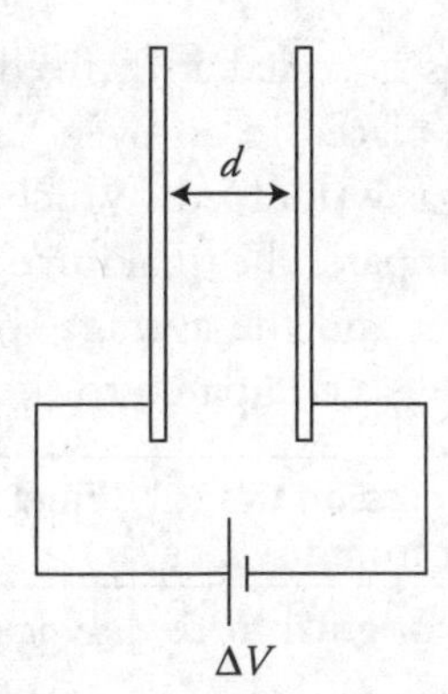

37. A capacitor, with parallel plates a distance d apart, is connected to a battery of potential difference ΔV as shown in the figure. The plates of the capacitor can be moved inward or outward to change the distance d. To increase both the charge stored on the plates and the energy stored in the capacitor, which of the following should be done?

(A) Keep the capacitor connected to the battery, and move the plates closer together.
(B) Keep the capacitor connected to the battery, and move the plates farther apart.
(C) Disconnect the battery from the capacitor first, and then move the plates closer together.
(D) Disconnect the battery from the capacitor first, and then move the plates farther apart.

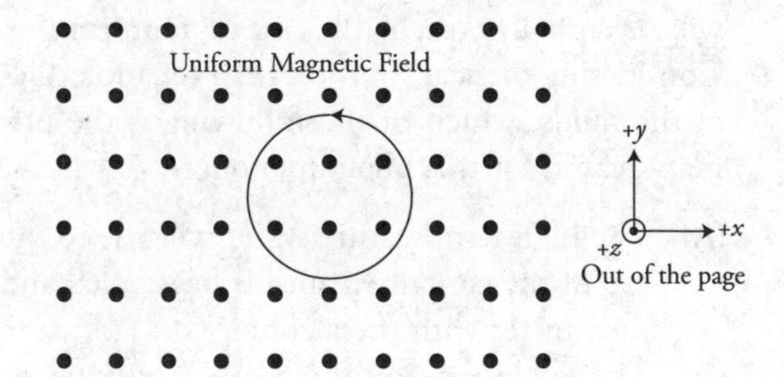

38. A loop of wire with a counterclockwise current is immersed in a uniform magnetic field that is pointing up out of the paper in the $+z$ direction. The loop is free to move in the field. Which of the following is a correct statement?

(A) There is a torque that rotates the loop about the x-axis.
(B) There is a torque that rotates the loop about the z-axis.
(C) There is a force that moves the loop along the z-axis.
(D) There is a net force of zero and the loop does not move.

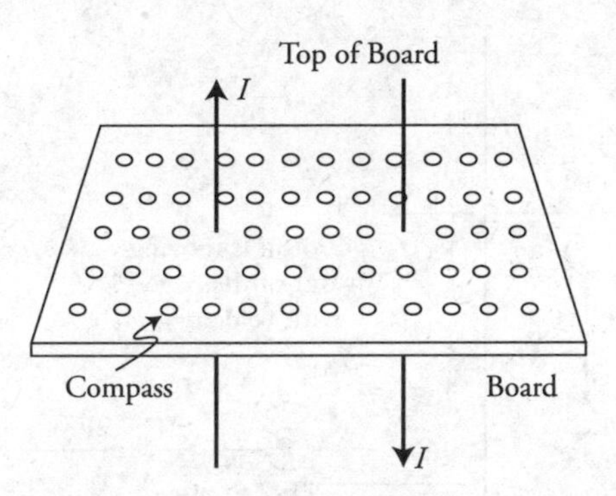

39. Two long wires pass vertically through a horizontal board that is covered with an array of small compasses placed in a rectangular grid pattern as seen in the figure. The wire on the left has a current passing upward through the board, while the right wire has a current passing downward through the board. Both currents are identical in magnitude. Each compass has an arrow that points in the direction of north. Looking at the board from above, which of the following diagrams best depicts the directions that the array of compasses are pointing?

(A)

(B)

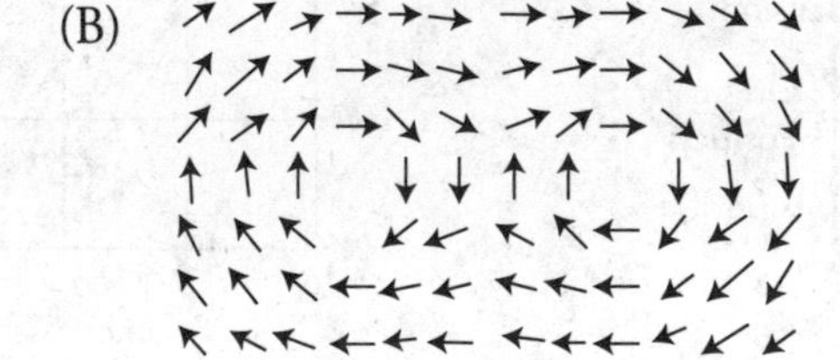

(C)

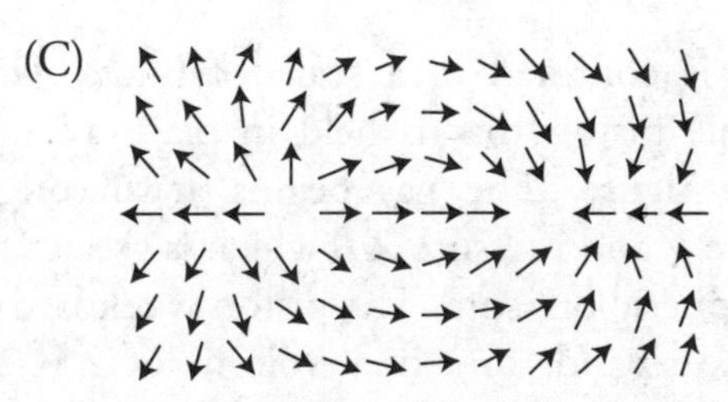

(D)

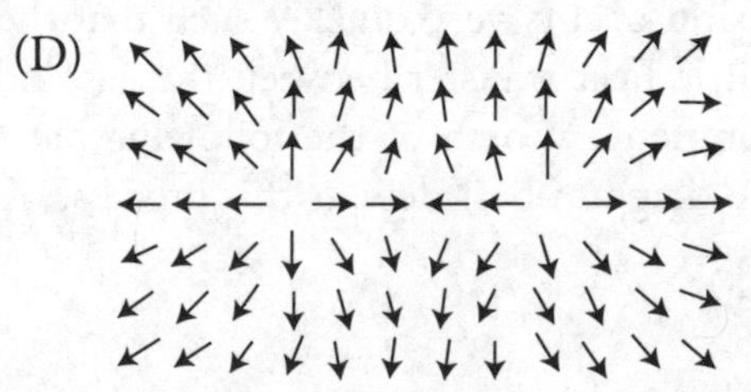

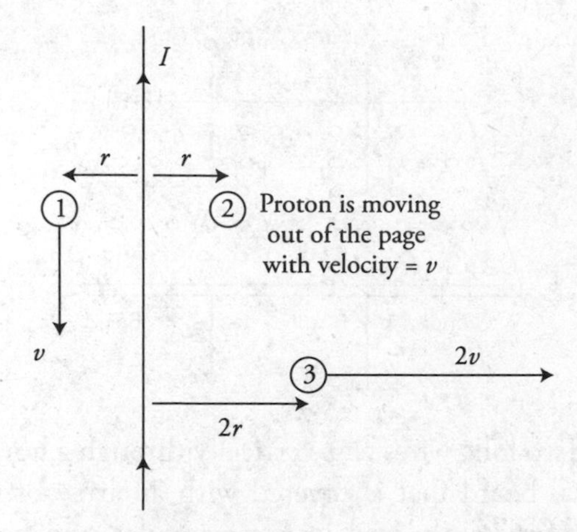

40. A long wire carries a current as shown in the figure. Three protons are moving in the vicinity of the wire as shown. All three protons are in the plane of the page. Proton 1 in moving downward at a velocity of v. Proton 2 is moving out of the page at a velocity of v. Proton 3 is moving to the right at a velocity of $2v$. Which of the following correctly ranks the magnetic force on the protons?

(A) $1 = 2 = 3$
(B) $1 = 2 > 3$
(C) $1 = 3 > 2$
(D) $3 > 1 > 2$

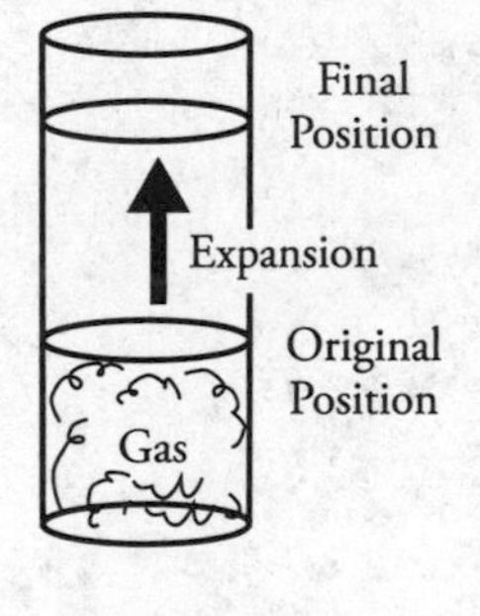

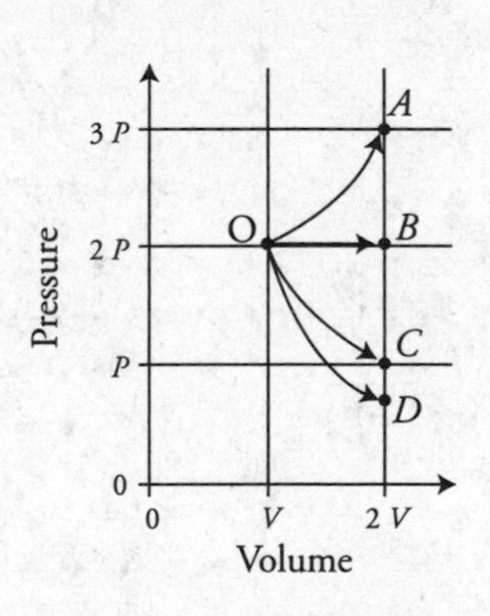

41. A gas is confined in a sealed cylinder with a movable piston that is held in place as shown in the figure. The gas begins at an original volume *V* and pressure *2 P*, which is greater than atmospheric pressure. The piston is released and the gas expands to a final volume of *2 V*. This expansion occurs very quickly such that there is very little heat transfer between the gas and the environment. Which of the following paths on the *PV* diagram best depicts this process?

(A) *A*
(B) *B*
(C) *C*
(D) *D*

42. An ideal gas is sealed in a fixed container. The container is placed in an oven, and the temperature of the gas is doubled. Which of the following correctly compares the final force the gas exerts on the container and the average speed of the molecules of the gas compared to the initial values?

	Final Force on the Container from the Gas	Final Average Speed of Molecules
(A)	Twice the original force	Twice the original speed
(B)	Twice the original force	The speed increases but will be less than twice the original speed.
(C)	The increase in force cannot be determined without knowing the area of the container.	Twice the original speed
(D)	The increase in force cannot be determined without knowing the area of the container.	The speed increases but will be less than twice the original speed.

43. When hot water is poured into a beaker containing cold alcohol, the temperature of the mixture will eventually reach the same temperature. Considering the scale of the size of the molecules of the fluids, which of the following is the primary reason for this phenomenon?

(A) The high temperature water will rise to the top of the container until it has cooled and then mixes with the alcohol.
(B) The molecules of the water continue to have a higher kinetic energy than the molecules of the alcohol, but the two liquids mix until the energy is spread evenly throughout the container.
(C) The hot water produces thermal radiation that is absorbed by the cold alcohol until the kinetic energy of all the molecules is the same.
(D) The water molecules collide with the alcohol molecules, transferring energy until the average kinetic energy of both the water and alcohol molecules are the same.

GO ON TO THE NEXT PAGE

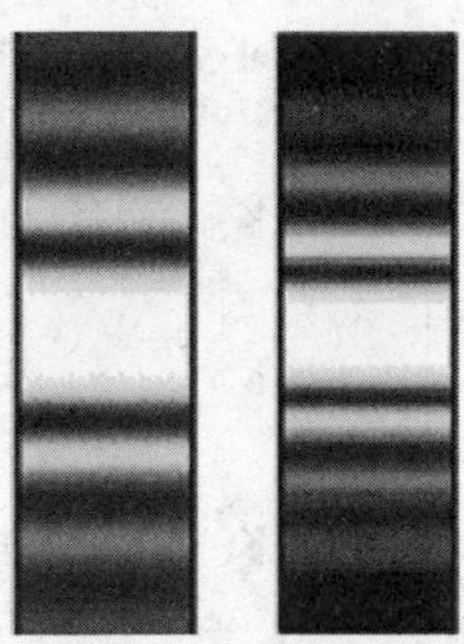

44. In an experiment, monochromatic light of frequency f_1 and wavelength λ_1 passes through a single slit of width d_1 to produce light and dark bands on a screen as seen in pattern 1. The screen is a distance L_1 from the slit. A single change to the experimental setup is made and pattern 2 is created on the screen. Which of the following would account for the differences seen in the patterns?

(A) $f_1 < f_2$
(B) $\lambda_1 < \lambda_2$
(C) $L_1 < L_2$
(D) $d_1 > d_2$

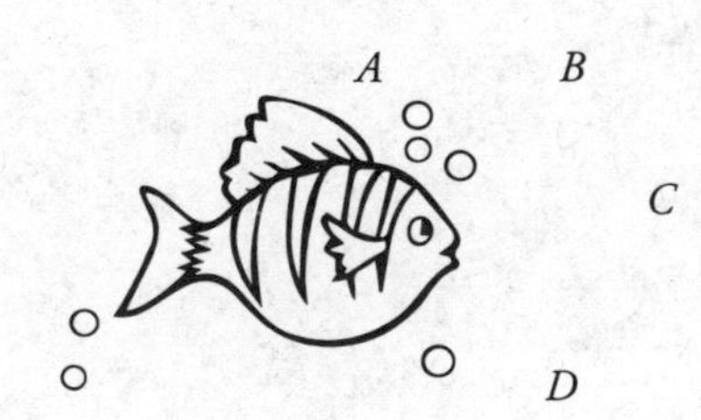

45. A bird is flying over the ocean and sees a fish under the water. The actual positions of the bird and fish are shown in the figure. Assuming that the water is flat and calm, at which location does the bird perceive the fish to be?

(A) A
(B) B
(C) C
(D) D

Questions 46–50: Multiple-Correct Items

Directions: Identify exactly two of the four answer choices as correct, and mark the answers with a pencil on the answer sheet. No partial credit is awarded; both of the correct choices, and none of the incorrect choices, must be marked to receive credit.

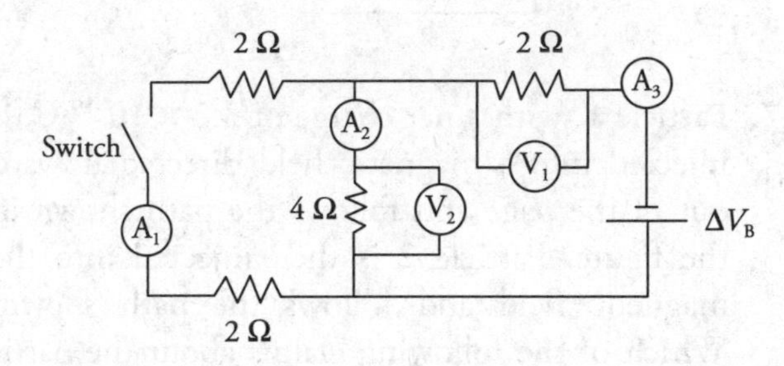

46. The circuit shown has a battery of negligible internal resistance, resistors, and a switch. There are voltmeters, which measure the potential differences V_1, and V_2, and ammeters A_1, A_2, A_3, which measure the currents I_1, I_2, and I_3. The switch is initially in the closed position. The switch is now opened. Which of the following values increases? **(Select two answers.)**

(A) V_1
(B) V_2
(C) I_2
(D) I_3

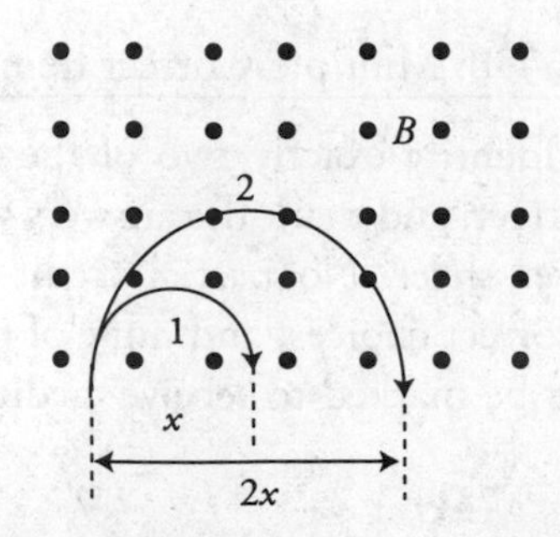

47. Particle 1, with a net charge of 3.2×10^{-19} C, is injected into a magnetic field directed upward out of the page and follows the path shown in the figure. Particle 2 is then injected into the magnetic field and follows the path shown. Which of the following claims about the particles would be a plausible explanation for the differences in their behavior? (**Select two answers.**)

(A) Particle 2 could have twice the energy of particle 1.
(B) Particle 2 could have twice the momentum of particle 1.
(C) Particle 2 could have half the charge of particle 1.
(D) Particle 2 could have half mass than particle 1.

	Initial Temperature	Final Temperature
#1	373 K	310 K
#2	273 K	310 K
#3	310 K	310 K

48. Three samples of gas in different containers are put into thermal contact and insulated from the environment as shown in the figure. The three gases, initially at different temperatures, reach a final uniform temperature of 310 K. Which of the following correctly describes the flow of thermal energy from the initial condition until thermal equilibrium? (**Select two answers.**)

(A) Heat flows from sample 1 to sample 2 during the entire time until thermal equilibrium of the system is reached.
(B) Heat flows into sample 2 only from sample 1 until both reach the equilibrium temperature of 310 K.
(C) Heat flows into sample 2 from both samples 1 and 3 until thermal equilibrium of the system is reached.
(D) Heat initially flows from sample 3 into sample 2 and then back from sample 2 into sample 3.

θ_1 (degrees)	θ_2 (degrees)
0	0
10	12
30	37
45	58
50	67

49. In an experiment, students collect data for light traveling from medium 1 into medium 2. The angles of incidence θ_1 and refraction θ_2 as measured from the perpendicular to the surface are given in the table. The data in the table supports which of the following statements? (**Select two answers.**)

(A) The index of refraction of medium 1 is approximately 1.2.
(B) Light travels slowest in medium 2.
(C) There are some angles at which the light will not be able to enter medium 2.
(D) Medium 1 and medium 2 are not the same material.

50. Which of the following phenomena can be better understood by considering the wave-particle duality of electrons? **(Select two answers.)**

(A) There are discrete electron energy levels in a hydrogen atom.
(B) Monochromatic light of various intensities eject electrons of the same maximum energy from a metal surface.
(C) A beam of electrons reflected off the surface of a crystal creates a pattern of alternating intensities.
(D) An X-ray colliding with a stationary electron causes it to move off with a velocity.

STOP: End of AP Physics 2 Practice Exam, Section 1 (Multiple-Choice)

AP Physics 2: Practice Exam 3

Section 2 (Free Response)

Directions: The free-response section consists of four questions to be answered in 90 minutes. Questions 2 and 4 are longer free-response questions that require about 25 minutes each to answer and are worth 12 points each. Questions 1 and 3 are shorter free-response questions that should take about 20 minutes each to answer and are worth 10 points each. Show all your work to earn partial credit. On an actual exam, you will answer the questions in the space provided. For this practice exam, write your answers on a separate sheet of paper.

1. (10 points—suggested time 20 minutes)

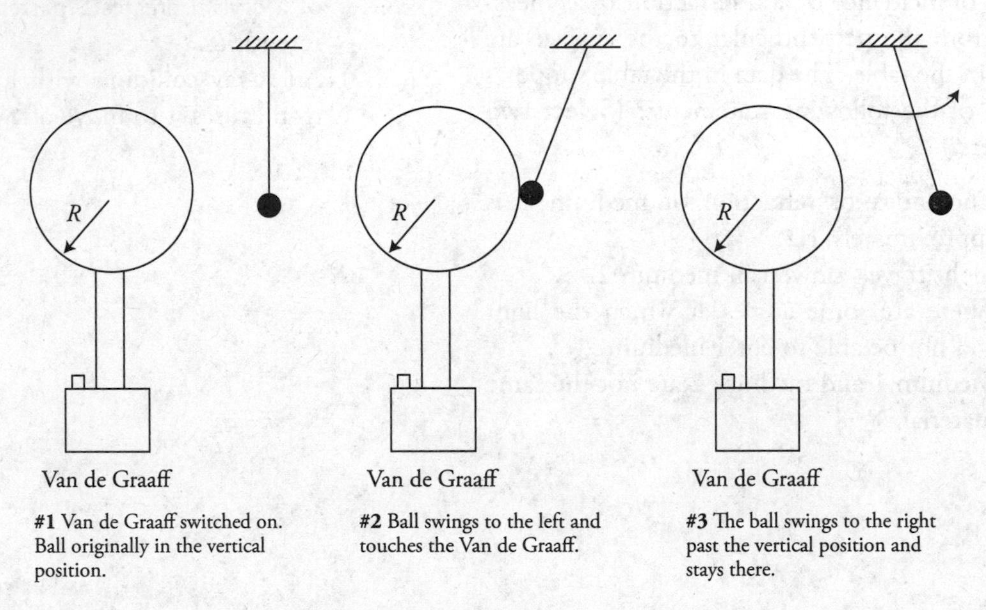

In a classroom demonstration, a small conducting ball is suspended vertically from a light thread near a neutral Van de Graaff generator as shown on the left of the figure. A grounding wire is attached to the ball and removed. Then the Van de Graaff generator is turned on, giving it a positive charge. After the Van de Graaff is turned on, the ball swings over toward, and touches, the Van de Graaff as shown in the middle diagram of the figure. After touching the Van de Graaff, the small ball swings away from the Van de Graaff toward the right and past the vertical position as shown in the figure at the right. The ball remains to the right of the vertical position.

(A) In a clear, coherent paragraph-length response, completely explain the entire sequence of events that cause the ball to behave as it does. Clearly indicate the behavior of subatomic particles and how any forces are generated.

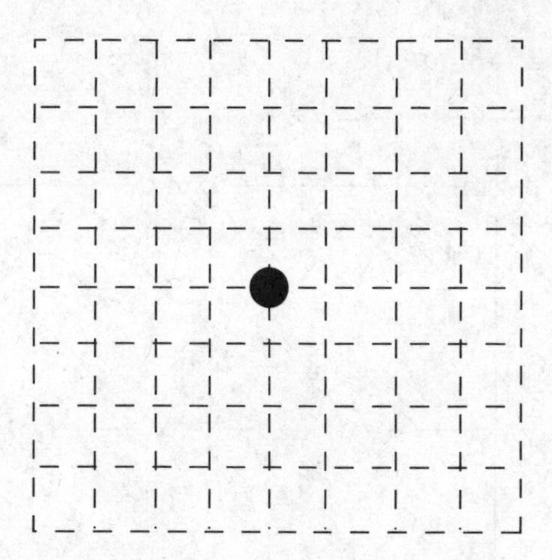

(B) The dot represents the small ball after it has swung away from the Van de Graaff when it is in its final position with an angle to the right of vertical. Draw a free-body diagram showing and labeling the forces (not components) exerted on the ball. Draw the relative lengths of all vectors to reflect the relative magnitudes of all the forces. (A grid is provided to assist you.)

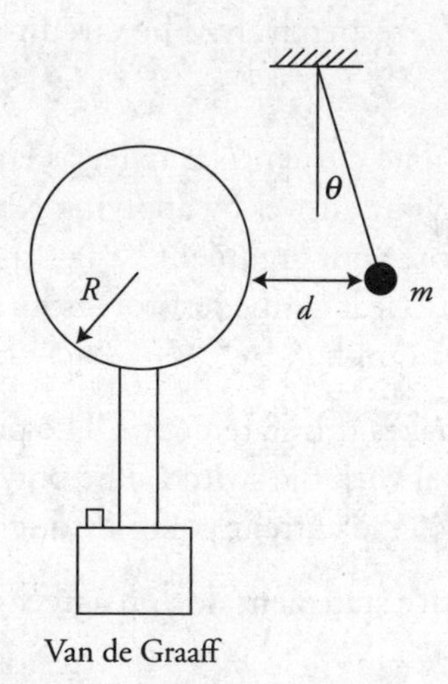

(C) In its final position the ball, mass m, is a distance d from the surface of the Van de Graaff generator and an angle θ from the vertical as shown in the figure. The Van de Graaff has a net charge of Q. Derive an expression for the magnitude of the net charge q of the small ball in its final position. Express your answer in terms of m, d, R, Q, θ, and any necessary constants.

2. (12 points—suggested time 25 minutes)

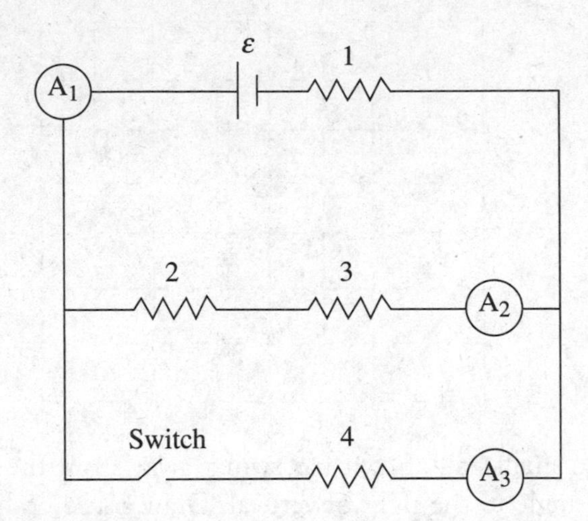

The figure shows a circuit with a battery of emf ε and negligible internal resistance, and four identical resistors of resistance R numbered 1, 2, 3, and 4. There are three ammeters (A_1, A_2, and A_3) that measure the currents I_1, I_2, and I_3, respectively. The circuit also has a switch that begins in the closed position.

(A) A student makes this claim: "The current I_3 is twice as large as I_2." Do you agree or disagree with the student's statement? Support your answer by applying Kirchhoff's loop rule and writing one or more algebraic expressions to support your argument.

(B) Rank the power dissipated into heat by the resistors from highest to lowest, being sure to indicate any that are the same. Justify your ranking.

The switch is opened. A student makes this statement: "The power dissipation of resistors 2 and 3 remains the same because they are in parallel with the switch. The power dissipation of resistor 1 decreases because opening the switch cuts off some of the current going through resistor 1."

(C) i. Which parts of the student's statement do you agree? Justify your answer with appropriate physics principles.

ii. Which parts of the student's statement do you disagree with? Justify your answer utilizing an algebraic argument.

The switch remains open. Resistor 4 is replaced with an uncharged capacitor of capacitance C. The switch is now closed.

(D) i. Determine the current in resistor 1 and the potential difference across the capacitor immediately after the switch is closed.

ii. Determine the current in resistor 1 and the potential difference across the capacitor a long time after the switch is closed.

iii. Calculate the energy (U) stored on the capacitor a long time after the switch is closed.

3. (10 points—suggested time 20 minutes)

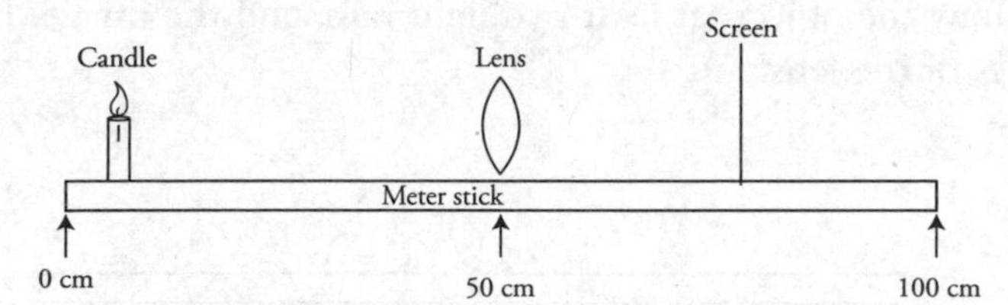

In a laboratory experiment, an optics bench consisting of a meter stick, a candle, a lens, and a screen is used, as shown in the figure. A converging lens is placed at the 50-cm mark of the meter stick. The candle is placed at various locations to the left of the lens. The screen is adjusted on the right side of the lens to produce a crisp image. The candle and screen locations on the meter stick produced in this lab are given in the table. Extra columns are provided for calculations if needed.

Candle location (cm)	Screen location (cm)
0	71
10	74
20	80
25	88
29	100

(A) Calculate the focal length of the lens. Show your work.

(B) Use the data to produce a straight line graph that can be used to determine the focal length of the lens. Calculate the focal length of the lens using this graph and explain how you found the focal length from the graph.

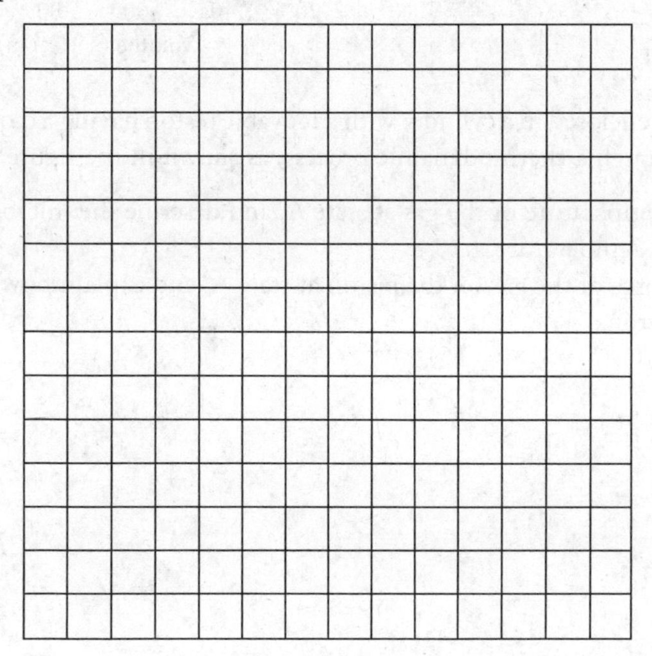

(C) Sketch a ray diagram to show how the candle would produce an upright image with a magnification larger than 1.0. Draw the object, at least two light rays, and the image. Indicate the locations of the focus on both sides of the lens.

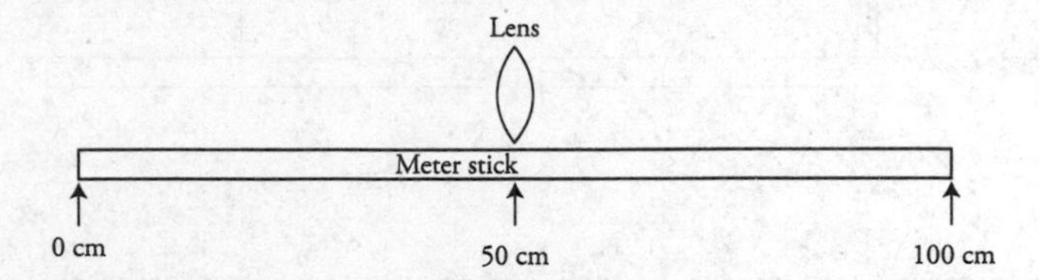

(D) A student says that virtual images can be projected on a screen. Do you agree with this claim? How could you perform a demonstration to support your stance with evidence?

4. (12 points—suggested time 25 minutes)

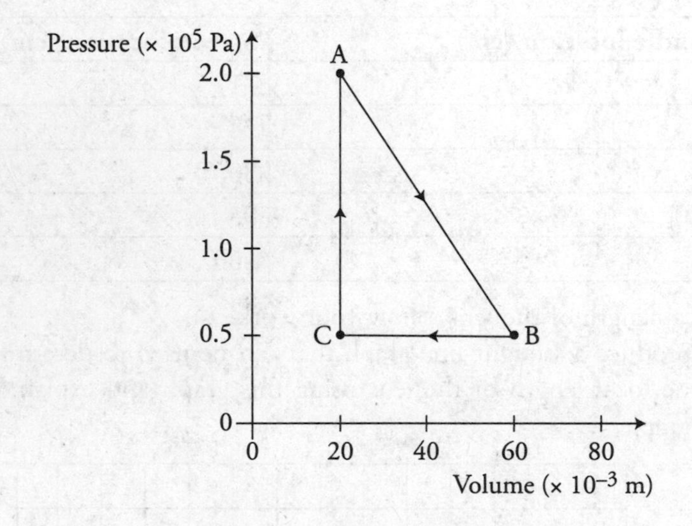

A mole of ideal gas is enclosed in a cylinder with a movable piston having a cross-sectional area of 1×10^{-2} m^2. The gas is taken through a thermodynamic process, as shown in the figure.

(A) Calculate the temperature of the gas at state A, and describe the microscopic property of the gas that is related to the temperature.

(B) Calculate the force of the gas on the piston at state A, and explain how the atoms of the gas exert this force on the piston.

GO ON TO THE NEXT PAGE

(C) Predict qualitatively the change in the internal energy of the gas as it is taken from state B to state C. Justify your prediction.

(D) Is heat transferred to or from the gas as it is taken from state B to state C? Justify your answer.

(E) Discuss any entropy changes in the gas as it is taken from state B to state C. Justify your answer.

(F) Calculate the change in the total kinetic energy of the gas atoms as the gas is taken from state C to state A.

(G) On the axis provided, sketch and label the distribution of the speeds of the atoms in the gas for states A and B. Make sure that the two sketches are proportionally accurate.

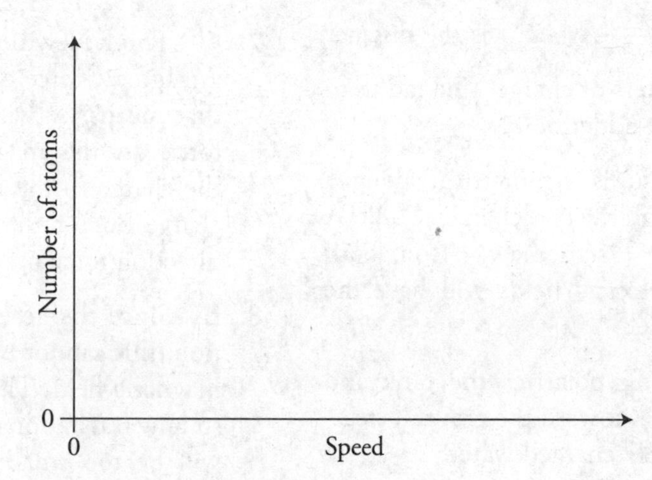

STOP: End of AP Physics 2 Practice Exam, Section 2 (Free Response)

Solutions: Section 1 (Multiple Choice)

Questions 1–45: Single–Correct Items

1. **D**—Since the magnitude of Q is much greater than (>>) the magnitude of m, the electric force will be many orders of magnitude larger than the force of gravity and we can neglect gravity. $|\vec{F}_E| = \frac{1}{4\pi\varepsilon_0}\frac{|q_1 q_2|}{r^2}$ where r is the distance between the two centers of charge. The radius of both spheres must be added to d.

2. **D**—The electric field is symmetrical along a horizontal axis through the two charges. Points A and D are the same distance away from both charges where the electric fields will have the same strength.

3. **C**—The negative ring polarizes the tank and drives negative charges toward the ground, creating a tank of positively charged water.

4. **B**—The electric field produced by a charged parallel plate capacitor is uniform in strength and direction everywhere between the plates away from the edges of the capacitor. Therefore, the electric force on the charges must be identical for locations A, C, and O. Near the edges of the capacitor, the field bows out a bit and is not quite as strong. This is consistent with answer choice B. Outside the capacitor the electric field is much weaker.

5. **D**—After the switch is closed and the capacitor has been connected to the circuit for a very long time, it has had time to fully charge and it will behave like an open circuit line. No current will bypass the middle resistor and, for all intents and purposes, the circuit looks just like it did before the switch was ever closed, with the exception that the capacitor now stores both charge and energy.

6. **C**—Conventional current will be moving around the wire in a counterclockwise direction, meaning that the current is traveling in the $+x$ direction for answer choices A and B and in the $-y$ direction for answer choices C and D. The magnetic field exits out of the north and into the south end of the magnet. This gives a magnetic field in the $+z$ direction for answer choices A and C and in the $-z$ direction for answer choices B and D. Using the right-hand rule for forces on a current-carrying wire gives us a magnetic force in the $+y$ direction for A, $-y$ direction for B, $+x$ direction for C, and $-x$ direction for D.

7. **A**—Think Newton's third law here! If we can get a magnetic force on the charge out of the page, that means we will have an equal but opposite force on the magnet into the page. Moving the charge to the right produces a force on the charge out of the page and thus an opposite-direction force on the magnet into the page.

8. **B**—Since the acceleration is constant in direction, this cannot be a collision or a force from a magnetic field. The electron mass is very small, so any gravitational acceleration on the election will be too small to see in a cloud chamber. However, a uniform electric field could easily produce this parabolic trajectory effect.

9. **B**—Buoyancy force is proportional to the displaced volume and the density of the fluid displaced. The density of the fluid is the same for each. The volume of the cube > volume of the sphere > volume of the cone.

10. **A**—Due to conservation of mass (continuity), the fluid must increase in velocity when the cross-sectional area decreases. Due to conservation of energy (Bernoulli), as the velocity of the fluid increases, the static pressure in the fluid decreases. This means the pressure on the bubbles will decrease, allowing them to expand in size.

11. **C**—The circumference of the balloon is related to the radius, and thus C^3 is proportional to the volume of the balloon. Temperature and volume of a gas are directly related. Therefore, as the temperature decreases, the circumference also decreases. Extrapolating the three data points will lead to a point on the graph where the circumference and thus the volume of the gas is zero. This temperature will be an experimental estimate for absolute zero.

12. **B**—To show ideal gas behavior, air must follow the relationships in the ideal gas law: $PV = nRT$. One variable should be plotted as a function of another, while the rest of the variables are held constant. With T held constant, $\propto \frac{1}{V}$. With P held constant, $\propto T$. With V held constant, $P \propto T$. Answer choice C is looking for the correct relationship but is using the wrong data set. Only answer choice B is looking for the correct relationship while holding the proper variable constant.

13. **C**—The largest final temperature will be at the location of the highest pressure times volume, PV location. The highest final of $6PV$ is along path C.

14. **B**—The point source model of waves tells us that waves display diffraction more prominently when the wavelength is about the same size or larger than the obstructions they are passing around. Sound can have a large wavelength similar in size to the corner it is bending around, while light has a wavelength that is much smaller. Thus light does not diffract very much going around corners.

15. **C**—The bug is walking from a position where the image is real, larger, and inverted toward $2f$, where the image will be real, inverted, and the same size as the bug.

16. **A**—First off, the student is seeing a smaller, upright, virtual image. This means this has to be a diverging lens with a negative focal length and a negative image distance.

Using the magnification equation:

$M = \frac{s_i}{s_o}, \frac{1}{2} = -\frac{s_i}{10 \text{ cm}}$ gives an image distance of –5 cm.

Using the lens equation:

$\frac{1}{f} = \frac{1}{s_i} + \frac{1}{s_o} = \frac{1}{-5 \text{ cm}} + \frac{1}{10 \text{ cm}}$, the focal length equals –10 cm.

17. **D**—This is a fission reaction where a large nucleus splits into two major chunks and released energy. Since energy is released in the reaction, conservation of mass/energy indicates that the final mass must be less than the initial mass. Answer choice A is a good distractor! Don't forget that the left side of the nuclear equation has a uranium nucleus *and* one neutron.

18. **D**—The positive and negative charges will receive forces in the opposite direction, creating a torque and causing it to rotate about its center of mass. When the bar is aligned with the electric field, it will already have an angular velocity and will overshoot the vertical alignment position. The process will repeat in the opposite angular direction, creating an oscillating motion that rotates the bar back and forth.

19. **A**—Answer choices B and D depict electric field diagram lines that begin on positive charges and end on negative charges. Electric field vectors are always perpendicular to the potential isolines. Answer choice C implies that there would be an electric field directed to the right or to the left between the two charges. This could be true only if the two charges have opposite signs. Two identical positive charges would produce a location of zero electric field directly between the two charges. This is implied by there not being any isolines in the middle of diagram A.

20. **D**—The resistance of the parallel set of three resistors on the far right is $\frac{2}{3}R$. Thus the total resistance of the circuit to the right of the battery is $\frac{5}{3}R$. The loop on the left has a resistance of $2R$. This means that the reading of $A_2 > A_1$. The current that goes through A_3 must be less than A_2 because it is in a branching pathway. Answer choice D is the only option that meets these requirements.

21. **C**— Ammeter A_2 is in the main line supplying the current to the right-hand side of the circuit. The total resistance of the circuit to the right of the battery is $\frac{5}{3}R$ (as described in the answer to question 20). This gives a total current passing through ammeter A_2 of:

$$I = \frac{\Delta V}{R} = \frac{V_B}{\frac{5}{3}R} = \frac{3V_B}{5R}$$

Using this current we can calculate the voltage drop through the resistor in the wire passing through ammeter A_2:

$$\Delta V = IR = \left(\frac{3V_B}{5R}\right)(R) = \frac{3V_B}{5}$$

This leaves the remaining voltage drop of $\frac{2}{5}V_B$ that will be read by the voltmeter.

22. **A**—$P = \frac{\Delta V^2}{R} = \Delta V^2\left(\frac{1}{R}\right)$, which means that the slope of the graph equals ΔV^2. The slope of the graph is approximately 9, which gives: $\Delta V \approx 3$ V.

23. **A**—We need a force on the proton beam pointed to the right. The velocity of the proton is forward or into the page. The top solenoid will produce a magnetic field pointing upward at the aperture. Using the right-hand rule for magnetic forces on moving charges, we can see that the proton will experience a force to the right in the upward magnetic field.

24. **B**—An induced current occurs when there is a change in magnetic flux through the loop:

$$\varepsilon = \frac{\Delta\phi}{\Delta t} = \frac{\Delta(BA\cos\theta)}{\Delta t}$$

This occurs only when the cart is entering the front edge and leaving the back edge of the field, because the flux area is changing. No current is induced while the cart is fully immersed in the magnetic field, because the magnetic field is completely covering the loop area; thus, the flux area is not changing.

25. **D**—The blocks are the same size yet sink to different depths, implying they have different masses and densities. The buoyancy force equals the weight of floating objects and thus cannot be the same. The boxes sink to different depths and static fluid pressure depends on the depth of the fluid.

26. **C**—x = (velocity)(time). From Bernoulli's equation we can derive that: $v = \sqrt{2gh_{above}}$ where h is the height above the hole to the top of the water. Time can be derived from kinematics: $t = \sqrt{\frac{2h_{below}}{g}}$ where h is the height below the hole to the ground. Multiplying these together gives the horizontal distance: $x = \sqrt{2gh_{above}}\sqrt{\frac{2h_{below}}{g}} = 2\sqrt{(h_{above})(h_{below})}$.
When the hole is moved to the new location, the height above is cut in half while the height below is doubled. Thus x remains the same.

27. **C**—nRT are all remaining constant. This means that PV = constant. Since the volume is decreased to ¼ of its original value, the pressure must have gone up by a factor of 4.

28. **B**—Both gases end with the same temperature and consequently end with the same average molecular kinetic energy. To have the same average molecular kinetic energy as carbon dioxide, hydrogen, with its smaller mass, must have a higher average molecular velocity.

29. **D**—The object begins at a distance beyond the focal length. This will produce an image that is inverted and real. As the lens is moved closer to the object, the image gets bigger and bigger. When the object is one focal length from the lens, no image will form. As the lens is moved even closer, the object is now inside the focal length where the image will be virtual and upright. This virtual image begins very large and decreases in size as we move the lens from 0.2 m to its final location of 0.1 m.

30. **C**—Light traveling from the water to the glass should bend toward the normal as it slows down in the glass. This eliminates answer choices B and D. The light will then travel from the glass to the air on the left where it will bend away from the normal because it is traveling fastest in the air. Answer choice A seems to show the angle in the air being the same as it was in the water, which cannot be true. The light travels faster in air than in water, which means the angle to the normal must be bigger than inside the aquarium. What about answer choice C? Since the light bends away from the normal entering air, it is possible that the angle of incidence between the glass and the air is beyond the critical angle, thus causing total internal reflection at the glass/air boundary; C is the best answer.

31. **B**—Increasing the brightness of the light increases only the number of photons not the energy of the individual photons. Thus, the number of ejected electrons goes up but their maximum energy and velocity will still be the same.

32. D—$E = hf = \frac{hc}{\lambda}$, therefore, wavelength is inversely proportional to the energy of the absorbed photon. The energy of the absorbed photon is equal to the jump in energy of the electron: $E_{final} - E_{initial}$. Electron B has the smallest energy jump, and electron A has the largest energy jump.

33. C—When S3 is charged negative, it will polarize the left two spheres. Since S1 and S2 are touching, S1 becomes negative and S2 positive during this polarization. When S1 and S2 are separated, they take their net charges with them. When the negative S3 touches the positive S2, their charges cancel each other out by conduction.

34. B—Note two things: (1) the force between pairs of charges must be equal and opposite and (2) due to a longer distance between them, the force between A and C is smaller than the forces between A and B or B and C. The forces are shown in the diagram. F_A must be larger than F_C. F_B must be larger than F_C.

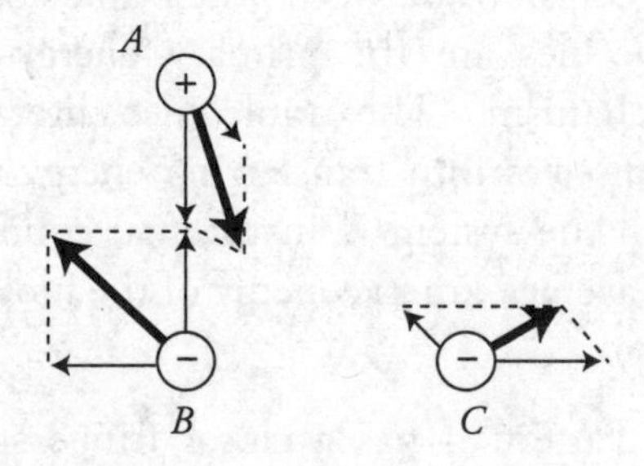

35. D—When the diameter of the wire is 1 mm the current is 0.1 A. When the diameter of the wire is doubled to 2 mm the current quadruples to 0.4 A. This means that when the diameter doubles, the resistance must be cut to one-fourth of its original value. Now let's look at what happens with the diameter triples to 3 mm. The current goes up nine times to 0.9 A. This means the resistance must have decreased to one-ninth of its original value when the wire had a diameter of 1 mm. Therefore, the resistance of the wire is inversely proportional to the diameter of the wire squared. There is no evidence from this data than the material is nonohmic or that the resistivity of the material changes.

36. A—Using Kirchhoff's loop rule it is easy to see that the voltages of A and B must be the same because both resistors are directly connected to the battery in a single loop. The voltages across A and B must equal the potential of the battery. The loop rule also shows us that in case C, there will be two resistors in any loop drawn between the plus and minus side of the battery. Therefore, the potential difference measured in case C must be less that of cases A and B.

37. A—Disconnecting the battery will ensure that the original charge Q on the capacitor cannot change, because there is nowhere for the charge to go and no way to add any new charge to the capacitor. Thus disconnecting the battery cannot be one of our choices. Decreasing the distance d between the plates will increase the capacitance C of the capacitor:

$$\uparrow C = \frac{\kappa \varepsilon_0 A}{\downarrow d}$$

Keeping the capacitor connected to the battery ensures that the voltage ΔV of the capacitor stays the same. This means that the change Q on the capacitor must increase as the capacitance C increases because the plates have been moved closer together:

$$\uparrow C = \frac{\uparrow Q}{\Delta V}$$

Since both capacitance C and charge Q are increasing, the energy stored in the capacitor must also increase:

$$\uparrow U_C = \tfrac{1}{2}(\uparrow Q)\Delta V = \tfrac{1}{2}(\uparrow C)\Delta V^2$$

38. D—Using the right-hand rule for magnetic force on a current-carrying wire, we can see there is a magnetic force directed radially outward on the wire loop that will try to expand the loop but will not move the loop.

39. A—Two background concepts: First, the magnetic field around current-carrying wire can be visualized using the right-hand rule by grasping the wire with your right hand with your thumb in the direction of the current. Your fingers will curl around the wire in the direction that the magnetic field rotates around the wire. Second, a magnetic dipole will rotate to align with the north end pointing in the direction of the magnetic field. With this in mind, looking at the board from the top, the magnetic field around the right wire will circulate clockwise and counterclockwise around the wire on the left. This is best depicted by answer choice A. Answer choice B shows clockwise circulation around both wires. Answer choices C and D depict fields pointing either away or toward the wires.

40. C—Using the right-hand rule for finding the magnetic field around current-carrying wires, we can see that the magnetic field points into the page on the right of the wire and out of the page on the left of the wire. Notice that proton 2 is moving parallel to the magnetic field from the wire. This means there will be no magnetic force on proton 2:

$$|\vec{F}_M| = |q\vec{v}||\sin\theta||\vec{B}| = |q\vec{v}||\sin 0||\vec{B}| = 0$$

Protons 1 and 3 are moving perpendicular to the field. Thus the sin θ equals 1.

$$|\vec{F}_M| = |q\vec{v}||\sin\theta||\vec{B}| = |q\vec{v}||\sin 90||\vec{B}| = q\vec{v}\vec{B}$$

We know that the magnetic field around a wire is:

$$B = \frac{\mu_0}{2\pi}\frac{I}{r}$$

Substituting this into the magnetic force on moving charges equation we get:

$$\vec{F_M} = q\vec{v}\vec{B} = q\vec{v}\left(\frac{\mu_0}{2\pi}\frac{I}{r}\right) = \left(\frac{\mu_0 qI}{2\pi}\right)\frac{\vec{v}}{r}$$

Therefore:

$$\vec{F_M} \propto \frac{\vec{v}}{r}$$

Since both the velocity and the radius from the wire are doubled for proton 3 compared to proton 1, the magnetic forces are the same on both protons.

41. D—There is no heat transfer to, or from, the environment. Therefore, this must be an adiabatic process: $Q = 0$. The gas is expanding ($+\Delta V$), which means that the work will be negative: $W = -P\Delta V$. From the first law of thermodynamics, we see that the change in internal energy of the gas must be negative: $\Delta U = Q + W$. Since the internal energy of the gas is directly related to the temperature of the gas ($\Delta U = \frac{3}{2}nR\Delta T = \frac{3}{2}Nk_BT$), the temperature of the gas must be decreasing. Finally, using the ideal gas law ($PV = nRT$), we can see that if the temperature is decreasing, the value of PV must also be decreasing. Only path D has a final PV value less than the initial value.

42. B—Doubling the temperature of a gas in a sealed container will double the pressure of the gas: $PV = nRT$. Doubling the pressure will also double the force, since the size of the container will stay the same: $F = PA$. Doubling the temperature of the gas will double the average kinetic energy of the gas molecules: $K = \frac{3}{2}k_BT$. But, kinetic energy is proportional to v^2. Therefore, the average speed of the molecules will only increase only by a factor of $\sqrt{2}$.

43. D—While there will be convection and radiation, the collisions between faster- and slower-moving molecules are the primary energy-transferring mechanism. The molecules literally collide themselves into transferring energy/momentum until the system is in thermal equilibrium and the average kinetic energy of the molecules is the same.

44. A—Pattern 2 has a closer fringe spacing than pattern 1. We could decrease the pattern spacing by simply moving the screen closer to the slit. Since that is not an option, answer choice C can be eliminated. Let's take a look at the interference pattern equation that models this behavior:

$$d\sin\theta = m\lambda$$

To get a pattern with a tighter spacing, we need to have a smaller angle θ. Rearranging the equation:

$$\sin\theta = \frac{m\lambda}{d}$$

The variable m is just the counter to find the angle for different order maxima and minima so we can neglect it. To get a smaller angle θ, we need to have a decreased λ or increased d. Neither of those is an option, so we can eliminate answer choices B and D. This leaves choice A as the correct answer. This makes sense because if we increase the frequency f of the light, the wavelength λ will decrease, making the pattern tighter, which is what we wanted!

45. A—Light traveling from water to air speeds up and will refract away from the normal. Draw several rays coming from the fish in the direction of the bird and backtrack the refracted rays to locate the image location.

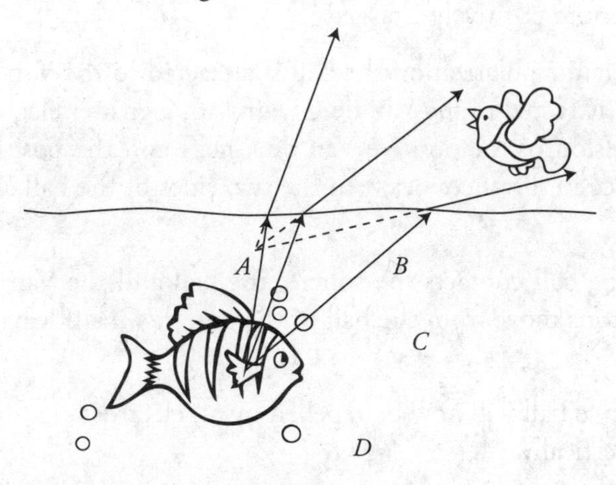

Questions 46–50: Multiple-Correct Items

(You must indicate both correct answers; no partial credit is awarded.)

46. B and C—With the switch closed, the equivalent resistance of the circuit is 4 Ω. When the switch is opened, the equivalent resistance of the circuit goes up to 6 Ω. This means that the current passing through A_3 decreases when the switch is opened, and that the voltage difference measured by V_1 must also decrease due to less current passing through the resistor. By process of elimination, the other two answer choices are correct. **Note:** This is not the only way to solve this problem, but seemed the fastest.

47. B and C—3.2×10^{-19} C is the charge of two electrons/protons. It is physically possible to have a charge exactly half this size as this would be the the net charge of a single electron/proton. Knowing that the magnetic force on the charge causes the particle to arc into a circular path, we can derive: $F_M = qvB = m\frac{v^2}{r}$, which gives us $r = \frac{mv}{qB}$. If the particle has twice the momentum or half the charge, it would turn in a circular arc with twice the radius.

48. A and D—Thermal energy always flows from high temperature to low temperature. Initially heat must flow into sample 2 from both samples 1 and 3. This will lower the temperature of sample 3 below the final equilibrium temperature of 310 K. Thus, as the temperature of sample 2 rises, heat will eventually have to flow back into sample 3 to bring its temperature back up to the equilibrium temperature of 310 K.

49. C and D—Since the angles in the two media are different, the speed of light must be different in the materials and the two media cannot be the same. The angle of refraction in media 2 is larger than the incidence angle, meaning that there will be a critical angle beyond which there will be total internal reflection. **Note:** We cannot assume that either of these two mediums is air ($n = 1$).

50. A and C—The discrete electron energy levels in hydrogen can be understood as being orbits of constructive wave interference locations for the electron. The alternating intensities seen in diffraction patterns are evidence of the wave nature of electrons reflecting off the atoms in the crystal. Both of the other choices are examples of the particle nature of electromagnetic waves.

Solutions: Section 2 (Free Response)

Your answers will not be word-for-word identical to what is written in this key. Award points for your answer as long as it contains the correct physics explanation *and* as long as it does not contain incorrect physics or contradict the correct answer.

Question 1

Part (A)

1 point—For indicating that the ball is originally neutral.

1 point—For explaining that the ball becomes polarized when the Van de Graaff becomes positively charged. During polarization some electrons in the ball are attracted to the left of the ball. This causes a charge separation with the left side of the ball more negatively charged and the right side of the ball more positively charged.

1 point—For indicating that due to charge polarization the ball is attracted to the Van de Graaff. The negative side of the ball is attracted to the Van de Graaff with a greater electrostatic force than the electrostatic repulsion of the positive Van de Graaff and the positive side of the ball because of the difference in distances between the two sides of the ball and the charged Van de Graaff.

1 point—For explaining that when the ball contacts the sphere, the ball and the Van de Graaff become the same charge. Electrons move from the ball to the Van de Graaff leaving the ball with a net positive charge.

1 point—For indicating that the positive ball will now be repelled by an electrostatic force to the right and will no longer hang vertically.

Part (B)

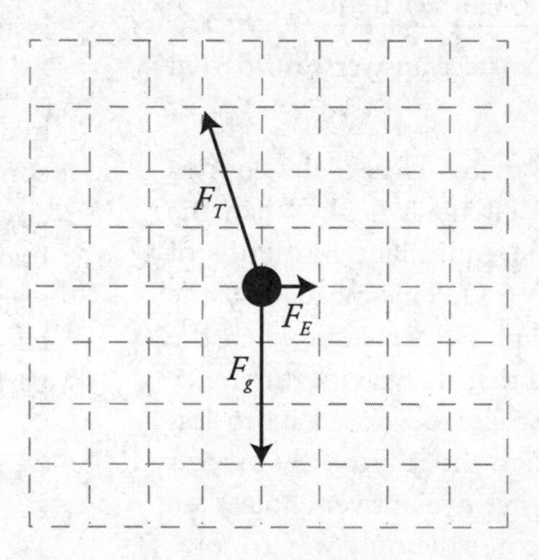

1 point—For drawing the electric force the same number of squares on the grid to the right of the ball as the tension force is drawn to the left of the ball.

1 point—For drawing the gravity force the same number of squares downward from the ball as the tension force is drawn upward from the ball.

Note: No points are awarded for incorrectly labeled forces. Deduct a point for each incorrect additional force vector drawn. The minimum score is zero.

Part (C)

1 point—For the correct expression of the electric force:

$$F_E = k\frac{qQ}{(R+d)^2}$$

1 point—For equating the horizontal component of the tension to the electric force *and* for equating the vertical component of the tension to the gravitational force.

There are two methods to do this:

Method #1

$$F_E = F_T \sin\theta$$

$$F_g = mg = F_T \cos\theta$$

Method #2

$$\tan\theta = \frac{F_E}{F_G} = \frac{F_E}{mg}$$

1 point—For the correct formula for the charge on the ball:

$$\tan\theta = \frac{F_E}{mg} = \frac{k\frac{qQ}{(R+d)^2}}{mg}$$

$$q = \frac{mg(R+d)^2\tan\theta}{kQ}$$

Question 2

Part (A)

Agree.

1 point—For correctly applying Kirchhoff's loop rule for the upper loop:

$$\varepsilon - I_2R - I_2R - I_1R = 0$$

$$\varepsilon = 2I_2R - I_1R$$

1 point—For correctly applying Kirchhoff's loop rule to the outer loop:

$$\varepsilon - I_3R - I_1R = 0$$
$$\varepsilon = I_3R - I_1R$$

1 point—For correctly using the two equations to show that I_3 is twice as large as I_2:

$$2I_2R - I_1R = I_3R - I_1R$$
$$2I_2R = I_3R$$
$$2I_2 = I_3$$

Part (B)

No points are awarded for the correct ranking of: $P_1 > P_4 > P_2 = P_3$.

1 point—For indicating that Power = I^2R and that all the resistors are the same. Therefore, the ranking is based on the current passing through the resistors.

1 point—For indicating that resistor 1 receives the most current as all the current must pass through it AND that resistor 4 receives more current than resistors 2 and 3 AND that resistors 2 and 3 receive the same current because they are in the same conductive pathway.

Part (C)

(i.) Agree that the power will go down for resistor 1.

1 point—For indicating that when the switch is opened, there is only one path left for the current to pass through. This means the total resistance of the circuit increases. The potential difference across resistor 1 will decrease, which will bring its power dissipation down as well.

(ii.) Disagree that resistors 2 and 3 are unaffected.

1 point—For deriving a correct expression for the original current passing through resistors 2 and 3:

$$I_2 = \frac{1}{3}I_1 = \frac{1}{3}\left(\frac{\varepsilon}{R_{total}}\right) = \frac{1}{3}\left(\frac{\varepsilon}{\frac{5}{3}R}\right) = \frac{\varepsilon}{5R}$$

1 point—For deriving a correct expression for the new current passing through resistors 2 and 3:

$$I_2 = I_1 = \frac{\varepsilon}{R_{total}} = \frac{\varepsilon}{3R}$$

The new current is larger than the old; therefore, the power dissipation goes up. (Note that this argument can be also be made using potential difference and would also receive credit.)

Part (D)

(i.) Immediately after the switch is closed, the capacitor acts like a short-circuit wire that allows the current to bypass resistors 2 and 3.

1 point—For indicating that the current through resistor 1 will be $I_1 = \frac{\varepsilon}{R}$, and that the potential difference across the capacitor is zero ($\Delta V_C = 0$).

(ii.) After a long period of time, the capacitor becomes fully charged and acts like an open switch in the circuit.

1 point—For indicating the current through resistor 1 will be: $I_1 = \frac{\varepsilon}{3R}$.

1 point—For indicating that the potential difference across the capacitor will be equal to that of resistors 2 and 3 combined, because the capacitor is in parallel with them: $\Delta V_C = \frac{2}{3}\varepsilon$. Note this can be stated in words or symbolically to receive credit.

(iii.)

1 point—For calculating the potential energy stored by the capacitor:

$$\frac{1}{2}CV^2 = \frac{1}{2}C\left(\frac{2}{3}\varepsilon\right)^2 = \frac{2}{9}C\varepsilon^2$$

Question 3

Part (A)

1 point—For correctly calculating an appropriate set of image and object distances.

For example: Knowing that the lens is located at 50 cm and using the first set of data, we get the following:

$$s_o = 50\text{ cm} - 0\text{ cm} = 50\text{ cm}$$

$$s_i = 71\text{ cm} - 50\text{ cm} = 21\text{ cm}$$

1 point—For correctly calculating the focal length of the lens:

$$\frac{1}{f} = \frac{1}{s_i} + \frac{1}{s_o}$$

$$\frac{1}{f} = \frac{1}{50\text{ cm}} + \frac{1}{21\text{ cm}} = \frac{0.68}{\text{cm}}$$

$$f = 15\text{ cm}$$

Part (B)

1 point—For explaining how to produce a straight line from the data and why the *y*-intercept equals $1/f$.

1 point—For plotting $1/s_o$ on the *x*-axis and $1/s_i$ on the *y*-axis and drawing a best fit line through the data.

1 point—For calculating the correct focal length using the *y*-intercept.

Example: The lens equation can be rearranged to produce a straight line:

$$y = mx + b$$

$$\frac{1}{s^i} = (-1)\frac{1}{s_o} + \frac{1}{f}$$

Thus, if we plot $1/s_o$ on the *x*-axis and $1/s_i$ on the *y*-axis, we should get a graph with a slope of −1 and an intercept of $1/f$.

From our graph, the intercept is 0.067 1/cm, which gives us a focal length of 15 cm (the same as part A).

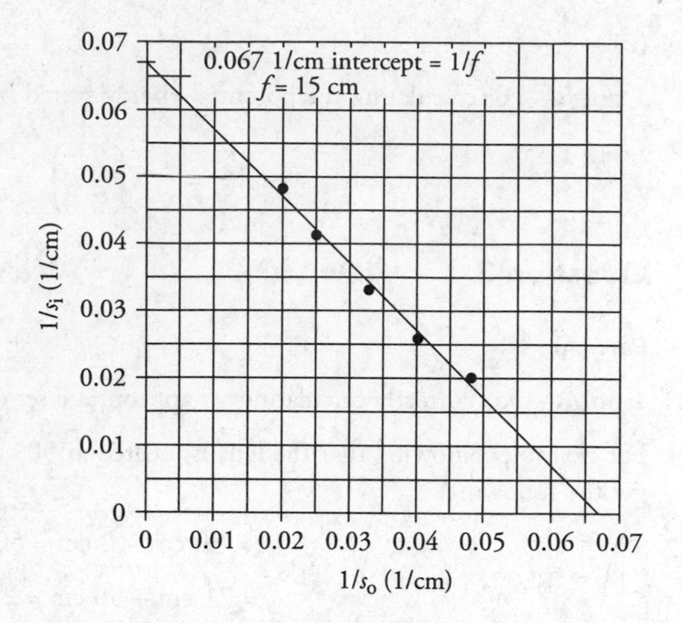

Part (C)

1 point—For drawing the candle or other object between the focus and the lens. This point cannot be awarded if the focal points are not designated on the drawing.

1 point—For drawing two correct rays from the object and passing through the lens. This point cannot be awarded if the focal points are not designated on the drawing. However, the point can be earned even if the object is not correctly located between the focus and the lens as long as the rays are correct for the object and lens placement.

1 point—For drawing a correct virtual image. The image should be upright, located at the intersection of the two outgoing rays, and be larger than the object.

Here is an example drawing:

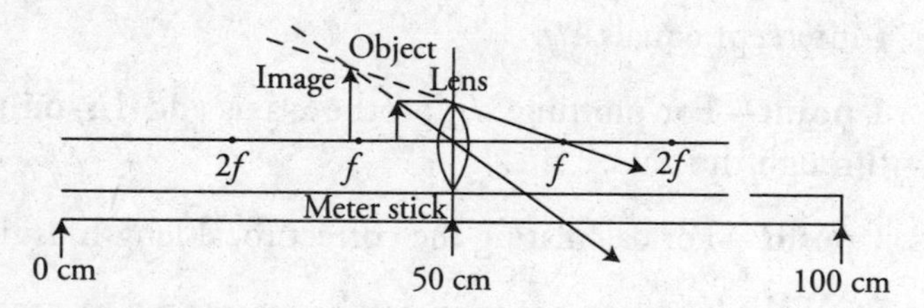

(An upright image will be virtual. The object will need to be between the lens and the focal point. Sketches will differ, and image locations will vary a bit depending on where the object is placed between the focal point and the lens.)

Part (D)

1 point—For disagreeing and stating that virtual images can be seen but cannot be projected on a screen.

1 point—For describing an appropriate demonstration to prove that virtual images cannot be projected on a screen.

For example: Produce a real image and show that it can be projected on a screen. Then create a virtual image and show that the image cannot be made to show up on the screen.

Question 4

Note that all the numbers in this problem are rounded to two significant digits because that is the accuracy of the data from the graph.

Part (A)

1 point—For the correct value of temperature with supporting equation and work: $PV = nRT$, $T = 480$ K.

1 point—For an explanation that the temperature of the gas is directly related to the average kinetic energy of the gas molecules.

Part (B)

1 point—For the correct force with supporting equation and work: $F = PA = 2{,}000$ N.

1 point—For an explanation of the mechanism that produces gas force on the piston.

For example: The gas molecules collide with the piston in a momentum collision that imparts a tiny force on the piston. The sum of all the individual molecular collision forces is the net force on the piston.

Part (C)

1 point—For indicating that the temperature of the gas is decreasing and explaining why this occurs.

For example: Since the PV value of the gas decreases, the temperature of the gas decreases in this process as indicated by the ideal gas law.

1 point—For indicating that the internal energy of the gas will decrease and explaining why this occurs.

For example: $\Delta U = nR\Delta T$ and the temperature is decreasing. Therefore, the internal energy of the gas also must decrease.

Part (D)

1 point—For indicating that the work in this process is positive because the process is moving to the left on the graph and that the thermal energy of the gas is decreasing because the temperature is decreasing.

1 point—For using the first law of thermodynamics to determine that heat is being removed from the gas.

For example: work is positive and the gas internal energy is decreasing. Therefore, using the first law of thermodynamics, $\Delta U = Q + W$, we can see that heat must be leaving the gas during this process.

Part (E)

1 point—For indicating that the entropy of the gas is decreasing because thermal energy is being removed in this process. This reduces the spread of the speed distribution of the gas, thus reducing disorder.

Part (F)

1 point—For the correct answer with supporting work:

$$\Delta U = \frac{3}{2} nR\Delta T$$
$$= \frac{3}{2} nR(T_A - T_C) = \frac{3}{2}(1 \text{ mole})\left(8.314 \frac{\text{J}}{\text{mole} \cdot K}\right)$$
$$\times (480 \text{ K} - 120 \text{ K}) = 4{,}500 \text{ J}$$

Part (G)

Based on the *PV* values, the temperature at point *A* is higher than that at point *B*. Thus, the peak for *A* must be at a higher speed than for *B*.

1 point—For both curves showing a roughly bell shape, and curve A having a higher average speed than curve B.

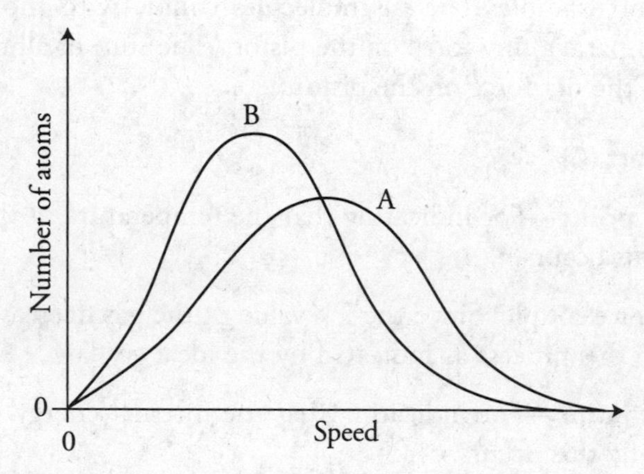

The area under the graphs must be equal because the number of molecules remains the same. This means the peak for *A* must be lower than that for *B*.

1 point—For curve A having a lower maximum than curve B, and both curves having roughly the same area beneath them.

How to Score Practice Exam 3

The practice exam cut points are based on historical data and will give you a ballpark idea of where you stand. The bottom line is this: If you can achieve a 3, 4, or 5 on the practice exam, you are doing great and will be well prepared for the real exam in May. This is the curve I use with my own students, and it is has been a good predictor of their actual exam scores.

Multiple-Choice Score: Number Correct ______________ (50 points maximum)

Free-Response Score: Problem 1 ______________ (10 points maximum)

Problem 2 ______________ (12 points maximum)

Problem 3 ______________ (10 points maximum)

Problem 4 ______________ (12 points maximum)

Free-Response Total: ______________ (44 points maximum)

Calculating Your Final Score

Final Score = (1.136 × Free-Response Total) + (Multiple-Choice Score)

Final Score: ______________ (100 points maximum)

Round your final score to the nearest point.

Raw Score to AP Grade Conversion Chart

Final Score on Practice Exam	AP Grade	
70–100	5	Outstanding! Keep up the good work.
55–69	4	Outstanding! Keep up the good work.
40–54	3	Great job! Practice and review to secure a score of 3 or better.
25–39	2	Almost there . . . concentrate on your weak areas to improve your score.
0–24	1	Don't give up! Study hard and move up to a better grade.

Words of Encouragement

Congratulations! You've made it through your AP Physics 2 class at school and this review book as well. You have worked hard and now you are ready. It is time to take the exam and show what you know. I'll leave you with these words of encouragement from Winnie-the-Pooh:

> There is something you must always remember. You are braver than you believe, stronger than you seem, and smarter than you think.
>
> —A. A. Milne, *Winnie-the-Pooh*

Believe in yourself. Be confident. Be bold. You are going to do great!

CB

Appendix

Constants
AP Physics 2 Equations

CONSTANTS

1 amu (unified atomic mass unit)	u	1.66×10^{-27} kg
mass of proton	m_p	1.67×10^{-27} kg
mass of neutron	m_n	1.67×10^{-27} kg
mass of electron	m_e	9.11×10^{-31} kg
electron charge magnitude	e	1.6×10^{-19} C
Avogadro's number	N_0	6.02×10^{-23} mol^{-1}
universal gas constant	R	8.31 J/(mol·K)
Boltzmann's constant	k_B	1.38×10^{-23} J/K
speed of light	c	3.0×10^{8} m/s
Planck's constant	h	6.63×10^{-34} J·s
	h	4.14×10^{-15} eV·s
Planck's constant · speed of light	hc	1.99×10^{-25} J·m
	hc	1.24×10^{3} eV·nm
vacuum permittivity	ε_o	8.85×10^{-12} C^2/N·m^2
Coulomb's law constant	$k = \frac{1}{4\pi\varepsilon_0}$	9.0×10^{9} N·m^2/C^2
vacuum permeability	μ_o	$4\pi \times 10^{-7}$ T·m/A
magnetic constant	k'	$\frac{\mu_0}{4\pi} = 1.0 \times 10^{-7}$ T·m/A
universal gravitation constant	G	6.67×10^{-11} N·m^2/kg^2
acceleration due to gravity at Earth's surface	g	9.8 m/s^2
1 atmosphere of pressure	atm	1.0×10^{5} Pa = 1.0×10^{5} N/m^2
1 electron-volt	eV	1 eV = 1.6×10^{-19} J

AP PHYSICS 2 EQUATIONS

Newtonian Mechanics

$$v = v_o + at$$

$$x = x_0 + v_0 t + \frac{1}{2} at^2$$

$$v_f^2 = v_0^2 + 2a(x - x_0)$$

$$F_{net} = ma$$

$$F_f \leq \mu F_N$$

$$a_c = \frac{v^2}{r}$$

$$\tau = F \cdot d$$

$$p = mv$$

$$\Delta p = \Delta F t$$

$$K = \frac{1}{2} mv^2$$

$$U_g = mgh$$

$$\Delta E = W = F \cdot \Delta x = Fd \cos \theta$$

$$P = \frac{W}{\Delta t} = \frac{\Delta E}{\Delta t}$$

$$P = F \cdot v$$

$$F = -kx$$

$$U_s = \frac{1}{2} kx^2$$

$$T_s = 2\pi \sqrt{\frac{m}{k}}$$

$$T_p = 2\pi\sqrt{\frac{L}{g}}$$

$$T = \frac{1}{f}$$

$$F_G = G\frac{m_1 m_2}{r^2}$$

$$U_G = G\frac{m_1 m_2}{r}$$

Electricity and Magnetism

$$|\vec{F}_E| = \frac{1}{4\pi\varepsilon_0}\frac{|q_1 q_2|}{r^2}$$

$$\vec{E} = \frac{\vec{F}_E}{q}$$

$$|\vec{E}| = \frac{1}{4\pi\varepsilon_0}\frac{|q|}{r^2}$$

$$\Delta U_E = q\Delta V$$

$$V = \frac{1}{4\pi\varepsilon_0}\frac{q}{r}$$

$$|E| = \left|\frac{\Delta V}{\Delta r}\right|$$

$$\Delta V = \frac{Q}{C}$$

$$C = \kappa\varepsilon_0\frac{A}{d}$$

$$E = \frac{Q}{\varepsilon_0 A}$$

$$U_C = \frac{1}{2}Q\Delta V = \frac{1}{2}C(\Delta V)^2$$

$$I = \frac{\Delta Q}{\Delta t}$$

$$R = \rho\frac{L}{A}$$

$$P = I\Delta V$$

$$I = \frac{\Delta V}{R}$$

$$R_S = \sum_i R_i = R_1 + R_2 + R_3$$

$$\frac{1}{R_P} = \sum_i \frac{1}{R_i} = \frac{1}{R_1} + \frac{1}{R_2} + \frac{1}{R_3}$$

$$\frac{1}{C_S} = \sum_i \frac{1}{C_i} = \frac{1}{C_1} + \frac{1}{C_2} + \frac{1}{C_3}$$

$$C_p = \sum_i C_i = C_1 + C_2 + C_3$$

$$B = \frac{\mu_0}{2\pi}\frac{I}{r}$$

$$\vec{F}_M = q\vec{v} \times \vec{B}$$

$$|\vec{F}_M| = |q\vec{v}||\sin\theta||\vec{B}|$$

$$\vec{F}_M = I\vec{\ell} \times \vec{B}$$

$$|\vec{F}_M| = |I\vec{\ell}||\sin\theta||\vec{B}|$$

$$\Phi_B = \vec{B} \cdot \vec{A}$$

$$\Phi_B = |\vec{B}|\cos\theta|\vec{A}|$$

$$\varepsilon = -N\frac{\Delta\Phi_B}{\Delta t}$$

$$\varepsilon = B\ell v$$

Fluids, Gases, and Thermal Physics

$$\rho = \frac{m}{V}$$

$$p = \frac{F}{A}$$

$$P = P_o + \rho g h$$

$$F_b = \rho V g$$

$$A_1 v_1 = A_2 v_2$$

$$\frac{\Delta V}{t} = Av$$

$$P_1 + \rho g y_1 + ½ \rho v_1^2 = P_2 + \rho g y_2 + ½ \rho v_2^2$$

$$\frac{Q}{\Delta t} = \frac{kA\Delta T}{L}$$

$$PV = nRT = Nk_BT$$

$$K_{average} = \tfrac{3}{2}k_BT \text{ and } K = \tfrac{1}{2}mv^2$$

$$W = -P\Delta V$$

$$\Delta U = Q + W$$

Waves and Optics

$$x = A\cos(\omega t) = A\cos(2\pi ft)$$

$$f = \frac{1}{T}$$

$$\lambda = \frac{v}{f}$$

$$\Delta L = m\lambda$$

$$d\sin\theta = m\lambda$$

$$n = \frac{c}{v}$$

$$n_1\sin\theta_1 = n_2\sin\theta_2$$

$$\frac{1}{s_i} + \frac{1}{s_o} = \frac{1}{f}$$

$$|M| = \left|\frac{h_i}{h_o}\right| = \left|\frac{s_i}{s_o}\right|$$

Quantum, Atomic, and Nuclear Physics

$$E = hf = \frac{hc}{\lambda}$$

$$p = \frac{h}{\lambda} = \frac{E}{c}$$

$$\lambda = \frac{h}{p} = \frac{h}{mv}$$

$$E = mc^2$$

$$K_{max} = hf - \phi$$